TELEPEN
10 0288861 X

AF598690

SCIENTIFIC PHILOSOPHY TODAY

BOSTON STUDIES IN THE PHILOSOPHY OF SCIENCE

EDITED BY ROBERT S. COHEN AND MARX W. WARTOFSKY

VOLUME 67

MARIO BUNGE

SCIENTIFIC PHILOSOPHY TODAY

Essays in Honor of Mario Bunge

Edited by

JOSEPH AGASSI

Boston University and Tel Aviv University

and

ROBERT S. COHEN

Boston University

D. REIDEL PUBLISHING COMPANY

DORDRECHT : HOLLAND / BOSTON : U.S.A.

LONDON : ENGLAND

Library of Congress Cataloging in Publication Data

Main entry under title:

Scientific philosophy today.

(Boston studies in the philosophy of science ; v. 67)
Includes bibliographical references and index.
1. Science—Philosophy—Addresses, essays, lectures. 2. Bunge, Mario Augusto. I. Agassi, Joseph. II. Cohen, Robert Sonné. III. Bunge, Mario Augusto. IV. Series.
Q174.B67 vol. 67 [Q175.3] 501s [501] 81-10686
ISBN 90-277-1262-X AACR2
ISBN 90-277-1263-8 (pbk.)

Published by D. Reidel Publishing Company,
P.O. Box 17, 3300 AA Dordrecht, Holland.

Sold and distributed in the U.S.A. and Canada
by Kluwer Boston Inc.,
190 Old Derby Street, Hingham, MA 02043, U.S.A.

In all other countries, sold and distributed
by Kluwer Academic Publishers Group,
P.O. Box 322, 3300 AH Dordrecht, Holland.

D. Reidel Publishing Company is a member of the Kluwer Group.

All Rights Reserved
Copyright © 1982 by D. Reidel Publishing Company, Dordrecht, Holland
and copyrightholders as specified on appropriate pages within
No part of the material protected by this copyright notice may be reproduced or
utilized in any form or by any means, electronic or mechanical,
including photocopying, recording or by any informational storage and
retrieval system, without written permission from the copyright owner

Printed in The Netherlands

EDITORIAL PREFACE

This volume is dedicated to Mario Bunge in honor of his sixtieth birthday.

Mario Bunge is a philosopher of great repute, whose enormous output includes dozens of books in several languages, which will culminate with his *Treatise on Basic Philosophy* projected in seven volumes, four of which have already appeared [Reidel, 1974ff.]. He is known for his works on research methods, the foundations of physics, biology, the social sciences, the diverse applications of mathematical methods and of systems analysis, and more.

Bunge stands for exact philosophy, classical liberal social philosophy, rationalism and enlightenment. He is brave, even relentless, in his attacks on subjectivism, mentalism, and spiritualism, as well as on positivism, mechanism, and dialectics. He believes in logic and clarity, in science and open-mindedness – not as the philosopher's equivalent to the politician's rhetoric of motherhood and apple pie, but as a matter of everyday practice, as qualities to cultivate daily in our pursuit of the life worth living. Bunge's philosophy often has the quality of Columbus's egg, and he is prone to come to swift and decisive conclusions on the basis of arguments which seem to him valid; he will not be perturbed by the fact that most of the advanced thinkers in the field hold different views. Though willing to alter his mind at the sight of a good argument, he holds that the best of public opinion, even when scientific and informed, need not be taken very seriously. He is, in short, emphatically autonomous in his judgment, swift or deliberate, well founded or not. He has won the respect and appreciation of many who share his opinions to a high or low degree, who enjoy or are critical of his brisk manner or who like him serious or who prefer him provocative.

His studies often explode multiple ambiguities, rectify logical errors, and point at classical defects such as dogmatism, subjectivism in science and in philosophy. His views on the conduct of research, his fallibilism, his view of science, are a constant assault on black-boxism, not to mention that his more specific doctrines of meaning and denotation, of things and their basic qualities, his theories of systems and their emergence, of life and mind, of causes and chance, and more, are bound to be judged among the splendid bodies of exact philosophical achievement of this century. May Bunge flourish, on and on!

We are grateful to Renate Hanauer and Carolyn Fawcett for their dedicated editorial assistance, and to Ms. Hanauer for translating the essay by Rudolf Haller.

Boston University Center for the Philosophy and History of Science

JOSEPH AGASSI
ROBERT S. COHEN

November 1980

TABLE OF CONTENTS

LEO APOSTEL

SOME REMARKS ON ONTOLOGY

1. INTRODUCTION

Many reasons lead us to evaluate Mario Bunge's philosophical work positively. His training as a physicist, added to his genuine philosophical interest, makes him appreciate the value of rigorous mathematical method, and simultaneously prevents him from concentrating on formal issues, forgetting the substantial ones. Few professional philosophers would dare to announce so comprehensive a work as his *Treatise on Basic Philosophy*, ranging over semantics, ontology, systems theory, epistemology, to ethics, seriously attacking traditional problems by means of modern methods. Yet, why are we philosophers at all if not because some such *Treatise on Basic Philosophy* is the purpose we are pursuing? We want to show our appreciation for his attitude which we fully share, by concentrating on volume III of his series, namely *Ontology I: The Furniture of the World*, and asking a series of questions about it. The questions would lead, if they are not mistaken, to a reordering of the material, to a tighter systematization and sometimes to locally different definitions and postulates. But we fully agree with the author that an ontology such as he offers here should be written; we want to take his pioneer effort seriously by looking at it in detail.

2. SOME REMARKS ABOUT BUNGE'S METHOD IN ONTOLOGY

Bunge states "the ontologist should stake out the main traits of the real world as known through science ... and should proceed in a clear and systematic way. He should recognize, analyze, and interrelate those concepts enabling him to produce a *unified* picture of reality" (p. 5). The difficulty we have here depends upon the meaning of the concept of unification. In the five chapters of the book we shall comment upon (reserving for another occasion the chapter about space-time), we find postulates about substance, form, things, possibility and change. These postulates are reached as generalizations from scientific procedure or result (with this we have no quarrel at all). They are introduced, however, as independent pieces in the architecture and few explanations are given, con-

J. Agassi and R.S. Cohen (eds.), Scientific Philosophy Today, 1–44.
Copyright © 1981 *by D. Reidel Publishing Company.*

necting them with other postulates and showing why we should prefer them to eventual alternatives. No *ontological* deductions at least are in evidence. *When* explanations occur, they are, curiously enough in a work so thoroughly opposed to subjectivism, related to the possibility of knowledge and of action. A few examples will show this:

(a) On p. 271, postulate 5.10 is justified as follows: every thing interacts with at least some other thing, while not interacting to the same extent with all other things. ["This partial interconnectedness of the parts of the world renders their study possible, for the study of every thing is partial and rests on the possibility of making contact with the thing" (p. 271)];

(b) On p. 17 when offered a list of ten ontological principles presupposed by scientific research, we meet again a curious mixture of principles *without* any justification; [then Bunge looks empirically at scientific research and tells us, for instance, that he observes that there are no forms that are not forms of things in this research or its result] and principles justified by means of *epistemological* remarks [M7, p. 17: "nothing comes out of nothing and nothing reduces to nothingness. If this were not so, we would make no effort to discover both the origin of new things and the traces left by things that have been destroyed." M10 is justified by a similar remark; M6 could be: if everything is knowable, and knowing can only occur through interaction, while interaction always presupposes change, everything changes; M8: "everything abides by laws" is explained as follows: "if there were no laws we would not utilize them in order to explain, foresee and do"].

Why do we point this out? *First*, because we think that if some type of epistemological or practical justification is offered for some basic principles or postulates, we are of the opinion that Bunge should either offer such justification for all of them, or show convincing reasons why such justification can be given only for some and not for others. *Second*, because these remarks, if taken seriously, clearly make volume III of the ontology dependent upon volume IV (a systems theory in which the concept of knowing and acting system can be introduced) and even upon later volumes on epistemology not yet published. *Third*, if taken seriously, as a result of these remarks, the pre- (or post-) Kantian attitude of an "ontology" that wants to talk about the universe without talking already about the knowing and acting subject is however lost. We fall back on the point of view of Kant's *Metaphysische Anfangsgründe der Naturwissenschaften* (even when we find it generalized by looking at the universe as not only determined by the conditions of the possibility of knowing but also by the conditions of the possibility of acting). This is (according to us) not incompatible with the possibility of another type of explanation of the postulates that would introduce a purely ontological justification for some of them. If our picture of reality can really be 'unified', then ontological explanations of its basic properties can be true, while simultaneously epistemological and praxiological explanations hold. But what would an ontological deduction look like? We cannot offer complete examples but only hints.

In chapter 3, Bunge asserts that only things (i.e., substances endowed with real properties) are real; in chapter 4, he asserts that real possibilities exist; and in chapter 5, he asserts that everything is continuously changing in some respect.

An ontological deduction would be proof of the fact that *if* there were non-things, reality could not be continuously changing, or there could be no real modalities; or that if there were reals not continuously changing, there could be real non-things, or no real modalities.

Without attempting to offer such relationships between the basic features of the world as they are described here, we think that we cannot really speak about a unification, but only about a general description. A general description is already much, but in view of the fact that Bunge himself tries to give justifications (and, wanting to be clear and evident, finds at his disposal only epistemological and praxiological ones) we think that he wants more; and we ourselves, true to traditional philosophy, certainly want more. If more is wanted, and if really a pre-Socratic or post-Kantian ontology is aimed at, then ontological deductions should be more prevalent in the work than they are at present.

A second methodological remark concerns Bunge's use of mathematics. We have no quarrel with his practice (he makes up the systems he needs, going along, and deviates from formerly known structures in ways clearly pointed out) but we have some quarrel with his theory. While conceding that "semantics or at least its application, does have metaphysical presuppositions" (p. 15 and the earlier volumes of the *Treatise*), he denies that logic and mathematics have them ["Deductive logic and pure mathematics . . . . are ontologically neutral" (p. 15)]. This leads him to the belief that classical set theory, expressed in second-order functional logic, is an effective tool for ontological constructions (and this attitude has some clear consequences in practice: for instance, without being the only reason for his radical rejection of modal logics, it seems to us to be one of the main reasons for this rejection). With this we cannot agree. The question "How can it be explained that there exists a system producing mathematics?" (we shall call it 'MacCullough's basic question') is one of the main questions of regional (*not* general) ontology and the question "What *in* reality makes mathematics applicable to it, while mathematics is not created in constant interaction with the external world?" is another basic question of general ontology.

If these two problems are – as they seem to us to be – genuine, then it could very well be the case (and we believe this to be true) that not classical mathematics, but rather some non-classical mathematics (for instance finitistic mathematics) would be the most appropriate tool for studying ontology (we think that both ontology and systems theory can offer arguments in favor of the view – and here again volume IV would have to determine volume III). But practically – and here Bunge is obviously completely correct – physics and the other advanced sciences are historically expressed in classical mathematics.

We shall thus have to proceed in a *non-linear* way that will, I suppose, displease Bunge (even though we do not think it is basically incompatible with his point of view). First we are 'mathematical opportunists' (p. 15 again). Then we develop ontology in this opportunistic fashion, until we can ask the basic question about the genesis and the applicability of mathematics (in systems theory, itself *dynamically* developed). Then we come eventually to a reformulation of mathematics. And in consequence of this reformulation we go back to ontology and reconstruct it accordingly. This method is the nonlinear method of constructing and changing the bases of the construction in consequence of its own development. If we have to find a classical name for this kind of procedure we can only call it dialectical. But we do not attach much importance to this name; we hope that Bunge himself will not attach too much importance to this name either but we would like to point out that this kind of progression is to be observed so often in the history of science that we can consider it as rather typical. In summary: we do *not* believe that we *can* remain "mathematical opportunists."

The two methodological remarks we just made refer (a) to the form and degree of systematization and unification of the ontology, and (b) to the use of its main tool: mathematics. We think that these two matters should be clarified if work is to progress in a satisfactory manner.

3. SOME REMARKS ON SUBSTANCE

The theory of substance is a theory about bare particulars, abstraction being made from all their qualities.

We have to ask about their status, and the reason for their presence; but first we have to present them, before being able to consider some alternatives.

Postulate 1.1 gives their characteristics:

1. The individuals considered belong to a set S, closed under an operation of composition $\circ$.

2. This composition is associative: $[x\circ(y\circ z)] = [(x\circ y)\circ z]$.

3. The operation $\circ$ is a binary operation.

Bunge adds to this *semi-group* the null individual, as the neutral element of the semi-group (this element in composition with an arbitrary element of S, either to the left or the right, yields the same arbitrary element all over again). After this addition the whole presents the *monoid* structure.

We have some questions about the relation between pages 26–27, and 28–29. Bare individuals are introduced as non-existent fictions (26), useful to understand things later on; but on 28 it is claimed that "S is the set of all substantial or concrete individuals." How can these two presentations be compatible? This question is not a useless semantical question but has deep consequences:

(a) according to us, if S is the set of concrete individuals it cannot contain the empty individual (that is certainly a complete fiction);

(b) moreover, if S is the set of concrete individuals, then association is a real relation, not a mathematical abstraction. The fact that we are studying ontology here and not the algebra of semi-groups or monoids, makes it necessary to understand composition as a real operation (although not one performed necessarily by organisms or man). Let us call this operation 'conglomeration' (the formation of a whole out of certain parts). If this is true, certain consequences follow that make it not as easy for us, as for Bunge, to accept the complete content of Postulate 1.1.

'$\circ$' is claimed to be commutative and idempotent '$x\circ y = y\circ x$' and '$x\circ x = x$.'

Bunge is well aware that these postulates do not express necessities. Thus, (a) when he has to consider the possibility that the same object comes in multiple copies (36–37) he considers '$x\circ x$' as different from 'x' and thus has tacitly to abandon idempotence; (b) he tells us (p. 29) that it would be 'interesting' to study non-associative and non-commutative algebras in ontology. But he simply *decides* not to do so, as if there were no ontological constraints to make this decision advisable or non-advisable.

We are of the opinion that he comes to this decision because, against his own intention not to be guided by formal considerations, he lets himself be influenced by them in order to reach a system closely similar to Leśniewski's (that mereology is the real inspiration of this chapter, anybody who consults Eberle's *Nominalistic Systems* could easily confirm). Let us, defending Bunge against Bunge, ask for (a) a concrete ontological interpretation of 'conglomeration', (b) a concrete ontological interpretation of the order '$x \circ y$', and (c) a concrete ontological interpretation of (and) (the groupings). Only if such interpretations were forthcoming, could we grasp the meaning of postulate 1.1 so swiftly introduced.

We shall presently come back to these matters, but already one conclusion imposes itself as the result of this request for a concrete ontological interpretation of the operations. Postulate 1.3 loses its force. This postulate, generalizing for n-ary sequences what association does for binary sequences, asserts that for any sequence $x1 \ldots xn$, its 'supremum' exists (i.e., its conglomeration or association). If a real whole is meant, we already have to introduce here a modality (in Bunge this occurs only in chapter 4) and we have to replace 1.3 by 1.3′: "For any set T of elements its supremum *can* exist." (The association of any two objects is possible but not real!)

Let us now return to the basic questions and try to give interpretations of association or conglomeration such that they are (a) associative and commutative, (b) associative and non-commutative, (c) commutative and non-associative, (d) non-associative and non-commutative.

Let the difference between '$x \circ y$' and '$y \circ x$' be the following: in the first case, x is separated from the rest of the universe (this idea presupposes the concept of *separation*, of *universe* and of *complement of an individual*), then later (this pre-time can be local), y is separated equally and both are finally bound together (a connection between x and y is established that may be visualized by vicinity or by a physical link). We consider the consequences of these separation actions. They will depend on the kinds or types of the links that have to be broken in order to isolate x and y, and on the relative strengths of these links or ties. Let us consider the following situation: (a) the kind or strength of a link between two objects depends upon the kinds or strength of the links between the other objects in the universe (or at least on some of them), and (b) the kinds and strengths of links between objects depend also on the objects involved in these links. If these two hypotheses are accepted, then it may be that first isolating x involves, in some (rare) cases, that the links between the rest (real complement of x) are so changed, that the succeeding isolation of y will admit a very strong linking of x and y (or a very varied and multiple or specific type of linking) while the reverse procedure, first isolating y, and then x, would not have the same consequence.

We realize that Mario Bunge will have the following answer ready: we are here introducing terms that cannot yet be defined: links, types of links, strengths of links, laws about interaction between links, temporal order (at least locally: first isolating x and then y or in reverse order) and moreover we are giving strong interpretations to 'association'. We are willing to concede all that. Our remark has however the following intent: if we are doing ontology, and not simply algebra, then exactly as we stated before (with reference to P I.3), from the beginning *modality* has to be present, and from the beginning *causality* and *pre-space* and *pre-time* have to be present (not the fully fledged realizations of these complicated concepts).

In the paragraph just finished, we introduced some considerations that made a non-commutative association operator plausible. Now we are going to introduce some others in favor of a non-associative one. Let us consider again the possibility of different types of strengths of associations. In the case of the presence of ( ), there is additional motive to do so if $x \circ (y \circ z)$, realistically interpreted, is the conglomeration of a whole $(y \circ z)$ with another object x. In the initial situation then, y and z are otherwise or more strongly tied to each other than x to y or x to z. To take a simple example: y and z are glued together, and x is tied to this whole. In the final situation $(x \circ y) \circ z$, x and y are linked in the way y and z were earlier. In a hierarchical universe (such as Bunge is describing) where wholes have properties different from those of their parts, it is to be expected that the associations of singletons, pairs, triads, n-tuplets in general will be different from level to level. Non-associativity might be the rule, not the exception. Moreover, we should remind the reader that even if the associativity rule were true, we should express the rule by means of a modality and not by means of a strict identity: if $[x \circ (y \circ z)]$ is the case, then it is possible for $[(x \circ y) \circ z]$ to be the case. Both cannot be simultaneously true if we interpret the ( ) in an ontological fashion. Our two examples show that non-commutativity and non-associativity can be due to different circumstances: the non-commutativity was due to the fact that the relationships between other parts of a whole may be altered by the separation of one part of the whole, while the non-associativity was due to the fact that various types and strengths of associations must in general be considered.

The reader might now get the impression that by means of weird and unrealistic examples we are simply breaking up the harmony of Bunge's useful algebra. We are not doing so, however. We think that we are preparing a possible justification of his first postulates by means of purely ontological considerations. We would be willing to defend the following assertion: if different types and strengths of association are present in the universe, there is at least part of this universe in which, considering only one type and one strength of association, and eliminating non-linear interactions, commutativity and associativity hold. This region will be one of the many levels of the universe. It will exist either as a logical consequence of a simplicity axiom (and we are not certain that simplicity axioms are purely ontological in nature) or as a consequence of a modal axiom: all possible types of levels are real (an axiom that according to Bunge would only occur much later but analogs of which have been considered both by the atomists and by Leibniz) or as a consequence of a partition axiom (the universe can be split up in parts in which strong interdependence holds while among these parts weaker or other interdependences hold) that could be derived as a systems-theoretical consequence from stability considerations *à la* Herbert Simon.

We are aware of the fact that this remark is not a deduction, but we claim that we owe it to Bunge to take his algebras ontologically seriously, and that we can only do so by means of preparing concrete interpretations for his statements, concrete interpretations that not only provide counter-examples but that, by providing counter-examples *in given circumstances*, reveal (and are indeed the only methods to reveal) the ontological reasons for the truth of his proposals in other selected contexts.

Moreover, the considerations of other algebras for substances can have some other interesting consequences. Page 34, theorem I.6, proves that the calculus of substances cannot have inverses (and, p. 35, Bunge recognizes that his semi-group of substances can never attain the group-structure). If we look for a moment at the reasons for this impossibility, we meet the following proof:

(1) Assume that '$x \circ y$' is identical to the empty individual (null)

(2) Now $x = x \circ \text{null}$ (by definition of null). By substitutivity of identity this entails

(3) $x = x \circ (x \circ y)$ (using the assumption).

(4) $x \circ (x \circ y) = (x \circ x) \circ y$ (by associativity)

(5) $(x \circ x) \circ y = x \circ y$ (by idempotence)

(6) $x \circ y = \text{null}$ (by hypothesis)

(7) By transitivity of identity: x equals the empty element.

The same proof can be given for y starting with $y = y \circ \text{null}$.

So, if two elements composed yield the null element, they are already null to begin with and we can never meet inverses.

We have encountered reasons to doubt associativity and idempotence. If these reasons bring us to the elimination of these postulates, the impossibility of inverses does not exist. If we do not have the null individual (and we do not see how we can have it), the existence of a complete inverse would have to be stated as follows: the composition of some x with some y reduces to the absence of x and y.

This leads us to a remark about 'identity'. Following an idea introduced by Lorenzo Peña in his doctor's thesis (Liège, 1979), we think that in ontology it is necessary, instead of the mathematical identity, to use a non-commutative 'reduction' operator. '$x = y$' is to be read as: x reduces to y (and the inverse is not therefore true). This non-symmetrical reduction relation introduces the part of the temporal relation we need to use at this early moment.

If our earlier considerations show us that it is not necessary to eliminate the inverse altogether, they also entail however that not all elements will have inverses (because in parts of our universe the associativity laws will be true). The problem arises then (we cannot solve it here but only introduce it), under what conditions there will be ontologically complete inverses. A more general problem would be: is it possible (and under what conditions) to introduce partial inverses? We define a partial inverse as follows: $(x \circ y) = z$, where z is a proper part of x. If $z \circ y$ again reduces to u (where u is a proper part of x) this would mean that y is a universal reducer. Then either x has an infinite number of elements or the repeated application of y is a complete inverse. Without associativity this will not entail however that y is already a complete inverse.

Before finishing our remarks on substance by some remarks on the theory of assemblies, we have to draw attention to a point of principle. If our variable ranges over a set of bare individuals (without properties) we have the problem of finding a *principium individuationis* (in order to have building blocks that can associate or conglomerate, we must be able to distinguish them). This can only be done by means of the association relation itself (properties being absent!).

We could consider the following principle of individuation: two elements are distinct if and only if they have different parts. This will not do however if there are atoms (i.e., basic elements) having no parts (or having as parts only themselves and the empty element). This individuation principle will indeed identify all the atoms and leave us with only one atom (and without the possibility to associate).

So we have to consider the complementary principle: two elements are different if and only if they are parts of different elements. Finally, however (if we follow Bunge), all wholes belong to one and the same whole: the universe.

Let us now consider, if the n-th level contains the maximal individual, the universe, the $n-1$th level. Its elements are all identical because there are no different wholes to which they belong. So the $n-1$th level is already no longer differentiated. By using the same reasoning recursively we conclude again that there is only one element (and the possibility to associate disappears).

Bunge is aware of the necessity of discussing the problem of individuation (see p. 54) but he does not seem to have seen that there is a real antinomy here (as one would expect, in a theory where Aristotelian matter (*hyle*) is used and yet a multiplicity of substances are introduced on the same level). So we can only eliminate the non-individuation by one of two principles of infinity: either there is no one world that all substances are part of (and then we can use the second principle of individuation), or there are no atoms (and then one can use the first). However, if we choose this way of escape, is it still possible to speak about a well-defined domain over which the individual variables range? And – what is more important – are we not contradicting two basic properties we want our ontology to have (a) a basic level such that all substances are composed of substances of this level, and (b) one universe of which all substances are part?

We do think that these postulates are needed and in consequence we do not think this escape to be possible. The only other escape, however, is to enrich the theory of substances by introducing on the very level of ontological construction more individualizing properties than those present here.

Three methods could be used to enrich the theory of substances: (a) either the first part of ontology (introducing the building blocks of the universe) could be made to contain more than one type of substance (*here* there is only one type of entity: substance).

(b) Or, though containing only one type of entity, it could contain more than one type of relation (*here* there is only one type of relation: association). Even in assembly theory where juxtaposition and superposition are introduced, they can be easily reduced to one type of relation by operations (present in Leśniewski and in Goodman).

We were already suggesting this when we thought that even on this level 'possibilities' and 'links' should appear.

(c) Finally, we could also introduce types of primary entities other than substances. The two obvious candidates are either fields or processes. Outside of empiricist work (e.g., Whitehead's) we have no pure theory of fields and processes, however, and we think one should be worked out.

We would have one very speculative argument in favor of such a change: if, in the general materialism we encounter in Bunge, we use building blocks to which we ascribe no intrinsic tendencies (or if we ascribe to them tendencies only later, tendencies that do not constitute the kernel of substances, then the purely passive entities so introduced do not obey the most fundamental criterion of existence: everything that exists has to be, at least potentially, active. One methodological principle of ontology is for us the following: if an ontology is constructed, introduce only existent entities as building blocks of this ontology. On the other hand, however, the part-whole relation and the decomposition of complex systems into simpler ones is not only methodologically central, but also ontologically: *the more complex forms of existence must be, not reducible to, but caused by, the simpler forms of existence.* Even a process or a field ontology should obey this maxim and, in as far as it obeys this maxim, it should contain a part nearly isomorphic to the theory of substances presented here.

We do not comment in detail on assembly theory because all the fundamental questions have been asked in association theory. Assembly theory is simply the consequence of the fact that if one feels free to introduce a supremum (a smallest upper bound) one is also led to introduce an infimum (a largest lower bound) and if one has an upper semi-lattice, one is led to complete it by a lower semi-lattice so as to obtain a distributive complemented lattice. These operations are well known, of a formal nature, and they do not contain much 'ontological meat'. The difficulty of introducing the intersection of non-intersecting substances is taken care of by the null individual; and once the null individual is introduced, one feels no scruples about considering the complement of an individual as an individual. But if objections have been formulated against the zero individual, they will obviously have to be repeated when such artificial constructions appear.

A more essential problem is the one briefly argued about on p. 52: "to us the most basic dichotomy in any set of objects is the one between physical objects and conceptual ones." Some entities are constructs, other entities are substances. Bunge's tendency to isolate mathematics from ontology leads him here, as already earlier, to distinguish between concepts that are constructs and material entities that are (or are composed of) substances. To the contrary, it seems to us that if we want to remain close to scientific practice, we must consider constructs as material entities, produced by organisms and used in material interactions between these organisms. Bunge's dualism leaves him completely free to develop a completely conventional theory about the foundations of mathematics. This attitude is definitely not enlightening about the practice or the history of mathematics, and, by Bunge's very criteria, ought not to be accepted.

At the end of his chapter on substance the reader is left with two questions. (a) The theory of substance is obviously an attempt to develop a fundamental theory of matter. If the category of matter is something more stable than the theory of matter currently in favor, it is to be asked whether and why this theory of substance is really the core of the stablest part of our theory of matter? (b) Moreover, this theory of matter obviously uses Łeśniewski's and Goodman's ontology (but without the constructive function in the foundations of mathematics the first has, or the epistemological meaning the second has). Is this the best tool that can be found in the service of this task? And if so, why? By asking these two questions, we are not denying the answers given by Bunge to them. We are only noting down instructions for future work.

4. REMARKS ABOUT THE THEORY OF FORM

In his second chapter, Mario Bunge wants to define the concept of a 'substantial property' (p. 58), specifying and individualizing a substance (introduced by the theory of substances, or a modified version of it, commented upon earlier). We are astonished by the fact that constructing the theory of properties is considered as dependent upon the construction of semantics in the first two volumes of the *Treatise on Basic Philosophy*. Surely the ontological point of view would lead the philosopher to make his semantics dependent upon his theory of systems, and not his ontology dependent upon his semantics? But, for whatever reason, this is not the order selected by the author. A second source of astonishment is the following: attributes or predicates, the well-known entities used in functional logic (in fact functions representing or mapping individuals upon propositions), certainly basically different from properties (so much is taken for granted and strongly asserted by Bunge himself), are used in a roundabout way to approach the theory of properties. Even if it is true that we know properties only by means of the predicates used to describe them (and this is not true: we can know them by practice and/or acquaintance), the same could be said with equal right for substances (and yet the theory of substances has not been presented as the theory of correspondents of proper names or sortal names). Still, if P is the set of properties and A is the set of attributes, then a function from P to the power set 2^A makes for any property P a collection $r(P)$ of attributes.

We are of the opinion that nothing substantial would be lost if this representation relation were not used in the theory of properties.

Let us for instance consider the first postulate about properties P2.1 (p. 63).

Instead of saying "any substantial property in general is representable as a predicate or propositional function of the form '$A/SxTx\ldots Z$' mapped on propositions including A", we might say "any *general* substantial property attributed (where the attribution relation must be defined) to any member of a given class of the set S of substances, is a function of this class and of other classes belonging to S" (we would think that 'is attributed' should be purely ontologically defined, and should refer neither to a thinking nor to a speaking subject). In the same way a *specific* substantial property will be attributed to a specific substance.

By not attacking the problem directly but only by means of the concept of 'representation', Bunge avoids the crucial problem of defining 'attributing' in an ontological way. He sees the specificity of 'properties': he stresses the difference between 'properties' and 'predicates' more strongly than most writers we know (and doing this seems to us to be entirely correct); but he does not tackle the basic problem in a direct fashion. In this, his attitude is determined by his asymmetric view of the reality of *bare* substances and *free* properties. "Substrate free forms. . . are just as imaginary as formless substances. But at least the fiction of the bare individual makes mathematical sense while that of a pure form does not" (64). We would deny this: each concept is as meaningless or (we would think) as meaningful as the other. Historically, the Platonist trend of mathematics makes free properties much more popular than bare substances.

The basic question we are confronted with is answered in Bunge's Definition 2.2. "*P* is a substantial property iff some substantial individual exists that possesses this property" (71). Supposing that we have at our disposition a sufficient definition of substantial individuals, then the basic difficulty is to understand what it means for such an individual to 'possess a property'. Either we have an explicit definition for this '*flatus vocis*' (synonymous to 'the belonging of a property to an individual' or to 'a property being attributed to an individual'), or we have an implicit definition at our disposal. We do not encounter an explicit definition in Bunge but we do encounter an attempt to give an axiomatic definition characterized by the following conditions:

Postulate 2.4: (a) for every individual *x*, and every property, *x* possesses the property or it is false that *x* possesses the property (properties are dichotomous and sharp);

(b) no two substantial individuals have the same properties (=possess the same properties);

(c) for every two compatible properties, there is at least one other property *R* (and there may be more) such that this third property *R* is possessed by all objects which simultaneously possess the two others. (The *R* is thus not unique and we are not supposed to know what the conjunction of two properties is.)

Negatively: this axiom (*c*) does not hold for disjunction or negation. There are still other postulates given for properties (lawfulness [2.7, p. 78]; infinite number of basic properties [2.8, p. 81]), but it is clear that they do not throw any further light on the concept of 'possessing a property' (we can conceive of a non-lawful universe where individuals possess properties and we can conceive of an infinite number of basic properties).

The classification of properties into intrinsic or relational, quantitative or qualitative, and primary or secondary is not included in the concept of 'possessing a property' and cannot enlighten us about what it means to possess a property.

We must then, in order to see what Bunge means, concentrate our attention upon (a), (b) and (c).

With reference to these three statements, we can discuss the degree of their correctness, given the concept of property we think about.

Insofar as Bunge does not give us any substantial explanation, let us try to construct our own; not arbitrarily, however, but trying to work towards the concept Bunge wants, so we suppose, to explain. Let us consider the fact that every individual substance will be completely specified (i.e., constituted as a unique and completely determined '*proprium quid*' by means of a number of properties (finite or infinite)). We must thus certainly accept the fact that a property is such that, in conjunction with others, it may uniquely specify, or determine, all real relations of this individual with all other individuals in the real universe. Real relations are ways of being linked or tied to other individuals, so that basically the properties of an individual are those conditions that completely determine its links or ties. The verb 'determine' again must have an ontological meaning and this meaning can only be causal. So that we end up with the following result: the properties of an individual are the causes of the real relations of this individual. And the real relations of this individual are the causal interactions (as agent or/and patient) of this individual with the rest of the individuals in its universe. If this intuition is correct, then the concepts of (a) real relation and (b) cause are presupposed by the concept of property.

We do realize that Bunge explicitly rejects such an explanation (p. 69) where he refuses to concede to Helmholtz and Ducasse that properties are, in fact, causal interactions or their foundations. Our impression is, however, that in the chapter on Form of Bunge's *Treatise* a battle is waged between ontology and formalism and that, even though the ontological point of view is more strongly represented here than anywhere else, still it is overcome by formalistic tendencies.

Ontological in spirit and intention is the distinction made (on pp. 101 and 102) between binding and non-binding relations. Binding relations 'make a difference' to their relata. (102): they exist between two substances if some changes in the one are accompanied by changes in the other. The mathematical (arithmetic or geometric) relations are however not binding (in contrast to the hydrogen bond or cultural influence). It can be argued, however, that if a relation is not binding or reducible to a set of binding relations, it cannot exist (and then it cannot be said to characterize its relata). The task of interpreting mathematical relations among ontological sets as binding relations can be undertaken if it is considered that, in general, mathematical characteristics have physical consequences. Bunge does not look in this direction, however. Once it is realized that ontological relations must be binding, the further question can be asked: can relations be reduced to intrinsic properties? It is claimed (on pp. 70–71) that all binary relations can be reduced to eventually infinite sets of intrinsic properties. Bunge is fully aware of the purely formal character of his argument (see p. 70) which is in effect the following: any $R(x, y)$ can be reduced to $P(x)$ if we take successively all constant values over which the variable ranges, and if in every specific case we consider then for $y = c$, R depends only on varying values of x. This is the purely non-ontological replacement of the set of couples (x, y), bound by a binding relation R, (the example is: x falls on y) by the syntactical set of triads for all $c1, c2, c3 \ldots R(x1, c1), R(x2, c1), R(xn, c1), R(x1, c2) R(x2, c2) \ldots$ and so forth. If this ordered set exists, it is on the metalevel again linked by a binding relation, and the reduction of relation to property is not achieved. The same negative result, *a fortiori*, is achieved if even on the metalevel this cannot be claimed to be true. But in no sense can it be asserted that the new relations so created have any right to be substituted for the relation we started out to analyze and understand.

The arguments seem to show (a) that relations are not reducible to properties in ontology, and (b) that all relations must be binding relations. The question (comparing our unexpected proposal to that of Bunge) remains: Why should one think that intrinsic properties should be reduced to relations?

The suggestion to do so is by no means tied to phenomenalism (rightly rejected by Bunge). It derives from taking seriously (as Bunge wants us to do) traditional disputes about properties.

It is indeed characteristic for an individual that it is characterized by a set of properties (if the individual is x, this set is called $p(x)$ in the present work) and it is characteristic for a property to have a scope: i.e., to be possessed by a set of individuals (if the property is P, the scope is called $S(P)$ in this work). It can be said that exactly the same property characterizes all the elements of its scope. Then this property is either present in as many identical copies as the scope's |numerosity| demands, or the identical property is to be ascribed to all members of its scope, and it is then an entity present in a large multiplicity of relational locations (if we already had space-time at our disposal we would say: present in many times and spaces). If we accept the principle of the identity of indiscernibles, then we reject the first eventuality; and if we accept that every thing that exists is an individual (has a specific relational place), then we reject the second eventuality.

Even if we do not reject the existence of universals (as we are inclined to do if we realize that the existence of inherent universals is as difficult to accept as the existence of free universals) in both Platonism and Aristotelianism, the 'possession of a property by an individual' must be a relation between a universal and an individual entity; and this relation on the one side must be a particular and on the other side must be a universal. If we really take seriously the concept of 'the scope of a property' we meet these problems and cannot escape from them in the easy way Bunge has preferred. Even though we share his two preferences (a) to avoid Platonism and (b) to let properties characterize the individuals, we think that the difficulties to be met are far more serious than he seems to consider them. The proposal to consider properties as binding relations which exercise causal forces compelling any given individual to be different from any given other individual presupposes that the problem of individuation of substances has been solved (and it is far from being solved as we saw in our earlier paragraph) but it is an attempt to take seriously the demand that any property, to be substantial, should be individual and yet should be an individualizing force simultaneously (the *partial* character of every individualizing force can be used in the future to explain that different individuals can share the same property). This expression would then mean that they are distinguished from a certain number of other individuals by means of the same set of causes, a subset of the total set of causes that characterize their complete specificity.

The reader can return, after hearing our defense of our thesis (a defense that is after all an attempt to avoid some of the formalism that remains in Bunge's account and to take seriously the ontological difficulties he has raised) to the typical characteristics of Bunge's 'possessing a substantial property'.

His three postulates follow from our definition: (a) The sharpness of the property-concept derives from the fact that it is an individualizing cause. (If fuzzy properties were allowed, their individualizing force would be weaker; yet we still believe that the intersection of a sequence of fuzzy properties could completely individualize a concrete substance but certainly more of them and more specific ones should be used.) The dichotomy postulate, even if not absolutely necessary, is the simplest way to realize the task we think properties are serving in reality (once more we come perhaps to a justification of something originally introduced by postulation). (b) If partial identifications must be able to build up a complete identification, the conjunction postulate should be true (and we consider it as one of the excellencies of Bunge's treatment that he has raised the question of this postulate at all). And finally. (c) no two individuals distinct from each other can be characterized by the same complex of individualizing forces.

A last confirmation of our proposal is postulate 2.7 (following definition 2.7): for any property P, if P is a substantial property, then there exists at least one other substantial property Q, such that for all x, $P(x)$ entails $Q(x)$ or $Q(x)$ entails $P(x)$. The entailment relation has here to be understood as a merely material implication; no conditions are imposed on the predicates and quantifiers of a law statement. For this reason we think that the conditions utilized are too weak (how could subjunctive conditionals be derived from pure material implications in non-trivial senses?). But we cannot enter here into a detailed discussion of the different properties of law statements. Let us only consider this problem solved; it is evident that if properties are individualizing forces (or causes) they must either entail other properties in the same or different substances, and exclude a third type of properties from the first or other substances. Once more the difficulties of the relations between particulars and universals are present, however: if all properties are antecedents or consequents of laws, and if laws are universal statements both in their predicates and in their quantifiers, then all properties are universals (and thus not real in the sense of the word we are using here). This is a serious difficulty: even an infinite set of unreals cannot transform unreality into reality – and laws about non-existent entities would then characterize in conjunction an existent entity. We can only overcome these difficulties if we allow the expressions we call 'laws' in science to be as different from real ontological laws as predicates are from substantial properties (in fact this distinction follows from Bunge's own). Real laws would be real existent relations between properties that are themselves relations (causal relations). We thus must admit (and this in contradiction to what Bunge tells us in his parag. 4.4 [pp. 98–99]) that there exist properties of properties and that these properties of properties are even extremely important (because they are needed to give real existence to the lawful character of properties). Far from being an idle formal creation the iteration of the property concept is inscribed in its very meaning. If property is an individualizing force, then properties entail and exclude other properties and are constituted by these entailment relations, hence, a property necessarily has properties.

We come to the end of this discussion of substantial properties by making two remarks: (a) about the weights of properties, and (b) about the similarity of individuals. Bunge (a) first computes indices of similarity and (b) afterwards defines the weights of properties. We agree with his exposition and consider both these concepts as very important. However, we would introduce the weights as codetermining the degrees of similarity, and we are not certain that his definition of weight is completely adequate.

The definition we have sketched for property makes it necessary to consider that properties have different weights (a property is essentially defined by its specific weights, i.e., (p. 94)) by considering the influences it has on other properties. We thus have an ontological reason to agree with Bunge's epistemological reason to distinguish properties as more or less essential (i.e., as having more or less weight: 94–95)

The general conditions for a weight function are presented as follows:

The weight function varies between 1 and 0 and obeys the following conditions:

(1) $W(P, P) = 1$. If P is necessary for Q, $W(P, Q)$ (the weight of P for Q) is greater than or equal to the weight of Q for P.

(2) If P and Q are incompatible, they have zero mutual weight.

(3) If P and Q are concomitant, their mutual weight is different from zero.

(4) For all Q different from P, the sum of the weights of all these Q for P is 1.

We cannot agree with conditions (2) and (3). Instead of (2), we would propose, "If $W(P, Q)$ is equal to $W(P, -Q)$, $W(P, Q)$ is zero." As to (3) we would propose adding the condition to its antecedent that there is no third R such that R has the weight 1 for both, then the consequent of (3) holds.

Even when all these conditions are satisfied, it would seem to be possible to define weights in function of circles of essentiality for a substance. We would consider of maximum essentiality, those properties that, for a given individual x, are not conditioned by any other property of x, while conditioning all other conditioned properties of x. The immediately following degree of essentiality would belong to properties of x only conditioned by properties of maximal degree, and conditioning all others not conditioned only by those of maximal degree. With reference to x, the weight of a property would then be conditioned by its degree of essentiality. The intrinsic weight of a property would be the average weight of this property taken over all substantial individuals possessing it.

We stress the fact here that the definition of law is the main tool we have to use in order to define this concept of weight (and that Bunge also, in his general conditions for weight, uses this concept of law in a central place).

If we have introduced the concept of weight we seem to be able to define the degree of similarity of two individuals by the number of their common properties, each of them weighted however by their intrinsic weights. This number can only have significance if it is finite, and Bunge considers each individual to have an infinite number of properties. In order thus to use his degree of similarity he has to relativize it completely by comparing two individuals only with reference to a finite subset B of properties (p. 86). This necessary relativization made us look for reasons to overcome the arbitrariness of B, and the most plausible reason we could find was introducing the rank of essentiality of properties before we introduce similarity measures, and taking as B the set of properties of sufficiently high rank (we would have to introduce an extra postulate: for every substantial individual, the number of properties of sufficiently high rank is finite).

If we remember our problem about the individuality of substances and the universality of properties, we could make our proposed partial solution more precise as follows: the fact that we recognize properties of properties as existent (in contradiction to Bunge) makes it possible for us to state that all properties of individuals are particular but that these properties have given degrees of similarities. This is naturally only an easier (though not necessarily easier) replacement of the problem, but it is now possible to apply our solution to higher levels of properties than to properties themselves. (In fact classical nominalism has had recourse to this stratagem a number of times, though – we point this out – without complete success, if no added stratagem, akin to our earlier ones, can be introduced.)

We come at the end of our reflections on Bunge's treatment of properties to the following conclusion: if we consider the definition he gives for 'system' (on p. 264 of his *Ontology*), and if we provisionally abstract the concept of 'history' not yet introduced, then, taking into account the strong interaction between the different properties of things, both internally (by means of the essentiality ranks and the weight concept) and externally (properties being finally, in order to avoid certain antinomies, as binding relations, real relations), we come again to the view that the fourth volume of Bunge's *Treatise on Basic Philosophy* is already presupposed by the third. The concept of system seems to be as basic as both the concept of individual and that of property.

We do realize that this is a rather radical change in perspective. Perhaps we have come to this conclusion too fast, and perhaps this added complication can be avoided. If it can, we would be grateful to be wrong; if it cannot, we can still continue to keep the main points of Bunge's presentation, but we have, as we announced, to introduce local changes. We hope that future discussion will throw light on this matter.

5. REMARKS ON THE THEORY OF THINGS

In this part of our paper we shall make some remarks on

1. Bunge's definition and algebra of things,
2. Bunge's definition of thing models (with reference to his characterization of constructs),
3. Bunge's definition of state and state space,
4. Bunge's concept of natural kind,
5. Bunge's definition of existence.

The reader will easily recognize that the remarks made here depend upon our earlier discussion; we shall try to be as succinct as possible.

5.1. Bunge defines a thing as the ordered couple of a substantial individual and the collection of its 'unarised' properties: $\langle x, p(x)\rangle$.

For these ordered couples, sufficient postulates and definitions are introduced to define the juxtaposition operation $(x \dotplus y = (x \dotplus y, p(x + y))$ and to give the set of things the structure of a commutative monoid of idempotents, with a maximum (the world), a minimum (the empty thing) and a supremum for all sets of juxtaposed things.

It looks strange to us that, for a second time, quite isomorphous axioms have to be introduced in order to yield the commutative idempotent structure for things without being able to use any earlier properties about substances and properties. Moreover the fact that $p(x)$ is a structure of sets of entities with a conjunctive operation defined on them enters nowhere into the definition of thing. The same objections formulated in our first paragraph against the juxtaposition or addition operations apply to the thing structure. Finally, it is obvious that while we are able to think about a thing as an ordered couple, we cannot identify things and ordered couples of a certain type.

5.2. Concepts like sets, relations, individuals, propositions and contexts are defined by Bunge (postulate 3.4, p. 117) as belonging to the class of constructs, and while no thing is a construct, all objects are either concepts or things.

We think that this methodological dualism (118) is as unacceptable as Popper's third world theory that Bunge explicitly rejects. The status of a non-real object, neither a word nor a brain process, and yet able to represent real things is unexplained (and we think unexplainable). Why not define concepts as the results or properties of human actions, products that indeed are different from words or brain processes but that are certain complex patterns of objects or brain processes? Such an attitude would be closer to Bunge's ontological stand and we would be able to give such patterns the properties needed to 'represent' other things.

But even when we had introduced such 'real concepts', we still would claim that things should be approached otherwise than by means of the study of their models only. A model of a thing X is an ordered couple constituted by a base set M, and a sequence of functions F, defined on M, each representing a property of x. If it is true (as claimed on p. 121) that the same individual may have any number of different models, we must be able to state by means of the 'representation' relation what modelled thing and model-thing have in common so that the one may represent the other and we must be able to construct a theory about the thing modeled *itself*. There is no reason to suppose that the properties of a set of models (even if adequate) of a thing are identical to the properties of a thing, just as there is no reason to suppose that we have necessarily to develop the theory of things by means of the theory of their representations. We recognize that the introduction of the representations is, for the physicist Bunge, a quite understandable move in order to show his awareness of the fact that 'reality as such' is not captured by any particular model of it. We share this epistemological caution, but we still think that in an ontology (if one dares to construct one at all) one has to overcome one's epistemological scruples and tackle the universe as we think (indeed only as we think) it is.

5.3 If a couple (M, F), where $f1, f2..fn$ are the elements of the function set F, is a representation of an object x, then the elements of F are called the state variables in the representation in question. Laws formulate restrictions about the values of the state variables and about their interrelations. The state space of the individual is the total set of values of its state space variables; the lawful state space of the individual is the subset of this state space that obeys the restrictions imposed upon it by the laws. A state as such in the representation selected is a complete set of values for all state variables (p. 127).

It is obvious that for this definition of 'state' the two essential problems are (a) the definition of the criteria to be imposed on a state function and (b) the definition of the criteria to be used to decide whether and how far we have a complete set of state functions.

In a more ontological way we could have defined a state space as constituted by the possible values of the essential properties of the individuals studied, and a specific state by a complete set of fully concrete values of the essential variables. The difficulty in saying what is a concrete value of a property, and of enumerating the essential properties, is neither larger nor smaller than the present one, but we would at least have remained on the object level. In a change or process ontology, the momentary state of a thing would only have the status of a non-existent limit. The state during a certain interval would then have to be the specific processes going on in the individual during this interval. The interval could be taken to be very short (even though we do not yet speak about infinitesimally short ones) and the velocity of the change could be very slow.

We do not study the problem (even though it is a formidable one) in order to find out whether it is possible for all state variables (or for all essential properties) to have simultaneously completely determinate values. We accept Bunge's non-operationalist views and we consider it to be an important assertion that this simultaneous precision can, even not operationally, still ontologically be presupposed.

We do recognize, however, that there exist ontologies in which individuals and properties will have various *degrees* and *forms* of reality. The Bunge ontology tacitly assumes that there exists only one form of reality (existence) and only two values for it (existence or non-existence). In an ontology with degrees and multiplicities of reality (and we still do not have convincing proof that it has neither scientific adequacy nor ontological reasons in favor of it) the concept of state would have to be defined in very different ways.

5.4 A beautiful result of Bunge's approach to individuals and properties is the one stated in a nutshell (on p. 141): "substantial individuals come in filters, while substantial properties come in ideals."

We fully agree with Bunge that any valuable ontology must give to its objects of study a clear algebraic structure. Our own difficulties make it less easy to reach this aim than his own views; and we recognize this to be an objection against our own misgivings. Still the points of view of ontological adequacy and those of algebraical elegance must be combined and at the present moment we can only regret that our own difficulties with the scopes of properties and the sets of properties of a given object do not allow us to come easily to similar results.

Bunge's preoccupation with ontology as distinct from formal algebra makes him introduce, immediately, new and fruitful concepts: '*kinds*' are sets of individuals sharing not one property but a set of properties and '*natural kinds*' are sets of individuals sharing a set of laws. (Laws are considered, as we remember, as properties.) Chemical or biological species will be examples of such natural kinds.

We think that the definition of these kinds is one of the most important contributions of Bunge's ontology. Once more (in par 3.4) Bunge reaches an elegant result about the algebra of kinds (theorem 3.5). We want to remark that more restrictions have to be imposed on both the concepts of 'kind' and of 'natural kind' than Bunge imposes. The difference between sets of individuals sharing one property and sets of individuals sharing two properties is not the crucial one that will transform the second type of sets into kinds. The several properties shared must be related by an intrinsic relation, and the series of kinds constituted must form a systematic whole presenting certain global properties. We think both requirements are needed. Mineralogy, botany, zoology, cristallography all show us that we have to introduce 'related properties', 'subordinate properties', 'superordinate properties', 'families of properties'. We are aware of the fact that it adds enormously to the difficulties to be overcome if we do this, but the importance of the introduction of kinds and natural kinds is great enough to compel us to go somewhat further in this direction than has been done here. The remark made about the concept of 'kind' has natural consequences for the concept of 'natural kind'. The hierarchies among more or less essential properties, and their relations with more or less universal or more or less *necessary laws* (we apologize for introducing this concept of 'necessary' that we are only to meet in the following chapter of the treatise but that we can interpret here as being more or less easily or simply deducible from the bases of a theory about the field – and thus as reflecting a more or less essential property of the world of things) would play a role in a more demanding concept of natural kind.

5.6 Having introduced natural kinds without having recourse to Platonist entities, Bunge has the right to be satisfied, when he states that his universe contains not only things but also mono- or pluri-specific populations having partially emergent properties. This result seems extremely desirable to us too. We had the occasion however, in our earlier remarks, of pointing out certain traces of formalism, Platonism, and subjectivism that, in due course of time, should have to be wiped out in order to be completely certain about the results announced on pp. 153–154.

Such a trace is Bunge's dualism (on page 157): An individual x will be said to exist conceptually if it exists in the conceptual world; it will be said to exist really if it exists in the world of things.

For us the conceptual world is part of the world of things; and the world of things cannot be characterized as it has been by the commutative idempotent monoid of the couples $(x, p(x))$. There are too many models of commutative monoids around, and even if we enrich the definition of Th by postulates about natural kinds (and so implicitly by postulates about real properties and their lawlike relations), we have indeed to introduce stronger versions of 'property' and of 'law' than Bunge is inclined to use. But once more, his introduction of existence as a predicate, on the one side, and his existence criterion (p. 160) on the other side ("an object other than the entire world exists really if it is shown to be connected to some real object other than itself"), point in the right direction. We would have liked the criterion (on p. 160) to be included in the definition of 'existence' itself. Certainly a danger of circularity would seem to exist: if existence is the power to influence or be influenced by existence, *definiens* and *definiendum* contain the same term. But either a closure property could be introduced (existence means the possibility to be connected to the maximal number of other existents – a type of Leibnizian materialism, introduced in Bunge's materialism), or the definition of existence could be made recursive (basic existence could be distinguished from derived existence), or finally the intensity and form of the connection could be taken into account.

Why, however, should the reader accept the challenge to follow us on such a dangerous trip? We think we could persuade him by pointing out how very few and very general the properties of the Th set are, such as they are defined in Bunge's *Ontology*, and how clearly one has to go in the direction just pointed out if one wants to introduce stronger properties for Th (our discussion of properties points in the same direction).

If account is taken of these remarks, we cannot completely agree with the remark on p. 152: "the view that the world consists in facts, or in events or processes, is incompatible with physical cosmology and fails to clarify the very notions of fact, event and process." We are not concerned with facts (they also appear to us as artefacts), but we think that many of our remarks go in the direction of a theory that identifies the 'furniture of the world' with 'thing-processes' (neither the thing-concept nor the process-concept being less basic than the other). The relational view of properties we came to; the binding character of relations (entailing causal interaction and thus process) seem to maximalize the common presence and interaction of dynamism and reism. We shall see later, when commenting on 'change', whether the claim that processes and events become ununderstandable when they are taken as basic is really to be accepted. However, the blunt assertion that process and event ontologies "are not in agreement with physical cosmology" is – to say the least – astonishing, if one thinks about the multiple rejections of the concept of thing found in present-day physics. We do not say that we agree with these rejections of the thing concept; quite to the contrary, we think Bunge is right; but his conclusion that process ontologies are not convergent with present-day cosmology is – so it seems to us – neither more true nor more false than the assertions made by others according to which a thing-ontology is not convergent with present-day physical cosmology.

6. REMARKS ON THE THEORY OF POSSIBILITY

The chapter on possibility in Bunge's *Ontology* is characterized by a positive proposal: the proposal to introduce 'real' possibilities.

Classical modal logic attaches modal operators as operators to propositions. In ontology such modalities have no use, if, as Bunge proposes, a strict distinction between concepts as constructs and things as reals is preserved. The modalities he introduces are then rather 'real modalities' (and in the first place 'real possibilities'). These possibilities are operators applied to 'real' entities. Now we have seen that the only real entities in this *Ontology* are things (substantial individuals endowed with real properties). We should thus expect the following three concepts: (a) possible things, (b) possible properties, (c) possible individuals. Bunge does not follow this course directly, however, but rather introduces first the concept of fact, and then the concept of 'possible fact'. This procedure can be defended (but it is astonishingly close to classical modal logic because, as so many authors have proposed, facts can be considered as correspondents of propositions); the other method (defining possible things first) seems, to us at least, closer to Bunge's central inspiration.

That modalities have to be introduced at all in a thing-metaphysics shows once more (as did already the introduction of weights for properties, and the introduction of natural kinds) the influence of Aristotle on Bunge's materialism. His argument in favor of the existence of real possibilities is the following one: if change as such is to be real, then future states must already be present in some way in earlier states (otherwise, change is necessarily a discontinuous shifting and jumping from state to state without the possibility of explaining the future by the past). One can agree with this argument, but one may ask whether the introduction of degrees of reality as such would not serve the same function as the introduction of the specific way of gradualizing reality that is the introduction of potentiality (= real possibility). We come back later briefly to this remark. It is however remarkable that a *modal materialism* (we find this mixture in Marxism also) is presented in a modern form. The originality of the approach shows itself once more.

We must ask, however, what is a fact, and what is the structure of the set of facts? A fact involving a thing (p. 169) is either a state of that thing or an ordered couple of states representing events.

The reader will remember that for us states are defined in a different way than they are in Bunge's work (either as complete sets of real properties, or as 'limits' of processes). But even in Bunge's approach states are very specific entities and possible states must be states coming into existence (if not, then the argument in favor of their introduction does not hold water). But, as happens more than once in the present *Treatise*, the author, instead of looking for the specificity of states and of changes of states, introduces very fast (and without any discussion (on p. 170)) (a) the set of possible facts as a Boolean algebra (neglecting both the meaning of 'fact' and the meaning of 'possibility') and (b), in his Postulate 4.1, he gives even to the set of facts the same structure as to the set of 'possible facts'. He has no intention of making the second move (as is shown by his discussion with Suppes, p. 171, and by his interpretation of the concept of fact, p. 170) but there remains the ambiguity in the expression of the postulate that should be eliminated. Even if this ambiguity disappears, however, we still have before us a universe of possible facts that has received a structure (Boolean algebra) without reference to the meaning of 'fact' as state or as change of state, and without reference to the meaning of 'possible' as a function of change. This procedure is, as far as we can see, not in accordance with the intention of constructing an ontology.

The actualization operator that applies to possible facts in order to make real facts of them has once more the same Boolean structure. Bunge's desire not to say that the union of any two real facts is a fact is, in reality, counteracted by the Boolean nature of the set F of facts and of the operator of actualization A.

For us, then, the structure of possible facts remains undetermined if we are not able to derive it from the properties of states and changes and from the properties of possibility as such.

In the present context it is obviously the last topic that must interest us most.

We have to expect that (a) possibility depends upon the changes that are occurring or that will occur but that (b) many possibilities remain unactualized. These two basic properties of possibility are taken into account by Bunge, when he compares Chrysippian possibility to real possibility as lawfulness.

Chrysippian possibility, in its historical form, defines 'x is possible' by means of 'x is real now or will be real in the future' (time is thus presupposed by Chrysippos, and not yet introduced by Bunge). Not being able to take this definition over in its literal form, Bunge defines a relation between facts "if fact $f1$ entails the inexistence of fact $f2$ then $f1$ hampers $f2$" and "a fact is possible if there does not exist an actual fact hampering it." This modified Chrysippian possibility as it stands cannot be accepted by us (because the actualization operator has purely conventional properties and is applied to a space that itself has purely conventional properties).

We think that the first part of our intuition about 'real possibility' is better taken care of as follows: (a) Let us consider a process. Later stages of a process are possible with reference to earlier stages (stages are not punctual entities but directed intervals), if earlier processes have existed the initial segments of which stand to final segments in the same relations as the already realized part of the process stands to the parts called 'possible'.

The concept of 'hampering' as introduced by Bunge is too simple, and both too weak and too strong: it is too strong because a fact may hamper another without preventing it altogether. We say that a process (instead of fact we use only processes; in Bunge's definition anyway parts of the world of facts) presents *tendencies* towards the prevention of other processes, events, or states. The concept of tendency will be commented upon later. The strength of these tendencies measures the degree of hampering of one process (or event or state) by other processes.

The degree of possibility of states, events, or processes is measured by the strengths of favorable tendencies compared to unfavorable tendencies. It thus seems to us that possibility, related to tendency, is by its very essence a measurable entity. This will be important later on (and allows us to prove something that Bunge is compelled to postulate). Having introduced a weaker concept of 'hampering' (as a function of changing processes), we see immediately that it is also a stronger one. Indeed while the relations between Bunge's 'facts' can be arbitrary, our own 'obstacles' are necessarily connected with each other by means of the tendencies figuring in their definition.

The remarks just presented make us define a possible thing as a thing such that reality tends in part towards it, and that those parts of reality preventing it are neither too overpowering nor too durable.

Now we must take care of the second part of the meaning of the possible: unactualized events can be possible. Bunge tries to give an account of this feature by introducing the concept of lawful event.

Even if the concept of law is adequately strengthened (in the direction we proposed) we would suggest however that it is not sufficient for defining real possibility to identify it with the lawfulness of a fact.

We would consider a process or event (rather than a state; or, if necessary, a state in our modified definition of state) to be really possible if it is simultaneously lawful *and* satisfies the conditions of our modified Chrysippian possibility.

A second reason to link these two conditions together is that either the lawfulness of facts is defined with reference to all true laws (and then it is operationally impossible to find out whether a fact is possible) or it is defined with reference to all lawlike accepted propositions at a given moment (and then this concept of possibility is extremely relative). We are not saying that we overcome this dichotomy between relativisation and loss of operationality by adding tendential possibility and lawful possibility in order to reach real possibility. But we add at least some factual features that are both less relative to the knowledge of the moment and still operationally useful. Moreover, this addition of tendential to nomological possibility has the advantage of overcoming a lack of symmetry between Bunge's treatment of necessity and possibility. His necessity

UNIVERSITY OF BRISTOL LIBRARY

is defined with reference to other facts. A fact is really necessary if it is really possible and moreover entailed by actualized facts (entailed we suppose according to the laws that define lawfulness). Given Bunge's definition of real possibility not referring to other facts, and his definition of real necessity referring to them, it is easy to understand that necessity cannot be defined for him by means of possibility. We, to the contrary, would preserve this connection present in other modal logics by considering a really possible process as a later stage of a process already actually going on. Moreover, this change in terminology allows us to disagree with the assertion (on p. 176) that "the prefix 'necessarily' is unnecessary." On the contrary, explaining an event is showing that a seemingly contingent event was necessary. We can keep this usage of actual science on the basis of our definitions of real possibility.

Bunge has however done the philosophical public the great service of taking real possibility seriously, even though we again disagree with him on points of detail.

We have followed him until now in his decision to consider 'real possibility' as a property of facts. Our only deviation has been to consider processes and events as the most eminent facts, and to use a definition of state that did not coincide with his.

Let us now return to his basic reism (which we combine with dynamism). As the eminently real objects are things, we ask ourselves what it means for a thing to be really possible.

We are inspired by examples (on p. 178) to give the following definition: *A* '$t \in \mathrm{Th}$' (a thing element of the set of things) is really possible if (a) all the parts that together make up t in one given decomposition of t exist (for at least one given decomposition); (b) there are other objects u in Th, with other parts such that the relations between the u parts are either identical or analogical to those between the t parts; and (c) there exists (or has existed) a process such that when applied to an existent t', the result of it would be t. We are not claiming that this definition constitutes a real guarantee for the existence of t in the future (if we may use that word) but this type of definition (and descriptions which might strengthen it and which the reader may easily introduce) could be used to define the possibility of states (a state is possible if a thing presenting that state is possible), of events, of processes and thus finally also of those doubtful categories joining states to events that are called facts.

This family of definitions, even though quite clearly related to the earlier family we studied, makes it unnecessary for us to believe that "empirical criteria of possibility are impossible" (179).

Having examined possibility, we must now follow Bunge's treatment of (a) causal disposition, (b) probability, and (c) chance propensity. We shall end our comments by making some observations on the relation between Bunge's theory of possibility and classical modal logic.

If $z = (x \dotplus y)$ is a thing composed of different things x and y, then P is a disposition or a causal propensity of x, if there is a property Q of y, such that $x \dotplus y$ has a property R possessed by neither x nor y, and if z possesses neither P nor Q. The prototype of this complex definition is: sugar and water each possess specific chemical structures, sugar put into water dissolves, and this is due to the solubility of sugar and the sugar dissolving efficiency of water, two properties only actualized when both are brought together. According to us the definition is not tight enough: (a) the simple addition of x and y must be strengthened by an addition of a binding relation between the two added things, and (b) the relations between R, P and Q must be such that R implies the existence of P and Q in the parts of z. If, moreover, P and Q imply the possibility (or rather real potentiality) of $x \dotplus y$, then according to the strength of this factual implication, we can say that there is not only a disposition for x to present Q but even a tendency for x to present Q. We have defined possibility by tendency earlier, and here we define tendency by possibility. This would commit us to a vicious circle, that can only be avoided by recursive definition (defining by basic tendencies less basic possibilities, themselves defining derived tendencies and so forth). The basic problem of tendency remains unsolved. When we enrich the definition of disposition in this way (and we think that the examples given entitle us to do so, and even compel us to follow this way), we must ask ourselves (even if we provisionally do not answer the question), whether all changes are not the realizations of preexistent dispositions and whether this assertion is not the adequate condition for realizing the possibility of real change (under preservation of thing-identity).

Bunge introduces probability in a very classical way as a non-negative, real valued function on the set of possible facts, completely additive and normed. Once more, the pragmatist Bunge and the ontologist Bunge enter into conflict. The ontologist, if he had his way, would have to ask: why should probability as a measure of degree of possibility be applicable? Are possibilities defined in such a way that they necessarily can and must be measured? If they are so defined, what in the very concept of possibility does compel us to state that the possibility-measure should have these properties? But the ontologist Bunge at certain times remains silent in front of the pragmatist Bunge who, aware of the great usefulness of probability in statistical mechanics and quantum theory, introduces the function *ad hoc*. Once this probability is introduced, it then receives an interpretation and naturally it receives the propensity interpretation. We again (as in most cases) agree here with Bunge: probability needs an objective interpretation on the one side, and not a phenomenal but an ontological one. The propensity interpretation is the only one fulfilling these conditions. But as the reader knows (one has but to look at Tuomela's reader on *Dispositions* to be aware of it), there are many propensity interpretations of probability. On p. 191, Bunge defines the probability of a state *s* for a thing *t* as the strength of the propensity of that thing *t* to be in that state *s*. The conditional probability is the strength of the propensity of that thing to reach a state *x* being in a state *y*. The concept of 'propensity' is not defined unless the concept of tendency is defined (see p. 192 "propensity or tendency"). Strangely enough, Bunge nowhere defines tendency. We made a short attempt to define it as a special case of disposition. We could elaborate this attempt as follows: if a thing *t* is in a state (or process – the distinction is not essential here) *i*, and if the possible histories that *t* can have following *i* contain in most cases the state *e*, then *t* in *i* has a tendency to get into the state *e* (measured by either the number of histories in which *e* occurs or, more adequately, the intensity of the forces actually together with *i* and driving *t* to the state *e*).

Here, however, the concept of tendency is defined by means of a concept of force that is more general than the usual concept of force, but that seems unavoidable at this point. We claim that, exactly as Bunge introduces a generalized energy (as the only property existent in all things and everywhere additive), he has to introduce a generalized force (to reach tendency, and by means of tendency, propensity). This is a fundamental remark, raising a problem that we cannot solve at the present moment. Let it be clear, however, that the definition must be an objective one; the probability function as introduced here is so clearly dependent upon a probability space (and this probability space has been so clearly defined as a Boolean algebra in a purely formal way) that we cannot consider the present introduction of probability as realizing the aim of finding an ontological probability (an aim obviously to be pursued).

We think that the introduction of forces to explain tendencies, and of tendencies to explain propensities, leads necessarily to a causal theory of probability. This is a consequence of great importance, because it makes it impossible for chance propensity to be an irreducible notion, as Bunge would have it. In fact his own definition of 'chance propensity' shows that he cannot defend such an irreducibility. On p. 197, in def 4.14, we consider again a composition z made up out of a multiplicity of things x and y. Q will be a chance propensity if (a) Q is representable non-trivially as a random variable or a probability distribution, and (b) y has a property R such that when the addition occurs, the probability that x acquires the individual property q (a value of Q) is equal to p. This definition of propensity offers us, in a clear circle that we can only consider to be vicious, twice the concept of probability presupposed (a) in the second condition (where y attributes to Q values with given probabilities) and in the first condition where the concept of random variable and probability distribution is used. Now how could probability be defined by chance propensity, if chance propensity did already presuppose probability? It is obvious that, to get out of such a circle, probabilities must be strengths of causal tendencies to realize certain states. These strengths, relative measures of forces, belong to the realm of causal propensities.

Bunge will presumably reject this analysis and claim that the interpretation of quantum theory entails the irreducibility of probabilities. We can only state (a) that we accept the objection but (b) that we consider this incompatibility to constitute an ontological argument against the irreducibility of random variables, *even* in quantum mechanics.

The statistical way of looking at the world seems to us to be basically derived while the causal method seems to be basically primary (even if by introducing the category of real possibility we have left behind us the program of Spinozistic determinism).

Only if we relate real possibility (i.e., causal possibility) to real probability can we define the one by the other, and this occurs when we take the propensity theory seriously, but it does not occur when we leave essential parts of it ('tendency', 'force') undefined or give them only an empirical and not an ontological interpretation.

We want to end these comments on Bunge's theory of possibility with some remarks. Bunge has the tendency to reject modal logic very strongly, both because of its syntactical properties and because of its semantic interpretation. One can easily understand this because modal logic has indeed neglected the ontological possibility Bunge is legitimately concerned with. However sometimes his rejections seem to be too strong: (a) on page 200, he tells us that "science has little use for absolute possibility." We agree to a certain extent but we want to add that both Georg Henryk von Wright and Nicholas Rescher have presented calculi of conditional modalities. Modal logicians are not as strongly identified with absolute modalities as Bunge thinks. (b) Since Kripke, modal semantics has been introduced with reference to relations between possible worlds, (202–204) and Bunge rejects these possible worlds for many reasons. We want to suggest (even though we do so with diffidence) that realistic definitions of possible worlds have been suggested. A first interpretation of a possible world presents it as a part of the real world (the uniquely systematized real world) that has sufficient closure and is sufficiently close to the totality to be considered as a world. In a second interpretation (due to Cresswell, among others), a possible world is a redistribution of the things present in the present world changing their external and not essential relations (we are not yet allowed here to speak about space, but Cresswell thinks about possible worlds as redistributions of matter over space). One could state this in a more general way: if laws are not deducible from facts nor facts from laws, then the application of the same laws to other boundary conditions or (in a more remote sense) of other, somewhat related, laws to the same boundary conditions would be the set of possible worlds. As long as the split between laws and boundary conditions is allowed to exist, even with reference to present-day science, it does not seem to be the case that possible worlds are entirely unconnected with reality.

But we do agree that real possibilities have to be defined with reference to the one and unique real world and not with reference to possible universes. To have undertaken this task is one of Bunge's great achievements.

7. REMARKS ON CHANGE

In this last and final part of our comment on *Ontology* we shall analyze:

1. The theory of processes and events;
2. The theory of frames;
3. The theory of interactions.

Our comments will be inspired by the main philosophical problem of change that is the following one: how can a thing both change and remain. This problem can be related to all the other basic questions: how can a thing change both quantitatively and qualitatively, both continuously and discontinuously, both in a deterministic and a random fashion? To be sure, Bunge is also inspired by these basic questions but he solves them in his usual masterful but *ad hoc* fashion: by introducing *ad hoc* definitions and postulates, imposing upon his universe the necessary wealth of structure so that it presents all these types of processes and changes.

Recognizing that his method has the advantage of clarity and simplicity, we, to the contrary, want to introduce as few postulates as possible and to understand (if it can be done at all) why things necessarily change, and why these changes take necessarily all these different forms. Even if we fail, the task is worth the work.

7.1 *The Theory of Processes*

We can consider a thing with a finite number of states (or a denumerably infinite number of states) and a thing with a non-denumerably infinite number of states. In general we do not think that infinity has an ontological privilege (quite to the contrary), but infinity can be here a model of fuzziness: a thing with a fuzzy set of states, states not completely discernible from each other, can best be modelled by a thing with a non-denumerable number of states.

For this reason we consider it preferable to introduce processes for things with this maximal number of states. We shall speak about special cases (events are only special processes) later. Here, for once, the mathematical structure will help. We can agree with Bunge that the set of processes is a category and it is a well-known result in category theory that the basis of the category can be replaced by sets of morphisms. This will help us to show that in a trivial way the theory of processes can become the basis for the theory of events, and of states. The fundamental question is – as stated – the connection between things and processes.

Let S_L be the set of lawful states of an object x. Bunge proposes to represent (if the power of $S_L(x)$ is non-denumerable) the evolution of x by an arbitrary representation of this set on itself. Not completely arbitrary, however, because the representation has to be a function. This allows the fusion of things, but not their fission (although Bunge does not discuss this consequence). Any function compatible with the laws on S is thus the representation of a possible process. One particular process will be partially represented by a triple (s, s', g) where s and s' are the initial and the final states of the process and g represents part of the transformation leading from initial to final state (and thus maps the intermediary states, as far as certain properties are concerned).

The set of processes (theorem 5.1, p. 230) is a category (Bunge expresses this result for events, but it holds in full generality for processes):

(a) the basis set of this category is the set of processes;

(b) processes can be composed by the following operation

$$\langle s, s', g\rangle^* \, c\langle s'', s''', h\rangle = \langle s, s''', h \circ g\rangle \text{ if } s' = s'',$$

and $h \circ g$ belongs to the set of G_L of lawful transformations (p. 231);

(c) there is an identity process (s, s, i_G) mapping a state on itself by means of the identity function of the set of functions;

(d) it can be proved that the composition c is associative.

We want to raise the following question: is any function an adequate representation of a process? (Must we not introduce conditions of continuity, differentiability, at least on intervals, analyticity while also imposing restrictions on the behavior of measures over the set of sets represented on each other by these functions?) We are convinced by looking at actual science, that the extreme generality of Bunge's decision, allowing arbitrary functions is going too far in one direction, but his choice of this alternative can be defended by the fact that we cannot immediately give a plausible set of restrictions, ontologically and not purely factually, derived for the admissible law-functions.

Having raised the question, we are not going to propose an answer at this point. Let us go back to the definition of process. We can then immediately define an event as a specific type of process: an indecomposable process. We call 'event' any process such that it cannot be represented in the theory we work in (or, stronger, in any adequate theory) as the composition of a process with another process (except if one of these processes is the identity process). Bunge, for the sake of clarity does first define event, and only later process. We, to the contrary, think it unavoidable to introduce process first. The reason is clear from Bunge's exposition itself: his definition of events by ordered pairs of states cannot distinguish between events and processes because of the fact that the intermediary states are left completely free.

If the set of processes is, as was presented, a category, then we can represent the set of states by identity morphisms of the category. This entails that formally we reduce events and states to processes. Let us remark that the theory of processes does not depend on any specific property of states. This entails that we could apply the theory of processes to the changed concepts of states we have been proposing in earlier parts of this paper. We could even apply it immediately to substantial individuals, properties or sequences of properties (defined themselves as causal functions, as we have suggested).

So far we have been able to follow up Bunge's suggestions, taking his own definitions as seriously as possible and deriving consequences from them that are more important from our point of view than from his.

Now we come to a more fundamental problem, however. On p. 221, Bunge solves, by a principle of 'nominal invariance', what is according to us (and according to philosophical tradition also, we think) the basic problem of change. His principle 5.1 tells us "a thing, if named, shall keep its name throughout its history, as long as the latter does not include changes in natural kind, changes which call for changes of name." The preservation of identity-in-change is thus simply, according to 5.1, a designation rule: "there are no self-identical things but only constant names helping us to keep track of the changes undergone by things." If this last statement means that there are no unchangeable things, we can only agree (indeed we wish to derive this as a consequence and not introduce it as a fact), but if there is no basic relation between the changing and the unchanging part of things, then the whole argumentation in favor of the introduction of modalities, earlier defended by Bunge, collapses and finally it becomes vague what it means to "keep track of one thing" if there are no things any more. We have come full scale from the emphasis on substantial individuals in the first chapter of the *Ontology* to the emphasis on processes we have defended from the beginning. But this is not merely a careless oversight. Bunge is here tackling a problem that has been with us at least from the *Parmenides* on. How, without invariance, can identity be preserved? This problem is indeed crucial (and to stress equally strongly both thing and process is one of the deep insights of Bunge's *Ontology*). But the task is not easy. It is certainly not so easy that we can hope to execute it by introducing a designation principle. There is a basic problem for ontology here: (a) We have to designate things and keep them relatively constant; if not our science is not possible. This is a Kantian remark, out of place in ontology; (b) we make it because if we cannot derive from ontological bases that such near invariance has a temporary existence, then our description of the world as given in science is basically different from our ontology, and the whole program of constructing a 'scientific ontology' (a program we basically adhere to) is condemned to fail.

We think, however, that our own conception of property as expressed above in our 'Comments on Form' is able to give a solution. If all properties are basically causal properties, and if causality, entailing interaction, entails change, then we can at least state that postulate 5.1 is, in an even stronger sense than it has on p. 218, a consequence of the meaning of the concept of property. However, if simultaneously the 'binding interactions' that are the properties, make the substantial individual that is their center different, then at least for a given sequence of processes, these necessary changes are identity-preserving. We have, in other words, defined our properties in such a way that they entail a concept of identity. This concept of ontological identity is basic, however, and typically different from logical identity (because it is accompanied by change of properties). Here the ranking of properties as more or less essential is evidently crucial. An individual is destroyed if it loses one property of the highest rank of essentiality.

Having offered this idea (not worked out at all and waiting for further analysis) as a possible solution for a problem that Bunge tries to solve "by cutting the Gordian knot" (as he is wont to do; perhaps rightly so if he is to reach a workable theory; but we cannot follow him, being less impressed by axiomatic decisions than by the natural development of a system of concepts), we are now ready to derive the truth of a certain number of Bunge's own postulates.

1. P5.2(p. 239): every realistic state function has at least one continuous component. We know that this is agreeable to us: every realistic state function has at least one fuzzy component.

We think that the fact that individuation occurs by means of interaction at all possible levels of reality prevents us indeed from having a universe with a punctual basis. Even the elements deepest down in the cosmological hierarchy are partially overlapping with certain others.

2. Postulate 5.5(p. 242): the rate and extent of change in anything composed of a finite number of things is bounded.

We do not derive this, as Bunge does, from a postulate on energy (pp. 240–242) but again from the principle of individuation: if the sequence of individualizing properties of a thing is complete, it must be finite (infinity must be incompleteness), and if it is finite each of these individualizing processes must exhaust itself in due course (because if it did not, it would be able to eliminate by its influence in some dimension all other individuals and thus be incompatible with the existence of a world containing multiple individuals).

3. The same type of reasoning gives us the consequence that all real processes are impure (i.e., are to be represented by a multiplicity of functions). This is the case because all real processes are concrete and because being concrete they are multi-faceted, in exactly the same way and

for exactly the same reason as concrete things are multi-faceted. But the definition of property by relation yields still another consequence: no process can be considered to touch one substantial individual only. This would, in a future version of an ontology, compel us to define processes always over mixtures of individuals and thus would introduce even a deeper type of impurity than the one we are talking about at present.

4. Because processes are concrete they are finite, and necessarily also have beginnings and ends. And because substantial individuals are concrete and thus finite, they necessarily show at least at some points (beginning and end) discontinuous changes.

This simple remark does not solve the problem of continuity and discontinuity, however. Before leaving this topic, we have to make two remarks: (a) Continuity and discontinuity in the mathematical sense of the word are not to be considered as synonymous to continuity and discontinuity in the ontological sense of the word. Bunge's strategy has unfavorable consequences here. (b) The fact that individuation occurs through interaction entails that in all processes contributing to individuation a continuous component (the interactional one) and a discontinuous component (the individuating one) are to be combined.

5. We cannot derive from our premises the infinity postulate that every process is preceded and followed by other processes. If it is to be justified it must be justified on other foundations.

6. Assume that we have a multi-level universe (as Bunge certainly wants us to have, and as follows from the existence of emergent properties – an existence we did not yet comment upon but that follows from the fact that a multiplicity of properties also presenting real interactions between each other must, when things are conjoined by binding relations and only then give rise to new and emergent – though dependent – properties). Then the processes occurring on these various levels must occur with different velocities; they must have more or less deep changes as their effects (the depth of changes is measured by the number of properties modified and by the rank of essentiality of these properties; for us every change is a qualitative change). But changes that have an impact on more essential properties (excluding the maximally essential ones, where changes entail the destruction of things), can be opposed to those that modify, even on superficial levels, properties the values of which can be ordered on a linear scale. Bunge's distinction between quantitative and qualitative change is otherwise defined; he uses the relation to coordinate

systems defined by frames of reference. We shall come back to frames of reference later on, but we have, without introducing them, sufficient means at our disposal to classify types of change quantitative or qualitative.

Bunge introduces an important postulate when he states that every real process presents at least one random component. As the reader knows already from our remarks on probability, we could not condone the classical definition of random component. We could, however, in the Kolomogorov manner, define a random process as a process whose description would necessarily have a complexity equal to that of the process itself. The concept of description is not an ontological concept and we cannot use it in ontology. If we accept the postulate however that every process is reflected (a) on all hierarchical levels or (b) has some impact upon, and submits to, the impact of all other processes, then we can say not strictly but, within a certain approximation, that every process has components of complexity as high as the set of interactions of all individuals with all individuals or (if Bunge's version is preferred) of all levels with all levels. This would show, without compelling us to accept randomness as such, that Bunge's postulate (as in almost all cases) has a kernel of ontologically derivable truth (without being necessarily defensible as it stands).

If change is not a discontinuous jump, something must be preserved during the change. This introduces into the definition of every individual changing thing the concept of stability (a changing thing being stable if the effects of the change remain within a certain subspace of the space of states, in Bunge's language).

Finally, we do not consider the problem of reversibility or irreversibility as a pseudo-problem, as Bunge does. It is one for him because he accepts (as physics does in general, but not everywhere) the distinction between laws and boundary conditions as logically necessary in all cases. If this distinction is taken as such to be without problems, then we can naturally say that even though laws are invariant under time-reversal, this does not entail processes to be reversible because the initial circumstances may be such that the laws entail irreversibility. How did these initial circumstances come about, however, in a universe supposedly completely governed by time-reversible laws? This basic problem of change has not been treated by Bunge, and it remains for us one of the basic riddles (even though he considers it to be without importance). Any theory of change must take it into account and a future version of the ontology should be centered on this difficulty. We do think however that we can already come, with the present concepts, to a strange conclusion: if the universe is to be a radically irreversible system then it must be a hereditary process (determined by its whole history).

We could go on making remarks about the interaction of various types of change. We do not want to take too much space, however. But we still want to make the following basic point: taking our (and Bunge's) commitment to dynamism and to reism seriously, the attempt to solve the problem thus created has suggested to us possible derivations (that have, to be sure, to be made more precise) for assertions that in Bunge's *Ontology* are introduced as purely isolated postulates.

7.2 *The Theory of Frames*

The concept of frame has played a central role from the moment Bunge has introduced his model-functions. We did not discuss it at that point because, as the reader will remember, we tried to do without these model functions (considering their introduction to be a step in a non-ontological direction). However it is remarkable that Bunge who uses them in effect from the very moment he introduces properties, does not define frames of reference until much later (model functions are with us from chapter 2 on, and yet frames are defined (even though continuously used to characterize states) only in chapter 3. We want to look into their definition, modify it to a certain extent and then derive the existence of frames of reference from ontological viewpoints.

Bunge's definition of frame of reference is to be found on p. 265, def 5.36: if X is a thing represented by a functional schema (M, F) and if f is another thing with states in $S(f)$, then f is a reference frame for X if and only if:

(1) $f \dotplus X$ is such that no binding relations exist between f and X;

(2) the domain M of the state functions of X equals the state space of f: $M = S(f)$.

We have somewhat changed his definition because Bunge uses the concept of history that we only want to introduce later. However the gist of the definition has remained the same. Before starting our discussion we want to make two minor remarks. (a) The absolute inexistence of interaction between f and X is presumably not necessary; let us only suppose that in comparison to most other interactions of X the (f, X) interaction is weak. (*b*) Moreover, the identity of the two state spaces is not necessary (and usually not possible) either. So let us only introduce a more or less strong correspondence (in the best case an isomorphism). As Bunge himself remarks, his treatment has the advantage of calling a reference frame a real thing (which, as such, is itself susceptible of change). We think we can even strengthen this advantage in one direction: we can define the concept of 'reference frame' without introducing functional schemata. Let x be a thing. Then a reference frame for x is any other thing t' such that there is no strong interaction between both, and the states of the one correspond sufficiently to the states of the other. The assertion that "things have no intrinsic states, but only states with reference to their reference frames" corresponds then to the following assertion: "things have only relational properties and these relational properties can only be individualized by their weak interaction with things having analogous relational properties." Or, if one wants another and in a certain sense dual version: "things have only relational properties that can be individualized by the effect-maps they leave on various other things having complexities analogous to their own" (in the second version of the criterion we abandon the lack of interaction of frame and thing in order to introduce a strong unilateral interaction between thing and frame; in the first version we compare the things emitting their effects as rays to similar things having similar rays). This 'ontological reference frame' concept confirms one of the assertions that, according to us, is present at various moments in Bunge's work but in a hidden way. Causal binding relations must be present from the very moment we introduce properties. We see here that his attempt to develop a *realistic relativism* (frames as real objects, and correspondences with frames as real relations) leads Bunge exactly in the same direction. He does not want to be led in that direction, however, and for that reason he immediately introduces the conceptual construction of the functional schema that lies, according to his dualism of thing and concept, outside of the world of real objects.

7.3 *Interaction and Causality*

Let us call the sequence of states that a thing runs through from its beginning to its end, the history of this thing. A history is thus for Bunge a member of the n-th Cartesian product of the state space of the thing with itself (for a duration of n moments). For us, who prefer processes as basics, a history of a thing is the ordered series of all processes the thing has participated in.

Now we define a connection between two different things. The addition + is used as connection by Bunge. We, when talking about real addition already introduced a real binding relation for addition to be possible, and for that reason we must presuppose the concept of action that Bunge is here trying to define. We come back to this point later on. Def 25.28: "a thing changes by itself if for any other thing $h(x + y) = h(x)$ union $h(y)$," is not applicable in our own ontology because addition always entails a link. We can however compare various strengths of links (measuring various degrees of change, measured by distances defined between states or between processes). We shall say that a thing changes relatively spontaneously if the type of change it undergoes, for any type of system y it is linked to by addition, is either the same for all y (and is thus non y-specific), or only concerns an nth part (part of a part of a part) of the thing x. (Here n can be conventionally or intrinsically defined.) If, after being linked, x follows another history according to the object it is linked to, while the converse is not true, then y acts on x without x acting on y (again this cannot be strictly true according to our definition of link, but again we can rephrase the definition by making the change in the y history small in comparison to the change in the x history).

Postulate 5.10: "everything acts on and is acted upon by other things" is for us evidently not a postulate but a consequence of our definition of property. Postulate 5.11 according to which everything changes spontaneously in some respects and under constraint in some others, is again a consequence of the fact that if the thing has to be a proper individuality expressing itself in its ways of change, some changes must be spontaneous (i.e., determined, if determined at all, by internal factors) while on the other side it must necessarily submit to the individuating 'repulsion' of many other things (forms of 'repulsion' a metaphor that constitutes the individuality of these other things).

Given the fact that the concept of real link or bond is so crucial for us (in opposition to Bunge's present treatment, but not, we trust, in opposition to the whole spirit of his work) we are convinced that early in the chapter on properties we should introduce an axiomatization of the concept of either action, or bond. The place of this crucial idea is not at the end of the doctrine of change but in the center of the doctrine of individuation.

We leave this problem open for the moment, referring the reader to our *Matière et forme*, volume 1, in which, though in a provisional way, we made an attempt to give such a characterization.

CONCLUSION

Bunge's ontology does not end with his doctrine of change. A very important chapter on space-time concludes the work. We do not feel called upon to comment upon it here, however, both because of the fact that no new ontological concepts are introduced and because of the fact that an adequate critical analysis of it would call for parts of general geometry, while Bunge feels free to use, here as elsewhere, the classical concepts of geometrical thought.

We have followed the *Ontology* for a long while. The reader ought not to think that our numerous proposals to change definitions and axioms are expressions of antagonism. On the contrary, we have accepted as thoroughly as we could the ontological point of view as separate from and independent of the epistemological point of view. Many of our deviating proposals derive from this commitment. We have also accepted his mereological (*pace* Bunge) materialism, his theory of things, his theory of natural kinds and essences, his possibilism and his dynamism. But everywhere it has been our aim to weed out as best we could the remainders of formalism, to stress both concretism and dynamism with equal force, to emphasize the importance of process and interaction as much as that of individuation, and to reduce the number of axioms by deriving some of them. This commentary cannot, however, reach the standards of explicitness and clarity of the original.

We should be very happy if our questions encourage others to enter into ontology and to rebuild it as the living science it once was and it deserves to be. Mario Bunge has had the courage to take a decisive step in that direction. Let others continue.

Rijks Universiteit Gent
Vrije Universiteit Brussel

MÁXIMO GARCÍA-SUCRE

A KIND OF COLLAPSE IN A SIMPLE SPACETIME MODEL

ABSTRACT. We have studied a kind of collapse that may occur to a spacetime in the framework of a simple model introduced in previous papers. All the concepts of our model are defined in terms of only two primitive concepts, namely those of preparticle and of membership relation of set theory. Preparticles are considered to be the most elementary components of matter. Particles are represented by sets of sets of preparticles. A point of a spacetime is represented by an equivalence class of points of crossing between particles such that all these points of crossing have the same structure. We introduce two postulates. Firstly, that two points of a spacetime appear distinct from each other only because they have different structures. Secondly, that the extension of a spacetime can be defined only with respect to a given collection $\mathscr{C}$ of physical systems (in the sense that only those points of a spacetime crossed over by particles belonging to physical systems entering in $\mathscr{C}$ are the points which contribute to the extension of this spacetime). We then show that if the extension of a spacetime is defined with respect to only one very large physical system, this extension vanishes.

INTRODUCTION

We will be mainly concerned here with a property of a certain kind of global fields (which we will call complete fields) within the framework of a simple model of physical systems, according to which any regular part of such a global field (necessarily covering completely this field) reduces itself to only one point. This property is related to the concept of a detector physical system which we have introduced in a previous paper together with the concepts of spacetime, reference frame, and energy, momentum, frequency and wavelength of a physical system in a given reference frame [García-Sucre 1979]. We have paid particular attention in defining 'detector physical system' starting from the same primitive concept in the terms of which all the other concepts of the theory are defined [García-Sucre 1975, 1978a, 1979]. In this way, detector physical systems are incorporated on the same footing as any other physical system. In fact, there are physical systems fulfilling certain conditions (to be specified later) related to the mutual modification that detector and detected physical systems produce on each other due to their interaction. Furthermore, interaction is described here as a purely physical process. We exclude any intervention of

J. Agassi and R.S. Cohen (eds.), Scientific Philosophy Today, 45–69.
Copyright © 1981 *by D. Reidel Publishing Company.*

mind in the detection process other than the interpretation of the resulting data and the proposition of how to perform the detection in question according to the theory in mind. [On this point we follow Bunge's point of view, as explained in *Foundations of Physics* [1967].] Also, we explain clearly what the interaction between detector and detected systems consists of, since the concept of interaction between physical systems is a derivative concept in our model, and it happens to be such that the interaction process can be completely characterized in terms of the most elementary components of a physical system [García-Sucre 1978a, 1979].

We will show below that the property mentioned above according to which the regular parts of complete fields reduce themselves to only one point has an important consequence for the framework of our model. Namely, that the extended character of spacetime (a global field) arises only when one reconstructs spacetime by assembling many regions of spacetime, each one of these being the region detected by one of *many* detector systems. If, on the other hand, we try to detect the whole of spacetime using only one detector physical system having a size appropriate to this detection, the extension of a spacetime (as it will be defined later on) reduces itself to a minimum value. We think that this property may be relevant to physical theories predicting different kinds of collapses in the physical world, since this property could be related to a kind of collapse somewhat different from those which have been considered usually in the literature [see for instance Patton and Wheeler 1975].

Finally, let us point out that our model can be considered a candidate to pregeometry in the sense that it provides a framework in which both geometrical and non-geometrical features of the physical world could arise as a consequence of the properties of the most elementary components of matter; these properties having no relation with the concepts of distance and place. [For a discussion on the necessity of a pregeometry, see Misner *et al.* 1970 and Patton and Wheeler 1975.] Let us point out that, in this sense, arguments have been given according to which geometry as a background for the whole of physical phenomena can no longer be accepted; something more elementary than geometry must be at the basis [Misner *et al.* 1970]. This feature gives an increasing interest to the foundation theories of space [see for instance Penrose 1971; Bunge and García-Máynez 1976], time [see for instance Noll 1967; Bunge 1967, pp. 93–101; García-Sucre 1975], and spacetime [see for instance Basri 1966; Finkelstein 1969, 1972a, 1972b, 1974, Finkelstein, Frye and Susskind 1974; Penrose 1967, 1968, 1975; García-Sucre 1979], since these theories do not assume either space and time, or a spacetime geometry, but instead try to obtain geometry as a consequence of the properties of more elementary entities. Despite the differences existing between these theories every one of them can be considered an effort in the search of a pregeometry. It will perhaps soon be answered which one is nearest to the more appropriate pregeometry, or even the question of whether the search for such a pregeometry is the right way to proceed in the present stage of fundamental physics. Yet, there are some indications in favor of the search for a pregeometry as the background of the whole of physics [Misner *et al.* 1970; Patton and Wheeler 1975]. In addition to this feature, the above mentioned theories try to answer the following questions of ontological interest: What is space? What is time? What is spacetime? One is thus trying to know better about the basic concepts of physics.

THE MODEL

In this section we are going to give first a brief account of our model which has been explained in more detail in previous papers [García-Sucre 1975, 1978a, 1979].

There are only two primitive concepts in our model, namely, that of preparticle and that of the membership relation $\in$ of set theory. In connection with the last concept, we adopt the whole of set theory in the version of Fraenkel and Bar-Hillel [1958]. On preparticles we say that they are unchangeable physical objects without internal structure. We consider preparticles as the most basic ingredients of the physical world [García-Sucre 1975, 1978a, 1979]. Since preparticles do not change they cannot be divided into parts. Therefore, preparticles are atoms in the etymological sense, i.e., indivisible. Yet, this atomistic hypothesis differs from usual atomism in that preparticles do not interact with each other.

In fact, we will see below in our model that the concept of interaction is a derivative one, which is well defined only when the referents are systems of particles, i.e., when one speaks of the interaction between systems of particles [García-Sucre 1978a]. We also speak of the interaction between particles when the interacting physical system is such that it contains only one particle. Since in our model preparticles are neither systems of particles nor particles, it follows that there is no sense in saying that two preparticles interact or that a preparticle interacts with a system of particles. The distinguishing trait mentioned above between particles and preparticles in connection with the concept of interaction leads to

UNIVERSITY OF BRISTOL

certain advantages with respect to a kind of paradox occurring in atomistic models of the physical world. This point has been discussed in the introduction of a previous paper [García-Sucre 1978a].

In our model, a particle is formed by preparticles in the following precise sense. Consider a set $B \equiv \{\alpha_i \mid i \in I\}$, where I is a finite set of labels, such that its elements α_i's are all the existing preparticles. We represent a particle as a subset of $P(B) - \phi$, where $P(B)$ and ϕ denote respectively the power-set of B and the empty set. On the other hand, we represent a physical system as a set whose elements are sets representing particles [García-Sucre 1975].

Let the set

$$p_i = \{a^i(x) \mid x \in X \text{ and } a^i(x) \in P(B)\} \tag{1}$$

represent a particle, where X is a finite set of labels. We say that the particle represented by p_i is an evolving or a nonevolving particle depending on whether the elements $a^i(x)$'s above, can or cannot be completely ordered according to the proper inclusion relation "$\subset$" [García-Sucre 1975]. Therefore an evolving particle is represented by a chain, since a chain is defined as a set whose elements can be completely ordered by the proper inclusion relation [Fraenkel and Bar-Hillel 1958. pp. 128–131].

An α state $s^i(x)$ of a particle represented by the set p_i above, Eq. (1), is given by

$$s^i(x) = a^i(x) - \bigcup_{x' \in X'(x)} a^i(x') \tag{2}$$

where $a^i(x)$, $a^i(x') \in p_i$ and every $x' \in X'(x)$ is such that $a^i(x') \not\supseteq a^i(x)$ [García-Sucre 1975]. We denote the set of α states of p_i as $\sum(p_i)$. Furthermore we consider that the α states are ordered according to the following rule: $s^i(x) < s^i(y)$ if and only if $a^i(x) \subset a^i(y)$, where $a^i(x)$ and $s^i(x)$, and $a^i(y)$ and $s^i(y)$, are related according to Eq. (2).

To illustrate the above definitions, consider a set p_i representing an evolving particle. This means that its elements $a^i(x)$'s are completely ordered by the proper inclusion relation. Also, its α states $s^i(x)$ are completely ordered in the set $\sum(p_i)$, since according to the ordering rule for α given above, this ordering is induced by the proper inclusion relation. A possible representation of an evolving particle is any finite curve and the elements of the particle are represented by sectors of the curve all going from the initial point of the curve to some point falling on the same curve (Fig. 1). In this way, all the elements of a particle so represented are ordered according to the proper inclusion relation. On the

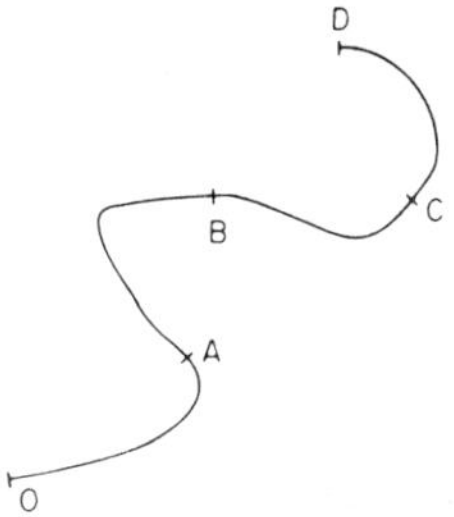

Fig. 1. We illustrate here an evolving particle p by the curve OD. The elements of the set representing p appear in the figure as the curve sectors OA, OB, OC, OD. All the elements of the set representing p which appear in the figure respectively as the curve sectors OA, OB, OC and OD can be ordered by proper inclusion relation ($OA \subset OB \subset OC \subset OD$). The α states of p are illustrated by the curve sectors OA, AB, BC and CD (see Eq. (2)). Consider, for instance, that every molecule of the track of ink of the curve OD stands for a preparticle.

other hand, from Eq. (2) it follows that the α states of p_i are the curve sectors determined by successive end points of the elements of p_i in this figure. Namely, the elements of p_i in Fig. 1 are illustrated by the curve sectors OA, OB, OC, and OD. The α states are represented by OA, AB, BC, and CD. Note that the elements of an α state are preparticles and that we have assumed that the total number of preparticles is finite. Therefore, although we represent an α state by a curve sector (e.g., OA on Fig. 1), this does not mean that the points over which OA passes represent preparticles. Instead, we can say, for instance, that every molecule entering in the track of ink going from O to D stands for a preparticle in Fig. 1. Let us point out here that the order of the elements of p_i and of its α states is completely specified once we specify the set p_i, since the order is induced by the relation $\subset$. Thus, the order specified in Fig. 1 assigning OA as the first element of p_i and OD as the last one, merely illustrates the order already specified by the structure of p_i with respect to the relation $\subset$. Following the same conventions as in Fig. 1, we illustrate in Fig. 2 two different cases of nonevolving particles. In Fig. 2a we illustrate the case of a nonevolving particle represented by a set corresponding to a graph with a tree structure. On the other hand, in Fig. 2b we illustrate the case of a nonevolving particle that is represented by a set having with respect to the relation $\subset$ the structure of a several-branch-graph such that these branches do not

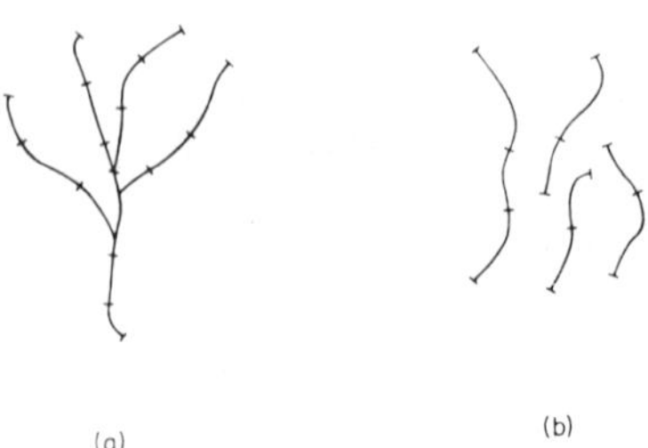

Fig. 2. In (a) we illustrate the case of a nonevolving particle represented by a set corresponding to a graph with a tree structure with respect to the proper inclusion relation $\subset$. In (b) we illustrate the case of a nonevolving particle corresponding to a several-branch-graph with the restriction that these branches do not cross each other. Any other possible case of a nonevolving particle is necessarily a "mixed" case between (a) and (b). If we illustrate a physical system of very many particles with the same conventions as for (a) and (b) above, we should obtain a figure having the aspect of very many figures as either Fig. 1 or Fig. 2 all together forming a network.

cross each other. According to our definition of nonevolving particles any "mixed" case between (a) and (b) in Fig. 2 also represents a nonevolving particle [García-Sucre 1975]. The points of crossing between branches in Fig. 2a arise because the corresponding α states share some or all preparticles, i.e, because the intersections between these α states are nonempty.

Since we represent physical systems by sets of sets representing particles, a diagram constructed with the same conventions used in Figs. 1 and 2 illustrating the case of a physical system may have the same aspect as those of these figures. Of course, if the physical system concerned is represented by a set of very many sets representing particles, the corresponding diagram may have the aspect of an extended and/or dense network. Furthermore, here and there one may have cuts of different sizes depending on the physical system with which we are concerned [García-Sucre 1975, 1978a]. We call each point of the network a point of crossing even if only one particle passes over this point. The most general case will be that of points of crossing where several particles meet each other. More precisely, we represent a point crossing of a physical system S as a pair $(s^i(x); \pi^i_x(S))$; where $s^i(x)$ is an α state of a particle entering in S and $\pi^i_x(S)$ is the set of all the sets representing particles of S which have α states yielding nonempty intersections with $s^i(x)$ [García-Sucre 1978a]. We also call $(s^i(x); \pi^i_x(S))$ a complex of S [García-Sucre 1978a]. We call $s^i(x)$ and $\pi^i_x(S)$ the center and the π-set of the point of crossing $(s^i(x);$

$\pi_x^i(S)$). The structure of $(s^i(x);\ \pi_x^i(S))$ will be that of a center where several branches meet or cross each other, except in the case where only one particle enters in the point of crossing. In the last case, the structure of the point of crossing will be that of a center over which one particle passes, or where a particle either starts or ends. This concept of point of crossing of a physical system is appropriate to describe the structure of a physical system [García-Sucre 1978a]. In this sense, we call structure-set $\sum\sum(S)$ of a physical system S the set to which all the points of crossing of S belong.

To characterize the structure of S by set $\sum\sum(S)$, is equivalent in our example above to characterizing the structure of a network by the structure of the knots of the network (the structure of such a knot can be characterized by the number of filaments meeting at this knot, the length of these filaments, etc.).

To be more precise, we shall start by defining what we understand by particles having the same structure (or similar particles).

We say that two particles p and p' have the same structure (or are similar), if there exists a one-to-one mapping Ψ between $\sum(p)$ and $\sum(p')$ which preserves the order in each of these sets of α states. Hereafter, we call the mappings Ψ's above, similarity mappings. On the other hand, when we have a nonevolving particle represented by a set p_n it may occur that there exists a proper subset $\tilde{p}_n$ of p_n such that all its elements are ordered by the proper inclusion relation $\subset$. In this case, we can follow the same ordering rule for α states of particles and thus order the α states corresponding to the subset $\tilde{p}_n$ given by Eq. (2). We denote such a set as $\sum(\tilde{p}_n)$ and it will be a completely ordered set. We call $\tilde{p}_n$ a branch of p_n, and, according to what we have said above, a branch of a nonevolving particle has the same properties as a set representing an evolving particle. Thus, in the same way as in the case of two particles we can say that a particle and a branch or that two branches have the same structure when the same conditions as those of similarity between particles are fulfilled.

Let us now denote the relation of similarity between particles and/or branches as R (e.g., when the particle p_i has the same structure as the branch $\tilde{p}_j$ of the particle p_j we write that $p_i R \tilde{p}_j$).

Consider two points of crossing $\sigma_i \equiv (s^i(x);\ \pi_x^i(S))$ and $\sigma_j \equiv (s^j(y);\ \pi_y^j(S))$ and let $\pi_x^i(S)/R$ and $\pi_y^j(S)/R$ be the quotient sets of the π-sets of σ_i and σ_j, with respect to the equivalence relation R. Thus, an element $\bar{p} \in \pi_x^i(S)/R$ is a set of particles and/or branches which are similar to each other. We say that the points of crossing σ_i and σ_j have the same structure or are similar to each other if there exists a one-to-one mapping Φ between the quotient sets $\pi_x^i(S)/R$ and $\pi_y^j(S)/R$ fulfilling the following conditions:

(i) Two equivalence classes $\bar{p}_i \in \pi_x^i(S)/R$ and $\bar{p}_j \in \pi_y^j(S)/R$ of evolving particles and/or branches of nonevolving particles in correspondence according to $\bar{p}_i \overset{\Phi}{\longleftrightarrow} \bar{p}_j$ are necessarily such that any $p_i \in \bar{p}_i$ has the same structure as any $p_j \in \bar{p}_j$.

(ii) For any two $p_i \in \bar{p}_i$ and $p_j \in \bar{p}_j$ there exists necessarily a similarity mapping Ψ which puts into correspondence α states $s \in \sum(p_i)$ and $s' \in \sum(p_j)$ having nonempty intersections with the centers $s^i(x)$ and $s^j(y)$ respectively.

We denote the similarity relation between points of crossing as $\sim$. The above definition of similarity between points of crossing is somewhat different from those introduced in previous papers [García-Sucre 1978a, 1979]. Here the one-to-one mapping Φ is between the quotient sets $\pi_x^i(S)/R$ and $\pi_y^j(S)/R$ instead of being between the π-sets $\pi_x^i(S)$ and $\pi_y^j(S)$. We have introduced this change for the sake of being consistent with our assumption below according to which two physical entities are distinct from each other only if they can be represented by sets having different structures.

Using again our example of the network above, consider a network such that its knots stand for the centers of the points of crossing belonging to the structure-set $\sum\sum(S)$. Furthermore, assume that the filaments of this network stand for equivalence classes of evolving particles and/or branches of nonevolving particles. Then we say that two points of crossing have the same structure if we can put their centers (knots) one upon the other in such a way that the filaments meeting at one knot reproduce exactly the filaments of the other knot.

We are prepared now to introduce the concept of a field produced by a physical system [García-Sucre 1978a, 1979]. The field $f(S)$ produced by a physical system represented by S is in turn represented by the quotient set $\sum\sum(S)/\sim$. Furthermore, the points of the field $f(S)$ are represented by the equivalence classes belonging to $\sum\sum(S)/\sim$.

The reasons to choose the set $\sum\sum(S)/\sim$ to represent a field (for instance, instead of $\sum\sum(S)$), have been given in previous papers [García-Sucre 1978a, 1979]. Let us point out here only the following aspects of the question. The points of a field are represented by equivalence classes x's to which points of crossing having the same structure belong. The connections between two points, say x and x', are established by the particles passing over the centers of at least two points of crossing, one of them necessarily belonging to x and the other point of crossing to x'.

We say that the points represented by x and x', where $x, x' \in \sum\sum(S)/\sim$, are connected by particles fulfilling the condition that we have just described above [García-Sucre 1978a]. The set $\sum\sum(S)/\sim$ can thus be seen as a set of points (equivalence classes), connected by sets representing particles which enter in the physical system producing the field under consideration. The specific way in which these points are connected depends upon the particles entering in the physical system represented by S. We can have points either connected by many particles or not connected at all. Thus, fields may present the aspect of a network with regions where the points are connected in nearly the same way and/or regions presenting strong irregularities in this concern, and finally regions where cuts may be present. As we have already stressed, all these features will depend upon the physical system producing the field under consideration.

On the other hand, it can be shown that we can define reference frames in our model in such a way that they can be thought to be inside the fields [García-Sucre 1978a, 1978b, 1979]. By means of such frames we can unambiguously ascribe space and time coordinates to points of a field, each frame corresponding to one of the possible ways in which space and time coordinates can be assigned to points. One definition of such a reference frame is the following [García-Sucre 1979]. Each α state of a particle can be interpreted as a stage of the history of this particle [García-Sucre 1978a]. Then, consider a subset τ of the evolving particles entering in S and take into account only the points of $\sum\sum(S)/\sim$ which are connected by the particles entering in τ. Also consider a subset σ of evolving particles connecting the particle entering in τ. Then, each evolving particle belonging to τ orders a subset of points of $f(S)$, ascribing to each such points a time coordinate. One can do this unambiguously if the particle p of τ under consideration passes only once over a point of $f(S)$ and if each α state of p has a nonempty intersection with centers of points of crossing entering in only one point of $f(S)$. Then, the order of the α states of p induces an order for the points of $f(S)$ over which the particle p passes, thus completely ordering these points of $f(S)$. The time coordinate of such a point will be the ordinal of the corresponding α state of $\sum(p)$. In this sense, the particles entering in τ play the role of clocks. The ordered set of points of $f(S)$ over which the particle p passes will be the trajectory of p in $f(S)$ [García-Sucre 1978a, 1979], which we denote as T_p^f. If the trajectories

T_p^f's of the particles p's entering in τ do not cross each other we can think of them as parallel oriented segments which are in turn ordered by evolving particles entering in σ. Assume for instance, that only one evolving particle p_σ belongs to σ. Furthermore, that p_σ crosses only once each trajectory T_p^f, where $p \in \tau$, and finally, that each α state of p_σ has a non-empty intersection with centers of points of crossing belonging to one or several points entering in only one trajectory T_p^f. This means that the particle p_σ induces an ordering in the set of trajectories T_p^f, $p \in \tau$, since p_σ is an evolving particle (i.e., $\sum(p_\sigma)$ is a completely ordered set). In short, the trajectories T_p^f, $p \in \tau$, may be viewed as vertically oriented lines which are ordered horizontally according to the order induced by the set $\sum(p_\sigma)$ in our example above. Furthermore, we may establish a one-to-one correspondence between the first elements of the trajectories T_p^f's, the second element of all the trajectories T_p^f's, etc. Now, denote R_f° the set of points of the field $f(S)$ which enters in the trajectories T_p^f with $p \in \tau$, i.e., $R_f^\circ = \bigcup_{p \in \tau} T_p^f$. Then, we call the triad $(R_f^\circ, \tau, \sigma)$ reference frame R_f inside the field $f(S)$, where R_f° may be either equal to $f(S)$ or to a proper subset of $f(S)$. We say that x is a point of R_f if $x \in R_f^\circ$. The particles belonging to τ and σ establish the only connections between points of $R^\circ \subseteq f(S)$ that we take into account in the construction of the reference frame R_f [García-Sucre 1979]. Note that even if we do not vary the set R_f°, by simply considering different sets τ's and σ's we obtain different reference frames. Moreover, this last case is what usually is assumed in Galilean and Lorentz transformations since when we use them we are concerned with the same points of a field independently of the reference frame under consideration, and what changes from one frame to another are the space and time coordinates ascribed to each of these points. Precisely the way in which coordinates are ascribed to points of $f(S)$ depends in our model upon the sets τ and σ. The example of the reference frame described above is a one-space-axis reference frame. Separations in space between two points of R_f are measured by counting the number of vertical lines separating the two vertical lines over which these points respectively fall. Separations in time are measured counting horizontal rows of points separating the two rows of points over which these two points fall respectively.

The generalization to cases of reference frames with several space axes is straightforward [García-Sucre 1978a, 1978b, 1979].

In short, fields may be such in our model that their points are connected in an exceedingly rich way and this feature opens the possibility of defining very many different reference frames inside the same field. On the other hand, the same richness in the connections between the points of a field makes it necessary to consider only a restricted number of connections in the construction of each reference frame in order to allow the ascription of different space and time coordinates to different field points. In other words, without such restrictions when considering connections between points it may occur for fields sufficiently rich in connections (which in turn depends upon the fact that in S enter very many evolving particles of each kind in relation to the number of α states) that for any two points of the field there always exists a short path (an evolving particle with few α states) connecting them [García-Sucre 1978a]. In this sense, any two points of a field sufficiently rich in connections may be considered to be very near to each other when all the connections between points are taken into account. Of course, this degeneracy with respect to the distance between points in such fields disappears when we define a reference frame inside the field. This is one of the reasons for the necessity of a reference frame in order to classify points of fields unambiguously, ascribing different coordinates to each of them.

The choice we have made to represent the field $f(S)$ by the quotient set $\sum\sum(S)/\sim$ is based upon the following postulate [García-Sucre 1979].

Any two points of a field are distinct from each other only because they have different structures.

The relation between this postulate and our choice above of representing fields by sets $\sum\sum(S)/\sim$'s can be understood by remarking that the equivalence relation $\sim$ stands between two points of crossing when they have the same structure. Therefore, each point $x \in \sum\sum(S)/\sim$ is completely characterized by its structure with respect to all other points of the same field. How this feature of our model relates to the usual way in which the points of a field are characterized in General Relativity has been discussed in a previous paper [García-Sucre 1978a, 1979]. On the other hand, note that our choice of $\sum\sum(S)/\sim$ to represent the field produced by the physical system represented by S makes it unnecessary to introduce any extra labels to characterize points of a field since they are completely characterized within a field by their respective structures.

Let $S_1, S_2, \ldots, S_n$ be a collection of sets representing a collection of physical systems. We call 'spacetime' that which is associated to this collection of physical systems to the field produced by the global physical system constituted by all these physical systems. We represent this spacetime by the set $ST(S) \equiv \sum\sum(S)/\sim$ where $S = S_1 \cup S_2 \cdots \cup S_n$ [García-Sucre 1979]. Therefore, spacetime is in our model a global field with respect to the collection of physical systems relevant to the problem at hand. Then, all that we have said about fields also applies to spacetime. In particular, reference frames can be seen again as entities defined inside a given spacetime which are appropriate to unambiguously ascribing space and time coordinates to the points of a spacetime.

Let $R_f = (R_f^\circ, \tau, \sigma)$ be a reference frame defined inside the spacetime $f = ST(S) = \sum\sum(S)/\sim$. We may now ask how the trajectory of a particle p will appear in the reference frame R_f. We have said that a point x of the field f belongs to the trajectory T_p^f of a particle p in the field f if at least an α state of p has a nonempty intersection with at least one center of a point of crossing entering at the point x. Furthermore, the points of T_p^f are ordered according to the ordering of the α states of the particle p in the set $\sum(p)$. Yet, we are now interested in the trajectory of p in the reference frame R_f and not in the field f. We denote this trajectory as $T(p; R_f)$ and it is equal to T_p^f in connection with the points of f belonging to them; but they differ in the ordering of these points. In the case of $T(p; R_f)$ such an ordering is induced by the time coordinates that the points of $T(p; R_f)$ have in R_f. We have chosen this convention since it corresponds well to the usual way in which the points of a trajectory of a particle in a reference frame are ordered according to the time readings of the clocks entering in this frame [García-Sucre 1978b]. Thus, the trajectory $T(p; R_f)$ will appear in R_f as a path of points of R_f ordered according to their time coordinates in R_f.

A second question that we may ask is the following: Where in a given reference frame R_f will the trajectory of particles chosen at random tend to appear more frequently? According to what we have said above a point of R_f is also a point of the trajectory $T(p; R_f)$ if an α state of p has a nonempty intersection with at least a center of one of the points of crossing entering at this point. Therefore, the larger the number of preparticles belonging to the union of all the centers of the points of crossing entering in a given point, then the probability that this point may belong to the trajectory of a randomly chosen particle is also larger. This feature of our model inclines us to define the intensity $I(x; f)$ of a field f at its point $x \in \sum\sum(S)/\sim$ as the number of preparticles belonging to the union of all the centers of the points of crossing belonging to x, i.e., $I(x, f) = N[\bigcup_{s(x) \in \sigma \in x} s(x)]$ [García-Sucre 1979]. If we adopt this definition, a definition of inertial frame immediately suggests itself [García-Sucre 1979]: the frame $R_f = (R_f^\circ; \tau; \sigma)$ is inertial if the intensity of the field f takes the same value at any point of R_f (recall that $R_f^\circ \subseteq f$ and that a point of R_f is a point that belongs to R_f°). According to the above definition, randomly chosen particles will tend to appear more frequently where the intensity of the field is larger. Or, in more figurative terms, it can be said that particles are "attracted" towards those points where the field under consideration has a larger intensity.

According to our definition above, the trajectory $T(p; R_f)$ of a particle p in a reference frame R_f is an ordered set of points of R_f. This set will be either partly or completely ordered depending on whether more than one point of $T(p; R_f)$ have the same time coordinates in R_f, or not. Also, the points of $T(p, R_f)$ may appear either distributed in a large region of R_f or concentrated in a small region of R_f. When the reference frame under consideration is inertial the most frequent situation will be that in which the trajectories of particles spread over the entire R_f. If a few preparticles enter in a particle p_i (which means according to Eq. (1) that the number of preparticles belonging to the union of all the elements $a^i(x)$'s of p_i is relatively small), the points of the trajectory $T(p_i; R_f)$ will probably appear sparsely, yet spreading over the entire R_f. In the opposite case in which very many preparticles enter in p_i the points of $T(p_i; R_f)$ tend again to be uniformly distributed in R_f but this time the density of points of $T(p_i; R_f)$ in R will be much higher than in the first case. Note, that even if only one preparticle enters in a particle p, the trajectory $T(p; R_f)$ may be a set to which much more than a point of R_f belongs. This can be understood by recalling that each point of R_f is a set of points of crossing having the same structure. On the other hand, the centers of points of crossing belonging to different points of R_f may share some preparticles, i.e., may be such that the intersection between these centers is nonempty. Therefore, one and the same preparticle may belong to several centers of points of crossing belonging to different points of R_f and thus the trajectory $T(p; R_f)$ of a particle p in which only one preparticle enters may be a set of more than one point [García-Sucre 1979].

We may now ask how the process of detection of a particle or more generally, of a physical system in a reference frame can be described in the framework of our model.

Let f and f' be two fields represented by the sets $\sum\sum(S)/\sim$ and $\sum\sum(S \cup S_d)/\sim$ respectively. Then, consider the reference frames R_f and $R_{f'}$ defined inside the fields f and f'. Assume, in addition, that R_f and $R_{f'}$ are defined in their respective fields by the same pair of sets τ and σ. In this respect, the frame inside f' more closely related to R_f (which is defined inside f) is precisely the frame $R_{f'}$. On the other hand, $R_{f'}$ may be different from R_f due to the interaction between the physical systems represented by S and S_d. To be precise, we say that the physical systems represented by S and S_d interact with each other if and only if the inequality $\Sigma\Sigma(S \cup S_d)/\sim \; \neq (\Sigma\Sigma(S)/\sim) \cup \Sigma\Sigma(S_d)/\sim$ stands [García-Sucre 1978a]. In this sense, $R_{f'}$ can be considered as the resulting reference frame when R_f is modified by the presence of the physical system S_d to yield together with S the resulting field f' inside of which $R_{f'}$ is defined. We state that the physical system S_d above is a detector in the field f' if there exists at least a physical system S_g which, not being localized in R_f for any interval of time larger than a given threshold δt, appears localized in $R_{f'}$ at least once for an interval of time larger than δt [García-Sucre 1979]. Though the idea of a physical system localized in a given reference is intuitively clear, let us state this concept precisely.

We have already defined what we understand by trajectory $T(p; R_{f'})$ of a particle p in $R_{f'}$. Similarly, the trajectory $T(S_g; R_{f'})$ of a physical system S_g in $R_{f'}$ is the union-set of all the trajectories $T(p; R_f)$ such that $p \in S_g$, and its elements are again ordered according to the time coordinates in $R_{f'}$. Then, S_g appears localized in $R_{f'}$ in a given interval Δt if in this interval of time all the points of the trajectory $T(S_d; R_{f'})$ appear concentrated closely around one of the time-axes of $R_{f'}$, i.e., closely around one of the vertical lines $T_p^{f'}$, $p \in \tau$, which we have mentioned in an example above illustrating our definition of reference frame. Note, that even if S_d is localized in $R_{f'}$ during the interval Δt it may be delocalized for any other interval of time in $R_{f'}$.

The way in which the localization of S_g occurs in passing from R_f to $R_{f'}$, by the agency of the detector physical system S_d may be understood as follows [García-Sucre 1979]. When we pass from the field f to the field f', these fields being respectively represented by $\Sigma\Sigma(S)/\sim$ and $\Sigma\Sigma(S \cup S_d)/\sim$, the particles of S_d when passing over the centers of points of crossing belonging to the equivalence classes belonging to $\Sigma\Sigma(S)/\sim$ may modify the structure of these points of crossing. On the other hand, recall that the points of a field are represented by equivalence classes of points of crossing having the same structure. Thus, the above-mentioned modification of the structure of points of crossing produced by the particles of S_d may yield as a result a redistribution of these points of crossing among the points of f to yield points of f'. The only other possibility is that due to the change of structure of the points of crossing, new points appear in f' having different structures from that of any point of f.

On the other hand, the points which belong respectively to $T(S_g; R_f)$ and to $T(S_g; R_{f'})$ depend upon which points of f and f' are sharing at least a preparticle with α states of particles belonging to S_g. (We say that a preparticle α_i enters in the point x of a field if α_i belongs to the union-set of the centers of all the points of crossing belonging to x.) We can thus see that the appearance of the trajectory $T(S_g; R_f)$ of S_g in R_f may be quite different from that of $T(S_g; R_{f'})$. In particular, $T(S_g; R_f)$ may appear delocalized in R_f for any interval of time larger than δt while this may not be the case for the trajectory $T(S_g; R_{f'})$ of S_g in $R_{f'}$ [García-Sucre 1979].

The fact that in our model a physical system which appears delocalized in a given reference frame may appear localized (at least for some intervals of time) in this reference frame once this reference frame has been modified by the intervention of a detector physical system, seems a good feature of our model. In this sense, it is well known that elementary particles lose the delocalized character that they may have when they are experimentally detected.

According to what we have said above, the question that then arises is what happens when the physical system S_g to be detected is much larger than a system constituted by one elementary particle. Could it not be that by means of a detector physical system S_d, the system S_g, despite being very large, may appear localized in $R_{f'}$ at least for one sizable interval of time? The answer to this question is that in our model such a localization may occur even for a very large system, but then the notion of localization itself loses part of its meaning. What occurs is that due to the fact that the trajectory $T(S_g; R_f)$ has very many points, one has to use a very large detector S_d in order to change the structure of very many points of crossing entering in the points of R_f and in this way localize S_g. On the other hand, as a consequence of a theorem proved below (see theorem 3), we will see that for a sufficiently large detector S_d not only the localization of S_g takes place but also the number of points of $R_{f'}$ becomes much smaller than the number of points of R_f. Therefore, although we have reduced the number of points in which S_g appears when passing from R_f to $R_{f'}$, the number of points of $R_{f'}$ itself has been reduced in such a way that we can no longer say that S_g appears in a small region of $R_{f'}$.

On the other hand, if we adopt here the definition of mass of a physical system S_g in a reference frame R_f introduced in a previous paper [García-Sucre 1979], according to which the mass is proportional to the number of preparticles entering in S_g, then the reduction in the number of points when passing from R_f to $R_{f'}$ could be interpreted as a kind of gravitational collapse. An argument of intuitive character in favor of this speculative idea could be that by using a very large detector physical system S_d we are considerably increasing the mass of the physical system producing the field f', represented by $\sum\sum(S \cup S_d)/\sim$, with respect to the mass of the physical system producing the field f, represented by $\sum\sum(S)/\sim$. As a result, the field f' contracts itself reducing its size with respect to the size of f (which in our terms is equivalent to saying that the number of points of f' is much smaller than that of f). This reduction in size equally affects the size of $R_{f'}$ with respect to the size of R_f since these reference frames are defined inside f' and f, respectively. We will see below that in an extreme case this reduction may be so strong that one ends up with a global field of vanishing extension.

We may now ask in which part of a spacetime $ST(S \cup S_d)$ the physical system represented by S_d participates. A natural answer to this question is that all those points of $ST(S \cup S_d)$ over which a particle of S_d passes constitute the part of $ST(S \cup S_d)$ in which S_d participates. We call such a set of points of $ST(S \cup S_d)$ the world of S_d in $ST(S \cup S_d)$ and we denote it as $W(S_d; ST(S \cup S_d))$. For a reference frame $R_{f'}$, where $f' = ST(S \cup S_d)$, such that all points belonging to $W(S_d; ST(S \cup S_d))$ also belong to $R^{\circ}_{f'}$, one has that $W(S_d; ST(S \cup S_d)) = \bar{T}(S_d; R_{f'})$, the bar over $T(S; R_{f'})$ meaning that $\bar{T}(S_d; R_{f'})$ is a plain set, i.e., a nonordered set.

On the other hand, we introduce the following definition:

We call the *n*-order-regular-part of the spacetime $ST(S)$, which we denote as $ST(S; n)$, to the set $\sum\sum(S_n)/\sim$, where S_n is the subset of S whose elements are all the particles p of S fulfilling that the α states of any $p \in S_n$ are all sets to which the same number n of preparticles belong.

Note that $ST(S; n)$ can be considered as a field since it is represented by a quotient set $\sum\sum(S_n)/\sim$. Therefore, we can define reference frames inside $ST(S; n)$, etc.

We now introduce the following postulate: The extension of a spacetime only has sense if it is defined with respect to a collection $\mathscr{C}$ of physical systems.

A precise definition of extension of a spacetime which is compatible with the above postulate is the following:

The *n*-order-extension of a spacetime $ST(S)$ with respect to a collection $\mathscr{C}$ of physical systems, which we denote as $EX(ST(S); \mathscr{C}; n)$, is given by

$$EX(ST(S); \mathscr{C}; n) = N[\bigcup_{Sd \in C} W(S_d; ST(S \cup S_d; n))], \qquad (3)$$

where $N[A]$ stands for the cardinal of the set A; C for the set whose elements are the set S's representing physical systems entering in the collection $\mathscr{C}$.

Note, that according to the above definition in the particular case where $S_c \equiv \bigcup_{S_d \in C} S_d = S$, it follows the simple relation

$$EX(ST(S);\ C;\ n) = N[ST(S;\ n)]. \tag{4}$$

However, when $S_c \not\subseteq S$ one has as a consequence of Theorem 3 below, that the extension $EX(ST(S); \mathscr{C}; n)$ may be very different from the number $N[ST(S; n)]$ of points of the n-order-regular-part $ST(S; n)$ of the spacetime $ST(S)$.

Note also our restriction in the above definition of extension of a spacetime $ST(S)$, according to which one must always refer to one of the regular parts of the spacetime $ST(S)$. This restriction allows the measurement of the extension of an entity as the number of its basic elements ("bricks") as long as these bricks are quite alike. The only difference that we admit between points of crossing entering in a regular part of a spacetime is with respect to their structure, since according to our first postulate this difference is what underlies the appearance of distinct points.

Two cases may arise with respect to the way in which regular parts distribute themselves in a spacetime. The first case occurs when none of the regular parts extend themselves over the whole spacetime under consideration: the regular parts form a kind of mosaic in the spacetime; these regular parts either overlap each other or not. The second case occurs when there exists a regular part which extends itself over the whole spacetime. This second case is compatible with either the case of a spacetime having only one regular part, or one having several regular parts but at least one of them extending over the entire spacetime under consideration and thus also over all the other regular parts of this spacetime.

To be precise, we say that a n-order-regular-part $ST(S; n)$ of the spacetime $ST(S)$ covers $ST(S)$ completely if all the preparticles entering in the physical system represented by S belong to the union-set of the centers of the points of crossing belonging to the points of $ST(S; n)$.

A spacetime $ST(S)$ is complete when it is represented by a set $\sum\sum(S)/\sim$ for which there exists a subset B' of the set B of all the preparticles, such that $p \in S$ if and only if $p \subseteq P(B')$. We have seen in a previous paper [García-Sucre 1979] that in this case, S is a lattice [Birkhoff 1961].

Complete spacetimes have the following property:

THEOREM 1. Any of the n-order-regular-parts $ST(S; n)$ of a complete spacetime $ST(S)$ covers the spacetime $ST(S)$ completely.

Proof. Let B' be the subset of B such that $p \in S$ if and only if $p \subseteq P(B')$. Consider the n-order-regular-part $ST(S; n)$ of $ST(S)$. This implies that $n \leqslant N[B']$. Then two cases arise: (a) $N[B'] = kn$, where k is an integer, or (b) $N[B'] = ln + r$, where l and r are integers, and $r < n$. In the first case, we can consider a partition $\tilde{B}_{kn}$ of B' in k subsets $b_i(i = l, k)$ all having the same number n of preparticles. Since, on the other hand, any $p \subseteq P(B')$ is such that $p \in S$, and $P(B')$ is by definition the set of all the subsets of B', there necessarily exists k evolving particles $p_i(i = l, k)$ such that their respective first elements are $b_1, b_2 \ldots b_k$. Furthermore, from Eq. (2) and recalling the ordering rule $s^i(x) < s^j(y)$ if and only if $a^i(x) \subset a^j(y)$, it follows that the sets $b_i(i = 1, k)$ are also the first α states of the evolving particles p_i with $i = l, k$. Therefore, according to our definition of point of crossing a number k of the points of crossing belonging to the structure-set $\Sigma\Sigma(S)$ have as centers precisely the sets of preparticles b_1, $b_2, \ldots, b_k$. On the other hand, we know that $\tilde{B}_{kn} = \{b_i \mid i = l, k\}$ is a partition of B' and therefore $B' = \bigcup_{i=1,k} b_i$. This, together with the fact that any center $s^i(x) \in \sigma \in \Sigma\Sigma(S)$ fulfills that $s^i(x) \subseteq B'$ (see Eq. (2) and recall that $p \in S$ if and only if $p \subseteq P(B')$) finally yield that the union-set of the centers of all the points of crossing entering in the points of $ST(S; n)$ is equal to B'. Therefore, the n-order-regular-part $ST(S; n)$ of the spacetime $ST(S)$ covers completely $ST(S)$.

The case (b) above, can be analyzed in the same way as we have done in the case (a), provided we consider a partition $\tilde{B}_{ln}$ of the set $B' - B_r$, where B_r is a set of r preparticles. Then, we start from the sets of preparticles b_1, b_2, b_l, b_r, where $b_i \in \tilde{B}_{ln}$, for $i = 1, l$, and where b_r is a set of n preparticles such that $b_r \supset B_r$ and $b_r \subset B'$. Thus the set b_r necessarily has a nonempty intersection with some of the sets $b_i \in \tilde{B}_{ln}$. Accordingly, $b_1 \cup b_2 \ldots \cup b_l \cup b_r = B'$ and following a similar argument as in case (a), we obtain that also in case (b) the n-order-regular-part $ST(S; n)$ covers completely $ST(S)$.

Finally, we can repeat the same argument as above for a regular part $ST(S; m)$ of any order m of the spacetime $ST(S)$, which completes the proof.

In order to prove Theorem 3 below, we need the following Theorem which we have proven in a previous paper [García-Sucre 1978a].

THEOREM 2. The intersection-set of any two of the α states, say $s^i(x)$ and $s^j(y)$, of an evolving particle is necessarily an empty set, i.e., $s^i(x) \cap s^j(y) = \emptyset$.

This theorem follows from Eq. (2) and the definition of an evolving particle (see theorem 1 in a previous paper [García-Sucre 1978a]).

THEOREM 3. The regular part $ST(S; n)$ of any order n of a complete $ST(S)$ reduces to a set of only one point.

Proof. If only one point of crossing belongs to $\sum\sum(S)$, it follows that only one element belongs to the quotient set $\sum\sum(S)/\sim$ and the theorem is trivially fulfilled. Then, assume that more than one point of crossing belongs to $\sum\sum(S)$ and that there exist at least two points of crossing $\sigma_i \equiv (s^i(x); \pi^i_x(S_n))$ and $\sigma_j \equiv (s^j(y); \pi^j_y(S_n))$ fulfilling that the α states of the particles belonging to $\pi^i_x(S_n)$ and to $\pi^j_y(S_n)$ have all the same number, say n, of preparticles. Therefore, the points of crossing σ_i and σ_j belong to either one or two points of the n-order-regular-part $ST(S; n)$ of $ST(S)$.

Consider the set $\sum(p_i)$ of α states of a particle or a branch $p_i \in \pi^i_x(S_n)$ given by

$$\sum(p_i) = (s^i_1, s^i_2, \ldots, s^i_t). \tag{5}$$

This set must be completely ordered since the elements of a $\pi^i_x(S_n)$ either represent evolving particles on a branch of a nonevolving particle having α states completely ordered according to the relation of proper inclusion $\subset$ (see the definition of point of crossing). Furthermore, any two $s^i_k, s^i_l \in \sum(p_i)$ in Eq. (5) fulfill $s^i_k \cap s^i_l = \emptyset$ for $k \neq l$ (see Theorem 2).

We will show that if $\sum\sum(S)/\sim$ represents a complete spacetime, then there always exists a particle or a branch $p_j \in \pi^j_y(S_n)$, which is similar to p_i.

Consider an arbitrary s^i_p belonging to $\sum(p_i)$. Then, any of the following cases may occur.

(i) $s^i(x) \cap s^i_p = \emptyset$ and $s^j(y) \cap s^i_p = \emptyset$

(ii) $s^i(x) \cap s^i_p \neq \emptyset$ and $s^j(y) \cap s^i_p \neq \emptyset$

(iii) $s^i(x) \cap s^i_p = \emptyset$ and $s^j(y) \cap s^i_p \neq \emptyset$

(iv) $s^i(x) \cap s^i_p \neq \emptyset$ and $s^j(y) \cap s^i_p = \emptyset$

When the case (i) holds we leave unaltered the α states s^i_p. In cases (ii), (iii) and (iv) we proceed as follows. Let $s^i_a, s^i_b, \ldots, s^i_g$ and $s^i_k, s^i_l, \ldots, s^i_n$, and $s^i_r, s^i_s, \ldots, s^i_z$ be the α states of p_i fulfilling respectively (ii), (iii) and (iv) above. Consider the intersection-sets $I_{ax} = s^i_a \cap s^i(x)$, $I_{ay} = s^i_a \cap s^j(y)$, $I_{bx} = s^i_b \cap s^i(x)$, etc. The intersection between any two of these I's sets is empty since according to Theorem 2 the intersection between any two α states of p_i is an empty set. Then, consider the sets $D_j = s^j(y) - s^i(x) - (I_{ky} \cup I_{ly} \cup \ldots \cup I_{ny})$ and $D_i = s^i(x) - s^j(y) - (I_{rx} \cup I_{sx} \cup \ldots \cup I_{zx})$. Notice that there always exists a D_i' having the same number of preparticles as D_i(i.e., $N[D_i] = N[D_i']$) such that $D_i' \subseteq s^j(y) - s^i(x)$. This is clearly fulfilled if $D_i = \emptyset$ since in this case one has $D_i' = \emptyset$, $N[D_i] = N[D_i'] = 0$, $D_i \subseteq s^i(x) - s^j(y)$ and $D_i' \subseteq s^j(y) - s^i(x)$. In the complementary case $D_i \neq \emptyset$, the property is also fulfilled as can be seen next. By definition of D_i, one has the result that $D_i \subseteq s^i(x) - s^j(y)$ since the α states s_r^i, s_s^i, $\ldots, s_z^i$ yield nonempty intersections with $s^i(x)$ (see the case (iv) above). On the other hand, since the points of crossing $\sigma_i = (s^i(x); \pi_x^i(S_n))$ and $\sigma_j = (s^j(y); \pi_y^j(S_n))$ belong to the points of the regular part $ST(S; n)$ it follows that $N[s^i(x)] = N[s^j(y)]$, and therefore $N[s^i(x) - s^j(y)] = N[s^j(y) - s^i(x)]$. Then, there exists a D_i' fulfilling that $N[D_i'] = N[D_i]$ and that $D_i' \subseteq s^j(y) - s^i(x)$. Following a similar argument as above we can also show that there exists a D_j' fulfilling that $N[D_j'] = N[D_j]$ and that $D_j' \subseteq s^i(x) - s^j(y)$.

We are prepared now to analyze the case (ii) above. Since $s_a^i, s_b^i, \ldots, s_g^i$ are α states of p_i, they yield empty intersections with any of the α states $s_k^i, s_l^i, \ldots, s_n^i$ or the α states $s_r^i, s_s^i, \ldots, s_z^i$ of p_i corresponding respectively to cases (iii) and (iv) above (see Theorem 2). Then, if we denote the set $s^i(x) \cup s^j(y) - D_i - D_j$ as D_k it immediately follows that $D_k \cap D_i = \emptyset$ and $D_k \cap D_j = \emptyset$. Furthermore, the set $P_{xy} = \{I_{ax}, I_{bx}, \ldots, I_{gx}, I_{ay}, I_{by}, \ldots, I_{gy}\}$ is a partition of a set $\tilde{D}_k$ such that $\tilde{D}_k \subseteq D_k$. (For instance, if any of the α states $s_a^i, s_b^i, \ldots, s_g^i$ yields an empty intersection with both $s^i(x)$ and $s^j(y)$, then $\tilde{D}_k = \emptyset$.) Consider now the set $D_k' = s^i(x) \cup s^j(y) - D_i' - D_j'$. Since D_i, D_i', D_j and D_j' are all included in $s^i(x) \cup s^j(y)$, $N[D_i] = N[D_i']$, $N[D_j] = N[D_j']$ and $D_i \cap D_j = D_i' \cap D_j' = \emptyset$, it follows that $N[D_k'] = N[D_k]$. Therefore, there exists a set $\tilde{D}_k' \subseteq D_k'$ having the same number of preparticles as $\tilde{D}_k$, *i.e.*, $N[\tilde{D}_k'] = N[\tilde{D}_k]$. Also, there exists a partition $P_{xy}' = \{I_{ax}', I_{bx}', \ldots, I_{gx}', I_{ay}', I_{by}', \ldots, I_{gy}'\}$ of $\tilde{D}_k'$ such that $I_{ax}', I_{bx}', \ldots, I_{gy}'$, relate themselves respectively to $I_{ax}, I_{bx}, \ldots, I_{gy}$ in the following way. Let us specify the relation between I_{ax}' and I_{ax}; (a) $N[I_{ax}'] = N[I_{ax}]$, and (b) $N[s^i(x) \cap I_{ax}] = N[s^j(y) \cap I_{ax}']$ and $N[s^j(y) \cap I_{ax}] = N[s^i(x) \cap I_{ax}']$. An identical relation can be established between I_{bx}' and I_{bx}, etc., until we exhaust all the elements of P_{xy} and P_{xy}' since $N[\tilde{D}_k] = N[\tilde{D}_k']$, $N[D_i] = N[D_i']$, $N[D_j] = N[D_j']$, $D_i \subseteq s^i(x) - s^j(y)$, $D_i' \subseteq s^j(y) - s^i(x)$, $D_j \subseteq s^j(y) - s^i(x)$ and $D_j' \subseteq s^i(x) - s^j(y)$. We then define the sets $s_a^{i'} = (s_a^i - I_{ax} \cup I_{ay}) \cup I_{ax}' \cup I_{ay}'$, $s_b^{i'} = (s_b^i - I_{bx} \cup I_{by}) \cup I_{bx}' \cup I_{by}'$, $\ldots$, $s_g^{i'} = (s_g^i - I_{gx} \cup I_{gy}) \cup I_{gx}' \cup I_{gy}'$. It is clear that

$N[s_a^i] = N[s_a^{i'}], \ldots, N[s_g^i] = N[s_g^{i'}]$, and that the intersection between any two of the sets $s_a^{i'}, s_b^{i'}, \ldots, s_g^{i'}$ is empty.

For the α states of p_i fulfilling the points (iii) and (iv) above, we proceed in the same way as above by considering the partitions $P_x' \equiv \{I_{rx}', I_{sx}', \ldots, I_{zx}'\}$ and $P_y' \equiv \{I_{ky}', I_{ly}', \ldots, I_{ny}'\}$ of D_i' and D_j' respectively. These partitions P_x' and P_y' relate themselves with the partitions $P_x = \{I_{rx}, I_{sx}, \ldots, I_{zx}\}$ and $P_y = \{I_{ky}, I_{ly}, \ldots, I_{ny}\}$ of D_i and D_j respectively, in such a way that $N[I_{rx}'] = N[I_{rx}], \ldots, N[I_{zx}'] = N[I_{zx}]$, and that $N[I_{ky}'] = N[I_{ky}], \ldots, N[I_{ny}'] = N[I_{ny}]$. This is always possible since $N[D_i] = N[D_i']$ and $N[D_j] = N[D_j']$. We then define the following sets: $s_k^{i'} = (s_k^i - I_{ky}) \cup I_{ky}', \ldots, s_n^{i'} = (s_n^i - I_{ny}) \cup I_{ny}'$, and $s_r^{i'} = (s_r^i - I_{rx}) \cup I_{rx}', \ldots, s_z^{i'} = (s_z^i - I_{zx}) \cup I_{zx}'$. According to these definitions one has the result that $N[s_k^{i'}] = N[s_k^i]$, etc., and that any two of the sets $s_k^{i'}, s_l^{i'}, \ldots, s_z^{i'}$ arising from the elements of $\sum(p_i)$, Eq. (5), have empty intersections with each other. This last point follows from the fact that any two elements of $\sum(p_i)$ have empty intersections between them, that $D_i' \cap D_j' = D_i' \cap D_k' = D_j' \cap D_k' = \varnothing$, and that any of the I''s sets above is an element of a partition of either D_i', D_j' or D_k'. Therefore, if two I'-sets belong to the same partition their intersection is necessarily empty by definition of partition. If they belong to two partitions, of say D_i' and D_j', then their intersection is again empty since $D_i' \cap D_j' = \varnothing$.

Consider now the completely ordered set

$$A = (s_1^{i'}, s_2^{i'}, \ldots, s_t^{i'}), \tag{6}$$

which has been ordered according to the order in which the elements of the set $\sum(p_i)$ appear in Eq. (5). Note that the elements of $\sum(p_i)$ fulfilling the point (i) above will be equal to the corresponding elements of A. Since the cases (i) to (iv) above cover all the possibilities the set A has the same number of elements as $\sum(p_i)$. We then consider the set

$$\tilde{A} = (s_1^{i'}, s_1^{i'} \cup s_2^{i'}, \ldots, s_1^{i'} \cup s_2^{\prime i} \cup \ldots \cup s_t^{i'}). \tag{7}$$

Since $s^i(x), s^j(y), s_1^i, \ldots, s_t^i \subseteq B'$ then every one of the elements $s_1^{i'}, s_2^{i'}, \ldots, s_t^{i'} \in A$ is also included in B'. Therefore, the set $\tilde{A}$ above fulfills that $\tilde{A} \subseteq P(B')$. Accordingly, $\tilde{A}$ represents an evolving particle p_j or a branch of a nonevolving particle of the physical system S_n, since according to Eq. (2) the α states of the particle represented by $\tilde{A}$ are represented by $s_1^{i'}, s_2^{i'}, \ldots, s_t^{i'}$ (see Theorem 1 of a previous paper [García-Sucre 1979]) and these sets have all the same number of preparticles. As a result, p_j fulfills all the conditions in order to be a particle of S_n. Furthermore, since

$p_i \in \pi_x^i(S_n)$ then at least one α state s_p^i of p_i in our argument above must be such that $s_p^i \cap s^i(x) \neq \emptyset$. Therefore, according to cases (ii) and (iv) above, one has by construction that at least one $s_p^{i'} \in A$ must fulfill that $s_p^{i'} \cap s^j(y) \neq \emptyset$. Since, on the other hand, the α states belonging to A in Eq. (6) have all the same number n of preparticles, the particle p_j represented by the set $\tilde{A}$ in Eq. (7) belongs to $\pi_y^j(S_n)$ and thus enters in the point of crossing σ_j above, which in turn belongs to a point of the n-order-regular-part $ST(S; n)$ of $ST(S)$.

Now, by making the comparison between Eqs. (5) and (6), one concludes that p_i is similar to p_j. Since we can repeat the same procedure for any particle belonging to $\pi_x^i(S_n)$, we can say that for any evolving particle or branch p_i entering in the point of crossing $\sigma_i = (s^i(x); \pi_x^i(S_n))$ there exists an evolving particle or branch p_j entering in $\sigma_j = (s^j(y); \pi_y^j(S_n))$ which is similar to p_i. Since, on the other hand, our reasoning does not depend upon which of the two points of crossing σ_i or σ_j we consider first, we also have the result that for any evolving particle or branch p_k entering in σ_j there exists an evolving particle or branch p_l entering in σ_i which is similar to p_k. This immediately leads to the result that there exists a one-to-one mapping Φ between the set of equivalence classes of similar particles or branches entering in σ_i and the set of equivalence classes of similar particles or branches entering in σ_j, i.e.,

$$\pi_x^i(S_n)/R \overset{\Phi}{\longleftrightarrow} \pi_y^j(S_n)/R, \tag{8}$$

such that Φ establishes a correspondence between equivalence classes having the same structure.

On the other hand, in our construction above when passing from the set $\sum(p_i)$ given in Eq. (5) to the set $\sum(p_j) = A$ given in Eq. (6), there is an exchange of roles between $s^i(x)$ and $s^j(y)$ in connection with their intersections with α states. Thus, the conditions on α states specified in point (ii) of the definition of similarity between points of crossing are also satisfied.

We then conclude that the points of crossing σ_i and σ_j are similar to each other, i.e., $(s^i(x); \pi_x^i(S_n)) \sim (s^j(y); \pi_y^j(S_n))$. In the same way, since we can follow the same reasoning for any other pair of points of crossing belonging to points of the n-order-regular-part $ST(S; n)$ of $ST(S)$, we conclude that all the points of crossing belonging to the points of $ST(S; n)$ are similar to each other. Accordingly, only one equivalence class of points of crossing belongs to $ST(S; n)$, i.e., $ST(S; n)$ reduces itself to a set of only one point. Finally, the same argument above applies to any other m-order-regular-part $ST(S; m)$ of $ST(S)$, the proof being thus completed.

A consequence of the above property of complete spacetimes relates to the property according to which the extension $EX(ST(S); \mathscr{C}; n)$ of a spacetime $ST(S)$ with respect to a collection $\mathscr{C}$ of only one physical system S_d having very many particles crossing over the particles of S, approaches the minimum possible value of 1. In order to see this, let us assume that the spacetime $ST(S)$ is not complete, but that the spacetime $ST(S \cup S_d)$ is complete. Thus, according to Theorem 3 above, and the fact that in C only S_d enters, one has from Eq. (3)

$$EX(ST(S); \mathscr{C}; n) = N[W(S_d; ST(S \cup S_d; n)] = 1, \tag{9}$$

since $N[ST(S \cup S_d; n] = 1$ (see theorem 3) and $W(S_d; ST(S \cup S_d; n)) \subseteq ST(S \cup S_d; n)$ by definition of $W(S_d; ST(S \cup S_d; n))$. Note that according to theorem 3 the above relation does not depend upon which of the regular parts of $ST(S \cup S_d)$ we consider, i.e., upon which of the possible values of n we choose in Eq. (9). Also, note that according to Theorem 1, any of the regular parts of a complete spacetime $ST(S \cup S_d)$ covers completely $ST(S \cup S_d)$. This means that the above relation given in Eq. (9) does not arise due to the fact that we are considering a small region of $ST(S \cup S_d)$.

Since a noncomplete spacetime $ST(S)$ may have very many points we say according to what we have discussed above that $ST(S)$ collapses when we try to define its extension with respect to a system S_d such that $ST(S \cup S_d)$ becomes a complete spacetime. Note that for any S there always exists a S_d such that $ST(S \cup S_d)$ is a complete spacetime. In order to see this, let B' be a subset of B (the set whose elements are all the preparticles) such that all the preparticles entering in the particles of the physical system represented by S are precisely the elements of B'. We can thus write $B' \subseteq B$. Then consider a set S_T for which one has $p \in S_T$ if and only if $p \subseteq P(B')$. Consequently, if we take $S_d = S_T - S$ it follows that $ST(S \cup S_d)$ is necessarily a complete spacetime. In the case where the physical system represented by S is relatively small, one has the result that $B' \subset B$ and $N[B'] \ll N[B]$. In this case the system S_d, such that $ST(S \cup S_d; n)$ collapses to only one point, may also be a relatively small system since $S_d = S_T - S$ and $p \in S_T$ if and only if $p \subseteq P(B')$ and $N[B'] \ll N[B]$. On the other hand, the case $B' = B$ corresponds to an extremely large physical system represented by S. Furthermore, the physical system represented now by $S_d = S_T - S$ (where now $p \in S_T$ if and only if $p \subseteq P(B)$) such that $ST(S \cup S_d)$ is a complete spacetime, may be either a small or large physical system depending on the measure that $\sum \sum(S)/\sim$ approaches a complete spacetime. Of course, we can also consider the case in which the physical system represented by S_d and with respect to which we define the extension of $ST(S)$ has a nonempty intersection with S. This case can be analyzed in the same way as we have done above when $S_d = S_T - S$.

We conclude that when we define the extension of a spacetime $ST(S)$ with respect to the only one physical system S_d, then in the extreme case corresponding to the largest S_d this extension reduces to its minimum possible value 1. Note that in this case any reference frame R_f defined within any regular part $f = ST(S; n)$ of $ST(S)$ also has only one point. This means that not only space separations collapse to zero but also time stops since in this case there exists only one trajectory T^f_p, $p \in \tau$, and this trajectory has in turn only one point. Also note, that in this case the physical system represented by S_d can be seen as a detector physical system since in passing from $f = ST(S; n)$ to $f' = ST(S \cup S_d; n)$ any trajectory in a reference frame $R_{f'}$ reduces itself to only one point.

The opposite case to the above one will occur when we define the extension of a spacetime $ST(S)$ which is not complete, with respect to a collection $\mathscr{C}$ of very many and nonoverlapping physical systems. In this case and according to Eq. (3) the extension $ST(S)$ is at least as large as the number of small physical systems entering in $\mathscr{C}$.

CONCLUSIONS

We have described in this paper a general trait of spacetime and more generally of fields, according to which these entities collapse when we define their extension with respect to only one very large physical system. The property according to which different types of collapses may occur in spacetime have been considered in the literature as a fundamental trait of spacetime [Patton and Wheeler 1975]. This feature described in our model can be added to other general traits of fields [García-Sucre 1978a], reference frames [García-Sucre 1978a, 1978b] and spacetime [García-Sucre 1979] that we have described in previous papers. This capability of our model of describing general properties of its referents is a typical trait of foundations of physics theories [Bunge 1977, p. 29].

Instituto Venezolano de Investigaciones Científicas (IVIC),
Centro de Física

REFERENCES

Basri, S.A.: 1966, *A Deductive Theory of Space and Time*, North-Holland, Amsterdam.

Birkhoff, G.: 1961, *Lattice Theory*, [*American Mathematical Society Colloquium Publications*, vol. 25], Providence.

Bunge, M.: 1967, *Foundations of Physics*, Springer, New York.

Bunge, M.: 1977, *International Journal of General Systems* **4**, 29.

Bunge, M. and García-Maynez: 1976, *International Journal of Theoretical Physics* **145**, 961–972.

Finkelstein, D.: 1969, *Physical Review* **184**, 1261–1271.

Finkelstein, D.: 1972a, *Physical Review* D **5**, 320–328.

Finkelstein, D.: 1972b, *Physical Review* D **5**, 2922–2931.

Finkelstein, D.: 1974, *Physical Review* D **9**, 2219–2231.

Finkelstein, D., G. Frye, and L. Susskind: 1974, *Physical Review* D **9**, 2231–2236.

Fraenkel, A.A. and Y. Bar-Hillel: 1958, *Foundations of Set Theory*, North-Holland, Amsterdam.

García-Sucre, M.: 1975, *International Journal of Theoretical Physics* **12**, 25–34.

García-Sucre, M.: 1978a, *International Journal of Theoretical Physics* **17**, 163–188.

García-Sucre, M.: 1978b, *Proceedings of the First Section of Interdisciplinary Seminars on Tachyons, Monopoles and Related Topics*, E. Recami (ed.), North-Holland, Amsterdam, pp. 235–246.

García-Sucre, M.: 1979, *International Journal of Theoretical Physics* **18**, 725–775.

Misner, C.W., K.S. Thorne and J.A. Wheeler: 1970, *Gravitation*, Freeman, San Francisco, pp. 1203–1209.

Noll, W.: 1967, 'Spacetime Structure in Classical Mechanics', *Delaware Seminar in the Foundations of Physics*, Springer, New York, 28–34.

Patton, C.M. and J.A. Wheeler: 1975, 'Is Physics Legislated by Cosmogony?', in *Quantum Gravity*, C.J. Isham, R. Penrose and D.W. Sciama (eds.), Clarendon Press, Oxford, pp. 538–605.

Penrose, R.: 1967, *Journal of Mathematical Physics* **8**, 345–366.

Penrose, R.: 1968, *International Journal of Theoretical Physics* **1**, 61–99.

Penrose, R.: 1971, '.....', in *Quantum Theory and Beyond*, T. Bastin (ed.), Cambridge University Press, Cambridge.

Penrose, R.: 1975, 'Twistor Theory, Its Aims and Achievements', in *Quantum Gravity*, C.J. Isham, R. Penrose and D.W. Sciama (eds.), Clarendon Press, Oxford, pp. 268–407.

RUDOLF HALLER

POETIC IMAGINATION AND ECONOMY: ERNST MACH AS THEORIST OF SCIENCE*

I

Mach's epistemological phenomenalism, the magnificent idea of constructing the world from those elements that are supposed to be nothing but sense impressions, seems to have left the scene of philosophical controversy as a loser.

Since Carnap's and later Goodman's [Carnap 1967; Goodman 1977; Frank 1938] attempts to erect a constitutional system as a rational reconstruction of the structure of the world based on subjective psychology revealed the unforeseeable difficulty of reducing physical to psychological phenomena, many people immediately and from the beginning seek to abandon the foundation of empiricism together with the epistemological foundation of the subjective psyche.

It is therefore not surprising that Mach's philosophy, always a philosophy of empirical science, lost that attraction and even fascination, which it possessed during the life-time of the great scientist and philosopher as well as within the Vienna Circle. With the verdict of the end of positivism and the historical judgment that somebody just killed logical positivism [Popper 1974], the relevance of Mach's philosophy seemed to have receded into the misty and ancient realm of the classics. Only in this way can the judgment of the last occupant of Mach's Vienna chair for the history and philosophy of the inductive sciences, Victor Kraft, be understood. In the postscript to the beautiful volume *Ernst Mach: Wegbereiter der modernen Physik* by K. D. Heller, Kraft wrote in 1964 that Mach surely was "a historical figure of far-reaching importance," but that his attitude towards epistemology has to be viewed as "outdated" at this time [Heller 1964]. This judgment is a misjudgment, a mistake, although it comes from a logical positivist. That is not to say that a point of view that is now outdated was wrong originally. On the contrary, it is typical for all scientific research and systematizing that it gets improved, perfected, outdated by further and new research. In that way the researcher contributes to decreasing the number of open

J. Agassi and R.S. Cohen (eds.), Scientific Philosophy Today, 71–84.
Copyright © 1981 *by D. Reidel Publishing Company.*

scientific questions of his time in his specialty or to raising new questions, research into which will broaden or at least promote knowledge. But even here the scientist has no guarantee. On the contrary, it is not uncommon that, with hindsight, his assumptions and convictions did not turn out to be fruitful, but rather unfruitful and even misleading.

If we listen to the judgment of Mach by contemporary philosophers and theorists of science, the assertion of mere outdatedness comes close to a euphemism. Rather, according to this one-sided judgment, it would have to be counted among the philosophically misleading and unfruitful teachings. There are three assertions about Mach's philosophy of science designed to support the accusation of obsoleteness: First, it is phenomenalist, second, it is narrowly instrumentalist-inductivist, and third, it is not adequate for science and its theories.

Before we form a judgment in this vague area of philosophical rumors and second-hand hearsay, it may be appropriate to call to mind Mach's basic ideas about the construction and change of scientific ideas and examine them in light of the controversy about philosophical theories in our day.

II. "SCIENCE EMANATES FROM LIFE"[MACH 1882b]

In contemplating scientific activity there is an advantage in proceeding at first from our life-world, the environment in which we live, in order to examine continuity and differences of familiar behavior in a familiar environment from this vantage point. In this life-world of ours, the experiential sphere in which we grow up, we learn just like other beings. Yet we are privileged, because of the possibility of receiving, so to speak, the experiences of other individuals, indeed the experiences of past generations, all at once, and of adapting them to our needs. This process is analogous to the process that transfers the 'experiences' gathered in the course of millions of years to living beings so that these experiences are available to future generations as inherited dispositions. Some people have viewed and characterized this very process as "knowledge producing," in which one generation genetically passes on to the next all information that is vitally necessary and promotes selection.

Man, however, belongs to two worlds insofar as he is given not only the 'information' of nature, but also that of the mind, or more precisely, language. "We are the same sort of thing as those in our physical surroundings and it is through ourselves that we become acquainted with them" [Mach 1976, p. 99]. The process of this acquisition of the conceptual apparatus is highly complicated and takes place initially without reflection and consciousness. If we become conscious at this stage of our learning process, which moreover is always practical, then what we find "within us is already a fairly complete world picture" [Mach 1976, p. 120]. But this conception of the world is, of course, not *a priori*, as might be suggested by our "instinctive knowledge," which inspires a great deal of confidence [Cf. Mach 1881]. This world picture is nothing other than the treasure of experiences which, just as with material wealth, concentrates the work of many individuals in one individual. "Not magic therefore," says Mach, "but economy." In science, too, the situation is basically the same. Only here the work of many is collected and unified without the necessity of exploiting anybody's energy, or, as Mach himself says: "without robbing others of anything" [Mach 1882a].

Mach's inexhaustible and often praised gift for reducing complex or opaque relationships to simple concepts has found a fruitful beginning in the concept of economy expressed here (the conception of a cunning of reason, as it would have been called elsewhere) for a theory of the history of the progress of knowledge. The value and extent of this conception are only now fully appreciated [Feyerabend 1970, p. 174 ff; Gargani 1979].

Mach does not tire of emphasizing the principle of economy, the economical nature of science as the principal characteristic of all investigation and achievement of knowledge of man. Again and again he points to the Austrian political economist Emanuel Herrmann, whom he met during his time at Graz, as the person who inspired this idea [Mach 1894, p. 209; 1910, p. 225]. While *Mach* wants to solve the problem of the dynamics of theories, i.e., the question of which principle determines change and progress of hypotheses and theories, with the aid of Herrmann's central concept of *economic theory*, namely the concept of economy, *Herrmann* himself strives to examine economic problems resulting from the principle of economy by means of the methodology of the *natural sciences*. Both scientists were convinced that a separation of man and nature can only be achieved artificially:

> The petty lines of division drawn between man and the rest of creation by superstition and false beliefs must dissolve in the light of science like fog in the sun and disappear; we know only *one* nature in which man finds a small and certainly quite insignificant place as a participating and cooperatively creative being.
>
> It is the very economy *of nature* (in the old and rotten sense of extra-human existence) which constitutes the most beautiful, most magnificent part of economic doctrine.

What is the saving in human labor, our buildings, our machines, compared to the economy according to which the vital organs of plants, and the sense organs and mental organs of animals, operate and create [Herrmann 1868, p. 21].

Out of the idea of the unity of the physical and the psychological, however, grew Mach's theory of elements, which was so often and readily dismissed as a sensualism that would lead to solipsism. Such an interpretation, however obvious, only too easily obscures the fact that Mach himself wanted to and did take a much more subtle view. Mach argues that a coin appears large and round or large and elliptical if viewed close up, and small and round or small and oval if viewed from a distance. This is to say that the 'effects' an object has are imparted in perception, and thus an object does not seem to be anything but a bundle of perceptions, a complex of perceptions. Taken at face value, Mach's position would seem nothing but a new version of Bishop Berkeley's epistemological idealism or idealistic empiricism. But only a superficial cliché of a specter of positivism can overlook the fundamental difference [Popper 1965, p. 172f]. While according to Berkeley only the fact of being perceived constitutes an entity and the 'esse est percipi' represents a metaphysical determination of what *is*, the question *whether* something is an object or not depends, in Mach's opinion, only on the *mode of observation* of the elements, which, however, are ontologically neutral: "It depends on the definite purpose of the observation whether we view *A* [i.e., an element, R.H.] as sensation or object" [Mach 1880]. Here Feyerabend, not always dependable because of his hedonistic tendency to reinterpret, is quite right when he states: Mach's "elements are not sensations" [Feyerabend 1970, p. 176; also Cohen, p. 132ff.]. It is rather the very concept of an element that was introduced *not only* as the irreducible part of the complex of sensations, *but also* as the irreducible part of a physical complex which is Mach's proper and independent safeguard against metaphysics.

In his confrontation with Max Planck – one of the few polemical writings of Mach – Mach clearly formulates his point of view which follows the natural world and is beyond idealism and realism. Here he asserts that it is experience which taught us that what at first sight appears to us to be immediately given is in fact not immediately and unconditionally given. Even ordinary people are aware of the influence of external circumstances as well as of subjective psychological and physiological conditions on our impressions of the world. This is even more true of scientific observations which, if they appear to be hallucinatory, must be corrected by the other senses or other persons if they are to have scientific, and therefore *social* value... Overestimating this exception [i.e., deception through hallucination, R.H.] easily leads to monstrous idealistic or even solipsistic systems.

And Mach adds a consequence which his interpretation seeks to avoid:

It would be strange if the experience of the world, because of its refinement, were to *abolish* [*aufheben*] itself and leave nothing of the world itself but unattainable phantoms [Mach 1910, p.235].

Here Mach opposes to the hypothesis of a shared world as the sum of the objects of experience the hypothesis that this world itself cannot be experienced independently of sense experience and the physiology of the living being. These, he believes, are the facts from which any theory of knowledge has to start. Of course, this is not yet the solution to the fundamental problem of epistemology. It is, nevertheless, a beginning the clarity of which is convincing and which is not surpassed by the so-called 'realists'.

Similar to Hertz and Boltzmann, Mach views the models which we create from our sense experiences, namely thoughts, as building blocks of a world which is to be perfected in its epistemological value by means of continuous and increasing adjustment. "We seek pictures of facts," Wittgenstein writes in his *Tractatus Logico-Philosophicus*, and the picture is composed of elements which represent the objects. It is a picture that itself is a fact and to which belongs the "representing picturing relation" so to speak as a feeler toward reality.[1]

Due to this similarity to Mach's conception and the identification of the elements with sense impressions a phenomenalist program in the Vienna Circle and especially in the early Carnap was bound to appear as a continuation of Mach, and the *Tractatus* was also interpreted in this vein. Also according to Mach the goal at first is: "The representation in thought of facts" [Mach 1976, p. 2]. However, while ordinary cognitive interest – or, as Mach puts it in the manner of his time, 'vulgar thinking' – pursues the conceptual completion of observations for the sake of practical cognitive purposes and is, therefore, wanting, scientific investigation is free from such considerations and therefore not only moves away from ordinary views, but constitutes one of them.

The representation of facts in thoughts and the adjustment of thoughts to facts by no means guarantees that the thoughts *agree with each other*. To put it differently: in constructing a scientific theory it is not only a necessary condition to use theoretical concepts; the concepts have to 'fit' with each other and be adjusted to each other. It is in this working out of the logical connections, which can be represented in their purest form in axiomatics, that Mach sees the essential difference between everyday thought and scientific investigation.

It is, however, a fact that "the facts are *not* compelled to conform to our thoughts"; "it is of course not possible to achieve perfect adaptation to every individual and incalculable future fact" [Mach 1976, p. 356, p. 355; cf. Mach 1894, p. 211]. *How* is it then that we adapt our thoughts?

The answer to this question shows anew the fruitfulness of a principle of economy if properly understood. If we did not have the means to connect the objects of the observable and the *un*observable world with the help of simplification and analogy, to reduce the pluralities to unities, we would senselessly stumble about in a disconnected realm of set pieces belonging to all kinds of possible worlds. If, on the other hand, we divide the complexes of facts into theoretical elements, or elements of thoughts, then we simplify, schematize, idealize. "The aim of research is the discovery of the equations which subsist between the elements of phenomena" [Mach 1894, p. 221, 264ff]. We must of course not expect that Mach is in a position to present a semantically clarified definition of these concepts. Brentano pointed out in a sharp and penetrating critique of Mach's *Erkenntnis und Irrtum* [*Knowledge and Error*] that both Mach's concept of similarity as partial identity as well as his concept of analogy were not sufficiently and satisfactorily analyzed.[2] And the physical equations express here "indeed a mathematically conceivable relation," but the logical scope of these relations is limited by experience. Those who so readily advocate and propagate the total theory-dependency of empirical concepts, such as observation, all too easily forget one of the pillars of their critical argumentation, namely, that there is no inductive necessity, or in Mach's words: "There simply does not exist a necessity other than a logical necessity, such as a physical necessity" [Mach 1896, p. 435]. Even if we assume that the simplest law corresponding to our experience holds, there is nevertheless no ground to assume that the simplest case will occur. Wittgenstein's formulation in the *Tractatus* is a clear echo of the Humean-Machian position:

A necessity for one thing to happen because another has happened does not exist. There is only *logical* necessity.

At the basis of the whole modern view of the world lies the illusion that the so-called laws of nature are the explanations of natural phenomena. [Wittgenstein 1922, 6.37 and 6.371]

Therefore no explanations. The logic of inquiry shows us that the adaptation of the thoughts to the facts is nothing else than their description. It was a credo of Mach's philosophy that to describe meant the concentration, the condensation of all experience, and that theories are nothing but the systematic summary [*Zusammenfassung*] and ordering of these experiences. But Mach did not characterize these descriptions as assertive, much less apodictic, judgments, as propositions which are either true or false. Rather he maintains that *all* empirical statements can only be hypothetical:

A proposition of natural science – Mach writes in *Knowledge and Error* – always has a merely hypothetical sense: if a fact A corresponds precisely to concepts M, then the consequence B corresponds precisely to concepts N; the two correspondences have the same degree of accuracy. Absolutely exact and perfectly precise and unambiguous determination of consequences from a presupposition in natural science, as in geometry, does not exist in sensible reality, but only in theory. [Mach 1976, p.356, cf. also p. 228].

Mach is of the opinion that mainly two reasons compel us to view descriptions not as assertive or apodictic, but as hypothetical judgments or statements: The first and deepest reason is that science provides an adequate picture of the facts and their conceptual real connections only as its *goal*. As inquiry or, as Mach says, as "*becoming* science," it moves in the realm of *conjectures*. Each hypothetical or theoretical value, therefore, is at first related to an area of particular circumstances with the hope of extending it to corresponding, similar and analogous cases. This "intended area of application" of a basic theoretical idea arises, so to speak, with the birth of the hypothesis or theory itself. From here the new problems grow which lead us to new hypotheses and so on. Viewed from the goal or retrospectively, hypotheses have a "self-destructive function" as Mach characterizes their tendency to change into the "conceptual expression of the facts," that is to say into successful adaptation to facts.

The second reason consists in the changeability of events and the changeability of opinions or conceptions of events and facts. Mach says in the introduction to the lecture "Die Geschichte und die Wurzel des Satzes von der Erhaltung der Arbeit" [The History and Root of the Principle of the Conservation of Energy] (1872): "History has made all; history can alter all," [Mach 1909; cf. also Höfler 1910 and Hiebert 1970]. If a change of conception did not have to be expected, it would be sufficient to store the accomplishments of the past in one's memory, or, even better, to relinquish this work to automatons. But just as there is no *necessary* sequence of events in nature itself, there is none in the history of nature or human ideas. Mach reminds us that there are hypotheses that last only moments, namely when we dismiss an idea at once. There are those, however, which persist for hundreds or thousands of years, such as Ptolemaic cosmology. "Absolute forecasts, consequently, have no significance in science" [Mach 1894, p. 223]. But despite this realization – "history made everything, history can change everything" – Mach did not become a proponent of an anarchic epistemology, as is common today among the so-called anti-positivists. And he is not ailing from the semi-scepticism of Critical Rationalism, the clichés of which he definitely opposes. Whoever really reads Mach will easily discover how much even those who are so eager to attack him owe him directly, even take from him.

I already mentioned at the outset that Mach emphasizes the principle of economy as the characteristic principle for distinguishing scientific from non-scientific ways of knowing. Most importantly and primarily, history, methodological conjecture, investigation and representation of an area of facts is to be preferred as scientifically justified to a merely accidental gain of knowledge. Thus the method of finding as well as the systematics of representation are subjected to the principle of economy. Or, as Pierre Duhem states in full agreement with and in the spirit of Mach: "The economy achieved by the substitution of the law for the concrete facts is redoubled by the mind when it condenses experimental laws into theories" [Duhem 1908, p. 24].

I am surprised that something in this controversial principle which is fundamental to it was evidently forgotten and could be rediscovered. The interpreters and opponents of the principle of economy overlooked the fact that in the rational theory of decision-making we proceed, of course, analogously to Mach. Someone who does not agree with the rules and norms of a rational theory of action, or of the realization of desires and purposes, strictly cannot be refuted any more than in the history of science a single theory could be falsified by exceptions. We would have to say of such a person that he is acting unreasonably, just as an anomaly disturbs the rational order of concepts in a system. However, the concept of rationality itself has by no means been clarified even to this day [cf. Haller 1981]. Furthermore, it is a relative concept which considers and refers to the kind of knowledge, or basis of information, somebody has at his disposal for his decisions and actions, the means he has at his disposal and the goals he pursues. As everyone knows, there are well-developed theories in which the expected utility, the worst or the best result possible among possible alternatives, can be calculated [Luce and Raiffa 1957, Stegmüller 1973, Hempel 1965, Spohn 1978]. These models and theories, as incomplete and irregular as they may be in the present state of investigation, seem to me to be a splendid confirmation of Mach's initial attempt [*Ansatz*] to investigate how the most economical way can be found which will be most useful in attaining a given goal, namely the cataloguing and systematization of all facts (and that, of course, is not limited to those already observed). Viewed in this way, the basic hypothesis of Mach's 'psychology and logic of investigation', namely economy, is nothing else than the basic hypothesis of the rationality of science, a rationality which is what makes understanding and comprehension possible.

At last at this point the question arises what role is to be given to imagination – mentioned in the title – in such a process which after all has its own genesis. Is not poetry above the facts, the comprehensive knowledge of which is the aim of investigation?

To be sure, the conjunction 'and' is patient, and with its help even the greatest inconsistency can be connected. What is the reason, then, that justifies placing poetic imagination next to the basic principle of the rationality of scientific investigation as if the structure of science rested on both these pillars?

The view expressed here is precisely that – following Mach – we need only these two principles in order to understand genesis, dynamics and the goal of scientific investigation. To wit, if the veil is pulled away from the birth of scientific hypotheses, a veil created especially by relegating heuristic to the realm of the incomprehensible, to that which cannot be rationally reconstructed [Popper 1961], then we could be unimaginative enough to attempt to discern behind it again the discredited induction. And we are disappointed if we do not find it. Thus we read in *Die Principien der Wärmelehre*:

> Those who have dealt with research or the history of research will have difficulty believing that discoveries are made according to the Aristotelian or Baconian scheme of *induction* (by enumerating identical cases). In that case discovery would be quite a cosy business [Mach 1896, p. 443].

What misconceptions in the estimation of positivism had to occur that *such* passages in Mach could be overlooked, missed or not read? And how much propaganda is necessary to give such mistaken judgments a measure of credibility! At the end, I intend to return to this point briefly, a point closely connected with the latest developments.

If, as we have heard, the activity of discovery cannot be a very cosy business because it is not possible to rely on past experiences in order to imagine new things, what then is it that enables us to see or to learn to see facts in a new light?

It is the work and role of the imagination which, according to Mach, is alone in a position to lead us away from what we habitually expect again and again. He who begins to let his imagination roam, leaves the shackles of everyday life behind him, plunges into the infinite realm of possibility and works on any worlds possible, no matter whether he enjoys it or not. Ultimately, he even leaves these worlds behind and crosses the line of what is *really* possible to the objects that are *logically* impossible. If the concept of the *perpetuum mobile* is an example of the former, every indirect *mathematical* and *logical* proof which shows the assumption of the opposite of what is to be proved to be absurd is an example of the latter. This is also the place for associations which lead us without any effort on our part to places which we would not reach without their seductive, capricious charm.

Without going into detail it may be appropriate to distinguish between two forms of imagination. One form of imagination continues the possibilities of actual events and thus makes it possible for us – afterwards, as with dreams – to remain in the everyday world with our wants and anxieties, even if the stream of its events has long flowed past us. The other rises, so to speak, *above* the events and creates a separate structure of relations where the intellect, unconscious of its actions, is quite capable of intervening and, out of the overlapping nets, to spin its own web, if I may use this image. Expressed less picturesquely, while the first mode follows the usual patterns of life and experience and substitutes only alternative course of events for the real ones, the latter mode constructs things, stories and structures, whose nature was hitherto unknown; this really is poetic imagination.[3]

Mach himself does not distinguish these two types, but simply assigns to imagination the role of *differentia specifica* for the animal organism of the human being. For him poetry is "the first step up from the everyday burden of life", [Mach 1976, p. 72] and poetic dreaming not only the very beginning of all mental development, but the source of completion and modification of experience as well, namely of that which exists as a fact.

Thus poetic imagination becomes the source for the formation of hypotheses and theories. What, after all, to remind ourselves, are the constituents of those facts that we subsume under laws? "Simplification, schematization, idealization," every theoretical concept also is a product of this disposition:

Every alteration appears as upsetting stability, as a dissolution of what until then existed together, abolishing the accustomed condition and so disturbing us and posing a problem, which drives us to look for a new connection and enquire into the cause. [Mach 1976, p. 205]

Abstraction and the activity of phantasy does the main work in the finding of new knowledge. The fact, emphasized by Whewell, that method can do little here, makes clear the mysterious feature that he thinks attaches to so-called 'inductive' findings. [Mach 1976, p. 236].

First, the result helps us to understand what always seems to have been difficult for other interpreters, namely, why Mach assumes that the *knowledge* of facts as well as *error* with respect to them have the *same* sources. From a phenomenological point of view, our imagination is *active* in both instances, either following the event and alternatively, or constructively and poetically. In this way it is possible for us in the empirical sciences to add to the experimental methods the important method of the thought-experiment, which the scientist shares with the builders of castles in the air, namely the novelists and science fiction writers [*technischen Utopisten*]. While it is hard to say how to achieve thought-experiments, Mach indicates a way: by means of paradoxes:

These [paradoxical situations – tr.] not only give us the best feeling for the nature of a problem since paradox is what makes a problem, but contradictory elements will not allow thought to rest. [Mach 1976, p. 143]

If knowledge and error therefore cannot be distinguished by means of their genesis, how then can truth be separated from falsehood, if everything is not to sink into the unknowable realm of mere conjecture?

Mach's answer to this question once more connects the two principles with his biological-evolutionary basic outlook [Capek 1968, p. 171ff] and with the realistic attitude of a great physicist and natural scientist who himself frequently erred. His answer is simple: only experience provides this distinction. He talks about an "intellectual risk," that one takes, and he asserts that the intellect of a genius differs from the normal intellect precisely because of "the certain expectation of the success of an intellectual move." In other words and expressed in theses, the following picture of the structure of scientific theories emerges:

(1) "The essential function of a hypothesis is that it leads to new observations and experiments, which confirm, refute or modify our

surmise and so widen experience." [Mach 1976, p. 176]

(2) The negative result of an experiment, that is, the falsification of a hypothesis must, however, "never be regarded as decisive" [Mach 1976, p. 157]

(3) An "unambiguous determination of consequences from a presupposition in natural science, . . . does not exist in sensible reality, but only in theory." [Mach 1976, p. 356]

(4) The result obtained through any quasi-inductive procedure has to be checked by deduction [Mach 1976].

(5) Science, like any other creation of man, is a product of history. History easily leads to prejudice, to clinging to old ways.

(6) Not only is there an evolution of life, but also of knowledge; knowledge itself promotes life.

(7) Only poetic imagination makes it possible to discover new problems and methods, the value of which always remains dependent on success either by adjusting to the existing theoretical structure or, sometimes, by breaking with it.

Contemplating these seven theses, which could easily be supplemented with many other compatible ones from Mach's works, is like opening a work in the theory of science of our century: All these thoughts can be found again almost without modification as the living accepted treasure of knowledge. And this thought can also be found in *those*, who elsewhere in their writings list Mach among those who contributed to a false understanding of science. Can there be a greater, more vivid success for a concept than that it is used and recognized after a hundred years even by its alleged opponents?

Universität Graz

NOTES

* Translator's Note: Throughout this essay the German term 'Phantasie' has been translated as 'imagination' rather than 'phantasy'.

[1] Wittgenstein [1922], 2.1 to 2.174; cf. Wittgenstein's early diary entry of October 15, 1914: "In the sentence we combine things, *on a trial basis*, so to speak; however, they do *not* have to be like that in reality." In Wittgenstein, [1960], p. 101.

[2] Cf. F. Brentano's 'Remarks about Ernst Mach's *Erkenntnis und Irrtum*'. Posthumous unpublished manuscript. The author hopes to publish Brentano's manuscript soon in *Studien zur Österreichischen Philosophie*, **6**, Rodopi, Amsterdam.

[3] Mach [1976], p. 112–113: "One can exercise phantasy in concepts too."

REFERENCES

Capek, M.: 1968, 'Ernst Mach's Biological Theory of Knowledge', *Synthese* **18**. Reprinted in R.S. Cohen and M.W. Wartofsky, eds., *Proceedings of the Boston Colloquium for the Philosophy of Science 1966/1968*, [*Boston Studies in the Philosophy of Science*, vol. 5], D. Reidel, Dordrecht, Holland, 1969.

Carnap, R.: 1967, *The Logical Structure of the World*, University of California Press, Berkeley. Tr. of R. Carnap, *Der logische Aufbau der Welt*, Weltkreis, Berlin, 1928.

Cohen, R.S.: 1968, 'Ernst Mach: Physics, Perception and the Philosophy of Science', *Synthese* **18**, pp. 123ff. Reprinted in R.S. Cohen and R.J. Seeger (eds.), *Ernst Mach: Physicist and Philosopher, Boston Studies in the Philosophy of Science*, vol. 6, D. Reidel, Dordrecht, 1970.

Dingler, H.: 1924, *Die Grundgedanken der Machschen Philosophie*, J.A. Barth, Leipzig.

Duhem, P.: 1908, *Ziel und Struktur der physikalischen Theorie*, tr. F. Adler, with a foreword by Ernst Mach, J.A. Barth, Leipzig. Engl. tr. P. Duhem, *The Aim and Structure of Physical Theory*, Philip Wiener, tr., Princeton University Press, Princeton, N.J., 1954. [Both tr. from the French *La théorie physique: Son objet et sa structure*].

Feyerabend, P.K.: 1970, 'Philosophy of Science: A Subject with a Great Past', in R.H. Stuewer (ed.), *Historical and Philosophical Perspectives of Science*, [*Minnesota Studies in the Philosophy of Science*, vol. 5], University of Minnesota Press, Minneapolis, pp. 172–183.

Frank, P.: 1938, 'Ernst Mach: the Century of his Birth', *Erkenntnis* **7**, 447ff.

Gargani, A.: 1979, *Wittgenstein tra Austria e Inghilterra*, Stampatori, Turin.

Goodman, N.: 1977, *The Structure of Appearance*, 3rd ed., D. Reidel, Dordrecht.

Haller, R.: 1981, 'Wie vernünftig ist der Common sense?', in H. Poser, ed., *Wandel des Vernunftbegriffs*, Alber, Freiburg/Br.

Heller, K.D.: 1964, *Ernst Mach: Wegbereiter der modernen Physik*, Springer, New York, Vienna, p. 168.

Hempel, C.G.: 1965, *Aspects of Scientific Explanation and Other Essays in the Philosophy of Science*, Free Press, New York.

Herrmann, E.: 1868, *Allgemeine Wirtschaftslehre*, vol. 1, part 1, J. Pock, Graz.

Hiebert, E.N.: 1970, 'Mach's Philosophical Use of the History of Science', in R.H. Stuewer, ed., *Historical and Philosophical Perspectives of Science*, [*Minnesota Studies in the Philosophy of Science*, vol. 5], University of Minnesota Press, Minneapolis, pp. 184–203.

Höfler, A.: 1910, 'Zur Geschichte und Wurzel der Machschen Philosophie', *Zeitschrift für den physikalischen und chemischen Unterricht* **23**, 1ff.

Luce, R.D. and H. Raiffa: 1957, *Games and Decisions*, New York.

Mach, E.: 1880, 'Tagebuchnotiz', p. 16, in Dingler [1924].

Mach, E.: 1881, 'Notizbuch, January 26', p. 15, in Dingler [1924].

Mach, E.: 1882a, 'Notizbuch February', p. 86, in Dingler [1924].

Mach, E.: 1882b, 'Notizbuch, April', in Dingler [1924].

Mach, E.: 1894, *Populärwissenschaftliche Vorlesungen*, J.A. Barth, Leipzig. Engl. tr. E. Mach, *Popular Scientific Lectures*, 3rd ed., rev. and enl., Open Court, Chicago, 1898.

Mach, E.: 1896, *Die Principien der Wärmelehre*, J.A. Barth, Leipzig.

Mach, E.: 1905, *Erkenntnis und Irrtum*, J.A. Barth, Leipzig.
Mach, E.: 1909, *Die Geschichte und die Wurzel des Satzes der Erhaltung der Arbeit*, Leipzig. Engl. tr. E. Mach, *History and Root of the Principle of the Conservation of Energy*, Open Court, Chicago, 1911.
Mach, E.: 1976, *Knowledge and Error* [*Vienna Circle Collection*, vol. 3], D. Reidel, Dordrecht and Boston. Tr. of E. Mach, *Erkenntnis und Irrtum*, 5th ed., J.A. Barth, Leipzig, 1926.
Mach, E.: 1910, 'Die Leitgedanken meiner naturwissenschaftlichen Erkenntnislehre und ihre Aufnahme durch die Zeitgenossen', *Scientia* **7**.
Popper, K.: 1961, *The Logic of Scientific Discovery*, Science Editions, New York. Tr. of K. Popper, *Logik der Forschung*, Vienna, 1935.
Popper, K.: 1965, 'A Note on Berkeley as Precursor of Mach and Einstin', in K. Popper, *Conjectures and Refutations*, London.
Popper, K.: 1974, 'Who Killed Logical Positivism?' in K. Popper, 'Autobiography', *The Philosophy of Karl Popper*, ed. P.A. Schilpp, La Salle, Ill., Open Court.
Spohn, W.: 1978, *Grundlagen der Entscheidungstheorie*, Scriptor, Kronberg.
Stegmüller, W.: 1973, *Carnap II, normative Theorie des induktiven Räsonierens*, Springer, Berlin. (His *Probleme und Resultate der Wissenschaftstheorie und analytischen Philosophie*. vol. 4: *Personelle und statistische Wahrscheinlichkeit*. vol. C.)
Wittgenstein, L.: 1922, *Tractatus Logico-Philosophicus*, Harcourt Brace, New York; Kegan Paul, French, Trubner, London.
Wittgenstein, L.: 1960, *Schriften*, vol. 1, Suhrkamp, Frankfurt.

PETER P. KIRSCHENMANN

SOME THOUGHTS ON THE IDEAL OF EXACTNESS IN SCIENCE AND PHILOSOPHY

I

The ideal of exactness occupies a peculiar place amid other requirements that one may lay down for intellectual work. Any attempt at assessing this place will have to struggle with the diversity of ways in which the label 'exact' has been used. Moreover, this distinctive label is mostly felt to endow whatever it is stuck upon with a special lofty appeal. In what to my knowledge is the only systematic study of the spectrum of its meanings, the adjective is called a " 'psychological' monster", which evokes at once a feeling of rigor and purity, and thus appears to legitimize any claim that is made in the name of exactness [König 1966]. Even opponents of one type of exactness might usurp the label for their own purposes, as when Goethe contrasted his idea of an "exakte sinnliche Phantasie" with the so-called exact sciences. The power of this word comes close to that of certain other terms in our modern vocabulary, such as 'freedom' and 'rationality', which are equally iridescent in their meaning, and which express ideals that also relate in peculiar ways to other values. As concerns rationality, especially scientific rationality, one may note here that exactness is often regarded as one of its ingredients.[1] In our intellectual history, each one of the three terms has frequently been employed to mark a new beginning, a purging of the mind or life from traditions considered obscurantist or stifling.

Exactness is primarily a formal ideal. There is no doubt that the label 'exact' has many a time been applied to argumentations or intellectual programs so as to make up for their lack of substantive ideas. Opponents of exact thought can find a welter of cases of such "exercises in empty exactness". Still, there are just as many good reasons for upholding the ideal of exactness. Its importance in the sciences, especially in mathematics and the natural sciences, is obvious, despite the controversy about whether its implementation is limited even in the area of science. The role the ideal should play in philosophy has been a matter of much greater dispute. To a large part, Professor Mario Bunge's work has been both a vigorous de-

J. Agassi and R.S. Cohen (eds.), Scientific Philosophy Today, 85–98.
Copyright © 1981 by D. Reidel Publishing Company.

fense of the view that the ideal can be generally implemented in all sciences and an impressive constructive advocacy of its indispensability and fruitfulness in many areas of philosophy. [See Bunge 1967a, 1967b, 1974, 1977, to mention only a few of his major works.]

In what follows I shall express or render some thoughts about a few issues concerning the ideal of exactness. Talking about exactness in some generality, as contrasted with analyzing, evaluating or even articulating specific exact methods or theories, has its particular risks and difficulties. Ironically perhaps, as we shall see, it is hard and may be even undesirable to be precise about exactness itself. Yet I find certain general questions about exactness intriguing enough to warrant some reflection. Above all, these are questions about the relationship of the ideal of exactness with complementary or other requirements or about possible limitations of its implementation. I shall begin, however, with a discussion of the notion of exactness itself. This will be followed by some brief comments on the delineation of useful from useless exactness. I shall then discuss the place of exactness in the so-called exact sciences. Finally, I shall do the same as regards philosophy, drawing upon Bunge's ideas about an exact metaphysics, which will be contrasted with Whitehead's contention that speculative metaphysics cannot be exact.

II

We could survey the diverse ways in which the term 'exact' has been and can be used and set up a catalogue of its many meanings [see König 1966]. We would state in it that in the expression 'exact method of measurement' the term means: yielding the one numerical value in question; in 'exact replica' it means: faithful in all detail, or isomorphic, to the original; in 'exact inference' the term means: logically deductive. We could then go on and seek to identify the general element shared by all the different concepts of exactness. One answer is that, in its most general sense, 'exactness' means: "definiteness (or, determinateness) qua definiteness."[2] An alternative way of specifying the common element, which will facilitate a discussion of some of its pecularities, is to articulate certain features of its relational structure. Dictionaries indicate this structure when giving as one of the primary meanings of 'exact' "perfectly conformed to a standard." This suggests the following definition.

DEFINITION 1. Some item is called *exact in a particular respect* if and only if it perfectly conforms to the corresponding standard, or, in a simple symbolism: for any x and s_i,

$$E_i(x) =_{\text{df}} PC(x, s_i), \text{ where } s_i \in S.$$

Certain refinements of this definition are obvious. The items that are possible candidates for the distinction 'exact' come in various kinds (as do the standards – a subdivision which I shall not specify in detail here). Further, only certain standards would be considered appropriately applicable to an item of any given kind. Thus one can lay down

DEFINITION 2. An item of some kind is judged *appropriately exact in a particular respect* if and only if, in addition to the conditions of Def. 1, the corresponding standard is appropriate for things of this kind, or briefly: for any kind K_j, and $x \in K_j$ and $s_i \in S$,

$$AE_{ij}(x) =_{\text{df}} PC(x, s_i) \ \& \ A(s_i, K_j).$$

A more problematic refinement is called for by the circumstance that we would not count all standards as standards of exactness. Supposing that we have reached a decision about which to call standards of exactness, we get

DEFINITION 3. An item is called *properly exact in a particular respect* if and only if, in addition to the conditions of Def. 1, the corresponding standard is a standard of exactness, or briefly: for any x and s_i,

$$PE_i^e =_{\text{df}} PC(x, s_i) \ \& \ s_i \in S^e \subset S.$$

Definitions 2 and 3 can be combined to

DEFINITION 4. An item of some kind is judged *properly and appropriately exact in a particular respect* if and only if the conditions of Defs. 2 and 3 are fulfilled.

As a last step for our purposes here, we can define a concept of being exact in all relevant (as concerns appropriateness and properness) respects, which may be called a concept of 'absolute' or 'complete' exactness (though I shall not make use of such qualifiers).

DEFINITION 5. An item of some kind is judged *properly and appro-*

priately exact if and only if it perfectly conforms to all standards of exactness that are appropriate for things of this kind.

Some comments on these definitions are in order. I shall begin by mentioning an example. Bunge calls a scientific predicate *exact* if it has an exact sense and a precise reference class [Bunge 1974, II, p. 147]. This characterization captures a concept of appropriate exactness (in two respects) inasmuch as predicates typically have (possibly precise) senses and reference classes. At the same time, it captures a concept of proper exactness (in two respects), since the conditions given clearly express standards of exactness. One question left is whether it also satisfies Definition 5. Bunge argues that, ideally, a scientific predicate should also have a definite extension ("truth set", or collection of objects for which it holds) [Bunge 1974, II, p. 135], although he does not wish to make this a condition of its exactness. Exactness, for him, has to do with constructs (predicates, propositions, theories) alone and not with relations external to constructs, such as truth valuations. Clearly, this philosophical choice of a particular subdivision of standards into those of exactness and others (which enters Definition 3) reduces the pretensions that sometimes are associated with the term 'exactness'.

A second question is whether Bunge's characterization is not circular, since it defines exact scientific predicate in terms of exact sense and precise reference class. I do not think so. These defining conditions could in their turn be analyzed along the lines given by our definitions, and so on, until ultimate standards of exactness are reached explicitly. As in the case of other norms, such an analysis is not an indefinite process either, and the resulting characterization could hardly be called circular in any straightforward sense. One should note, too, that a standard of exactness is not the same as an exact standard: talk of an 'inexact standard of exactness' is not a contradiction in terms. When referring to some standard as a 'standard of exactness' one is using the term 'exactness' analogically (in one particular sense of 'analogical use'). This analogical use does not indicate a definitional circularity; it rather points to the methodological circle which consists in the two-way traffic between (the refinement of) judgments of a certain kind and the formulation of the standards which are the basis of those judgments.

We can of course ask whether our definitions provide us with an exact concept of exactness. 'Exact' is an evaluational, not a scientific predicate; so we cannot use the above-mentioned conditions to answer this question.

Yet, do our definitions not imply at least that 'exact' can be used in an univocal sense? Being a general evaluational term, 'exact' does not meet this possible minimal requirement of exactness. It is incurably analogical (in a sense different from that referred to above), and our definitions cannot hide this fact. On the level of judgments, we speak of exact measuring techniques and exact theories; clearly, 'exact' is not being used in the same sense in the two cases. Our definitions may at first suggest that we can avoid this ambiguity by distinguishing exactness in one respect from exactness in another respect. But we are still faced with the question of how we can speak of exactness in both cases, or, how we can take a certain set of diverse standards as being standards of exactness, and how we can add to it new ones. We do so by analogy. Even the relationship between standards and items to be judged, the relation of "perfect conformity" as we called it, holds in different kinds of cases only in an analogical, not in an univocal sense. 'Conforming to' and, even more so, 'perfect' are analogical expressions themselves. The formulation of an exact (unequivocal) general concept of exactness seems to be neither possible nor, precisely for the sake of its generality, desirable. But it is both possible and desirable to have precise concepts of specific forms of exactness.

III

Given specific concepts of exactness designed to suit particular cases one will be able to identify a great variety of misapplications of the term 'exact', misuses of exact methods or ill-founded demands for exactness. Our schemes of definitions can provide merely a general guidance in surveying a part of this variety, namely those cases which have to do with the meaning and use of the expressions characterized by them. Thus, relying on Definition 3, we would say that the term 'exact' was being misapplied or, more specifically, improperly applied to some item if none of the standards to which this item conformed was a proper standard of exactness. The remaining cases have to do with the implementation of exactness rather than the use and misuse of the term 'exact'; our definitional schemes do not concern them directly. These cases raise the issue of whether and how, roughly speaking, "useful" exactness can be delineated from "useless" exactness;[3] indirectly, they will of course also have consequences for the meaning and use of the term 'exact'.

It should be noted, though, that in the notion of appropriate exactness, circumscribed in Definition 2, the meaning of 'exact' and the question of useful exactness can be closely intertwined. This is inevitably so in the many cases where the items considered are of an instrumental or functional value, as in the case of concepts, statements, theories or methods. In such cases it makes little sense to try to distinguish an appropriate intrinsic exactness of the item considered from its suitable exactness for extrinsic purposes. Concepts simply are meant to be useful for certain purposes; and if they are exact, their exactness is to be serviceable to those purposes. Otherwise, the exactness is useless and, as a consequence, the concept itself might become useless. Though no clear line can be drawn between intrinsic, appropriate exactness and extrinsic, useful exactness in the case of instrumental items of this kind, we could decide to distinguish instead between an exactness that is generally appropriate to them and whatever additional exactness they might be given to make them especially useful for particular purposes. In general, as we saw, we can require of a scientific predicate that it have an exact sense and a precise reference class. Yet, in the case of comparative concepts, for instance, one may for many purposes demand in addition that such concepts be transformed into quantitative concepts so as to increase their usefulness. Clearly, one critical issue involved in such a distinction is the determination of the types of exactness that are generally appropriate for the instrumental or functional item considered. The most profitable way of discussing this issue would be by examining claims and counter-claims bearing on particular cases.

This is also true, I think, of the question of delineating "useful" from "useless" exactness. The problem of finding general criteria of demarcation for the usefulness of exactness is relatively uninteresting and, most certainly, impossible to solve. It is much more interesting to examine specific claims concerning exactness in particular contexts (which is what I shall attempt to do in the sequel). It is more interesting to see why certain types of exactness may be regarded as useful or useless for particular purposes. Even so, a few general remarks concerning judgments of the usefulness of exactness will not be amiss.

One reason for our contention that the general demarcation problem is hardly solvable, is that exactness can legitimately be judged useful or useless in so many different ways. Judgments of useless exactness, for instance, can be articulated in a variety of nuances: such exactness may be castigated as "unsuitable", "vacuous", "idle", "fake", "misleading", "stultifying" or "impeding" the attainment of other, perhaps essential, goals. Again, without trying to draw any sharp lines, one can say that these judgments will in part concern cases where the type of exactness is intrinsically inappropriate to the items in question, while others – which no doubt make up the greater part – will have to do with cases where a type of exactness is useless for a particular purpose. Furthermore, there are other judgments of useless exactness which involve some confusion. These are cases where one may also want to speak of "useless" or "fake" exactness, in which requirements in addition to exactness are insufficiently met and in which this fact may even be disguised by the very presence of exactness. In such cases it is not so much exactness taken by itself which is useless or fake, but rather the item that has been given an exact form. For example, we tend to speak of fake exactness in the case of a mathematically well-articulated, but untestable or factually empty, theory in some empirical science.[4] There need not be anything that is fake about such mathematical exactness. But it may well have been implemented for the mere purpose of making the theory appear scientific. Clearly, in this case it is much rather the theory itself, or its supposed scientific character, which, strictly speaking, is fake (although we will then have good reason for calling the implemented mathematical exactness "a fake").

IV

Logic and mathematics are the fields in which certain types of exactness are realized in a paradigmatic fashion: precise language, strictly defined concepts, unambiguous propositions, rigorous deduction, fully axiomatized theories. All these features can more readily be achieved in these formal sciences than in non-formal sciences; they facilitate reproducibility and decidability, in short, maximum control over their subject matter. The important role and the virtues of exact methods in the empirical sciences, especially the physical sciences, are equally undeniable: we cannot imagine these sciences to exist without their use of mathematical theories, quantitative methods, precise observation and measurement techniques.[5] I shall restrict my comments here mainly to the ideals of quantification and mathematization in the empirical sciences.

Critics eager to show that there are fundamental limitations to the exact scientific method are wont to argue that science deals only with what can be quantified and measured. This is not the case, as Bunge has pointed out in several places [Bunge 1967a, I, p. 60, and 1967b, pp. 36f]: empirical science cannot dispense with such non-quantitative concepts as planet or solubility, and not all quantities used in its theories are empirically measurable. Conversely, this means that the empirical sciences will not even try to quantify or measure everything. To be sure, the use of quantitative concepts and theories with the aim of numerical precision may be considered exactness at its best. But many branches in mathematics, which can also find application in the empirical sciences, have nothing or little to do with numbers. Mathematization, and thus the exactness that comes with its use, is not restricted to quantification. There are other proper standards of mathematical exactness besides those of quantification, and many of them can be appropriate to empirical theories. This leaves us with the more general question of whether, in the empirical sciences, there are limitations to mathematization, including quantification as just one possibility.

We can, following A.N. Whitehead, regard mathematics as "the intellectual analysis of types of pattern" [Whitehead 1964, p. 117]. If, as appears to be the case, everything that exists possesses some pattern, we can conclude that the scope of mathematization is in principle unlimited. Accordingly, Whitehead was indeed very optimistic about the capacity of mathematics to go on conquering fields little or not yet touched by it [Whitehead 1961, p. 213]. Bunge is equally optimistic, yet for a somewhat more restricted reason [Bunge 1967a, II, p. 204]. It is our concepts, our ideas concerning a subject matter, not the subject matter itself, that gets mathematized or quantified; thus there should not remain any insurmountable barriers to mathematization. This view is compatible with Whitehead's, although there is a significant difference. Bunge's view, it seems to me, invites us to take a greater freedom in exploring possible mathematical treatments. But it does so by de-emphasizing the fact that the objects studied must have features that lend themselves to mathematical conceptualization. As concerns quantification in particular, an additional argument is given by Bunge for the view that its use can be extended far into the sciences of man [Bunge 1967a, II, pp. 203f]. For instance, fear can be quantified, because it can be correlated with adrenal level in blood. The significance of this and other correlations, already found or still to be found, is beyond dispute. But if this approach were to be taken as the primary path to the formulation of exact concepts and theories, it might come into conflict with the rather liberal approach just mentioned.

Mathematization, and also the reported optimism about its prospects, concerns only certain aspects of a subject matter studied: its pattern that can be articulated in an abstract way, or the formal structure of concepts and theories developed in studying it. In other respects, every intellectual discipline is at the same time governed by other goals or ideals. It is an almost trivial truth that exactness, even if it is mathematical exactness, is not enough; it is a means to achieve more substantive goals. Mathematics itself is no exception. G.H. Hardy indicated the ideals which, over and above that of exactness, govern this field by saying that mathematicians strive to come by serious (as opposed to trivial) theorems, that is, theorems which contain significant ideas, or ideas having a certain generality and depth [Hardy 1967, p.103].

The primary aim of empirical science is knowledge of the world. At its best, it tries to achieve this goal by means of comprehensive theories of great depth, which includes high explanatory and predictive power [cf. Bunge 1967a, I, passim, esp. pp. 506ff]. The theories must be well testable or tested; otherwise they can lay no claim to being factual knowledge. Mathematization not only satisfies requirements of formal exactness; it also facilitates the achievement of these goals by enhancing, for instance, a theory's deductive power and testability [cf. again Bunge 1967a, I, pp. 474ff]. In order to deliver these services, mathematical tools must be adequate to the particular object of study and purpose at hand. Thus, in the empirical sciences, the ideal of exactness is assigned a definite place within a network of checks and balances set up by other ideals or dominant goals. The great mathematized theories of physics can count as harmonious realizations of this complex of ideals. Most scientific research does not come close to achieving such a fortunate balance; in theoretical physics, mathematical refinement is frequently being pursued with little regard for, or maybe even at the expense of, explanatory value and testability.

Many of the ideals governing empirical science are analogous to those in mathematics. One obvious difference is this: the ultimate criterion for whether the goal of empirical science has been reached is the actual observational or experimental test and not the mathematical form of a theoretical construction. The difference is also exhibited by another fact, which Bunge especially has emphasized and constructively taken into account: an adequate axiomatization of a factual theory (a name which he prefers to 'empirical theory' for good reason) should, in addition to postulates characterizing its mathematical structure, include semantic assumptions specifying its factual meaning [see, e.g., Bunge 1967b; 1967a I, pp. 104ff, and II, pp. 16ff]. A difficulty here is that this specification cannot aspire to the same degree of exactness as that of the mathematical formalism.[6]

V

Many philosophical flags have carried the label 'exact'. I shall not attempt to give a survey of what was meant by it or of the aspirations and accomplishments that it covered.[7] For Bunge, 'exact philosophy' signifies "philosophy done with the explicit help of mathematical logic and mathematics". In philosophy, too, exactness is "a means for enhancing clarity and systemicity, hence control" as well as "the possibility of rational scrutiny", but it "should not be taken for an end" [Bunge 1973b, p. v]. Here, too, there are other ideals and dominant goals. There is the general ideal that philosophy tackle interesting, that is, genuine and deep problems [Bunge 1973b, p. v]. Yet Bunge has also adopted and in part executed a specific goal: the development of a scientific philosophy.[8] It is to consist in a system of interconnected theories, articulated in logical or mathematical form, which are both compatible with and relevant to science. By definition, then, scientific philosophy will be exact. Its relationship with science, of course, is not the same for all its parts. Semantics, inasmuch as it is restricted to semantics of science, is *about* science: it studies concepts like meaning, reference, truth in their relevance for scientific constructs [Bunge 1974, I, 'Preface']. On the other hand, scientific ontology (or metaphysics), whose goals are "to analyze and to systematize the ontological categories and hypotheses germane to science",[9] is continuous or overlapping with science:

> There is no gap between good metaphysics and deep science: every scientific theory with a wide scope may be regarded as metaphysical, and every ontological theory that brings together and generalizes scientific results, or else occurs in the background of an axiomatized scientific theory, qualifies as scientific.[10]

I have found it intriguing that Whitehead, surely not an enemy of logical or mathematical methods, reached quite different conclusions about the ultimate worth of exactness in, and the possible scientific character of, philosophy: "My point is that the final outlook of Philosophic thought cannot be based upon the exact statements which form the basis of the special sciences. The exactness is a fake."[11] This is the ending of the one of his last two essays, while the other ends as follows.

> But consciousness proceeds to a second order of abstraction whereby finite constituents of the actual thing are abstracted from that thing. The procedure is necessary for finite thought, though it weakens the sense of reality. It is the basis of science. The task of philosophy is to reverse this process and thus to exhibit the fusion of analysis with actuality. It follows that Philosophy is not a science. [Whitehead 1964, p. 121].

There are many substantive and also fundamental differences between Whitehead's process metaphysics and Bunge's thing or system ontology. I cannot discuss them here in detail.[12] One reason for these differences as well as for their different evaluations of the possible exact and scientific character of philosophy can be found in the fact that they assign different tasks to philosophy. Some of the passages quoted above have given an indication of this difference. Science always involves an abstraction from concrete actuality (which Whitehead called "a second order of abstraction" in contrast to what is involved in the actualization of something, namely, an abstraction from all the factors that also could have entered its actualization). Philosophy, according to Whitehead, should not proceed to even higher levels of abstraction, but ought to focus on the concrete constitution of what there is. The type of exactness appropriate to scientific theory is thus inappropriate to metaphysical theory. For Bunge, metaphysics is to be an extension of science: its basic categories and postulates are identical with the most general concepts and principles underlying or being used in science (which to Whitehead are prime examples of abstractions). Its goal is to develop a system of interrelated theories characterizing these categories and postulates in an exact manner, whereas for Whitehead the aim of what he calls "speculative metaphysics" is "to frame a coherent, logical, necessary system of general ideas in terms of which every element of our experience can be interpreted." [Whitehead 1957, p. 4] Scientific knowledge is accorded a central role, since it is an important element in our experience; yet it is just one among the data for metaphysical speculation. Whitehead, then, sought to obtain his fundamental concepts by generalizing categories of experiences (as when he identified "actual entity" with "occasion of experience"), and not by generalizing the basic concepts of science.

There also are differences as to the formal qualifications required of a metaphysical system. For Bunge, scientific ontology is to be a system of mutually consistent theories. Theories themselves are hypothetical-deductive systems, and they are exact if they are given a definite mathematical structure. Both these features enhance systemicity. In this way, the ideals of exactness and systemicity are thus conceived as being partly interwoven: systematization can be achieved better if exact, mathematical tools are being used. Whitehead's emphasis is on the requirement that the system of fundamental ideas be coherent, which means that they presuppose each other and would be meaningless in isolation. They are thus not characterizable by a definite mathematical structure of their own or by distinct mathematized theories. For Whitehead, then, metaphysics cannot derive its systematic character from mathematical form or have the kind of exactness envisaged by Bunge. There is agreement on the requirement that the metaphysical system be logical. Yet, while Bunge regards logical notions as metaphysically neutral, they must for Whitehead find their places in the system itself. Finally, Whitehead requires that the system be necessary. I understand him to mean by this that its universal validity should become manifest in and through the system itself, primarily through its interpretative power. This is not so much a formal requirement as one that concerns the evidence of the system. For Bunge, its primary evidence derives from its close contact with science. This means, first, that the final court of appeal in factual science, empirical test, is given a highly significant, if indirect and remote, influence in decisions on metaphysical matters. It also means a strong reliance on the self-critical power of science as regards its own basic assumptions, which thus are not considered as standing in need of a largely autonomous critique by philosophy.

Simplifying somewhat, one can attribute the two diverging assessments of the possible exact and scientific character of metaphysics to differing emphases placed on one feature of the world in its relation to human knowledge. For Whitehead, whatever actually exists is in infinitely many ways, physical and non-physical, related to other factors. No finite proposition or theory, which could then be expressed in precise statements, can therefore be adequate in an ultimate sense to what actually is. Science, which cannot help but proceed by way of abstraction, can arrive only at half-truths. For Bunge, too, there are no totally isolated things; but their relative isolation permits a stepwise approach, where each step in the conceptual reconstruction is capable of exact statements. While this is not denied by Whitehead, he takes it as typifying science as regards both its method and its limitations. If Whitehead is correct there will ultimately be limits to philosophizing in an exact manner. This is all the more the case when exactness is equated with the use of a given method such as the use of mathematical means, since for him the search for a metaphysical system includes the search for the appropriate metaphysical method. Yet the extent to which exact or scientific philosophy nonetheless can fruitfully be done will, it seems to me, remain an open question. It is worth noting here that both authors always stressed the tentative nature of their work in metaphysics.

The question about possible limitations of exactness in philosophy can remain in the background for Bunge because of his doctrine that this ideal applies to our conceptual tools alone. We have seen this doctrine exemplified in his view about mathematization in science. Earlier we saw it illustrated in his decision not to include the possession of a definite extension among the conditions of exactness for scientific predicates. The remaining conditions were that it had an exact sense and a definite reference class. Yet in general, the ideal of exactness is conceived even more narrowly: "to *exactify* a factual construct consists in either revealing or assigning its formal component" (the other component being its factual interpretation or "content") [Bunge 1974, II, p. 32, also p. 18; on form and content cf. I, pp. 115ff]. Thus, the exactification of a concept, proposition or theory, which is to be carried out by means of logic and mathematics, concerns only formal sense alone. In addition we can mention here that Bunge stresses the point that our conceptual constructs are human artifacts, "total fictions", though indispensable ones [Bunge 1974, I, pp. 13, 26ff]. These views, I take it, hold for philosophical constructs as well. Thus, Bunge has both sufficiently restricted the aspects of philosophical thought to which exact methods are to be applied and considerably enlarged the freedom of using diverse logical or mathematical structures for philosophical ends.

The primary specific challenge for an exact philosophy will always be to find and use logical and mathematical tools which satisfy the more substantial requirements and goals. They must be adequate to the purpose at hand, suitable for the exhibition of interesting formal aspects of important philosophical concepts and theories, helpful in clearing up or solving significant philosophical problems. Just as the issue of possible limitations of metaphysical exactness, judgements on these matters will remain themselves philosophically controversial. It is hard to imagine that there could be exact philosophical theories which in their general recognition would vie with the great achievements of theoretical physics which harmoniously satisfy the several ideals of factual science. Concomitantly, there scarcely are indications that the analysis of philosophical problems or concepts would stimulate the development of new branches of mathematics most suitable for the analysis at hand, in the way much theoretical work in physics and some philosophical work in previous ages have done. Such observations should not, however, be turned into an argument against the use of ready-made exact tools for philosophical purposes. Their appropriateness must simply be evaluated afresh in any particular application. In a most persistent and exemplary fashion, Professor Bunge has contributed greatly in the search for appropriate exact forms in which to cast philosophical thought.

Vrije Universiteit Amsterdam

NOTES

[1] For a discussion of this point, with focus on more recent developments in the philosophy of science, see my [1977].
[2] "Bestimmtheit qua Bestimmtheit", König [1966], p. 101.
[3] I am obliged to one of the editors for urging me to dwell on this issue explicitly.
[4] Bunge offers an example of such a theory in [1967a] I, pp. 477f.
[5] Cf. Bunge [1967a] I, pp. 474ff, and II, p. 202. For a few more details on this and the following points see also my [1977].
[6] Cf. also my review of Bunge [1973a] in *Philosophy of Science* **42** (1975), 94–96.
[7] For an overview of the meaning diversity see again König [1966], esp. pp. 81ff.
[8] For his characterization of scientific philosophy see the "General Preface to the *Treatise*" in any of the volumes and also the "Introduction" in vol. III.
[9] Bunge [1974] III, p. 12 (italicized in the original).
[10] *Ibid.*, p. 24.
[11] Whitehead [1964], p. 104. The last sentence is set apart as a paragraph of its own.
[12] See Bunge [1977], passim (for precise references look under "Whitehead" in the Index of Names). [See also L. Apostel's contribution to this volume – Eds.]

REFERENCES

Bunge, M.: 1967a, *Scientific Research*, vol. I and II, Springer, New York.
Bunge, M.: 1967b, *Foundations of Physics*, Springer, New York.
Bunge, M.: 1973a, *Philosophy of Physics*, D. Reidel, Dordrecht.
Bunge, M.: 1974, *Treatise on Basic Philosophy*, vol. I and II, D. Reidel, Dordrecht.
Bunge, M.: 1977, *Treatise on Basic Philosophy*, vol. III, D. Reidel, Dordrecht.
Bunge, M. (ed.): 1973b, *Exact Philosophy*, D. Reidel, Dordrecht.
Hardy, G.H.: 1967, *A Mathematician's Apology*, Cambridge University Press, Cambridge.
Kirschenmann, P. P.: 1977, *Exactness*, Vrije Universiteit, Amsterdam.
König, G.: 1966, *Der Begriff des Exakten*, A. Hain, Meisenheima. G.
Whitehead, A.N.: 1957, *Process and Reality*, Harper, New York.
Whitehead, A.N.: 1961, *The Interpretation of Science*, A.H. Johnson (ed.), Bobbs-Merrill, Indianapolis.
Whitehead, A.N.: 1964, *Science and Philosophy*, Adams, Paterson, New Jersey.

WŁADYSŁAW KRAJEWSKI

ON HYPOTHESES AND HYPOTHETICISM

1. INDUCTIVISM, DEDUCTIVISM, HYPOTHETICISM

The traditional method of induction, both enumerative and eliminative, is of a qualitative nature. It leads to the discovery of qualitative laws of different types: causal, structural, evolutionary, etc. Some related methods, more important for the sciences, are of a quantitative nature. Statistical reasoning leads to the discovery of empirical statistical laws. Interpolation and extrapolation of measurements leads to the discovery of functional dependencies between directly measured magnitudes. These methods can be considered to be generalizations of induction. They can be called inductive methods in a broader sense.

Inductivism is a view that only the methods described above are used in the empirical sciences or, at least, that they are the main methods of these sciences. According to this view, science always starts from facts (observations, measurements) and passes from them to laws.

Inductivism was founded in the seventeenth century by Francis Bacon. It prevailed in the methodology of science in the nineteenth century, especially in Great Britain (J. Herschel, J. S. Mill, A. Bain). It was also propagated by all forms of positivism in other countries (Comte, Mach, the Vienna Circle).

Inductivism in methodology is closely connected with empiricism in the theory of knowledge. An opposite view was held by rationalism (apriorism). All adherents of this view in philosophy – from Descartes, Leibniz, Kant, and Hegel to the neo-Kantians and phenomenologists – claimed that the fundamental Laws of Nature, or at least some of them, can be inferred by deduction from basic philosophical ideas. According to this view, only simple empirical laws can be discovered by induction. Hence, deduction is more important for science than induction. This view can be called *deductivism*.

However, there were other adversaries of inductivism among philosophers of the nineteenth and twentieth centuries. They showed the crucial role of hypotheses in science. In order to explain phenomena, scientists invent hypotheses, deduce empirical consequences from them and then test

J. Agassi and R.S. Cohen (eds.), Scientific Philosophy Today, 99–109.
Copyright © 1981 *by D. Reidel Publishing Company.*

them by experiments. This method has been called the hypothetico-deductive method (H-D). The view that it is the main method of science, in any case at the advanced stages of its development, can be called *hypotheticism*.

The British philosophers of the nineteenth century, W. Whewell and W.S. Jevons, can be considered the founders of hypotheticism. They emphasized the role of inventiveness and creativity in science in the formation of hypotheses. Friedrich Engels also opposed one-sided inductivism and pointed out the role of hypotheses in science, though he did not develop this subject. At the beginning of the twentieth century, the French conventionalists H. Poincaré and P. Duhem criticized extreme empiricism and stressed, though with some exaggeration, the role of creativity, especially of conventions, in science. In the thirties, K.R. Popper became the main adversary of inductivism. He called his position 'deductivism.' However, in contradistinction to the apriorists, Popper speaks about the deduction not from *a priori* statements but from hypotheses; he stresses the necessity to criticize hypotheses. Therefore, the name 'hypotheticism' is much more adequate at this point.

All hypotheticists consider experience to be an ultimate criterion of hypotheses. Therefore, they are methodological empiricists. However, they stress the role of reason and deduction. Therefore, their standpoint in the theory of knowledge can be called 'rationalized empiricism'.

* * *

A great controversy started when Popper published his classical work [1935]. Popper and his school were on one side, Carnap and his followers – among them many eminent logicians – on the other side. This controversy is unusual because it is maintained only by one side.

The Popperians claim that there is a basic contradiction between them and the Carnapians. The latter are attacked sharply for different shortcomings: they mistakenly think that the process of the discovery of laws starts from facts which are generalized by induction, that induction justifies a law, that it leads to truth, that the growth of science is a simple accumulation of truths. In other words, they are blamed for the faults of inductivism, justificationism, verificationism, cumulativism. The Popperians, to the contrary, insist that scientists start from hypotheses (hypotheticism), deduce empirical consequences from them (deductionism), try to refute them (falsificationism). When a hypothesis cannot be falsified, it is corroborated; however, this does not justify it; we cannot even speak about a probability of a corroborated hypothesis, it can always fail (fallibilism). The growth of science is not an accumulation of truths (anticumulativism), it consists of constant refutations of falsified hypotheses, it is a 'permanent revolution' [cf. Agassi 1966].

The Carnapians answer that they do not accept the terms 'inductivism', 'hypotheticism', etc.; they deny the existence of any essential contradictions between them and the Popperians at all. Their arguments are as follows. Everybody accepts the role of hypotheses in science, the use of the H-D method, the fact that some hypotheses are accepted and others are refuted. The difference between the two sides is rather terminological. The Popperians use the term 'induction' in a narrower sense, the Carnapians in a broader sense. For the latter, every fallible method in science, including the H-D method, is an induction. W.C. Salmon has said during the 1965 London discussion on inductive logic that every "ampliative and nondemonstrative inference" is an induction [Lakatos 1968, p. 28]. According to this standpoint, the single nonterminological difference is the difference in the field of interests. The Carnapians investigate only logical relations between sentences, especially relations between hypotheses and facts; they try to evaluate the probability of a hypothesis with respect to some set of facts, i.e., to state its degree of confirmation, etc. They do not consider the real process of discovery in science, the temporal relations between different steps, etc. According to Bar-Hillel, we are dealing with a 'division of labor': the Carnapians concentrate on a 'rational synchronic reconstruction of science', whereas the Popperians are "mostly interested in the diachronic growth of science" [Lakatos 1968, p. 69].

I cannot agree with these arguments. Methodology should consider all methods leading to truth and not only purely logical relations. This does not mean that I accept the methodology presented by Popper, in any case in his classical work. I cannot agree with his denial of induction: different kinds of induction, mentioned at the beginning of this paper, play an essential role in science, though not as important as the H-D method. Moreover, I cannot agree with Popper's one-sided falsificationism. Falsification plays a great role in science; however, a positive result has a greater value. It may be called confirmation or corroboration – it does not matter. Even when the concept of probability in the strict sense is inapplicable here, we must search for the way to evaluate different degrees of confirmation.

As M. Bunge has rightly said, we must choose a '*via media*' between extremes of inductivism and deductivism [Bunge 1963, p. 151]. The same view is held by I. Niiniluoto and R. Tuomela: the Carnapian inductivism exaggerates the role of induction and does not see the role of theory, both in the identification of facts and in the testing of hypotheses; the Popperian deductivism does not see the role of induction and does not want to recognize the fact that highly corroborated hypotheses are reliable (though not absolute in any sense) and that predictions based upon them are rational [Niiniluoto and Tuomela 1973, pp. 200–203].

The methodology elaborated in Popper's classical work is an extreme version of hypotheticism. Later on, Popper and his school developed towards a more moderate and reasonable version of hypotheticism which recognizes the role of both positive and negative results of testing. It does not identify the falsification of a hypothesis (i.e., the finding of facts incompatible with it) with its repudiation because there is often a possibility of its preservation by means of new auxiliary hypotheses. In general, an old theory is given up only when a new theory is created which explains facts better than the old one. Even when we give up a theory we usually retain some of its elements. Therefore, the growth of science is not a permanent revolution which simply overthrows all subsequent theories, but a progress, despite revolutions, a passing from worse theories to better ones, which come increasingly close to the truth. These ideas are contained in the newer papers by Popper [1972] and especially in papers by his disciple, the late Imre Lakatos [1970].

We can add that in the developed sciences the method of idealization plays a great role; therefore we can speak about idealizing hypotheticism. When we also take into account the role of rivalry between different hypotheses (which has been put forward by Lakatos) we can speak about pluralistic idealizing hypotheticism. All these concepts are introduced in my English book [1977] and I shall not discuss them here.

2. THE HYPOTHETICO-DEDUCTIVE METHOD

We shall now attempt to provide a more thorough consideration of the H-D method, i.e., the process of the creation and testing of hypotheses.

A general sketch of this process is presented by C.G. Hempel in his excellent popular book [1966]. The book begins from the analysis of Semmelweiss's discovery of the cause of childbed fever. Semmelweiss took into account various hypotheses concerning its cause, he deduced empirical consequences from them ('test implications' in Hempel's terminology) and then executed experiments to test them. This is a general method of the empirical sciences. Hempel puts forward the view that we cannot start from the facts, as 'narrow inductivists' claim, because without a hypothesis, we do not know, which facts should be collected. There are no facts relevant to a problem, there are only facts relevant to a given hypothesis. One should deduce 'test implications' from the hypothesis and test them in experiments. The negative result of experiments leads to the falsification of the hypothesis, and multiple positive results – to its confirmation. Later on, the author considers factors which determine the degree of confirmation.

Hempel describes the H-D method although he does not employ this term. He calls this method "inductive in a broader sense", in contradistinction to the "narrow inductive method." The name is not important, of course. In any case, Hempel stresses very strongly the fact that we cannot induce hypotheses from facts, we must invent them. Hypotheses are 'happy guesses' which often require 'great ingenuity.' Hempel cites Whewell and Popper here [Hempel 1966, p. 15]. Hempel's standpoint is a consistent hypotheticism, even a 'militant hypotheticism," because his book is directed against narrow inductivism.

Some authors point out that we start not from hypotheses but from facts which need explanation. N.R. Hanson criticizes both the adherents of induction and the adherents of the H-D method. According to him, the former are right claiming that investigation always starts from facts but are wrong when they claim that facts are generalized by induction. The latter are right claiming that hypotheses must be invented for the explanation of facts but are wrong when they claim that a scientist starts from hypotheses. He starts from facts to be explained, analyzes them and then invents hypotheses [Hanson 1958, p. 70]. This is true but I think we have just a more complete account of the H-D method here, not a 'third way.'

W.C. Humphreys points out that science tries to explain not all facts but only anomalies, i.e., facts which cannot be explained by existing theories. A scientist starts from the identification of anomalies, i.e., he must ascertain that given facts indeed cannot be explained by the known theories; only then does he create new hypothetical theories [Humphreys 1968, Ch. I]. An anomaly can often be removed by auxiliary hypotheses added to the old theory. Sometimes we see that this procedure is impossible, i.e., that a given fact is indeed an anomaly, only after the creation of a new theory [cf. also Pietruska-Madej 1978]. In any case, we always start from the identification of anomalies. This identification is often trivial (e.g., the high mortality in Semmelweiss's clinic), but in many cases it is more complex, and the problem, whether a fact is an anomaly, may give rise to a controversy (Humphreys presents different examples from the history of physics).

Hence, we have the following steps in the process of discovery:

1. The identification of an anomaly, i.e., of facts which need explanation.

2. The creation of an explanatory hypothesis.

3. The deduction of empirical consequences from this hypothesis.

4. Experiments testing these consequences.

5a. The acknowledgment of this hypothesis when the result is positive (confirmation).

5b. The refutation of this hypothesis when the result is negative (falsification).

In the last case, steps 2–5 must be repeated.

This scheme is a simplification. The real process is more complex. First of all, empirical consequences usually do not follow from the tested hypothesis alone but from a conjunction of this hypothesis and certain auxiliary assumptions. Later on, in contemporary physics and other developed sciences, there are additional complications. The experimental test must be preceded by theoretical tests. When the hypothesis is a project of a mathematically elaborated theory, we must examine its coherence. Then, we must examine whether it is compatible with accepted, well-confirmed knowledge. In contemporary physics a new fundamental theory, usually more general than its predecessor, must be so constructed that the old theory (after reinterpretation) turns out to be a limiting case of the new one. This postulate is known as the correspondence principle [cf. Krajewski 1977]. In less fundamental domains of physics and in other natural sciences, an examination is necessary in order to determine whether the new hypothesis is compatible with the accepted fundamental theories of physics. In other words, a new hypothesis must pass a test of internal coherence, a test of external coherence, and only then experimental tests [cf. Such 1975].

In contemporary science, a hypothesis increasingly often is an idealizing one. In this case the whole process is still more complex. The hypothesis must pass a procedure of factualization before experimental tests [cf. Krajewski 1977, Ch. 2]. The experimental data often have to be transformed too, they have to be interpreted in theoretical terms, in order to undergo test [cf. Bunge 1970]. We shall not discuss these questions in detail. In any case, it is clear how far we are from simple induction and from simple deduction.

3. SIMPLE AND DEEP HYPOTHESES

The hypotheticists focus on hypotheses. The inductivists answer that they also take hypotheses into account. This is true. Moreover, every case of inductive reasoning uses hypotheses, makes hypothetical assumptions. But what is their nature? Induction by enumeration uses a hypothesis about the class of objects which have the given feature (when we say that each A is B, we may make the class A broader or narrower). Induction by elimination uses a hypothesis about the set of factors which can be the cause of the given event. The method of interpolation uses a general postulate of simplicity and a hypothetical assumption about the shape of the functional dependence in the given case. Statistical reasoning uses a hypothesis about the probability of an event in certain conditions, etc. However, all these hypotheses deal with the observed facts, with direct experience. We shall call all hypotheses of this kind 'simple hypotheses'.

When Carnap and his followers examine the problem of the probability of a hypothesis, they almost always take simple hypotheses into account, which are expressed in observational terms. When they evaluate a hypothesis that an object belonging to class A belongs also to class B, they consider observational terms A and B, even when they take dispositional predicates into account (e.g., 'soluble in water'). We may add that Popper in *The Logic of Scientific Discovery* considered mainly simple hypotheses too. However, such hypotheses do not occupy the central place in science.

All great discoveries in science consist in confirmation of deep hypotheses. Most of them concern unobserved things, hidden mechanisms of observed processes. Of course, some observational consequences always follow from these hypotheses – otherwise they would not be scientific – but the deduction of these consequences consists of many steps and needs different auxiliary assumptions. In many cases it is not easy to find the consequences which can be tested by experiment. Their finding needs invention, ingenuity.

There is an abundance of examples of great discoveries based on deep hypotheses: that sound is a longitudinal wave, that light is a transverse wave, that white light is complex, that heat is chaotic molecular movement, that electric current is an ordered movement of electrons or ions, that a molecule is a system of atoms with a definite space-structure, that the valency of a chemical element depends upon the number of electrons on the external atomic orbit, that infectious diseases are caused by bacteria or viruses, that genetic information is contained in the molecules of DNA which have the shape of the double helix, that this information is transported by the molecules of the RNA, etc.

None of these discoveries could have been made by any type of induction. They needed ingenuity, imagination, courage to create hypotheses which surpass the observed phenomena. As we have noticed, invention is necessary not only in the course of the creation of hypotheses but also in the course of deduction. None of the above-mentioned hypotheses entails empirical consequences alone, for in all cases different auxiliary hypotheses are necessary. The planning of experiments which test these hypotheses needs invention, too.

Deep hypotheses, when formulated, usually cause different emotions: astonishment, distrust, often resistance, and later enthusiasm and admiration. An inductive conclusion can hardly cause such emotions.

When we take the language of scientific theories into account, we may say that deep hypotheses are statements formulated by means of theoretical terms. Many authors oppose the dichotomy 'theoretical-observational', pointing out the theoretical component of all terms. We shall not discuss this controversy. However, even when a dichotomous division is impossible, it is obvious that some terms are more, and some less, theoretical. The deeper a hypothesis is, the more theoretical terms are necessary for its expression. True, a measure of the degree of depth is hardly possible.

Many contemporary methodologists point to these problems using different terminology. The book by Hempel mentioned above includes a special chapter on the role of theories and theoretical explanation. The author claims that theories "afford a deeper and more accurate understanding of the phenomena in question," because they interpret them as "manifestation of entities and processes that lie behind or beneath them" [Hempel 1966, p. 70]. We see that Hempel has surmounted all inductivist and positivist limitations. However, he does not criticize Carnap. Other authors do. As we have mentioned before, Niiniluoto and Tuomela criticize Carnap and other inductivists for neglecting the role of theories and for comparing single hypotheses with facts. Bunge has made a similar criticism in the course of the 1965 London discussion. At that time Bar-Hillel replied that Bunge's phrase "inductivist and anti-theoretic philosophy of Carnap" is irresponsible because "few philosophers have contributed more to an understanding of the function of theories in science than Carnap and Hempel" [Lakatos 1970, p. 284]. Why did Bar-Hillel identify Carnap with Hempel? Bunge has made no criticism of Hempel!

I want to point out that I hold Carnap to be an eminent logician and methodologist in our century. His merits are especially great in the examination of the probability of hypotheses. However, he usually confined himself to simple empirical hypotheses, hence, to elementary parts of science. True, in the last period of his life Carnap introduced theoretical concepts into the range of his considerations. Sometimes he spoke about the theories of contemporary physics. Nevertheless, he still discussed the possibility of eliminating theoretical concepts by means of a "Ramseyification" [cf. Carnap 1966]. Hence, he did not see the epistemological role of deep hypotheses which reveal the internal mechanism of natural processes.

* * *

All idealizing laws are hypotheses which – usually after a factualization - are confirmed by experience. They are deep hypotheses although they do not always reveal the internal mechanisms. Newton's laws of classical dynamics, e.g., do not concern the internal structure of matter. Nevertheless, they are deeper than Kepler's laws or Galileo's law of free fall because Newton revealed the idealizing conditions which must be assumed in order to obtain Kepler's laws or Galileo's law.

K. Popper quotes these examples in his 1957 paper 'The Aim of Science' which is a great step forward in comparison with *The Logic of Scientific Discovery* because in this paper he passes to consideration of fundamental processes of the growth of science. According to him, science seeks "to probe deeper and deeper into the structure of our world" [Popper 1972, p. 196]. Popper points out the necessity of idealizing assumptions (he calls them 'false assumptions'). In order to deduce Kepler's laws from classical mechanics, we must assume that there are no other planets (which disturb the movement of the planet considered). In order to deduce Galileo's law, we must assume that the earth's radius is infinite (only then is the earth's acceleration constant during the fall). Newton's theory corrects Kepler's and Galileo's laws while explaining them. In general, when "a new theory of a higher level of universality successfully explains some older theory *by correcting it*, then this is a sure sign that the new theory has penetrated deeper than the older ones" [Popper 1972, p. 202]. Only here has Popper come to a genuinely historical and realistic approach to science, and definitively surmounted any inductivism and positivism.

* * *

Many authors still hold the term 'inductivism' to be useless and misleading. I shall cite two examples of papers published recently in Poland.

Halina Mortimer calls the term 'inductivism' into question. She writes that Popper and his followers usually describe inductivism in a caricature, presenting only 'narrow inductivism' in Hempel's terminology [Mortimer 1978]. However, it is not a caricature. Narrow inductivism was represented by the Vienna Circle, and even if it were true that all its former participants have changed their minds, narrow inductivism is still represented by many scientists. They often consider it trivial to say that empirical sciences start from facts and then generalize them, inducing laws from facts. Some postgraduate students of the natural sciences told me that reading Hempel's book (I require students to read it before the examination in philosophy at Warsaw University) has given them new horizons because they were convinced hitherto that they should collect facts without any 'bias'.

Evaldas Nekrašas also criticizes the Popperian concept of inductivism. He writes that when we call a view that universal statements are obtained from singular statements by means of enumerative induction, the name 'inductivism' is empty [1978]. However, as we have seen, this name is not empty. In any case, many scientists are inductivists in this sense. Nekrašas also says that a Bayesian and a Popperian can have identical views about preference among rival hypotheses. I agree. The fault of the Carnapians (or Bayesians) lies, in my view, not in wrong methodological advice but in overlooking the most important problems of the methodology of science.

Warsaw University

REFERENCES

Agassi, J.: 1966, 'Revolutions in Science, Occasional or Permanent?', *Organon* **3**, 47–61. Reprinted in Agassi, *Science and Society*, D. Reidel, Dordrecht and Boston, 1981.

Bunge, M.: 1963, *The Myth of Simplicity*, Prentice Hall, New York.

Bunge, M.: 1970, 'Theory Meets Experience', in H. Kiefer, M. Munitz (eds.), *Contemporary Philosophical Thought*, vol. 2, Albany.

Carnap, R.: 1966, *Philosophical Foundations of Physics*, Basic Books, New York.

Hanson, N. R.: 1958, *Patterns of Discovery*, Cambridge University Press, Cambridge.

Hempel, C. G.: 1966, *Philosophy of Natural Science*, Prentice Hall, New York.

Humphreys, W. C.: 1968, *Anomalies and Scientific Theories*, Freeman Cooper, San Francisco.

Krajewski, W.: 1977, *Correspondence Principle and Growth of Science*, [*Episteme*, vol. 4], D. Reidel, Dordrecht.

Lakatos, I. (ed.): 1968, *The Problem of Inductive Logic*, North-Holland, Amsterdam.

Lakatos, I.: 1970, 'Falsification and the Methodology of Scientific Research Programmes', in *Criticism and the Growth of Knowledge*, I. Lakatos and A. Musgrave (eds.), Cambridge University Press, Cambridge.

Mortimer, H.: 1978, 'Karl Popper's Logic of Scientific Discovery', (in Polish), *Studia Filozoficzne* **4**.

Nekrašas, E.: 1978, *Analysis of the Concept 'Inductivism'*, (in Polish), *Studia Filozoficzne* **7**.

Niiniluoto, I., and R. Tuomela: 1973, *Theoretical Concept and Hypothetico-Inductive Inference*, D. Reidel, Dordrecht.

Pietruska-Madej, E.: 1978, 'Anomalies and Their Role in the Growth of Science', in W. Krajewski, E. Pietruska-Madej, J. Zytkow (eds.), *Relations between Theories and the Growth of Science* (in Polish), Ossolineum, Wrocław.

Popper, K. R.: 1935, *Logik der Forschung*, Springer, Vienna. Engl. tr. *The Logic of Scientific Discovery*, Basic Books, New York, 1959.

Popper, K. R.: 1972, *Objective Knowledge*, Clarendon Press, Oxford.

Such, J.: 1975, *Is There an Experimentum Crucis*? (in Polish), PWN, Warsaw.

J. LAMBEK

THE INFLUENCE OF HERACLITUS ON MODERN MATHEMATICS

Among the pre-Socratic philosophers we know that Thales, Pythagoras, Zeno and Democritus were involved in mathematics in one way or another. In fact, Pythagoras coined the very word "mathematics." Heraclitus does not appear to have occupied himself with mathematical questions; so how can he be said to have influenced modern mathematics? The answer to this question will take us on a small detour.

What has survived of Heraclitus' thought is contained in a number of pithy statements. I shall only mention two:

> "You cannot step into the same river twice",

and

> "Men do not know how what is at variance agrees with itself; it is an attunement of opposite tensions, as in the bow or in the lyre".

Of all the Greek philosophers, Heraclitus has perhaps been the most influential in the twentieth century. When Mao Tse-tung refers to the "law of contradiction in things" as "the unity of opposites," he is quoting indirectly from Heraclitus. How did this come about?

The nineteenth century German philosopher Hegel developed the ideas of Heraclitus into his theory of dialectics, viewing the universe as a sort of divine debating society in which opposing ideas are forever struggling to produce a synthesis. Marx and Engels took up Hegel's dialectics and gave it a materialistic basis. Marx himself had studied Democritus for his doctoral dissertation, comparing him with Epicurus, and mentioned Heraclitus only marginally. However, both Engels and Lenin acknowledged that dialectic materialism originated in the thought of Heraclitus, as has been eloquently argued and amply documanted by Kessidi.

At first it seems reasonable to expect to find the dialectical process in history, economics, biology and perhaps other sciences. But mathematics is supposed to deal with eternal and unchanging objects such as the number 5. Indeed, when later dialecticians rephrased the Heraclitian doctrine of continual change by saying that "a is not always equal to a,"

J. Agassi and R.S. Cohen (eds.), Scientific Philosophy Today, 111–121.
Copyright © 1981 *by D. Reidel Publishing Company.*

this appears to contradict a fundamental axiom of mathematics and is, in fact, not taken seriously by practicing mathematicians in communist countries.

However, Heraclitus' doctrine of the unity of opposites has occasionally been illustrated by examples from mathematics. Lenin asserted somewhere – and I quote from memory – that subtraction is the antithesis of addition, with arithmetic as the synthesis, and that integration is the antithesis of differentiation, with calculus as the synthesis. I did find a relevant quotation by Mao Tse-tung:

> The classification of scientific studies is based precisely upon the particular contradictions inherent in their objects .... For example, positive numbers and negative numbers in mathematics, action and reaction is mechanics,....

Other mathematical illustrations were not so fortunate. I recall one Marxist illustrating the change of quantity into quality by pointing out that the number 2 is even, instead of being twice as odd as the number 1.

I spent my sabbatical 1965–6 in Zürich, where I had many conversations with the young American mathematician, Bill Lawvere. We kicked around the idea that an interesting illustration of dialectic contradictions could be found in the "adjoint functors" of modern mathematics, which had recently been popularized by Peter Freyd in his book on Abelian categories. While the basic idea behind adjoint functors is fully contained in the "universal mapping problem" of Nicolas Bourbaki, their more symmetric presentation by Kan in 1958 relied on the notion of a category invented by Eilenberg and MacLane.

To me the connection between adjoint functors and dialectical contradictions was not much more than an amusing analogy, but to Bill Lawvere it was of profound significance and pervaded the whole structure and development of mathematics. Having been influenced by General Semantics in his youth, he was now becoming attracted to Marxism-Leninism and came to see the dialectical process everywhere in mathematics. When asked point blank years later he would not admit that adjoint functors were the only manifestations of contradictions in mathematics. However, in this essay, I shall confine myself mainly to this narrower view.

In his 1970 address on 'Quantifiers and Sheaves' to the International Congress of Mathematicians in Nice, Lawvere said the following:

> The unity of opposites... is essentially that between logic and geometry, ... We first sum up the principal contradictions of the Grothendieck-Giraud-Verdier theory of

topos in terms of four or five adjoint functors... When the main contradictions of a theory have been found, the scientific procedure is to summarize them in slogans which one then constantly uses as an ideological weapon for the further development and transformation of the thing. Doing this for "set theory" requires taking into account that the main pairs of opposing tendencies in mathematics take the form of adjoint functors....

Lawvere's ideas influenced a host of mathematicians working in the area of category and topos theory. Some of his students also adopted his ideological orientation. To show how even the philosophical vocabulary caught on, let me only quote a sentence in a review by Myles Tierney of an article by Anders Kock, which appeared in *Mathematical Reviews* in 1972: "No doubt, however, the reviewer has again failed in the struggle to grasp firmly the essential nature of the contradiction between "abstract" and "general"."

Again, in a recent paper, distributed in 1977, Henry Crapo writes: "The essential notion is that of a unity (of opposites) in a binary relation...." However, his notion of "unity of opposites" is quite different from the interpretations that will be discussed in the present essay. Let these examples suffice and let us turn to explain what the mathematical illustrations are all about.

I shall begin by explaining the notion of adjoint functors in a setting that does not presuppose an understanding of categories. Certain categories have always been known, namely preordered sets. A *preordered set* $\mathscr{A} = (A, \leq)$ is a set A together with a reflexive and transitive relation $\leq$ on A, that is, a binary relation satisfying

$$a \leq a$$

and

$$(a \leq a' \wedge a' \leq a'') \Rightarrow a \leq a''$$

for all elements a, a', a'' of A. We shall not assume the usual antisymmetry rule, although many of our examples do satisfy this law. We may write $a \cong a'$ to mean that $a \leq a'$ and $a' \leq a$.

Given two preordered sets $\mathscr{A} = (A, \leq)$ and $\mathscr{B} = (B, \leq)$, by a *functor* $F: \mathscr{A} \to \mathscr{B}$ we understand an order preserving function from A to B so that

$$a \leq a' \Rightarrow F(a) \leq F(a')$$

for all $a, a' \in A$.

Given two functors $F\colon \mathscr{A} \to \mathscr{B}$ and $G\colon \mathscr{B} \to \mathscr{A}$, we say G is *right adjoint* to F if

$$F(a) \leq b \Leftrightarrow a \leq G(b)$$

for all $a \in A$ and $b \in B$. This notion is not new; in fact, such a pair (F, G) used to be called a *Galois correspondence*.

Let us draw some obvious conclusions from this setup. Consider the functors $GF\colon \mathscr{A} \to \mathscr{A}$ and $FG\colon \mathscr{B} \to \mathscr{B}$. Of these the former is a *closure operation*, that is,

$$a \leq GF(a)$$
$$GFGF(a) \leq GF(a),$$
$$a \leq a' \Rightarrow GF(a) \leq GF(a')$$

for all $a, a' \in A$. The latter satisfies the dual properties:

$$FG(b) \leq b,$$
$$FG(b) \leq FGFG(b),$$
$$b \leq b' \Rightarrow FG(b) \leq FG(b')$$

for all $b, b' \in B$; we might call it an *interior operation*. Moreover, the functions F and G set up a biunique (up to $\cong$) correspondence between the *closed* elements of A, that is, the elements of A for which $GF(a) \leq a$, and the *open* elements of B, that is, the elements b of B for which $b \leq FG(b)$. Thus, the closed elements of A form a preordered subset $\mathscr{A}_0 = (A_0, \leq)$ of $\mathscr{A}$ and the open elements of B form a preordered subset $\mathscr{B}_0 = (B_0, \leq)$ of $\mathscr{B}$, and $\mathscr{A}_0$ is equivalent with $\mathscr{B}_0$ in the sense that

$$GF(a) \cong a, \ FG(b) \cong b$$

for all $a \in A_0$ and $b \in B_0$.

The following picture should throw light on the situation:

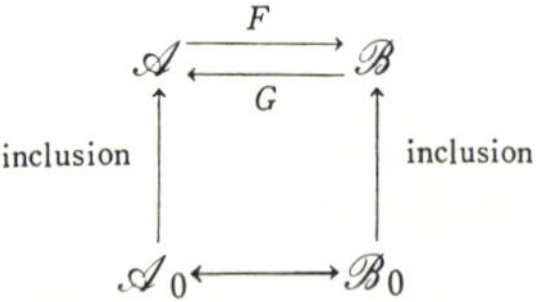

The same picture will apply to categories in place of preordered sets later in this essay.

There are at least two ways in which we can interpret adjoint functors as exemplifying the dialectical process.

Interpretation I. We may think of the pair (F, G) as establishing a contradiction between $\mathscr{A}$ and $\mathscr{B}$. The equivalence between $\mathscr{A}_0$ and $\mathscr{B}_0$ is then the unity of opposites. (This interpretation becomes even more convincing if we replace $\mathscr{B}$ by $\mathscr{B}^{\text{op}} = (B, \geq)$.)

Interpretation II. We may think of F as the thesis; its right adjoint G then becomes the antithesis. In this interpretation we should perhaps think of the adjunction itself as the synthesis. (This appears to be Lawvere's favorite interpretation.)

We shall illustrate these interpretations with a number of examples, dealing first with preordered sets, later with other categories.

EXAMPLE 1. Take $\mathscr{A}$ and $\mathscr{B}$ to be the ordered set of nonnegative integers, both with the natural order. Let

$$F(a) = \text{the } a\text{-th prime number when } a > 0,\ F(0) = 0;$$

$$G(b) = \text{the number of primes} \leq b.$$

Then $GF(a) \leq a$ holds for all positive integers a, while $b \leq FG(b)$ holds then and only then when b is a prime number. The "unity of opposites" here describes the biunique correspondence between positive integers and prime numbers.

Many examples arise from a binary relation $R \subseteq X \times Y$ between two sets X and Y. Take $\mathscr{A} = (\mathscr{P}(X), \subseteq)$ as the ordered set of subsets of X, ordered by inclusion, and take $\mathscr{B} = (\mathscr{P}(Y), \supseteq)$ as the ordered set of subsets of Y, ordered by the converse of inclusion. Put

$$F(a) = \{y \in Y \mid \forall_{x \in a}\ (x, y) \in R\},$$
$$G(b) = \{x \in X \mid \forall_{y \in b}\ (x, y) \in R\}$$

for any $a \subseteq X$ and $b \subseteq Y$. This situation is called a *polarity*. It gives rise to a lattice isomorphism between the lattice $\mathscr{A}_0$ of "closed" subsets of X and the lattice $\mathscr{B}_0$ of "closed" subsets of Y. (Note that open elements of $\mathscr{B}$ are closed subsets of Y, because $b \leq b'$ here means $b \supseteq b'$.)

Examples of polarities abound in mathematics; let me only mention two.

EXAMPLE 2. Take X to be the set of points of a plane, Y the set of half-planes, and write $(x, y) \in R$ for $x \in y$. Then, for any set a of points,

$UF(a)$ is the intersection of all half-planes containing a, in other words, the *convex hull* of a. Hence $GF(a) \leq a$ if and only if a is convex. For any set b of half-planes, $FG(b)$ is the set of all half-planes which contain the intersection of all planes in the set b. The "unity of opposites" here asserts that there are two equivalent ways of describing a convex set: by the points on it or by the half-planes containing it.

Our next example is a bit more technical and should be skipped by readers not familiar with the theory of rings.

EXAMPLE 3. Given a commutative ring C, take X as the set of elements of C, Y as the set of prime ideals of C. Again we write $(x, y) \in R$ for $x \in y$. Then $GF(a)$ is the intersection of all prime ideals containing the set a. In commutative algebra this is shown to be the set of all those $x \in X$ for which $x^n \in a$, for some natural number n, and is called the *radical* of a. On the other hand, the closure operation FG on the set of all subsets of Y makes Y into a topological space, in fact a compact space, called the *spectrum* of C. The "unity of opposites" here describes a biunique correspondence between "radical" ideals of the ring (ideals which coincide with their radical) and closed subspaces of its spectrum.

So far we have followed the first interpretation; but, according to the second interpretation, we could say that the above resolves a contradiction between a commutative ring and its spectrum.

Another example of a Galois correspondence is taken from elementary logic.

EXAMPLE 4. Take $\mathscr{A}$ and $\mathscr{B}$ both as the preordered sets of formulas of the propositional calculus, where $\leq$ is understood to mean entailment and is usually written $\vdash$. Given any formula p, we define $F: \mathscr{A} \to \mathscr{B}$ and $G: \mathscr{B} \to \mathscr{A}$ by

$$F(a) = p \wedge a, G(b) = p \Rightarrow b.$$

They form a pair of adjoint functors, because $a \vdash p \Rightarrow b$ if and only if $p \wedge a \vdash b$. It is easily seen that $GF(a) \leq a$ if and only if $p \Rightarrow a \vdash a$ and $FG(b) \leq b$ if and only if $b \vdash p$. (Here $p \Rightarrow a \vdash a$ means the same as $\vdash p \vee a$ classically, but not intuitionistically.) According to the first interpretation, we would thus be interested in the correspondence between formulas which entail p and formulas a for which $\vdash (p \Rightarrow a) \Rightarrow a$, but I don't know the significance of this correspondence. In fact, I believe that

Lawvere, who first studied this pair of adjoint functors, would see it as exemplifying the dialectical process according to the second interpretation as follows.

Proceeding from the functor F to its right adjoint G, we progress from the basic notion of conjunction to the more sophisticated antithetical notion of implication, thus introducing a new idea, albeit implicit in the old one. The synthesis of these two ideas is the system of logic embodying conjunction and implication, essentially the positive intuitionist propositional calculus, also known as Heyting algebra. (I am here neglecting the symbols T, $\perp$ and $\vee$.)

From now on we shall assume that $\mathscr{A}$ and $\mathscr{B}$ are categories and that (F, G) is a pair of adjoint functors between them. As in the special case of preordered sets, we obtain an equivalence between full subcategories $\mathscr{A}_0$ and $\mathscr{B}_0$, where $\mathscr{A}_0$ consists of all objects A of $\mathscr{A}$ for which the canonical arrow $A \to GF(A)$ is an isomorphism, while $\mathscr{B}_0$ consists of all objects B of $\mathscr{B}$ for which the canonical arrow $FG(B) \to B$ is an isomorphism. The meaning of all these terms will be explained in the appendix. Again we have two dialectical interpretations, as in the special case of preordered sets.

EXAMPLE 5. $\mathscr{A}$ is the category of Abelian groups and $\mathscr{B}$ is the opposite of the category of topological Abelian groups. Let K be the compact group of the reals modulo the integers. For any abstract Abelian group A, $F(A)$ is the group of all homomorphisms of A into K, with the topology induced by K. For any topological Abelian group B, $G(B)$ is the group of all continuous homomorphisms of B into K. Then $\mathscr{A}_0$ is $\mathscr{A}$, while $\mathscr{B}_0$ is the opposite of the category of compact Abelian groups. According to our first interpretation, we have here a contradiction between abstract and topological Abelian groups, the "unity of opposites" being the well-known Pontrjagin duality between abstract and compact Abelian groups.

EXAMPLE 6. $\mathscr{A}$ is the category of rings and $\mathscr{B}$ is the opposite of the category of topological spaces. For any ring A, $F(A)$ is the topological space of homomorphisms of A into the ring of integers modulo 2. For any topological space B, $G(B)$ is the ring of continuous functions of B into $\{0, 1\}$, with the ring structure inherited by that of $\{0, 1\}$. Then $\mathscr{A}_0$ is the category of Boolean rings and $\mathscr{B}_0$ is the opposite of the category of zero-dimensional compact Hansdorff spaces. According to our first interpretation, we have here a contradiction between algebra (the category of rings) and topology (the opposite of the category of topological spaces), the "unity of opposites" being the well-known Stone duality.

EXAMPLE 7. In a manner similar to Example 6, we study the contradiction between (the category of) Banach algebras and (the opposite of the category of) topological spaces and obtain the well-known Gelfand duality as the "unity of opposites."

The details of Examples 6 and 7 have been worked out in collaboration with Basil Rattray. There are numerous other examples of the same nature, let me just mention one more.

EXAMPLE 8. $\mathscr{A}$ is the category of presheaves on a topological space X, $\mathscr{B}$ is the category of spaces over X. There is a pair of adjoint functors between these categories which induces an equivalence between sheaves and local homomorphisms.

In Examples 5 to 8, the first interpretation was useful. In the following example, which is analogous to Example 4, the second interpretation is more interesting.

EXAMPLE 9. $\mathscr{A}$ and $\mathscr{B}$ are both the category of sets (or any Cartesian closed category). Given any object C, we consider the functors F and G determined by $F(A) = C \times A$ and $G(B) = B^C$. These are a pair of adjoint functors in view of the correspondence between arrows $C \times A \to B$ and $A \to B^C$. Here exponentiation should be viewed as the antithesis of Cartesian multiplication, and the struggle between the two produces essentially a Cartesian closed category. (We have ignored the terminal object.)

Up to this point we have considered mathematical illustrations of the Heraclitian principle of the unity of opposites. His doctrine of continual change has been less influential, but it too has recently entered the thinking of Bill Lawvere. In his talk at the Bristol Logic Colloquium of 1973, he advocated the view that "the theory of topoi ..... is a basis for the study of continuously variable structures, as classical set theory is a basis for a study of constant structures." The idea is that a sheaf of rings, for example, may be looked upon as a variable ring, or perhaps as a ring of variable quantities. He quotes Engels as saying that "the advance from constant quantities to variable quantities is a mathematical expression of the advance from metaphysics to dialectics". The surprising thing is that results about constant sets when proved intuitionistically carry over to variable sets, that is to say, to sheaves.

McGill University, Montreal

APPENDIX

A *category* consists of two classes, the class of *arrows* and the class of *objects*, together with two mappings between them, called *source* and *target*. We write $f: A \to B$ as shorthand for: source$(f) = A$ and target$(f) = B$. Moreover, there is given for each object A an arrow $1_A: A \to A$, called the *identity*, and a *composition* of arrows thus:

$$\frac{f: A \to B \quad g: B \to C}{gf: A \to C}.$$

Finally, the following equations are postulated:

$$1_B f = f, f 1_A = f, (hg)f = h(gf),$$

for all $f: A \to B$, $g: B \to C$ and $h: C \to D$.

A *functor* $F: \mathscr{A} \to \mathscr{B}$ between categories sends the objects of $\mathscr{A}$ to objects of $\mathscr{B}$ and the arrows of $\mathscr{A}$ to arrows of $\mathscr{B}$ so that

$$\frac{f: A \to B}{F(f): F(A) \to F(B)}$$

and

$$F(1_A) = 1_{F(A)}, F(gf) = F(g)F(f).$$

A *natural transformation* t: $F \to G$ between two functors from $\mathscr{A}$ to $\mathscr{B}$ sends the objects of $\mathscr{A}$ to arrows of $\mathscr{B}$ so that $t(A) : F(A) \to G(A)$, for all objects A of $\mathscr{A}$, and so that

$$G(f)t(A) = t(A')F(f),$$

for all arrows $f: A \to A'$ in $\mathscr{A}$, as is illustrated by the diagram:

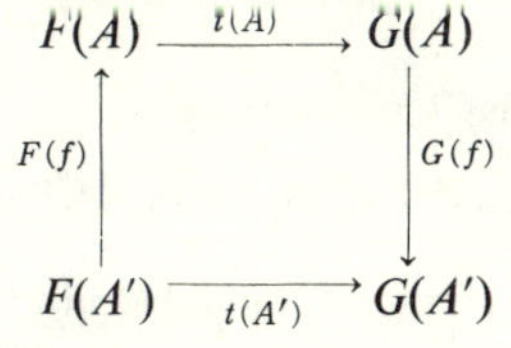

$$\begin{array}{ccc} F(A) & \xrightarrow{t(A)} & G(A) \\ {\scriptstyle F(f)}\uparrow & & \downarrow{\scriptstyle G(f)} \\ F(A') & \xrightarrow[t(A')]{} & G(A') \end{array}$$

We say that $G: \mathscr{B} \to \mathscr{A}$ is *right adjoint* to $F: \mathscr{A} \to \mathscr{B}$ if there are given natural transformations $\eta: 1_{\mathscr{A}} \to GF$ and $\varepsilon: FG \to 1_{\mathscr{B}}$, where $1_{\mathscr{A}}$ and $1_{\mathscr{B}}$ are the identity functors on $\mathscr{A}$ and $\mathscr{B}$ respectively, such that, for all objects A of $\mathscr{A}$ and B of $\mathscr{B}$,

$$G(\varepsilon(B))\eta(G(B)) = 1_{G(B)}, \varepsilon(F(A))F(\eta(A)) = 1_{F(A)},$$

that is, the following composite natural transformations are both identities:

$$G \xrightarrow{\eta G} GFG \xrightarrow{G\varepsilon} G, \quad F \xrightarrow{F\eta} FGF \xrightarrow{\varepsilon F} F.$$

We say that the adjoint pair (F, G) determines an *equivalence* between $\mathscr{A}_0$ and $\mathscr{B}_0$ if $\eta(A)$ and $\varepsilon(B)$ are isomorphisms (that is, have inverses) for all A in $\mathscr{A}_0$ and all B in $\mathscr{B}_0$.

The following categories are mentioned in the text.

When $\mathscr{A} = (A, \leq)$ is a preordered set, we may regard it as a category whose objects are the elements of A and whose arrows $a \to b$ are the pairs (a, b) for which $a \leq b$. In particular, a topological space gives rise to the preordered set of open subsets of X, which is then viewed as a category.

Other categories mentioned in the text are displayed in this table:

Category of	Objects	Arrows
sets	sets	functions
topological spaces	topological spaces	continuous functions
Abelian groups	Abelian groups	homomorphisms
topological Abelian groups	topological Abelian groups	continuous homomorphisms
rings	rings	homomorphisms
Banach algebras	Banach algebras	norm decreasing homomorphisms
presheaves on X	functors $X^{\mathrm{op}} \to \mathrm{Sets}$	natural transformations
spaces over X	continuous functions $f: Y \to Z$	continuous functions $h: Y \to Y'$ so that $f'h = f$

REFERENCES

Crapo, H.: 1977, 'Unities and Negation: On the Representation of Finite Lattices', preprint.

Eilenberg, S. and S. MacLane: 1945, 'General Theory of Natural Equivalences', *Trans. Amer. Math. Soc.* **58**, 231–294.

Freyd, P.: 1964, *Abelian Categories*, Harper and Row, New York.

Kan, D.N.: 1958, 'Adjoint Functors', *Trans. Amer. Math. Soc.* **87**, 294–329.

Kessidi, T.: 1973, *Les origines de la dialectique matérialiste (Héraclite)*, Les éditions du progrès, Moscow.

Lambek, J. and B. A. Rattray: 1979, 'A General Stone-Gelfand Duality', *Trans. Amer. Math. Soc.* **248**, 1–35.

Lawvere, F.W.: 1969, 'Adjointness in Foundations', *Dialectica* **23**, 281–296.

Lawvere, F.W.: 1971, 'Quantifiers and Sheaves', in *Actes du Congrès International des Mathématiciens, Nice 1970*, vol. 1, Gauthier-Villars, Paris, pp. 329–334.

Lawvere, F.W.: 1975, 'Continuously Variable Sets, Algebraic Geometry = Geometric Logic', *Logic Colloquium '73. Proceedings of the Logic Colloquium,* Bristol, July 1973, H.E. Rose and J.C. Shepherdson (eds.) [*Studies in Logic and the Foundations of Mathematics* **80**], North-Holland, Amsterdam, pp. 135–156.

Mao Tse-tung: 1962, 'On Contradiction', in *An Anthology of His Writings*, Mouton, New York.

Russell, B.: 1965, *History of Western Philosophy*, George Allen and Unwin, London.

Tierney, M.: 1972, 'Review of A. Kock, "Strong Functors and Monoidal Monads," ' *Math. Reviews* **43**, 6282.

HUGUES LEBLANC

FREE INTUITIONISTIC LOGIC: A FORMAL SKETCH

The inference rule

$$\text{If } \vdash A(t), \text{ then } \vdash (\exists x)A(x) \tag{1}$$

and its dual

$$\text{If } \vdash (\forall x)A(x), \text{ then } \vdash A(t) \tag{2}$$

are of course sound only if t in $A(t)$ denotes something or other. So, prior to 1959, the singular terms owned by logic, be it classical logic or intuitionistic logic, were presumed to be denoting ones. The presumption was overtly made by some writers. It was not by others, and as a result many a novice must have substituted for 't' in (1) – (2) a nondenoting term, thereby perpetrating a *non-sequitur*. But nondenoting terms do play a large role both in ordinary and in scientific discourse, and a logic that accommodates them eventually proved indispensable. One was supplied in 1959 by Jaakko Hintikka [1959], and another supplied independently and in the very same year by Theodore Hailperin and me [Leblanc and Hailperin 1959].

First-order logics *with* identity, they both renounced (1) – (2) in favor of the weaker, but more welcome, rules.

$$\text{If } \vdash A(t) \text{ and } \vdash (\exists x)(x = t), \text{ then } \vdash (\exists x)A(x) \tag{1*}$$

and

$$\text{If } \vdash (\forall x)A(x) \text{ and } \vdash (\exists x)(x = t), \text{ then } \vdash A(t), \tag{2*}$$

where '$(\exists x)(x = t)$' is often shortened to read '$E!t$'. Hintikka further renounced the theorem

$$(\forall x)A(x) \supset (\exists x)A(x), \tag{3}$$

which presupposes that *there exists something or other.*[1] Otherwise, the logics in [Hintikka 1959] and [Leblanc and Hailperin 1959] were *classical as opposed to intuitionistic in spirit.*

Saul Kripke in his 1963 paper on modal logics with quantifiers [Kripke 1963] renounced (3), as Hintikka had done. He further flirted with what

J. Agassi and R.S. Cohen (eds.), Scientific Philosophy Today, 123–138.
Copyright © 1981 *by D. Reidel Publishing Company.*

came to be called "free logic" when allowing bound variables to draw their values from *one* possible world and free ones to draw theirs from the union of that world and various *other* possible worlds. However, truly free modal logics probably date from the mid-seventies [see Leblanc 1976, Chapters 8–9], and the first free intuitionistic logic may be that in [Leblanc and Gumb 1982], a paper written by Raymond D. Gumb and me in the fall of 1978. That logic, to be known as $\text{IQC}^*_=$, reflects in its way Michael Dummett's contention that in intuitionistic logic

> a proof of $(\exists x)A(x)$ is a proof of some statement of the form $A(t)$, *together with a proof that the object denoted by the term t belongs to the domain*; and a proof of $(\forall x)A(x)$ is a construction of which we can recognize that, when it is applied to any term t *and to a proof that the object denoted by t belongs to the domain*, it yields a proof of $A(t)$ [see Dummett 1977, p. 43].

Indeed, read '$(\exists x)(x = t)$' where Dummett had 'the object denoted by the term t belongs to the domain', and you get (1*)–(2*).

As the subscripted '=' signals, $\text{IQC}^*_=$ is a first-order logic *with* identity. Free first-order logics *without* identity are quite intriguing, though. Karel Lambert sketched a classical one in 1963 [Lambert 1963]. Borrowing from [Leblanc and Gumb 1982] and from a paper that Robert F. Barnes and I wrote in 1979 [Barnes and Leblanc 1982], I provide here an intuitionistic counterpart of Lambert's logic, to be known as IQC*. In the absence of '=', neither of (1*)–(2*) can be had, and one must be content with free variants of the theorems underlying (1)–(2). These theorems are of course

$$A(t) \supset (\exists x)A(x)$$

and

$$(\forall x)A(x) \supset A(t),$$

and their free variants respectively run, as in Lambert:

$$(\forall y)(A(y) \supset (\exists x)A(x))$$

and

$$(\forall y)((\forall x)A(x) \supset A(y)).$$

The semantics I outfit IQC* with, is of the *truth-value* sort: i.e., quantifiers are interpreted substitutionally and truth-value assignments are used in place of models. Readers who wince at this are invited to translate offending definitions and theorems into *model-theoretic* idiom. (They may, incidentally, find the task a rather arduous one, models being far less tractable than the truth-value assignments I employ.) As in earlier writings, I have *two* runs of *individual letters*. One run, to wit:

$$x_1, x_2, x_3, \ldots,$$

serve as *bound variables*, and bound variables *only*. The other run, to wit:

$$p_1, p_2, p_3 \ldots,$$

can be thought of as *free variables*, hence as place-holders for (*possibly nondenoting*) singular terms. They can alternatively be thought of as (*possibly nondenoting*) *singular terms*. If the latter interpretation be favored, then the predicate parameters of IQC* should be thought of as *predicates* (rather than *free predicate variables*). Using two runs of individual letters pays, as the cliché has it, "handsome dividends" both at syntactic and at semantic junctures.

This paper, as the subtitle warns, is but a *sketch*. So definitions and theorems are stated in full, but, for lack of space, proofs are generally omitted. The reader who hankers for them should consult [Leblanc and Gumb 1982] and [Barnes and Leblanc 1982]. And the sketch is essentially a *formal* one. Those who, as they read Section 2, would welcome extra motivating asides should consult [Leblanc and Gumb 1982] (and possibly [Kripke 1963] from which some of Section 2 stems).

1. SYNTACTIC MATTERS[2]

The (primitive) *signs* of IQC* will be

(a) for each d from 0 on the aleph$_0$ predicate parameters of degree d

$$f_1^d, f_2^d, f_3^d, \ldots,$$

(b) the aleph$_0$ individual parameters

$$p_1, p_2, p_3, \ldots,$$

with 'p_1' known as the alphabetically first individual parameter of IQC$\underset{=}{*}$, 'p_2' as the *second*, 'p_3' as the *third*, etc.,

(c) the aleph$_0$ individual variables

$$x_1, x_2, x_3, \ldots,$$

(d) the logical operators '$\sim$', '$\supset$', '&', '$\vee$', '$\forall$', and '$\exists$',[3] and

(e) '(', ')', and ', ',

and the *formulas* of IQC* will be all finite strings of signs from (a) – (e).

Predicate parameters will be referred to by means of 'F^d'; individual parameters by means of 'P' and sets thereof by means of 'π'; individual variables by means of 'X' and 'Y'; individual signs in general (i.e., individual parameters and individual variables) by means of 'I'; formulas by means of 'A', 'B', and 'C', and sets thereof by means of 'S'; A being a formula, the set of the various individual parameters occurring in A by means of '$\pi(A)$'; S being a set of formulas, the set of the various individual parameters occurring in members of S by means of '$\pi(S)$'; and, A being a formula, $I_1, I_2, \ldots, I_m$ ($m \geq 0$) being distinct individual signs, and $I'_1, I'_2, \ldots, I'_m$ being individual signs not necessarily distinct from one another nor from $I_1, I_2, \ldots, I_m$, the result of simultaneously substituting I'_1 for I_1, I'_2 for I_2, $\ldots$, I'_m for I_m in A (will be referred to) by means of

$$((A)(I'_1, I'_2, \ldots, I'_m/I_1, I_2, \ldots, I_m)).$$

Note: Given the following conventions the inference rules (1)-(2) would now run

$$\text{If} \vdash A(P/X), \text{ then} \vdash (\exists X)A$$

and

$$\text{If} \vdash (\forall X)A, \text{ then} \vdash A(P/X).$$

and theorem (3) would run

$$(\forall X)A \supset (\exists X)A.$$

To resume my definitions, the *atomic well-formed formulas* of IQC* will be all predicate parameters of degree 0 *plus* all formulas of the sort $F^d(P_1, P_2, \ldots, P_d)$, where F^d is a predicate parameter of non-zero degree d and $P_1, P_2, \ldots, P_d$ are (not necessarily distinct) individual parameters; and the *well-formed formulas* of IQC* will be the atomic ones just defined *plus* all formulas of any of the following three sorts:

(a) $\sim A$, where A is a well-formed formula,

(b) $(A \supset B)$, $(A \ \& \ B)$, and $(A \vee B)$, where A are B are (not necessarily distinct) well-formed formulas, and

(c) $(\forall X)A$ and $(\exists X)A$, where – for any individual parameter P – $(A)(P/X)$ is a well-formed formula.[4]

'Well-formed formula(s)' will hereafter be abridged 'wff(s)' and outer parentheses will be dropped whenever clarity permits.

Finally, π and π' being (not necessarily distinct) sets of individual parameters and **R** being a function that maps π into π', I shall call **R** a *relettering function* for IQC* ; A being a wff and **R** a relettering function defined over $\pi(A)$, I shall understand by the **R**-*relettering* $\mathbf{R}(A)$ of A the wff $A(\mathbf{R}(P_1), \mathbf{R}(P_2), \ldots, \mathbf{R}(P_m)/P_1, P_2, \ldots, P_m)$, where $P_1, P_2, \ldots, P_m$ $(m \geq 0)$ are in alphabetic order the various members of $\pi(A)$; and, S being a set of wffs and **R** a relettering function defined over $\pi(S)$, I shall understand by the **R**-*relettering* $\mathbf{R}(S)$ of S the set consisting of the **R**-*reletterings* of the various members of S.

The *axioms* of IQC* will be all formulas of any of the following eighteen sorts:

A1. $A \supset (B \supset A)$
A2. $(A \supset (B \supset C)) \supset ((A \supset B) \supset (A \supset C))$
A3. $(A \supset B) \supset ((A \supset \sim B) \supset \sim A)$
A4. $A \supset (\sim A \supset B)$
A5. $A \supset (B \supset (A \,\&\, B))$
A6. $(A \,\&\, B) \supset A$
A7. $(A \,\&\, B) \supset B$
A8. $A \supset (A \vee B)$
A9. $B \supset (A \vee B)$
A10. $(A \supset C) \supset ((B \supset C) \supset ((A \vee B) \supset C))$
A11. $(\forall X)(A \supset B) \supset ((\forall X)A \supset (\forall X)B)$
A12. $A \supset (\forall X)A$
A13. $(\forall Y)((\forall X)A \supset A(Y/X))$
A14. $(\forall X)(\forall Y)A \supset (\forall Y)(\forall X)A$
A15. $(\forall X)(A \supset B) \supset ((\exists X)A \supset (\exists X)B)$
A16. $(\exists X)A \supset A$[5]
A17. $(\forall Y)(A(Y/X) \supset (\exists X)A)$
A18. $(\forall X)(A(X/P))$, where A is an axiom.[6]

And, for any wffs A and $A \supset B$, B will count as the *ponential* of A and $A \supset B$.

A being a wff and S a set of wffs, I shall understand by a *proof* of A from S in IQC* any finite column of wffs such that (i) every entry in the column is a member of S, an axiom, or the ponential of two earlier wffs in the column, and (ii) the last entry in the column is A. I shall say that A is *provable* from S in IQC* (for short $S \vdash A$) if there is a proof of A from S; and I shall say that A is *provable* in IQC* (for short, $\vdash A$) if $\phi \vdash A$.

2. SEMANTIC MATTERS[7]

By a *truth-value assignment* for IQC* I shall understand any function from (the set of) the atomic wffs of IQC* to $\{T, F\}$; by a *parametrically relativized and indexed truth-value assignment* for IQC* I shall understand any triple of the sort $\langle\alpha, \pi, r\rangle$, where α is a truth-value assignment, π is a (*possibly empty*) set of individual parameters, and r is a real number; and by a *truth-value triple* for IQC* I shall understand any triple of the sort $\langle K, \langle\alpha, \pi, r\rangle, R\rangle$, where K is a nonempty set of parametrically relativized and indexed truth-value assignments, $\langle\alpha, \pi, r\rangle$ is a member of K, and R is a binary relation on K such that

(a) R is reflexive and transitive, and

(b) if $R(\langle\alpha', \pi', r'\rangle, \langle\alpha'', \pi'', r''\rangle)$ for (not necessarily distinct) members $\langle\alpha', \pi', r'\rangle$ and $\langle\alpha'', \pi'', r''\rangle$ of K, then

(b1) $\pi' \subset \pi''$ and

(b2) for any atomic wff A, $\alpha''(A) = T$ if $\alpha'(A) = T$.

In such triples as $\langle\alpha', \pi', r'\rangle$, $\langle\alpha'', \pi'', r''\rangle$, etc., (i) α', α'', etc., correspond to the models of standard semantics, (ii) π', π'', etc., consist of the parameters which – as regards α', α'', etc. – hold place for denoting terms (or, should 'p_1', 'p_2', 'p_3', etc. be thought of as terms, consist of the terms which – as regards α', α'', etc., – do denote), and (iii) r', r'', etc., allow (for sheer technical convenience) one and the same parametrically relativized assignment to turn up under different guises, say, once with r' as index, once with r'', etc. And in such triples as $\langle K', \langle\alpha', \pi', r'\rangle, R'\rangle$, $\langle K'', \langle\alpha'', \pi'', r''\rangle, R''\rangle$, etc., (i) K', K'', etc., correspond to Kripke's sets of possible worlds in [Kripke 1965] and (ii) R', R'', etc., correspond to Kripke's accessibility relations between the members of K', those of K'', etc. As for (b1) – (b2) in my characterization of R, they match well-known restrictions of Kripke's on his accessibility relations. Given that $R(\langle\alpha', \pi', r'\rangle, \langle\alpha'', \pi'', r''\rangle)$, (b1) requires that, say, the terms denoting as regards α' also denote as regards α'' and (b2) requires that any atomic wff true on α' be also true on α''.[8]

Turning (as this last remark invites us) to the question of truth, let A be a wff (atomic or not) and $\langle K, \langle\alpha, \pi, r\rangle, R\rangle$ be a truth-value triple. I shall say that A is *true* on $\langle K, \langle\alpha, \pi, r\rangle, R\rangle$ if

(a) in the case that A is atomic, $\alpha(A) = T$,

(b) in the case that A is a negation $\sim B$, B is not true on $\langle K, \langle\alpha', \pi', r'\rangle, R\rangle$ for any member $\langle\alpha', \pi', r'\rangle$ of K such that $R(\langle\alpha, \pi, r\rangle, \langle\alpha', \pi', r'\rangle)$,

(c) in the case that A is a conditional $B \supset C$, B is not true on $\langle K,$

$\langle\alpha', \pi', r'\rangle, R\rangle$ or C is for any member $\langle\alpha', \pi', r'\rangle$ of K such that $R(\langle\alpha, \pi, r\rangle, \langle\alpha', \pi', r'\rangle)$,

(d) in the case that A is a conjunction B & C, both B and C are true on $\langle K, \langle\alpha, \pi, r\rangle, R\rangle$,

(e) in the case that A is a disjunction $B \vee C$, at least one of B and C is true on $\langle K, \langle\alpha, \pi, r\rangle, R\rangle$,

(f) in the case that A is a universal quantification $(\forall X)B$, $B(P/X)$ is true on $\langle K, \langle\alpha', \pi', r'\rangle, R\rangle$ for any member $\langle\alpha', \pi', r'\rangle$ of K such that $R(\langle\alpha, \pi, r\rangle, \langle\alpha', \pi', r'\rangle)$ and any member P of π', and

(g) in the case that A is an existential quantification $(\exists X)B$, $B(P/X)$ is true on $\langle K, \langle\alpha, \pi, r\rangle, R\rangle$ for at least one member P of π.

Then, where A is a wff and S a set of wffs, I shall declare A *valid* (for short, $\models A$) if A is true on every truth-value triple; and I shall take S to *entail* A (for short, $S \models A$) if, for any truth-value triple $\langle K, \langle\alpha, \pi, r\rangle, R\rangle$ and any one-one relettering function **R** such that **R** is defined over $\pi(S \cup \{A\})$, $\mathbf{R}(A)$ is true on $\langle K, \langle\alpha, \pi, r\rangle, R\rangle$ if $\mathbf{R}(B)$ is true for every member B of S.[9]

3. SOUNDNESS THEOREMS

Proof that IQC* is sound under the semantics in Section 2 is relatively easy. You first prove the following generalization of restriction (b2) on R.

THEOREM 1. Let $\langle K, \langle\alpha, \pi, r\rangle, R\rangle$ be a truth-value triple for IQC* and $\langle\alpha', \pi', r'\rangle$ be any member of K such that $R(\langle\alpha, \pi, r\rangle, \langle\alpha', \pi', r'\rangle)$. If a wff A of IQC* is true on $\langle K, \langle\alpha, \pi, r\rangle, R\rangle$, then A is true on $\langle K, \langle\alpha', \pi', r'\rangle, R\rangle$.

With this result on hand, you go on to show that

THEOREM 2. If A is an axiom of IQC*, then $\models A$,

and

THEOREM 3. If two wffs A and $A \supset B$ of IQC* are true on a truth-value triple $\langle K, \langle\alpha, \pi, r\rangle, R\rangle$ for IQC*, then B as well is true on $\langle K, \langle\alpha, \pi, r\rangle, R\rangle$. And given these two results, you then go on to show that

THEOREM 4. Let the column made up of $A_1, A_2, \ldots, A_p$ constitute a proof of A from S in IQC* and let $\langle K, \langle\alpha, \pi, r\rangle, R\rangle$ be a truth-value triple for IQC* on which every member of S is true. Then, for each i from 1 through p, A_i is true on $\langle K, \langle \alpha, \pi, r \rangle, R \rangle$.

Hence, with S taken to be ϕ:

THEOREM 5. If $\vdash A$, then $\models A$. (Weak Soundness Theorem for IQC*)

Suppose, on the other hand, that S is not empty. It is readily shown that if $S \vdash A$, then for any relettering function $\mathbf{R}$ defined over $\pi(S \cup \{A\})$ there exists a proof of $\mathbf{R}(A)$ from $\mathbf{R}(S)$ whose entries exhibit only individual parameters from $\pi(S \cup \{A\})$. Hence, by dint of Theorem 4:

THEOREM 6. Let $S \vdash A$; let $\mathbf{R}$ be a relettering function for IQC* that is defined over $\pi(S \cup \{A\})$; and let $\langle K, \langle \alpha, \pi, r \rangle, R \rangle$ be a truth-value triple for IQC* on which $\mathbf{R}(B)$ is true for every member B of S. Then $R(A)$ is true on $\langle K, \langle \alpha, \pi, r \rangle, R \rangle$.

Hence:

THEOREM 7. If $S \vdash A$, then $S \models A$. (Strong Soundness Theorem for IQC*)

4. COMPLETENESS THEOREMS

That IQC* is complete under the semantics of Section 2 follows from the completeness of $\text{IQC}^*_=$, one of the logics studied in [Leblanc and Gumb 1982], and from a result in [Barnes and Leblanc 1982] concerning provability in $\text{IQC}^*_=$ and IQC*. So a good look at $\text{IQC}^*_=$ is in order.

$\text{IQC}^*_=$ has one more (primitive) sign and one more sort of atomic wffs than IQC* does: the sign '=' and the atomic wffs $(P = P')$, where P and P' are (not necessarily distinct) individual parameters. In the version that I employ here $\text{IQC}^*_=$ also has four more axiom schemata than IQC* does, to wit:

A19. $P = P$

A20. $(\forall X)(\exists Y)(Y = X)$

A21. $P = P' \supset (A \supset A(P'/P))$, where A is atomic

A22. $P = P' \supset (A(P'/P) \supset A)$, where A is atomic

In the presence of *A19–A22*, three axiom schemata could be dispensed with: *A13*, *A14*, and *A17*; but to ensure that $IQC^*_=$ is an extension of IQC* in the most straightforward sense of the word I retain them. The reader may wish to note that

$$A(P/X) \supset ((\exists X)(X = P) \supset (\exists X)A)$$

and

$$(\forall X)A \supset ((\exists X)(X = P) \supset A(P/X))$$

are provable in $IQC^*_=$ and hence that (1*)–(2*) are available as derived rules of inference of $IQC^*_=$.

To prevent confusion, incidentally, I shall write '$S \vdash_= A$' for 'A is provable from S in $IQC^*_=$', and '$\vdash_= A$' for 'A is provable in $IQC^*_=$'. The result I shall borrow from [Barnes and Leblanc 1982] is to the effect that if $S \vdash_= A$, then $S \vdash A$, so long as the members of $S \cup \{A\}$ are wffs of IQC*, and hence that $IQC^*_=$ is what is called a *conservative extension* of IQC*.

Five extra syntactic notions will be needed below, and two semantic ones.

(1.1) Let A be a wff of $IQC^*_=$ and π be a set of individual parameters. I shall say that A is *in* π if every individual parameter occurring in A belongs to π.

(1.2) Let S be a set of wffs of $IQC^*_=$. I shall say that an individual parameter P is *foreign* to S if P does not occur in any member of S; I shall say that S is *infinitely extendible* if aleph_0 individual parameters are foreign to S; I shall say that S is *consistent* if there is a wff A of $IQC^*_=$ that is not provable from S in $IQC^*_=$; and I shall say that S is *saturated* if

(a) S is consistent,

(b) if $S \vdash_= A$, then A belongs to S,

(c) if $A \vee B$ belongs to S, then so does at least one of A and B, and

(d) if $(\exists X)A$ belongs to S, then so does $A(P/X)$ for at least one individual parameter P such that $(\exists x_1)(x_1 = P)$ belongs to S.

(2) Let $\langle \alpha, \pi, r \rangle$ be a parametrically relativized and indexed truth-value assignment for $IQC^*_=$,[10] and $\langle K, \langle \alpha, \pi, r \rangle, R \rangle$ be a truth-value triple for $IQC^*_=$.[11] I shall say that $\langle \alpha, \pi, r \rangle$ is *identity-normal* if

(a) For every individual parameter P, $\alpha(P = P) = T$,

(b) for every atomic wff A of $IQC^*_=$ and any two individual parameters P and P', $\alpha(A) = \alpha(A(P'/P))$ if $\alpha(P = P') = T$, and

(c) for any two individual parameters P and P', P belongs to α if P' does and $\alpha(P = P') = T$.

And I shall say that $\langle K, \langle \alpha, \pi, r\rangle, R\rangle$ is *identity-normal* if every member of K is identity-normal.

With this last notion on hand, require the truth-value triples on which A must be true *plus* those on which $\mathbf{R}(A)$ must be true if $\mathbf{R}(B)$ is for every member B of S to be identity-normal, and the definitions of validity and entailment on p. 129 will suit IQC$^*_=$.

Proof that $S \vdash_= A$ if $S \models_= A$ calls for four lemmas, some of them suggested to Gumb and me by a 1968 paper of Richmond H. Thomason [Thomason 1968]. You first establish that

THEOREM 8. Let S be a set of wffs of IQC$^*_=$ that is infinitely extendible; let A be a wff of IQC$^*_=$ such that not $S \vdash_= A$; let π^∞; consist of aleph$_0$ individual parameters foreign to $S \cup \{A\}$; let π^+ be $\pi(S \cup \{A\}) \cup \pi^\infty$; and suppose the complement of π^+ is of cardinality aleph$_0$. Then there exists a set S^+ of wffs of IQC$^*_=$ such that

(a) $S \subset S^+$,

(b) $\pi(S^+) = \pi^+$,

(c) S^+ is infinitely extendible and saturated, and

(d) though $\pi(A) \subset \pi^+$, A does not belong to S^+.

The theorem, as the reader may have guessed, is the *intuitionistic* counterpart of one Leon Henkin used in his 1949 paper to show QC (the classical quantificational calculus of order one) complete [Henkin 1949].

You then go on to establish that

THEOREM 9. Let S be a set of wffs of IQC$^*_=$ that is infinitely extendible and saturated, and let A be any wff of IQC$^*_=$ in $\pi(S)$. Then there exists an identity-normal truth-value triple $\langle K, \langle \alpha, \pi, r\rangle, R\rangle$ for IQC$^*_=$ such that A belongs to S if and only if A is true on $\langle K, \langle \alpha, \pi, r\rangle, R\rangle$.[12]

The two theorems yield:

THEOREM 10. Let S be a set of wffs of IQC$^*_=$ that is infinitely extendible and A be a wff of IQC$^*_=$ such that not $S \vdash_= A$. Then there exists an identity-normal truth-value triple $\langle K, \langle \alpha, \pi, r\rangle, R\rangle$ for IQC$^*_=$ such that

(a) every member of S is true on $\langle K, \langle \alpha, \pi, r\rangle, R\rangle$, but

(b) A is not true on $\langle K, \langle \alpha, \pi, r\rangle, R\rangle$.

Note for proof of Theorem 10 that by virtue of Theorem 8 there exists a superset S^+ of S meeting conditions (c) – (d) in Theorem 8. Hence by virtue of Theorem 9 there exists an identity-normal truth-value triple $\langle K, \langle\alpha, \pi, r\rangle, R\rangle$ for IQC$^*_=$ such that, for every wff B of IQC$^*_=$ in $\pi(S^+)$, B belongs to S^+ if and only if B is true on $\langle K, \langle\alpha, \pi, r\rangle, R\rangle$. But every member of S, belonging as it does to S^+, is sure to be in $\pi(S^+)$; whereas by virtue of (d) in Theorem 8 A, though in $\pi(S^+)$, does not belong to S^+. Hence there exists an identity-normal truth-value triple $\langle K, \langle\alpha, \pi, r\rangle, R\rangle$ for IQC$^*_=$ on which every member of S is true but A is not.

Taking ϕ as S delivers:

THEOREM 11. If $\models_= A$, then $\vdash_= A$. (Weak Completeness Theorem for IQC$^*_=$).

Now for nonempty S. The next theorem brings relettering functions into the picture and permits proof of Theorem 13.

THEOREM 12. Let S be an arbitrary set of wffs of IQC$^*_=$ and A be a wff of IQC$^*_=$ such that not $S \vdash_= A$. Then there exists a one-one relettering function $\mathbf{R}$ for IQC$^*_=$ defined over $\pi(S \cup \{A\})$ and an identity-normal truth-value triple $\langle K, \langle\alpha, \pi, r\rangle, R\rangle$ for IQC* such that

(a) for every member B of S, $\mathbf{R}(B)$ is true on $\langle K, \langle\alpha, \pi, r\rangle, R\rangle$, but

(b) $\mathbf{R}(A)$ is not true on $\langle K, \langle\alpha, \pi, r\rangle, R\rangle$.

For proof let $\mathbf{R}$ be the relettering function such that, for each i from 1 on, $\mathbf{R}(`p_i\text{'}) = `p_{2i}\text{'}$. Then $\mathbf{R}(S)$ is infinitely extendible But, as is easily verified, $S \vdash_= A$ if $\mathbf{R}(S) \vdash_= \mathbf{R}(A)$. Hence not $\mathbf{R}(S) \vdash_= \mathbf{R}(A)$. Hence by virtue of Theorem 10 there exists an identity-normal truth-value triple $\langle K, \langle\alpha, \pi, r\rangle, R\rangle$ for IQC$^*_=$ such that (a)–(b).

Hence:

THEOREM 13. If $S \models_= A$, then $S \vdash_= A$. (Strong Completeness Theorem for IQC$^*_=$).

To obtain the analogues of Theorems 11 and 13 for IQC*, suppose from now on that S consists exclusively of wffs of IQC* and A is a wff of IQC*. It follows by definition that

$$\text{If} \models A, \text{ then} \models_= A \text{ and hence} \vdash_= A \tag{4}$$

and

$$\text{If } S \models A, \text{ then } S \models_= A \text{ and hence } S \vdash_= A. \tag{5}$$

So we merely need guarantee that, whether S be empty or not,

$$\text{If } S \vdash_= A, \text{ then } S \vdash A, \tag{6}$$

and the trick is done.

Proof of (6), though easy, runs to considerable length. By way of a sketch, let the following column **C** of wffs of IQC$^*_=$, to wit:

$$\begin{array}{l} A_1 \\ A_2 \\ \cdot \\ \cdot \\ \cdot \\ A_p \end{array}$$

constitute a proof of A from S in IQC$^*_=$. If '=' does not occur in any A_i ($i = 1, 2, \ldots, p$), then of course $S \vdash A$. If, on the other hand, '=' does occur in one or more of $A_1, A_2, \ldots, A_p$, then [Barnes and Leblanc 1982] shows how to translate each A_i into a wff $T_C(A_i)$ of IQC* such that

(a) if A_i is an axiom of IQC$^*_=$, then $\vdash T_C(A_i)$ and

(b) if A_i is the ponential of two earlier entries A_g and A_h (say $A_h = A_g \supset A_i$) in **C**, then $T_C(A_i)$ is the ponential of $T_C(A_g)$ and $T_C(A_h)$.

Since any member of S that figures in **C** is a wff of IQC* and hence is its own T_C-translation and (ii) so is A $(= A_p)$, the column

$$\begin{array}{l} T_C(A_1) \\ T_C(A_2) \\ \cdot \\ \cdot \\ \cdot \\ T_C(A_p) \end{array}$$

is sure to constitute a proof of A from S in IQC*. Hence (6).

The translation process that turns A_i into $T_C(A_i)$ will be sketched in the Appendix,[13] and proof that $T_C(A_i)$ meets (a)–(b) above is in [Barnes and Leblanc 1982]. So by virtue of (4) and Theorem 11:

THEOREM 14. If $\models A$, then $\vdash A$. (Weak Completeness Theorem for IQC*).

And by virtue of (5) and Theorem 13:

THEOREM 15. If $S \models A$, then $S \vdash A$. (Strong Completeness Theorem for IQC*.)

Proof of either theorem can doubtless be had without resort to Theorems 11 and 13, indeed without resort to IQC$^*_=$, but – I suspect – would be rather arduous.

APPENDIX

The process whereby, given (i) a set S of wffs of IQC$^*_=$ and a wff A of IQC$^*_=$, (ii) a column **C** of wffs of IQC$^*_=$ that qualifies as a proof A from S in IQC$^*_=$, and (iii) an entry A_i in **C**, A_i is translated into a wff $T_C(A_i)$ of IQC* that meets conditions (a)–(b) on p. 134, is as follows. I presume that at least one entry in **C** contains a predicate parameter; if none does, preface **C** with the axiom '$(\forall x_1)(f_1^1(x_1) \supset (f_1^1(x_1) \supset f_1^1(x_1)))$', and you will have a proof of A from S of the sort considered here. I also use '$(A \equiv B)$' as short for '$((A \supset B) \& (B \supset A))$', with A to be known as the *left component* and B the *right component* of $A \equiv B$. I dub *foreign* to **C** any individual sign that does not occur in any entry in **C**. And by a *quasi-wff* of IQC$^*_=$ I understand any formula of IQC$^*_=$ that is not well formed but would be if it contained individual parameters at one or more places where it contains individual variables.

Let $P_1, P_2, \ldots, P_k$ $(k \geq 0)$ be all the individual parameters, and $X_1, X_2, \ldots, X_m$ $(m \geq 0)$ be all the individual variables, that occur (in one or more entries) in **C**; and let $Q_0, Q_1, Q_2, \ldots, Q_m, Q_{m+1}, Q_{m+2}$ be $m + 3$ distinct individual parameters, and X_{m+1} and X_{m+2} be 2 distinct individual variables, foreign to **C**.[14]

Following Barnes [Barnes and Leblanc 1982], I first show how to translate a formula of IQC$^*_=$ of the sort $(I = I')$ (I and I' arbitrary individual signs) into $T_C(I = I')$, and then any formula A of IQC$^*_=$ of four further sorts into $T_C(A)$.

A. Given any formula $(I = I')$ of IQC$^*_=$

(1) form for each predicate parameter F_j^d $(d, j > 0)$ that occurs (in one or more entries) in **C** all atomic wffs of the sort $F_j^d(P'_1, P'_2, \ldots, P'_d)$, where $P'_1, P'_2, \ldots, P'_d$ are (not necessarily distinct) individual parameters from the list $P_1, P_2, \ldots, P_k, Q_0, Q_2, \ldots, Q_{m+2}$,

(2) for each atomic wff A obtained by dint of (1), form all quantifica-

tions of the sort $(\forall x_1)(\forall x_2) \ldots (\forall x_h)\ ((A \equiv A)(X_1, X_2, \ldots, X_h/Q'_1, Q'_2, \ldots, Q'_h))$, where $Q'_1, Q'_2, \ldots, Q'_h$ are all the individual parameters among $Q_1, Q_2, \ldots, Q_m$ that occur in A,

(3) in each quantified biconditional obtained by dint of (2), replace Q_0 in the left component of the biconditional by I, replace Q_0 in the right one by I', insert ampersands between the resulting wffs, and insert enough parentheses to ensure that the resulting formula is well formed. The result is $T_C(I = I')$.

B. Given any wff or *quasi-wff* A of IQC$\underline{\underline{*}}$ that is not of the sort $(I = I')$, take $T_C(A)$ to be:

(i) A itself if A is atomic,

(ii) $\sim T_C(B)$ if A is of the sort $\sim B$,

(iii) $(T_C(B) \supset T_C(C))$ if A is of the sort $(B \supset C)$, and

(iv) $(\forall X)T_C(B)$ if A is of the sort $(\forall X)B$

It is clear that $T_C(A)$ is a wff of IQC*, no matter the wff A of IQC$\underline{\underline{*}}$. Hence so is $T_C(A_i)$ no matter the entry A_i in **C**. And, as announced earlier, it can be shown that $\vdash T_C(A_i)$ if A_i is an axiom of IQC$\underline{\underline{*}}$ and that $T_C(A_i)$ is the ponential of $T_C(A_g)$ and $T_C(A_h)$ if A_i is that of A_g and A_h. So (6) on p. 133.[15]

Temple University

NOTES

[1] At this juncture one frequently quotes

$$(\exists x)(A(x) \vee \sim A(x)) \qquad (3')$$

rather than (3). But I preferred a formula which unlike (3′) holds in standard versions of intuitionistic logic and does presuppose that *something or other exists*. Whether proof of formulas such as (3) and (3′) is blocked in [Hintikka 1959] has been questioned by some. The issue, however, is a marginal one here. What matters is that Hintikka meant to renounce (3) and that in this paper as in other papers of mine subsequent to [Leblanc and Hailperin 1959] I do renounce it.

[2] Much of this material in Section 1 is borrowed by way of [Leblanc and Gumb 1979] from [Leblanc 1976].

[3] Following Church, Kleene, and a great many others I use the same connectives in intuitionistic as in classical logic.

[4] Because of (c) formulas in which identical quantifiers overlap are not well formed, a theoretically insignificant departure from standard procedure but one which vastly simplifies substitutions.

[5] With $A \supset (\forall X)A$ and hence A presumed here to be a wff, X is sure not to occur in A. The same remark applies to X in *A18*.

[6] *A18*, which enables one to dispense with Generalization as a rule of inference, stems from [Fitch 1948]. The dual

$$(\exists X)(\exists Y)A \supset (\exists Y)(\exists X)A$$

of *A14* is easily had as a (meta) theorem as Peter Woodruff brought to my attention. Over the last few years I repeatedly tried to prove *A14*, but repeatedly failed. Kit Fine – I understand – has proof that the schema is independent of the rest, and I yield to him on the matter. The presence of *A14* in the above list of axiom schemata is jarring, but jarred one has to be, it seems.

[7] Section 2 owes some to Kripke, as already acknowledged, and to several other writers, among them P.H.G. Aczel and M.C. Fitting. The necessary references are in [Leblanc and Gumb 1982].

[8] For an intuitive justification of (b1)–(b2), see [Leblanc and Gumb 1982].

[9] Merely requiring that, for any truth-value $\langle K, \langle \alpha, \pi, r\rangle, R\rangle$, A be true on $\langle K, \langle \alpha, \pi, r\rangle, R\rangle$ if every member of S is, would not do, as readers familiar with truth-value semantics are sure to know. Under these circumstances $\{f_1^1(p_1), f_1^1(p_2), f_1^1(p_3), \ldots\}$ would entail '$(\forall x_1)f_1^1(x_2)$': if '$f_1^1(p_i)$' is true on $\langle K, \langle K, \pi, r\rangle, R\rangle$, then '$f_1^1(p_i)$' is true on $\langle K, \langle \alpha', \pi', r'\rangle, R\rangle$ for any member $\langle \alpha', \pi', r'\rangle$ of K such that $R(\langle \alpha, \pi, r\rangle, \langle \alpha', \pi', r'\rangle)$. There are countless **R**'s, on the other hand, such that (a few quotations omitted to enhance readability) '$\mathbf{R}(f_1^1(p_1))$', '$\mathbf{R}(f_1^1(p_2))$', '$\mathbf{R}(f_1^1(p_3))$', etc. are true on $\langle K, \langle \alpha, \pi, r\rangle, \mathbf{R}\rangle$, but '$(\forall x_1)f_1^1(x_1)$' (='$\mathbf{R}((\forall x_1)f_1^1(x_1))$' is not: the function **R** such that $\mathbf{R}(p_i) = p_{2i}$ is a case in point. For further details on this point, see the comments following Theorem 12.

[10] With '$\mathrm{IQC}^*_=$' substituted here for 'IQC*', understand α in $\langle \alpha, \pi, r\rangle$ to be defined for every atomic wff of $\mathrm{IQC}^*_=$.

[11] With '$\mathrm{IQC}^*_=$' substituted here for 'IQC*', understand K to consist of parametrically relativized and indexed truth-value assignments for $\mathrm{IQC}^*_=$.

[12] As the reader may expect, α assigns T to all and only the atomic wffs of $\mathrm{IQC}^*_=$ that belong to S, and π consists of every individual parameter such that $(\exists x_1)(x_1 = P)$ belongs to S.

[13] The translation $T_C(A_i)$ of A_i stems from [Leblanc 1971], where, alas, it was inaccurately described. Proof of (a)–(b) will also be found in [Woodruff 1982], a paper written at the same time as but independently of [Barnes and Leblanc 1982].

[14] The role played by Q_0 will become obvious in (3). The extra individual signs Q_1, Q_2, X_{m+1}, and X_{m+2} help to show that certain generalizations of *A21* – *A22*, when translated as in **A** – **B**, are provable in IQC*.

[15] Thanks are due and most gratefully extended to Raymond D. Gumb, with whom [Leblanc and Gumb 1982] was coauthored, and to Robert F. Barnes, Jr., with whom [Barnes and Leblanc 1982] was. This paper was written while I held a research grant from the Foundation for the Advancement of Interdisciplinary Studies and was in residence at Dalhousie University, Halifax.

REFERENCES

Barnes, R. F., Jr. and H. Leblanc: 1979, 'Identity-Elimination in Various Free Quantificational Logics', in *Studies in Epistemology and Semantics*, Haven Publishing Corporation, New York.

Dummett, M.: 1977, *Elements of Intuitionism*, Clarendon Press, Oxford.

Fitch, F.B.: 1948, 'Intuitionistic Modal Logic with Quantifiers', *Portugaliae Mathematica* **17**, 113–118.

Henkin, L.: 1949, 'The Completeness of the First-Order Functional Calculus', *The Journal of Symbolic Logic* **14**, 159–166.

Hintikka, J.: 1959, 'Existential Presuppositions and Existential Commitments', *Journal of Philosophy* **56**, 125–137.

Kripke, S.A.: 1963, 'Semantical Considerations on Modal Logic', in *Modal and Many-Valued Logics, Societas Philosophica*, vol. 16, Helsinki, pp. 83–94.

Kripke, S.A.: 1965, 'Semantical Analysis of Intuitionistic Logic I', in *Formal Systems and Recursive Functions. Proceedings of the 8th Logic Colloquium*, Oxford 1963, J.N. Crossley and M.A.E. Dummett (eds.), North-Holland, Amsterdam, pp. 92–130.

Lambert, K.: 1963, 'Existential Import Revisited', *Notre Dame Journal of Formal Logic* **4**, 288–92.

Leblanc, H.: 1971, 'Truth-Value Semantics for a Logic of Existence', *Notre Dame Journal of Formal Logic* **12**, 153–168.

Leblanc, H.: 1976, *Truth-Value Semantics*, North-Holland, Amsterdam.

Leblanc, H., and R.D. Gumb: 1982, 'Soundness and Completeness Proofs for Three Brands of Intuitionistic Logic', in *Studies in Epistemology and Semantics*, Haven Publishing Corporation, New York.

Leblanc, H, and T. Hailperin: 1959, 'Nondesignating Singular Terms', *Philosophical Review* **68**, 239–43.

Thomason, R.H.: 1968, 'On the Strong Completeness of the Intuitionistic Predicate Calculus', *Journal of Symbolic Logic* **33**, 1–7.

Woodruff, P.: 1979, 'Eliminating Identity in Free Logic', in *Studies in Epistemology and Semantics*, Haven Publishing Corporation, New York.

FERNAND LEMAY

SOME LESSONS IN THE SUN

I. IDENTIFYING THE SET S

Here is a set (Fig. 1).

A pupil says that it looks like a sun – the lines are rays.
So that is what we call it: a SUN.

II. THE CORE OF THE SUN

What can be said about this set?

. . . There are lines in all directions – some long and some short – some black – two green . . .

Certainly the lines (SEGMENTS) differ in colour, in length and in direction. But have they not anything in common?

They all pass through the same point.

(Since the segments are fixed it is not strictly true to say that they pass through a point. Inaccuracy of this sort is common and does not matter as long as it is understood.)

Would it make sense to say that the SUPPORTS of the segments pass through a special point?

Could you add new segments to the sun? How many?

An infinity, asserts a pupil. Thus the picture cannot be completed. But nothing stops us from imagining the final COMPLETE SUN and it is to this that we will always refer. Fortunately a few segments or even just the centre are enough to represent it.

Now that we know what a sun is and can imagine it should we not

J. Agassi and R.S. Cohen (eds.), Scientific Philosophy Today, 139–150.
Copyright © 1981 *by D. Reidel Publishing Company.*

write out an official description defining it so that we do not have to keep on remembering it?

As the construction of definitions had become for various reasons a very popular activity, someone soon offers to start framing one.

SUN: several supports passing through a centre.

Marie does not find this quite right and modifies it:

SUN: several segments whose supports pass through a centre.

But had we not said that we were going to think of the complete sun? . . .

Gradually the pupils elaborate a text that satisfies them. It becomes official by being duplicated, distributed to all and added to the mathematical dictionary which other investigations had led us to compile. This dictionary is the arbiter in discussion. If at any time an entry in it is no longer sufficiently discriminating then it has to be improved.

III. PATHS THROUGH THE SUN

Show a path joining the two green segments—using for instance an elastic band stretched to start with to the length of one of the segments.

Such a path could be represented by an appropriate number of segments suggesting various positions reached during the transformation from the initial position to the final one (Fig. 2).

That is fine but now I would like a path WHOSE COMPONENTS ARE ALWAYS CONTAINED IN THE SUN (Fig. 3).

This first slightly cautious suggestion is followed by more and more adventurous ones (Fig. 4).

IV. BARRICADES

Here is a red barricade (Fig. 5).

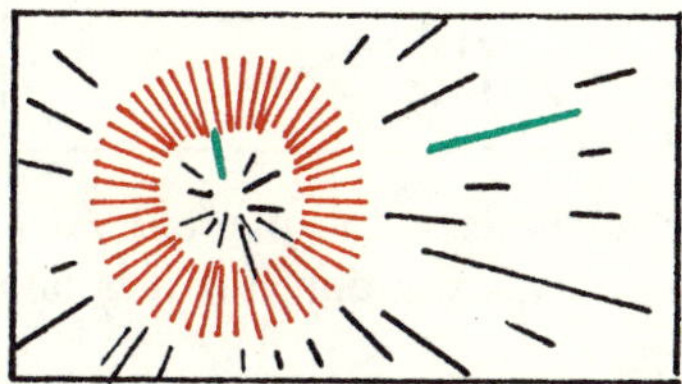

Is this obstacle going to stop one green segment finding the other by moving along a path that can be traced in the sun itself?

Anyone who has immediately seen a way THROUGH the red segments needs to be reminded that the figure was only a sketch – the CROWN should be complete. (How delicate is communication – a casual remark can here force the framing of an official definition of a crown.)

Before trying an actual solution the situation has to be explored in imagination. At first there will be disagreement about the existence of a path joining the two green segments. After discussion it happens that opinions converge to an agreement (Fig. 6).

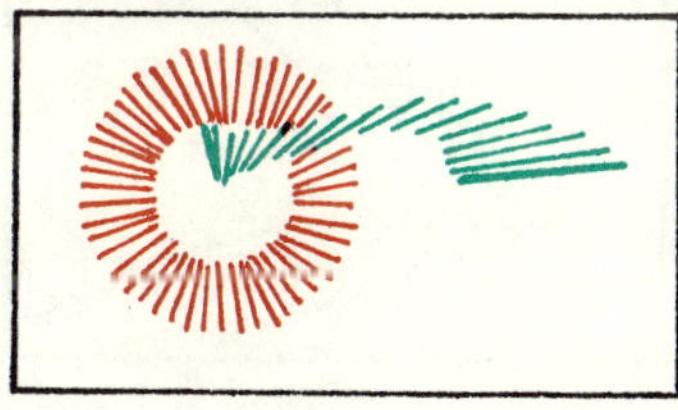

The discovery of a path joining the two green segments shows that there is a path made out of segments not lying in the crown. ($\text{Path} \cap \text{Crown} = \varnothing$).

V. THE BEAM

Having shown that the crown is permeable by segments of the sun we still have the problem of constructing an impenetrable frontier which would separate the two green segments – a frontier which would intersect all paths joining the two segments.

Someone believes at first that four POLICEMEN placed round a green segment would stop it escaping (Fig. 7).

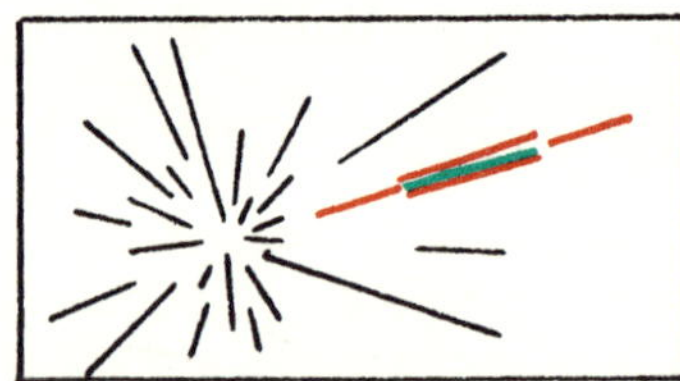

Eventually another suggests colouring red ALL the segments other than the two green ones. This solution, for it is one, seems unnecessarily radical.

Then a sort of comet is suggested; this is made out of all the segments contained in a certain sector (Fig. 8).

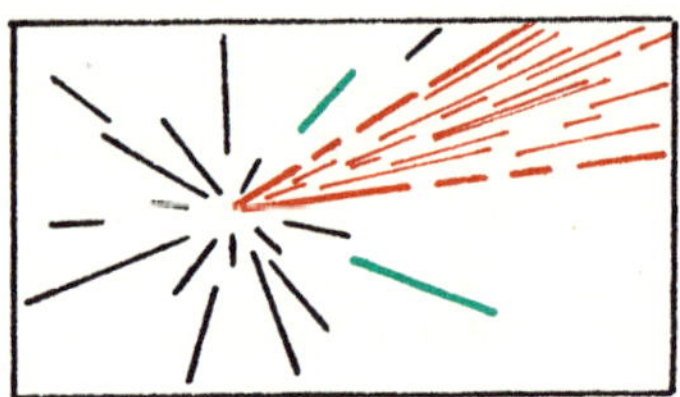

But it is soon recognised that despite its strength this obstacle does not cloud the whole horizon (Fig. 9).

As various ways of modifying the latter barricade arise there is also an interest in finding the most economical barricade.

Threading the needle, as it were, a girl invents the BEAM, namely the set of segments supported by a particular line (Fig. 10).

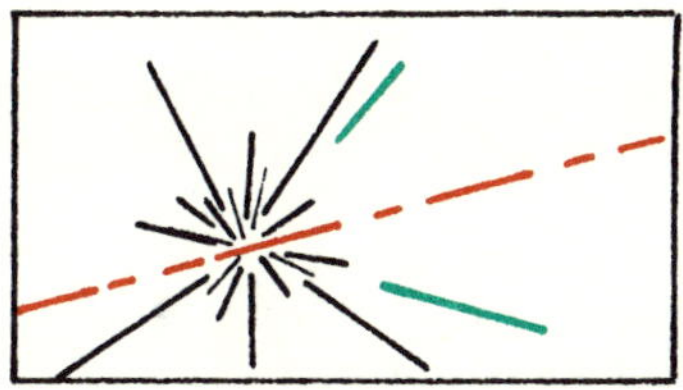

VI. CROSSING THE BEAM

Among the suggestions then there is this very thin barricade, the beam (Fig. 10). Does this set make an efficient frontier? In other words, are all the paths joining the green segments intercepted by this set?

Once again opinion is divided; the class falls into two nearly equal groups. Those who believe this barricade is efficient have the task of seeking amongst themselves an argument that might convince their opponents. The others have to invent a path crossing the beam which they would be able to display to the whole class.

After some time each group comes back enthusiastically convinced they are right. *It is impossible to cross it*, says the spokesman for the first group, *since in passing from one side to the other the green segment has to take the direction of the beam and at that moment it will lie in the beam.* Unmoved by this, the spokesman for the other group presents a subtle path invented by two members of the group (Fig. 11).

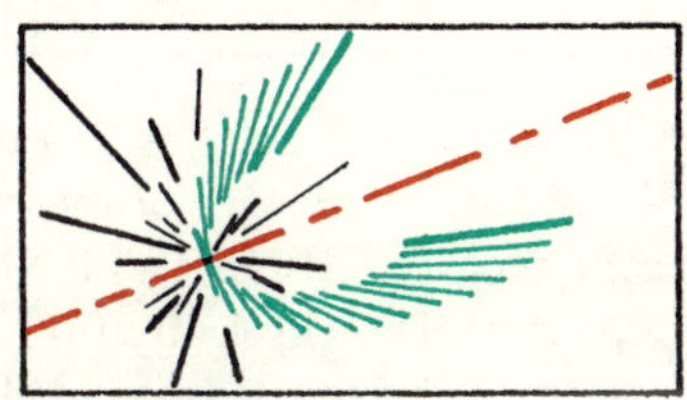

He shows however that TWO appropriately placed beams would provide an impenetrable obstacle (Fig. 12).

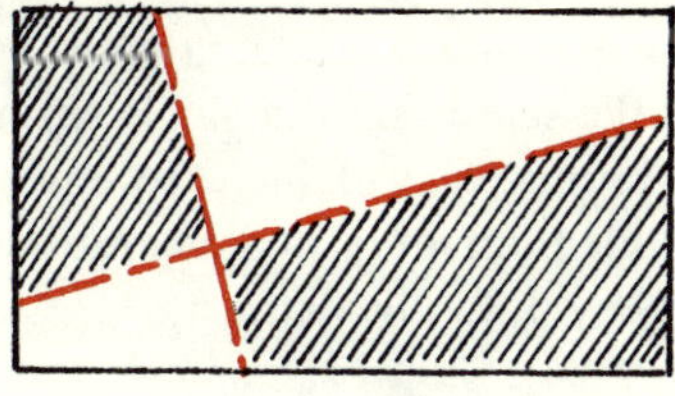

(But the members of the first group are now no longer interested in understanding where they might have gone wrong ...)

VII. SEEING A STRATEGY

The last barricade is only useful as long as it can be inserted between the green segments. But there are some cases where this is not possible (Fig. 13).

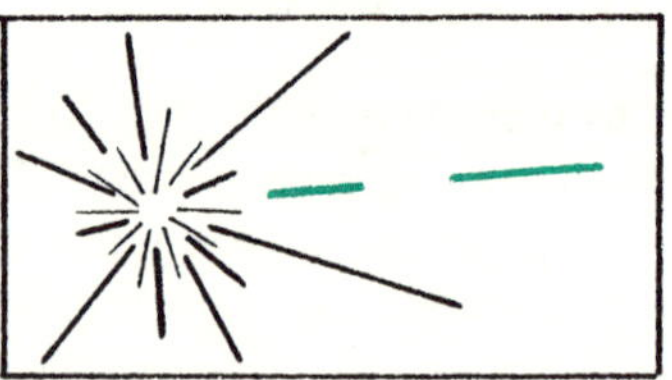

A new frontier has to be invented for this case. It does not take long before someone invents the following set – spontaneously called AN ECLIPSE (Fig. 14).

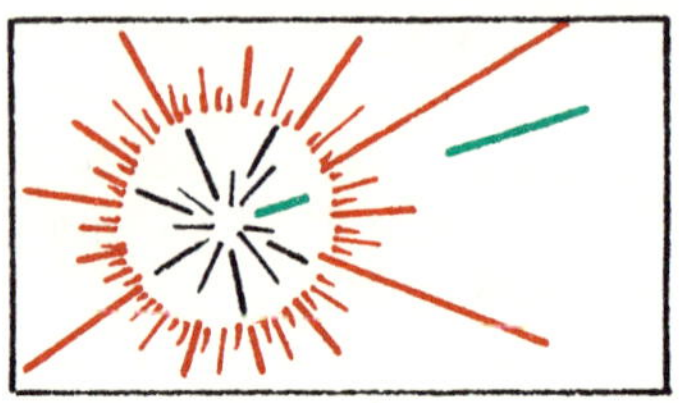

But is it really an impenetrable obstacle? A girl doubts it and asserts that she sees a route. She reduces the length of the moving segment until it becomes a point! Yet if such DEGENERATE SEGMENTS are permitted then they also occur in the eclipse itself.... Thus the class gets involved in defining the term 'segment'. It is decided that the extremities of a segment must be distinct. (This is the eventual definition but it was difficult to reach because the class felt it necessary to make the meaning of the word 'line' more precise.)

Not only is the eclipse permitted as a frontier but it is also recognised to be a THIN frontier in the sense that if a single segment were removed at least one path would be made available. Moreover the inventor of the eclipse will have provided an oral demonstration of this. (Certainty creates the need for proof – J. L. Nicolet.)

It may rightly be supposed that the terms eclipse, frontier and so on will eventually appear in the dictionary for inevitably they arise in the course of involved disagreements. Meanwhile there is nothing to stop us undertaking this task in any case.

VIII. SABOTAGE OF THE ECLIPSE

Pursuing the strategy of sabotage let us submit the case of two partially superimposed segments as shown (Fig. 15).

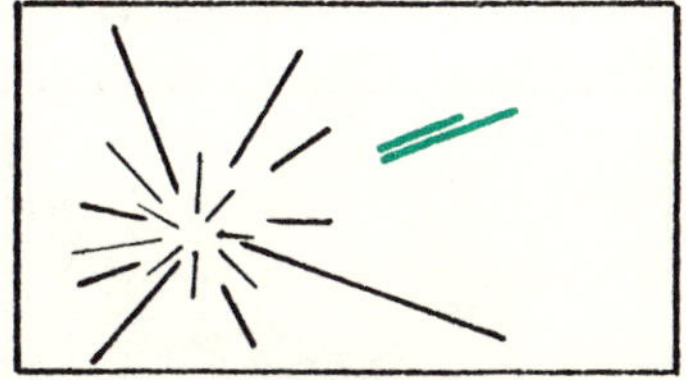

It is now no longer possible to separate these two segments with beams or with an eclipse and new forms of frontiers must now be looked for. Concentrating on the fact that any path joining the two green segments must contain some segments of every intermediate length will produce the notion of a frontier that prohibits one of these lengths (Fig. 16).

This new frontier, THE STAR, is a subset of the sun made out of segments of some previously chosen length.

This is indeed only one solution among many – one might have thought for instance of an inverted eclipse whose segments lie inside the underlying circle.

IX. SABOTAGE OF THE STAR

Notice another special case which will exclude the use of the star – two green segments of the same length (Fig. 17).

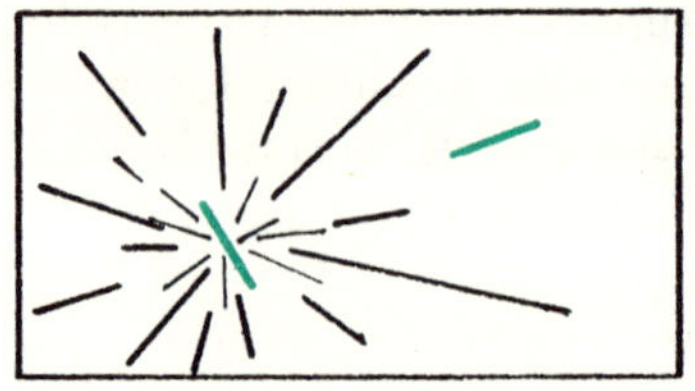

Though it is possible to call on a pair of beams or an eclipse we can interpose the elegant SPARKLER, namely that part of the sun made out of segments having one extremity at the centre (Fig. 18).

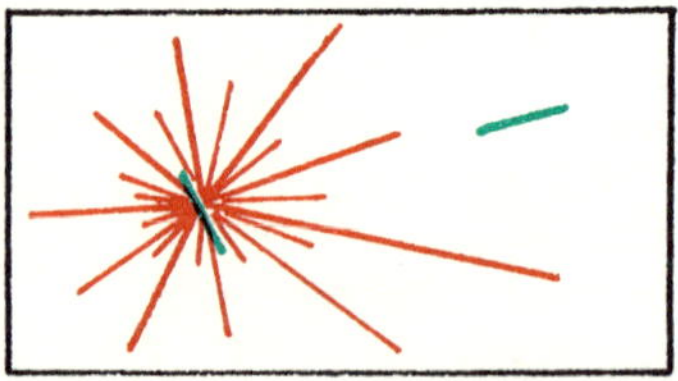

The sparkler – as indeed the star, the eclipse or the pair of beams – splits the sun into two isolated parts.

X. DESCENDING TO THE EMPTY SET

Sabotage is no longer so important since there are now available a variety of frontiers that can split the sun whatever the given pair of distinct green segments.

The study of the various paths joining two segments of the sun can be repeated for each of the new sets which have appeared as frontiers.

Here, by way of illustration, is a 'descent' starting from the sun (Fig. 19).

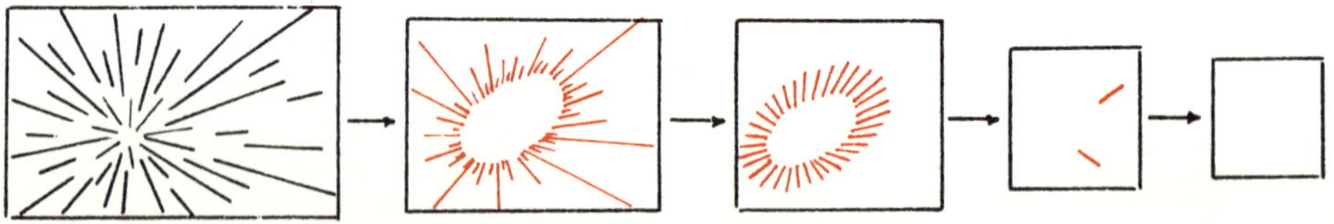

But this is only one among many as the following scheme suggests (Fig. 20).

SUN

BEAM
STAR
ECLIPSE
SPARKLER

WAVE
CROWN
CANDLE

SPARSE $\longrightarrow \varnothing$

XI. POINCARÉ'S PROBLEM

Just as the 'atomic matter' of conventional geometry merges into complexes that are called geometrical figures so does the basic matter of the universe of segments permit the construction of such geometrical objects as suns, stars, sparklers, beams, eclipses and so on. The children explored about twenty such sets and gave them delicious names One day I wrote a very serious letter to the thirty-two children in the class, daring to propose that they establish a classification of such sets and of those that remained to be invented.

The empty set would have to be the sole member of the first category.

The next category, to be designated $\mathscr{C}_0$, would contain, besides the empty set, the SPARSE SETS, that is to say those in which segments are already isolated from each other without introduction of any frontier.

There would then be a category $\mathscr{C}_1$, containing sets which could be separated from each other by frontiers taken from the previous category $\mathscr{C}_0$. (This category certainly contains the empty set and all sparse sets but are there other sets in it?)

There would follow a category $\mathscr{C}_2$, made up of sets which could be separated from each other by frontiers taken from $\mathscr{C}_1$. And so on – to form larger and larger categories.

$$\ldots \mathscr{C}_n \supset \mathscr{C}_{n-1} \ldots\ldots \mathscr{C}_3 \supset \mathscr{C}_2 \supset \mathscr{C}_1 \supset \mathscr{C}_0 \supset \{\varnothing\}$$

This classification was conceived by Henri Poincaré (1854–1912), one of the greatest mathematicians of our time. The sets which are in $\mathscr{C}_n$ without being in the preceding categories are said to be OF DIMENSION n (There is no need to define a dimension for the empty set.)

XII. ASCENDING TO THE UNIVERSE

Through frontiers we have descended from the sun as far as the empty set. The latter serves as a frontier for the sparse sets and in turn each set could be considered to be a frontier to larger sets, for example to the set of all plane segments, the UNIVERSE.

Given the freedom to move about in the whole universe, the beam seems such a weak obstacle that it can be surmounted almost absent-mindedly (Fig. 21).

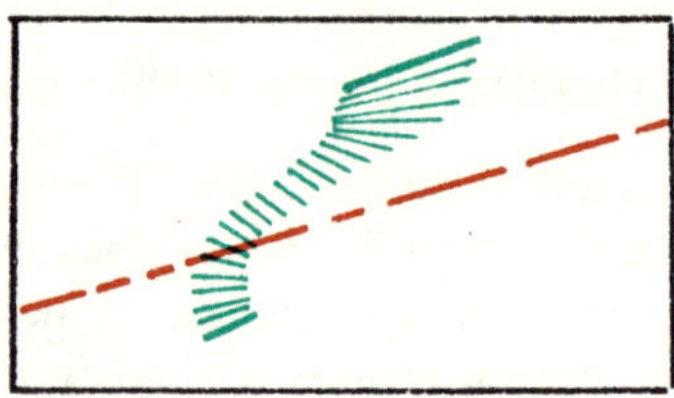

We can pass just as easily under the stars (Fig. 22),

or slip through the eclipse (Fig. 23),

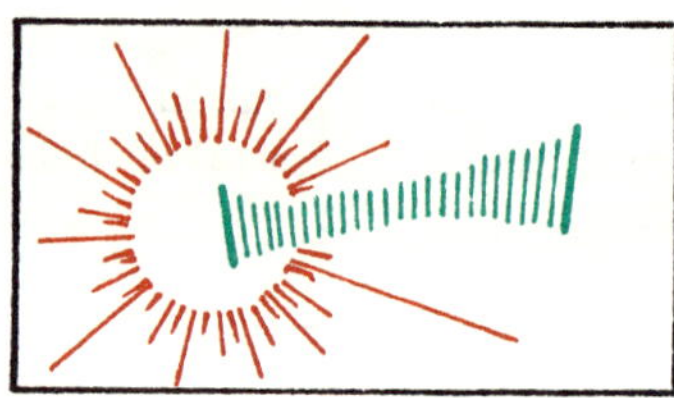

or even escape the sparkler (Fig. 24).

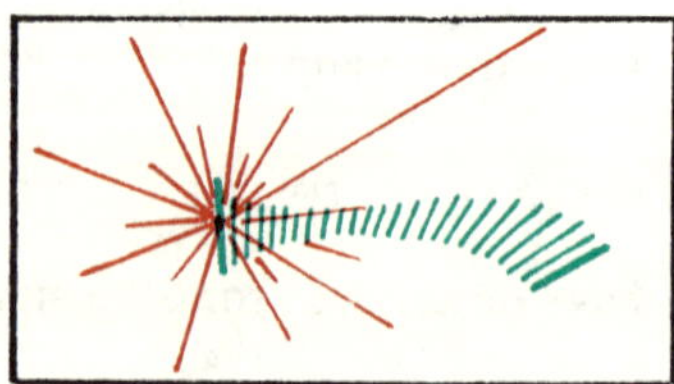

But all these paths could be intercepted by an adequately situated sun (Fig. 25).

Nevertheless the sun is not impenetrable (Fig. 26).

But as some pupils point out the rays of two suns double their effect and do finally just trap the segments (Fig. 27).

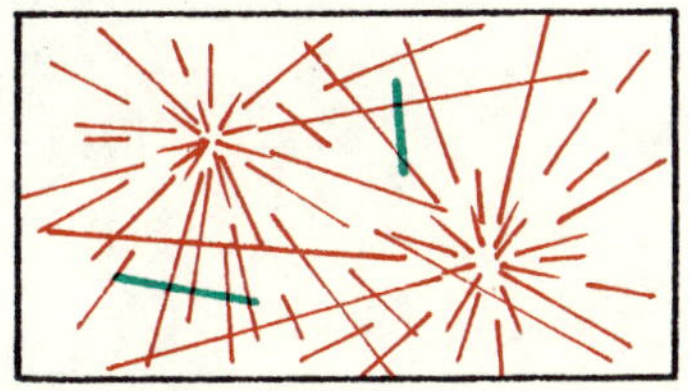

XIII. THE DIMENSION OF THE UNIVERSE

These were only 'some lessons in the sun'. Many others have been evoked for the sun could have been sabotaged and after working at various tasks and sharing information the class might have made some lessons in the SNOW, in A STORM, or in A GALAXY ...

	SNOW	BEAM		
	STORM	STAR	WAVE	
UNIVERSE	SUN	ECLIPSE	CROWN	SPARSE SET → {∅}
	GALAXY	SPARKLER	CANDLE	
	....			

Since the various descents to $\emptyset$ tally, the universe (of line segments in the plane) is then seen to be of dimension four.

Université Laval, Québec

HANS LENK

INTERPRETATIVE ACTION CONSTRUCTS

In this short contribution* I intend to reveal by way of epistemological argumentation as well as by providing some examples of how the extensive discussion in the analytical philosophy of action bears on social scientific approaches to the analysis of sport including in particular the social psychological analysis of motivation, the sociological analysis of the normative orientation of sporting behavior, etc.

Let me begin by roughly sketching out some main lines and articulation points as well as different positions in the analytic philosophy of action and by describing a new interpretative model of an action theoretical approach. This new approach seems to be apt to dispense with most difficulties, anomalies and even antinomies of the traditional action theoretic models underlying social scientific analyses of action. In particular, one result is that purely behavioristic approaches do not satisfy the aims of a full-fledged social-scientific analysis of meaningful actions guided by social norms, values, conventions, symbols, etc.

In the analytic philosophy of action mainly three different controversies characterize the state of the discussions to date: These are the controversies between

(1) causalists vs. logical intentionalists,
(2) particularists vs. generalists (or repetitionists),
(3) pluralists vs. reductionists.

Let me give a brief glimpse of these three controversies.

Re (1).

Whereas the logical intentionalists (e.g., Melden, C. Taylor, R. Taylor, v. Wright) think that there is a logical connection between the concept of an action and the concept of its motivating intention inasmuch as the intention and the action itself cannot be described independently of one another from a logical point of view, causalists (such as Chisholm, Danto, Davidson) think that there are logically independent internal causes contingently causing the action.

* Invited paper read at the 9th World Congress of Sociology, 1978.

J. Agassi and R.S. Cohen (eds.), Scientific Philosophy Today, 151–157.
Copyright © 1981 *by D. Reidel Publishing Company.*

The new interpretational and component theoretic approach to be outlined shortly will lead to the result that both views on either side of this dichotomy, if taken strictly, are wrong, but that both are right to a certain degree. Logical intentionalists are right in claiming that there is a conceptual bond between the description of an action and the description of its "grounds," "reason," "motivation," and/or its mental components in general, insofar as the set of these components cannot be conceived of as a logically independent cause of the action. As components they are part and parcel of the action description. However, single components can and will depend on independently describable, say, amongst others, physiological events which are possibly constructed as causally though not logically necessary conditions of a movement correlated to an action.

Re (2)

Whereas particularists, such as Brand, Danto, and Davidson conceive actions as single and singular, unrepeatable concrete events which can be uniquely identified in space and time, generalists like Chisholm and particularly Goldman think that in single action events only general action properties or action types are exemplified. Chisholm, e.g., interprets actions as abstracts, i.e., as repeatable, instantiable, not localizable and timeable entities which are dependent on statements, i.e., descriptions (this also holds true, by the way, for events and facts).

If you choose a descriptive constituent or component approach, this controversy turns out to be a bit terminological in character. The relata of descriptions of an action and their components may be accorded to a single event-point in time and space though their description has to take into account general expressions of action properties, types, without which identification of the action will be impossible.

Re (3)

Whereas reductionists (as again among others Chisholm, Danto, Davidson) claim that there is only a unique sort of entities (namely 'bodily movements') characterizing an action and that each statement about actions can be reduced to statements about such primitive movements, pluralists (primarily Goldman) would necessarily accord different actions to different action descriptions, even if there is only one and the same movement to which both descriptions refer.

One can easily understand that both positions, if taken strictly, are led into grave difficulties. Neither the one-sided reductionist position only dealing with physical movements nor the proliferating world of abundant ontological types of actions infringing on Occam's razor principle would cope with the ramified variations of the reality of actions and their connection with conceptualizations and descriptions.

A descriptive or interpretational approach, however, can cope with these difficulties without resorting to either too simple or too abundant an ontology of action entities.

Now, what are the main suggestions of such a descriptive or interpretative (interpretational) approach combined with a component model like the one recently developed by Thalberg [1977]? In his book *Perception, Emotion, and Acton* Thalberg suggests to conceive actions not as movements and additional mental acts (e.g., acts of willing) but to understand the mental phenomena as intertwined with the action and its identification as a necessary ingredient, a component or constituent part, of the action. (Constitutive parts cannot be causes, nevertheless they can be correlates of, say, physiological, causes.) The set of components, so to speak, defines and identifies the action as such.

Yet Thalberg never distinguishes between sorts of components of an action: The physical or physiological components are on a par with psychic or psychological or even social elements. And what is their ontological status? To be sure, they would have to have one if taken in such a naive and direct supposition.

Well, one has to add a semantical dimension to avoid these ontological difficulties and further to really cope with exigencies of the social sciences. Actions are not plainly physical movements. This was always known by all except by hard-core reductionists and materialists. But what would remain if you subtract the physical movement from an action, what has to be added to a physical movement to make an action? – A famous question of the later Wittgenstein's *Philosophical Investigations* (§ 621). Wittgenstein answered: "nothing" – i.e., nothing which is existent in an ontological sense, no extra psychic act like a special act of will or so (Prichard). But certainly a plain physical movement (e.g., a reflex) *per se* is not an action. Therefore, something has to be added. But not necessarily something physical or ontologically mental, no additional act which would indeed beg the question again. (To be sure, there may be – and usually are – special nerve impulses, etc., material and physiological components, distinguishing, e.g., reflexes from an intentional action, but (a) these added factors are not themselves internal acts – and (b) they are dependent on a chosen interpretation or even are matters of fact, not of action.)

In short: *An action is not an ontological entity but an interpretative construct, a semantically interpreted entity: Actions are semantics-impregnated.* They can be conceptually analyzed only on a semantic level, they are not concepts of the object language but theoretical concepts referring to interpretations, perspectives, conceptualizations – they are interpretational constructs of at most observable movements. It is the interpretation or the description which has to be added to make an action out of a plain physical movement and to make it identifiable as such. Thalberg's components are of different sorts – some of which are based on a semantic dimension – and they are interpretative.

Now, to take an example from the realm of sport or sport-like behavior: Whether an agent performs a javelin throw as an action of spear hunting, warfare, sport, or other conventional, social or political action, certainly depends on the social setting and the interpretation or the description by the agent himself as well as by observers, participant or not. In medieval times, even the claim of a vassal for ground estate was sometimes measured by the applicant's ability to throw a stone or a spear to the utmost border of his future acquisition. In early Roman antiquity the priest's throw of a spear at a special place outside the city wall was a decisive symbol of declaring war.

In each of these cases the agent's physical movement may have displayed the same external form. At least, let us assume this for the sake of argument. The differences and the various assignments to a particular realm of action clearly depends on the socially impregnated definition of the situation, the social context and environment with all their norms, rules, traditions, values, frames of reference, reference groups playing a decisive role already in perceiving, and all the more in actively orienting, reacting and acting.

For the agent as well as for the observing partner and also for the observing, analyzing scientists' actions are therefore interpretative constructs consisting of constituents which are in part elements of the object language proper but which are in part and not in the least dependent on theoretical perspectives or even on metalinguistic concepts. *Action concepts are theoretical concepts of an interpretational character.*

The component theory developed by Thalberg has to be modified in order to take this theoretical-interpretive character of action concepts into account. The component theory therefore has to be enlarged toward *a descriptive-interpretational constituent theory of action*. It is only this way that the modified component approach can cope with the methodological difficulties and anomalies mentioned above.

Even the possible identification and classification of actions prior to their explanatory analysis is dependent on a descriptive frame delineated amongst others by constitutive rules.

Let us take up again our example from sport – throwing the javelin. Even regarding this seemingly simple uncontroversial action narrowly correlated to the agent's respective physical movement, social criteria of definition of the action necessarily tie in. There are conventional social rules of equivalence, substitutivity, equality, etc., which decide whether or not a movement is admissible as a javelin throw: to throw the javelin with the left instead of the right arm is certainly admissible. This would figure within the range of the admissibility class of the sport action, conventionally defined by the rules of the International Amateur Athletic Federation. Yet, no movement of throwing the javelin with, say, one's leg, if possible at all, would count as the respective admissible action. The thrower would be disqualified at once. Some years ago we had the telling discovery which did not become an innovation, although it could have done so, of a Basque javelin thrower who threw the javelin after turning around like a discus thrower – and shattering the world record by far. After considerable discussion, but nevertheless quickly, the IAAF banned the then admissible form of movement by *ex post facto* changing the rules. The Basque style was banned. It was not considered an admissible action of sport any longer. The admissibility class of javelin throwing was restricted by deliberate convention. In contradistinction, the respective restriction of intelligent innovations in sport was not executed in high jump regarding the new style of the Fosbury flop. (Maybe, this was due to the fact that there was not such a shattering supremacy of the new style regarding the level of performance as in the case of the Basque javelin thrower.)

Certainly, in sports games the impact of conventional rules of definitional and normative character with regard to the admissibility of actions are still far more obvious than in the cases of relative elementary, quasi-"natural" track and field events. The conventional character of the off-side rule in soccer need not be stressed at all.

The upshot of all theoretical and methodological remarks and of the simple examples so far is that all actions are interpretational constructs, perspectival, contextual and conceptual, even and all the more if we take one of the most differentiated approaches in the philosophy of action, namely a component approach (such as Thalberg's), avoiding the traditional dichotomies between causalists vs. intentionalists and pluralists vs. reductionists. All that has been said is clearly relevant for any social scientific analysis of action whatsoever, in particular also for the sociological, social psychological, e.g. research on motivation, and especially for the social philosophical analysis of actions and their methodological problems.

One may easily derive some corollaries regarding the realm of our example like the following:

Kinesiology and biomechanics are certainly very valuable and indispensable scientific disciplines for the analysis of sporting actions, but they certainly cannot cover the whole range of possible important questions connected to this phenomenon which has, among others, ineradicable social roots. Comprehensive studies of sport action have to exceed the realms of biomechanics and behaviorist approaches in general. Indeed, not even the objects of such a discipline can be identified in a pure behaviorist manner, but they can only be characterized by means of, in the last analysis, social and social philosophical concepts.

A purely behaviorist approach will *a fortiori* not do in social scientific analysis, although the increase of precision and the manipulability of variables makes behavioral and even behaviorist analysis a heuristically fruitful tool to begin with.

However, the long-range productiveness of such a methodologically restricted model should not be exaggerated nor even overestimated – and this is true because of and due to thoroughly detailed methodological analysis, indeed, if you take into account the philosophical foundation, constitution and analysis of the seemingly hardcore behaviorist and/or positivist concepts themselves.

These methodological considerations seem to point, indeed, to a necessary integration of the different branches of social sciences including philosophical disciplines. In bridging the boundaries of these for the most part neatly separated realms or "stakes of claims" in the interscientific territory we have unfortunately not come very far – at least so far. Yet, the interesting problems – at least from a social philosophical point of view – seem to loom in this interdisciplinary, intertheoretical and intermethodological realm. And the lucid though still sketchy analysis of nomo-pragmatic statements and rules of action by Bunge [1967, II, p. 134] is not only relevant here, but should also be understood in such an interpretational perspective.

Universität Karlsruhe

REFERENCES

Brand, M.: 1976, 'Particulars, Events, and Actions', in M. Brand and D. Walton 1976, pp. 133–157.

Brand, M. and D. Walton (eds.): 1976, *Action Theory*, D. Reidel, Dordrecht.

Brand, M. (ed.): 1970, *The Nature of Human Action*, Scott, Foresman, Glenview, Ill.

Bunge, M.: 1967, *Scientific Research*, vol. I and II, Springer, New York.

Chisholm, R.M.: 1969, 'Some Puzzles about Agency', in K. Lambert (ed.), *The Logical Way of Doing Things*, Yale University Press, New Haven, London, pp. 199–217.

Chisholm, R.M.: 1970a, 'Events and Propositions', *Nous* **4**, 15–14.

Chisholm, R.M.: 1970b, 'The Structure of Intention', *Journal of Philosophy* **67**, 633–47.

Chisholm, R.M.: 1971, 'On the Logic of Intentional Action', in R. Binkley, R. Bronaugh and A. Marras (eds.), *Agent, Action and Reason* (Fourth University of Western Ontario Philosophy Colloquium, 1968), University of Toronto Press, Toronto, pp. 38–80.

Chisholm, R.M.: 1972, 'Problems of Identity', in M. Munitz (ed.), *Identity and Individuation*, New York University Press, New York, pp. 3–80.

Danto, A.C.: 1963, 'What We Can Do', *Journal of Philosophy* **60**, 435–445.

Danto A.C.: 1965, 'Basic Actions', *American Philosophical Quarterly* **2**, 141–48; reproduced in N.S. Care and C. Landesman, 1968, pp. 93–112.

Danto A.C.: 1973, *Analytical Philosophy of Action*, Cambridge University Press, Cambridge.

Davidson, D.: 1963, 'Actions, Reasons, and Causes', *Journal of Philosophy* **60**, 685–700.

Davidson, D.: 1969, 'On Events and Event Descriptions', in J. Margolis (ed.), *Fact and Existence*, Blackwell, Oxford, pp. 74–84.

Davidson, D.: 1970a, 'Events as Particulars', *Noûs* **4**, 25–32.

Davidson, D.: 1970b, 'The Individuation of Events', in N. Rescher, *et al.* (eds.), *Essays in Honor of Carl G. Hempel*, D. Reidel, Dordrecht, pp. 216–234.

Davidson, D.: 1971, 'Agency,' in R. Binkley, R. Bronaugh and A. Marras (eds.), *Agent, Action and Reason* (Fourth University of Western Ontario Philosophy Colloquium, 1968), University of Toronto Press, Toronto, pp. 3–25.

Goldman, A.I.: 1970, *A Theory of Human Action*, Prentice-Hall, Englewood Cliffs, N.J.

Goldman, A.I.: 1971, 'The Individuation of Action', *Journal of Philosophy* **68**, 761–774.

Thalberg, I.: 1971, 'Singling out Actions, their Properties and Components', *Journal of Philosophy* **68**, 781–787.

Thalberg, I.: 1972, *Enigmas of Agency*, Allen and Unwin, London.

Thalberg, I.: 1977, *Perception, Emotion and Action. A Component Approach*, Blackwell, Oxford.

Wittgenstein L.: 1960, *Philosophische Untersuchungen* in L. Wittgenstein, *Schriften*, Suhrkamp, Frankfurt, pp. 279ff.

von Wright, G.H.: 1974, *Erklären und Verstehen*, Athenäum, Frankfurt.

For an extensive elaboration of the interpretation approach see:

Gebauer, G.: 'Überlegungen zu einer perspektivischen Handlungstheorie', and Lenk, H.: 'Handlung als Interpretationskonstrukt', in H. Lenk (ed.), *Handlungstheorien-interdisziplinär I*, Fink, Munich, 1979, pp. 351–371 and pp. 279–350.

LYNN M. LINDHOLM

IS REALISTIC HISTORY OF SCIENCE POSSIBLE?

*A Hidden Inadequacy in the New History of Science**

A recent well-known aphorist, Imré Lakatos, told us that the history of science is a caricature of its rational reconstruction.[1] If this is indeed the case, which do we prefer: do we want a rational reconstruction which is an idealization? do we want a history which is a senseless, even somewhat ridiculous rendition of events? Brought to such a pass, we would also surely ask: must the choice be limited, as the aphorist paradoxically implies, to an idealization and a caricature? Perhaps instead, a realistic account of the history of science is possible, a rendition of the past which is both accurate and theoretically respectable. Is such a history possible? In my view, the aphorist's paradox cuts much deeper than many who admire the excellent and sophisticated work of modern historians of science may suppose. Let me explain.

Historians of science since E. A. Burtt and Alexandre Koyré have, it seems clear, made remarkable steps toward more realistic and historically sensitive reconstructions of the scientific past. They admit that scientists may be mistaken, may concern themselves with old-fashioned views, philosophical views, even metaphysical views – intellectual preoccupations formerly ridiculed as caricatures of proper scientific concerns. Yet, I will argue, even these historians have not escaped the aphorist's paradox. The new history of science in its most critical and adventurous branches still does not make room for the influence on scientists of confused or incoherent metaphysical ideas. Such influences are ignored or dismissed in the reconstructions because, I propose, it appears impossible to accommodate them without rendering the science a caricature. Yet, the resulting distortions and idealizations seriously inhibit our ability to understand science. That the aphorism challenges even the best history of science shows that we need to stretch farther our views of the legitimate versus the caricature in the area of rationality itself. In the end, we must, I think, acknowledge that rationality in science may change from era to era just as theories do.

J. Agassi and R.S. Cohen (eds.), Scientific Philosophy Today, 159–186.
Copyright © 1981 *by D. Reidel Publishing Company.*

SECTION ONE: THE MERIT AND VULNERABILITY OF THE NEW HISTORY OF SCIENCE

Modern historians of many orientations now share the progressive historiographic policy of trying to avoid hindsight in the reconstruction of history, even the history of science. Of the many groups – economic historians, external historians, philosophical historians, cultural historians – who try to present the history of science in historically sensitive, realistic contexts, those who follow E. A. Burtt and Alexandre Koyré in reconstructing intellectual problem situations seem to me still the best and most critical. I will present some of the merits of this Burtt-Koyré tradition in part (a) of this section. That the tradition is nevertheless vulnerable to a specific version of the aphorist's challenging paradox will be evident, I hope, when I analyze I. B. Cohen's eulogy on Koyré in part (b).

(a) The Impact of Burtt and Koyré

Traditional history of science presented a bleak picture to critical historians: especially narrow ahistorical bias, including a stubborn commitment to anachronism, seemed to characterize efforts in the field [Agassi 1963, pp. 1–15; Kuhn 1968, p. 76]. Could this be corrected? Could it be corrected without abandoning the focus on "science" altogether? Particularly progressive moves toward a solution to this problem were made by E. A. Burtt in his classic *Metaphysical Foundations of Modern Science* of 1925 [1925/1932] and by Alexandre Koyré in his *Études galiléennes* of the '30s as well as in his other numerous works on the scientific revolution [Koyré 1935–37, 1968a, 1968b, 1968c, 1968d]. Both authors interpreted the need for a new history of science as the need for the proper acknowledgement of the theoretical side of science including theoretical change. By recognizing theoretical change within science, they offered one important solution to the problem: one reconstructs realistically by reconstructing the now hidden intellectual problem situation. The role of ideas in science got a new character; particularly, the status of ideas, even metaphysical ones, was rehabilitated.

More important for present purposes, Burtt and Koyré insisted upon appreciating theoretical traditions and conflicts between theories as they were understood in their time, when controversies about them raged. They presented developments as disputes couched in the intellectual context (often metaphysical) of their own day and age. The fruitfulness of the approach Burtt and Koyré thus developed has been quite remarkable. Historical work inspired by policies such as theirs has had a notable liberalizing impact on traditional theories of science: the approach allows history to show that out-of-date, false, even metaphysical ideas have been vital to the advancement of science.[2]

A brief review of the development of this branch of the new history of science would touch the following points. Early philosophers of science – I have in mind particularly Sir Frances Bacon, René Déscartes, and Robert Boyle – reacted against the abuse of metaphysics represented at the time by Aristotelian casuistry and Pythagorean mysticism. Later, this reaction developed into an anti-metaphysical, anti-speculative tradition utterly unrelated to the ills initially opposed; it reached a stage of near obscurantism with the positivist philosophers of the present century. For them, speculative metaphysics was the logical contrast to science. Speculation – particularly metaphysical speculation – was held by most thinkers during this period to be dangerous and irrelevant to the scientific enterprise.

A great deal of work in the history of science reflected this antispeculative anti-metaphysical philosophy; and, of course, a great deal had to be left out in reconstructions of the past in order not to present actual scientists as caricatures of this ideal [Kuhn 1968, pp. 74–75; Agassi 1963, pp. 1–22]. Scientists were congratulated for basing theories on fact, for being independent of all superstitious predecessors, for proving this or that doctrine in the field. Whatever did not fit this scheme was ignored or disparaged; it was emphasized that scientists qua scientists did not concern themselves with metaphysics.

Unfortunately for historians who tried to accommodate history to this anti-metaphysical philosophy of science, metaphysical concerns involve important questions about the nature of the world, the organization of the universe, and man's place in it. Metaphysical discussions include many of the most bold and far-reaching in our intellectual tradition. In trying to present science as basically inimical to, or at least aloof from, such concerns, historians found themselves in awkward positions. Inasmuch as they were successful, science began to look narrow and pedestrian, arrogantly authoritative, not at all part of an intellectual adventure. Scientists appeared at best dogged, ant-like creatures, at worst weird schizophrenics who were inexplicably half completely modern and half completely medieval. Thus, Lord Keynes was in a quandary: was Newton the first of the scientists or the last of the magicians [Keynes 1921]?

The words of Burtt and Koyré constituted a vigorous attack on these historically insensitive, traditional representations of science. They uncovered elaborate, powerful, and intellectually rewarding systems of thought among metaphysical traditions; they brought them forward as relevant, even essential, to the scientific endeavor. Burtt and Koyré – each of them – revealed the sophistication and deep insight into nature that made up the Neo-Platonic-Pythagorean worldview; they drew out the imaginative and penetrating nature of its idea of number, the aesthetic beauty as well as the intellectual breadth of its geometric world plan. Each traced into Newton's science the captivating and fruitful tradition of atomistic metaphysics. In general, their works showed very persuasively how important and revolutionary these allegedly woolly, wispy, incoherent notions had been.

Burtt's and Koyré's detailed, exciting reconstructions of classical metaphysical-cum-scientific ideas and their historical development were creations with many significant facets. By their own account, however, a major point in their histories was the lesson that *ideas*, the conflict of ideas, and the intellectual problems created by such conflicts, were the forces which impelled scientific advance – not practical ends, economic interests, or the social position of scientists. Also the ideas and theories which make up this intellectual background of science are influential *as ideas*, as theoretical speculative notions – not as facts, not as formulae for calculation, measuring devices, nor as research policies. Rather, the metaphysics in science plays its essential role as substantive theoretical speculation about the world.[3]

To elaborate, both Burtt and Koyré illustrated persuasively, with vivid historical accounts, that most of the important events in the history of modern science become inexplicable if their metaphysical background is ignored or if that background is given a degenerate interpretation as a set of empty calculating devices. The metaphysics must be acknowledged to have played its role qua informative metaphysical speculation in order to provide the elements that allow us to explain vital aspects of scientific work. These include the appearance of intellectual problems and concentrated attention on some problems rather than others, the inspiration for specific scientific ideas as well as their very articulation, the psychological and intellectual motivation behind criticism of certain ideas, even the formulation of criteria for scientific success! All of these intellectual activities need the contemporary metaphysical context to be properly understood.[4]

Through the Burtt-Koyré approach, science becomes more admirable and exciting through association with the realistic theories of nature provided by metaphysics and at the same time less admirable and exciting just because they become thereby less stable and quite unjustified. So, the approach of Burtt and Koyré both rescues and threatens some traditional views of science; it seems to make the history both more and less a caricature at the same time.

Notwithstanding my admiration for this new tradition in history of science, I want to draw attention to a hidden inadequacy in it – one which, I think, lessens significantly its historiographic and its philosophical power and which the aphorist pinpointed better than he knew. The weakness involves an assumption of clarity: the old masters are presented as if their writings were as clear and consistent as they became after historical reconstruction; the reconstructed history consistently precludes from the thought and work of scientists any confused or incoherent views.

To explain, careful consideration of the program implicit in the works of Burtt and Koyré allows one to see its vulnerability to an uncritical assumption of maximal clarity. For Burtt and Koyré, a coherent metaphysical system leads to, or gives rise to a coherent scientific system which conforms to it in some way. They do not say what conformity means in this context [Agassi 1964, p. 204]. Yet, leaving that aside for a moment, they do present the metaphysical tradition as intellectual, powerful, explanatory, certainly sensible. Indeed, they often emphasize its coherence and the rationality of the developments in the science which it guides. The impressive feel to the picture should be no surprise: after all, the recovery of metaphysics from oblivion and ridicule as nonsense is what Burtt and Koyré aimed at. I want to emphasize how powerfully they fulfilled their aim. Their accounts leave one with the impression that the metaphysical systems involved in the formation of modern science are intellectually respectable, even awesome in intellectual breadth. One is also left with the impression that the science which conforms to this metaphysics does not suffer from the association – at least not in intellectual cogency.[5] Indeed, their accounts leave open the possibility that a significant separation is maintained between the coherent science and the coherent metaphysics which has guided its development. In any case, both, it seems, remain uncompromised partners in the growth of our understanding of the world.[6]

The misleading element in this picture is its smoothness, its "unproblematic" finished quality. One sees a picture of developments that are admirable, reassuring, untroubled by any irrational fringe to the metaphysics. But this is an idealization. Further research – much of it inaugurated by these same historians – has illustrated that the metaphysics which presented itself to late Renaissance thinkers such as Copernicus, Galileo, Kepler, Harvey, or even Newton, was not solely an intellectually refined and straightforwardly explanatory geometrical worldview, nor a fruitful atomistic-corpuscularian theory. Rather, it was both of these together with an abundant admixture of Pythagorean number mysticism, the religious cosmology and an orientation toward natural magic from a bastard offspring of Jewish and Neo-Platonic thought – the Hermetic-Cabalist tradition – as well as degraded versions of Aristotelian hierarchical ideologies.[7] Of course, neither Burtt nor Koyré hid these matters actively; mainly, their mission was different – to show that the intellectually powerful side of the metaphysics does exist and was essential to the growth of science.

Nevertheless, as they stand, the new histories tend to camouflage the less intellectually respectable character of the traditions they disclose. The price of this weakness has not been estimated as it should be; and it is far from negligible. For, surely one of the main achievements of the early scientific theorists was precisely the clarification and refinement of the medieval metaphysical legacies. This achievement must be obscured or missed altogether if the metaphysical traditions are initially presented as a smooth, coherent set of theories. The idealization also allows the historical accounts to protect, whereas they would otherwise challenge, our contemporary ideas of rationality and the necessity for its historical stability.

(b) Koyré, I. B. Cohen, and the Assumption of Clarity

That the approach developed in the works of Burtt and of Koyré is indeed vulnerable to the distortion which I have just described may be confirmed in some degree by examining I. B. Cohen's commemoration of Koyré's place in the profession [Cohen 1966]. Cohen is himself one of the most eminent and respected followers in this tradition of new history of science; his understanding of its character is important testimony.

In my opinion, however, Cohen's eulogy of Koyré reflects from another angle the crucial equivocation which weakens the Burtt-Koyré approach. The equivocation concerns the question: can one do credit to a thinker's ideas without idealizing them, particularly with respect to their clarity and coherence? I will try to show below that in describing Koyré's approach, Cohen equates seeing an argument as it appeared in its own time with making such an argument coherent and valid. The equation is *not* made explicitly. Nevertheless, Cohen's report strongly suggests it; he does not deny the equation, and no alternative is apparent. Let me elaborate.

After saying that Koyré refused to give black and white judgments of theories by seeing them in terms of modern textbooks, Cohen describes Koyré's alternative as follows:

> Professor Koyré always attempted to find what the basis might possibly be of the particular argument, theory or set of concepts which he encountered . . . . Professor Koyré was interested [not like those who judge on modern standards of scientific truth] first and foremost to find out the degree to which men of science of the past may have been "right" in terms of their own time, not ours [Cohen 1966, p. 161].

This is a direct declaration of the position I say mistakenly supports the assumption of clarity; viz. that to understand a thinker in his own time, one must understand him to be right in his own time.

Cohen does not, however, *declare* that historical reconstruction is reconstruction of a view as correct. One finds only a vagueness that represents an equivocation. The vagueness concerns whether "right in their own time" is to be contrasted merely with "right in our time on our standards" or whether it is also to be contrasted with "wrong (although perhaps reasonable) in their own time on their standards" or even "unreasonable (although perhaps understandable) even in their own time on their standards." That is, Cohen equivocates exactly at the significant point: where an explication of " 'right' in one's own time" is called for.

At other times, Cohen does take a more critical position. For example, in the paragraph following that cited above, Cohen reports that when Koyré analyzed and reconstructed historical events, he tried to find out "whether or not" a particular idea "*followed or did not follow*" [my emphasis] from the views of the time [Cohen 1966, p. 161]. Looking closely, however, one can see that here also there is more equivocation than is superficially apparent. For example, if a view followed coherently from other contemporary opinions, were contemporary authors aware of the connection? Of its import?

More significant, one can see that in Cohen's account the range of contemporary views from which superficially strange ideas might or might not follow is so broad and so deep as to make the second possibility mentioned above – that the seemingly strange ideas were in fact not clear or coherent even in their own time – very remote, to say the least. The range includes, in Cohen's words, "either the explicit *or implicit* [my emphasis] assumptions of the author or general propositions of the age in which that author lived" [Cohen 1966, p. 161]. Thus, the first part of the policy statement indicating that one may be critical of the coherence of ideas within their own time is virtually withdrawn: one can hardly fail to find some predecessor from which a particular idea follows or to which it coheres when implicit as well as explicit assumptions of the author and general propositions of the age are at hand.

Furthermore, we must combine that range of options in making a thinker's ideas consistent with Cohen's additional note that "in some cases it was necessary to study hard and deep to discover such preconceptions" [Cohen 1966, p. 161]. These preconceptions are – the reader should remember – the means by which one can reconstruct the apparent caricature of a rational coherent view into the rational idealization it really is! No one would want to argue against penetrating and extensive historical probing. Yet, the danger here should be suspected by now: namely that, in the fashion of medieval scholastic text study, these options provide so much latitude that they are tantamount to a policy of never declaring a theory unclear, or incoherent in its own time.[8] They nearly guarantee as well that the policy can be followed regardless of historical facts!

Cohen caps his review of Koyré's policies by noting that the latter's analyses were often summarized with congratulations to the thinker in question for being "right": "Readers of Koyré's writings are familiar with his oft repeated comment after such an analysis: *And he was right*!" [Cohen 1966, p. 161].

I think Koyré's sense of the importance of conflict generally prevented his misapplying such a policy, but that does not mean he solved the theoretical difficulty it raises. Nor does it mean that Koyré's followers and admirers do not struggle with that difficulty in their own work.

To give credit to Cohen's more critical understanding of Koyré's policies, I want to note that he cites Koyré to say one must place works studied in their intellectual milieu and resist the temptation to make the thoughts of past thinkers *more* accessible than they should be by putting them into modern language which *clarifies* them only at the cost of deforming them. Koyré is then cited referring to the often obscure, and even confused thoughts of thinkers of the past [Cohen 1966, p. 161]. This appears to be a warning that clearly indicates both authors are aware of the dangers of unjustified streamlining of views through the adoption of an assumption of clarity.

Yet, even this direct warning is more equivocal than it seems. There is no answer to the question: are the thoughts we are warned not to streamline confused only to those of us who see them outside their full intellectual surroundings, only to those who are unfamiliar with the era and therefore are tempted to read modern meanings into the alien thoughts? Or, were these ideas understood in the confused manner at the time; were they perhaps confusions even to the thinker who advocated them? Just what status can a view that is confused on our standards, in their time have: *can* it be clear and coherent then if it isn't now!

The warning cited above is certainly not a declaration on Cohen's part that the ideas in question must have been clear in the proper native setting. But, it also does nothing to authorize or encourage any other interpretation. The anachronistic modernized interpretation makes the ideas more clear *to us*; one still does not know if it is possible to conclude that the views explored were confused at the time, *to them*! And, if the answer was yes, were they still scientific views? [Cohen 1966, p. 161]

Cohen's warning against idealization thus misses the mark slightly but significantly; historians do not know if they *can* declare a view unclear or incoherent in its own historical context. As I illustrate and explain below, they fear that this means the development in question must then return to the status of a caricature of "science," alternatively that such a verdict will indict the historian, not the history.

SECTION TWO: CASE STUDIES IN THE NEW HISTORY OF SCIENCE

The new history of science has added important dimensions to our understanding of science; aspects of history never before discussed in relation to science have been uncovered and used to fill out our picture of the scientific endeavor. A significantly improved and more realistic idea of the nature of science has been made possible in this way. I discuss below three particularly noted and influential contributions to this developing tradition.

The works are *Giordano Bruno and the Hermetic Tradition* by Frances Yates, 'Newton and the Pipes of Pan' by J. E. McGuire and Pyrali Rattansi, and 'Johannes Kepler's Universe: Its Physics and Metaphysics' by Gerald Holton.

Let me introduce the discussions briefly. The works of Yates are by now classic. She has opened up the entire field of the possible relation to Renaissance science of the renegade arm of Neo-Platonism, the Cabalistic-Hermetic tradition. Yet, I will argue, she tends to present only those elements of that tradition which fit the assumption of clarity even while indicating further material. Rattansi and McGuire have written one of the most famous and prestigious single essays in the post-war literature in the history of science. In it they delare they are merely showing that Newton took seriously the more esoteric ideas which may be found in his classical Scholia. I disagree. I think they also endeavor, though they do not declare this to be their aim, to present Newton's interpretations of esoteric wisdom as clear and coherent. Their own evidence offers ample reason to doubt the conclusion they suggest. It is, at any rate, a conclusion that should not be imported implicitly – as is currently done by the profession – but openly declared and discussed. Gerald Holton has more than a study of Kepler and other historical figures to his credit. I will not, however, discuss his variant of the Burtt-Koyré view. Let me only discuss his most influential study, his attempt to eliminate the traditional picture of Kepler as an intellectual schizophrenic. This I like, of course. Yet Holton also presumes, and like the others he does so implicitly, that to do so he must reconstruct Kepler's multi-leveled worldview as coherent and intellectually unobjectionable.

In this view of cases, therefore, I aim at illustrating my criticism of the new, superior historiographic policy of trying to avoid hindsight; although a guide to more adventurous and more realistic history of science, it nevertheless involves distortions of its own which limit our understanding of science and so need to be surpassed.

(a) The Assumption of Clarity in Frances Yates' Bruno

Frances Yates' work *Giordano Bruno and the Hermetic Tradition* [1964] is particularly important to any discussion of how to achieve a realistic history of science. In this work, Yates presented the neglected and rather arcane Hermetic-Cabalist branch of Neo-Platonic philosophy as an essential element in the backgrounds of Pico della Mirandola, Giordano Bruno,

Kepler and others. She thereby made it possible – even necessary – to consider the possible influence on scientists of ideas from this tradition. The Hermetic-Cabalist tradition is an esoteric blend of exceptionally bizarre and questionable intellectual movements including Jewish mysticism, Neo-Platonic cosmological and theological speculation, numerological prescriptions, and theories on natural magic. Clearly, this tradition is a set of ideas more arcane and farther outside the conventional view of the scope of science than any previously considered; the likelihood of idealizing science in this context or of seeing another caricature of reality increases. I will argue that her pioneering achievement is enhanced if we reject her remnant idealization of the Hermetic-Cabalist tradition since it falsifies her own evidence and betrays her own historical purposes.

In arguing such a case, however, one cannot appeal directly to stated historiographic principles. For, as an historian, Yates – like Rattansi, McGuire, and in part Holton as well – joins a long line of distinguished but monotonously uncritical predecessors and does not bother with historiography. So many otherwise perceptive historians embrace the astonishing faith that getting down to work is necessarily more fruitful and less controversial than discussing policy. In fact, although Yates has published several books and one or two major articles and although she has some pretensions to uncovering a significant and disturbing element in the history of science, her historiographic remarks are few and distressingly muddled.[9] Let me review her major book and go into specifics.

As many readers will know, Frances Yates' treatment of Giordano Bruno, the late-Renaissance thinker who was burned at the stake by order of the Inquisition in 1600, has inaugurated a great deal of discussion on the character of science and scientific growth. In the preface to that work, Yates describes a tradition distinct from Renaissance Humanism, the Hermetic-Cabalist tradition. Although unrecognized or neglected by many commentators, interest in this tradition was alive among notable intellectuals in exactly the period when science would also have been incubating [Kristeller 1961, 1972].

Yates connects the Hermetic-Cabalist orientation to the life and thought of Giordano Bruno. Bruno has been considered important in the history of science as the outspoken early advocate of the Copernican hypothesis that the sun was the center of the solar system, and the prophet who foresaw the modern cogency of theories on the plurality of worlds and the infinity of the universe. In many conventional histories, he was credited with an enlightened belief in the latest results of experimental science and with scientific martyrdom: he was the clear-sighted thinker who fell before the superstitious and dogmatic. Other historians, who acknowledge more fully the range of Bruno's theoretical views, dismissed him as a mere metaphysician. Thus, Yates' thorough and persuasive planting of him in this alien contemporary context makes her work a paradigmatic attempt to break through the aphorist's paradox: will the Bruno of history be an idealization or a caricature?

To reconstruct Bruno's history realistically, Yates' program suggests, one must escape both these limited alternatives and place Bruno in the Hermetic-Cabalist context of the day as a rather wild philosophical magician. In Yates' book, Bruno becomes a Magus – a philosophic-mystic seer in the style of alchemists and astrologers. His metaphysical views are blended with utopian hopes for magical transformations; his attitudes become just the sort of metaphysical speculations which were most castigated by traditional philosophers as anti-scientific, dangerous nonsense – not the clear expressions of a refined modern cosmology.

In this context, Bruno's support of Copernicus – and his other *avant-garde* scientific stances – remains both genuine and central to his life and thought, yet, it also becomes part of the strategy of an ambitious Magus who seeks the true cosmological pattern on which to work mystico-numerological spells. This is a startling change of image; Yates herself considers her portrait of Bruno revolutionary [Yates 1964, p. x].

Of course, Yates' presentation remains revolutionary only to the extent that she consistently avoids a retreat to the traditional dichotomy. Dismissing Bruno's metaphysics from consideration as part of "real" science, because its context taints it, would be such a retreat. Unfortunately, Yates is not stalwart here, and readers are never sure if (a) Bruno and his ideas have simply lost caste for being placed in the context that makes sense of them as wild metaphysics connected to magic, or if (b) Yates has discovered something new about science [Yates 1964, pp. 395–97, 450–451, 455].

The equivocation on Bruno is the pattern in Yates' work. For, in my analysis, Yates tries to honor an assumption of clarity in her efforts to reconstruct the Hermetic-Cabalist tradition in its relation to early modern science. She systematically projects the idea that the intellectual developments in question, to the extent that they have any connections with real science, are clear and coherent. When they are not clear or coherent, they may be seen to lose caste in her presentation and to become part of the "prehistory" of science or part of the "motivation" of scientists. As a general result, she hopelessly muddles her stories and contradicts herself as to her own major claims.

Although I cannot discuss them here,[10] a list of important points of equivocation and hesitation in Yates' *Bruno* would include the following: Were Pico della Mirandola and the Hermetic-Cabalist tradition infused with magic or not? She may have to say yes to make a connection with science, maybe no: she says yes and no and refers to further work [Yates 1964, pp. 88–90, 93, 99, 157]. Was the Hermetic-Cabalist tradition considered bizarre or respectable by the clerical establishment? She needs to say both one and the other in order to have a connection with science; she does.[11] Was Kepler a scientist because he cultivated Hermetic-Cabalist notions or because he abjured them? Again, Yates spends many pages setting up a "yes" answer only to say, in the end, "no" [Yates 1964, pp. 442–443, 447]. This time she defers to the idea that real science is just mathematics, which, if taken seriously, is a declaration of bankruptcy: if a scientific theory is merely a formula for calculation, then its accompanying metaphysics – clear or not, Cabalistic or Aristotelian – can have no place within science. Although both her research program and her historical research come to suggest the opposite conclusion, whenever she is forced to challenge the idea that science involves only clear ideas, she capitulates: the Hermetic-Cabalist tradition influenced "pre-scientific" ideas; it served merely as "motivation" for the "will" [Yates 1964, pp. 449–452], and not as an intellectual guide or as direct inspiration.

An uncritical caution in the criticism of esoteric views which includes a tendency to refer to further work is a significant tool in Yates' historiographic workshop. It is also used in all the other cases discussed in this section and will be the focus of critical discussion in Section Three. Perhaps, Yates regularly suggests, not enough is yet uncovered of this obscure and difficult Renaissance tradition to enable us to appreciate it properly. The hesitation in drawing unfavorable conclusions is particularly tempting for Yates since she is indeed working with much more incomplete and fragmentary sources than many other historians. The Hermetic-Cabalist tradition was, in fact, mainly an oral tradition. Thus, the tendency to rebuild it as clear or consistent may seem only fair. Scholars in her position are prompted to assume that any lack of clarity or inconsistency is the result of gaps in the historical record rather than evidence for the incoherence of a theory [Yates 1967, pp. 269–270].

The ease with which one may, however, through such understandable but misguided maneuvers, bypass apparent inconsistencies is an overlooked weakness in this historiographic program, one particularly important in reviewing Yates' studies. For, where no semblance of the proper coherence or sensibility may be maintained in the intellectual developments she is trying to reconstruct, Yates makes this policy a radical endorsement of the clarification program: she omits discussion and remarks on the need for further work in that particular area.[12] How does one decide when, on the one hand, the material under examination is simply confused and when, on the other, one needs to study it further to reveal and appreciate its clarity and coherence? Such questions do not appear in Yates' books. There is no hint of the aphorist's sophisticated sense that reconstruction is a problematic task involving controversial principles.

Yates' work features not only Bruno's thought but also debates between the Rosicrucian theorist Robert Fludd and the astronomer, Johannes Kepler. This research is complemented by D. P. Walker's studies of Fludd and other late-Renaissance figures on the borderline between science and philosophy [Walker 1958a, 1958b]. Walter Pagel, an extremely bold and insightful historian of biology, has produced similar studies revealing the wilder sides of the metaphysics of William Harvey, the early physiologist who discovered the circulation of the blood [Pagel 1967]. Thus, Yates' work represents a growing movement.

All these studies have critics, of course,[13] but *prima facie* the research leads in the direction of a hitherto unnoticed connection between esoteric thought and the development of science. If she did not attenuate it with qualifications due to the assumption of clarity, Yates' research could be a valuable, disturbing contribution to the development of a less idealized picture of science. Her research indicates that the metaphysics most popular with science-minded intellectuals of the Renaissance consisted of vague, confused notions, not even solely philosophical or cosmological, but partly religious, partly ethical, mystical, even magical. One could say that whereas Burtt and Koyré rescued metaphysics by emphasizing the more refined and respectable systems and views, Yates' work reveals the underside of those same systems and shows them to be far more problematic than suspected, both in themselves and as the supposed intellectual intimates of developing modern science. At present, however, Yates' work tends to submerge rather than raise the interesting problems since she veers, as the aphorist has maddeningly outlined, between an idealization and a caricature.

(b) The Assumption of Clarity in 'Newton and the Pipes of Pan'

'Newton and the Pipes of Pan' by J. E. McGuire and P. Rattansi [1966] is a much praised and notable historical paper which concerns the interaction of metaphysics and science. The 'Pipes of Pan' deals with the "classical" Scholia, i.e., commentary, which Isaac Newton appended to the last edition of his *Principia* and in which he presents quite extravagant religious beliefs and alchemical interests including a view of electricity as a vital force as well as conjectures as to the source of the pure wisdom of the ancients – the *prisca sapientia*. Not surprisingly, these concerns and opinions had traditionally been pushed under the rug by admirers of Newton's science.

Rattansi and McGuire undertake to end this disregard and dismissal by reconstructing the opinions and the problems of the Scholia in terms of the traditions of their day. The issues and opinions Newton expressed there, they say, were for him an integral part of his scientific enterprise. Newton took the more esoteric researches of the classical Scholia seriously: he quite genuinely "considered the arguments and conclusions of the Scholia an important part of his philosophy" [McGuire and Rattansi 1966, p. 108]. This position, the authors claim, was not "curious and anachronistic" except when seen in an inappropriate historical context such as that provided by "the generally-accepted view of the intellectual milieu in late seventeenth- and early eighteenth-century England" [McGuire and Rattansi 1966, p. 126]. According to that misleading historical interpretation, the era was secular, positivistic, or otherwise inimical to spiritual concerns. Such an interpretation makes the climate of opinion seem much closer than it actually was to our modern scientific one. But, seen more accurately "in the [true] intellectual environment of his century," Newton's Scholia research "was a legitimate task" [McGuire and Rattansi 1966, p. 138]. These passages appear innocuous in themselves, but as the essay is developed, they form part of an implicit program to present Newton's esoteric problems and opinions as *in fact* reasonable, and as consistent with his scientific worldview, as Newton and some of his contemporaries thought they were. Let me explain this criticism more slowly.

Rattansi and McGuire claim explicitly that Newton took the concerns of his classical Scholia seriously. They go beyond this claim, however; they take great pains to present the theories Newton cultivated as clear and unproblematic on modern standards when reconstructed in the proper context and to present his extended worldview (extended to include both his neglected esoteric researches and his well-known science) as uncompromised intellectually when reconstructed in the proper context. That is, not only did Newton take his program seriously, his program was serious. The care in reconstruction yields valuable historical insight. Yet the authors refrain from any discussion of the policies guiding their presentation and of the possible limits of those generous policies.[14] And in the absence of discussion and given the careful presentation, the approach comes to incorporate an uncritical assumption of clarity.

A specific example of how the authors' policy of reconstructing Newton's program becomes the suggestion that Newton was right to take seriously his Scholia is revealed in the statement below:

> The apparent contradiction between such a traditional Neo-Platonic philosophy [in the Scholia] and the stern inductivism of the *Principia* dissolves when we examine more closely how Newton modified the "mechanical" philosophy of nature which was current earlier in the century.[15]

To paraphrase: Newton seems to be making a mess of his scientific worldview only if one assumes he was an inductivist as well as a strict mechanist of the earlier Galilean-Cartesian variety. Now, the closer examination which McGuire and Rattansi provide reveals that in their view the contradiction between Neo-Platonism and inductivism cum mechanism was dissolved because Newton "restricted" the mechanical philosophy "especially in its pretensions to knowledge of the natural world" [McGuire and Rattansi 1966, p. 125]. This amounts to the claim that Newton surrendered his inductivist philosophy – namely, its claim to realism. This is too drastic a way to ease the conflict given the historically relevant realist trends of the time. Besides, it obviously plays havoc with the theory that Newton molded his science and his spiritual alchemy into one worldview – a theory also propounded by McGuire and Rattansi [1966, pp. 126–127]. They leave this apparent contradiction unresolved.

By the same line of argument: Newton failed to solve the ontological problem of causation but this failure was "less significant than his attempt to investigate it through a unique combination of methods." This unique combination of methods, however, was

> rigorously inductive philosophy, using controlled experiment and elaborate mathematics, complemented by an historical approach, reconstructing the *prisca sapientia* of the laws of God's agency in the world [McGuire and Rattansi 1966, p. 125].

To paraphrase: Newton's inductivism led to his metaphysical worldview. The point is that one cannot tell whether the authors wish to say that this program was *sensible* as well as "unique." They seem to agree with Newton in his opinion that his attempt – the inductive search through *prisca sapientia* of "the laws of God's agency in the world" into the ontological problem of causation – was well conceived; they seem to see no intellectual confusion or muddle in the views. This is proved by illustration, and the illustration fails; thus the authors end up grossly inconsistent.

One might think that Rattansi and McGuire prefer to call Newton's methods "unique" because they want to restrain themselves and avoid labeling those views "irrelevant" or "ridiculous" or "out-of-date." Indeed, as representatives of the new history of science, they no doubt welcome such a change of orientation and want to stress their support of it by avoiding any adjective with a disparaging connotation. This restraint may be admirable in some contexts. In this context, however, their own label, "unique," is either meaningless or yields the suggestion that Newton's methods were unexceptionable not only according to some standards – particularly the standards of the time – but likewise on modern standards.

Yet another example of the position McGuire and Rattansi take with respect to Newton's wilder views is reflected in their choice of the word "legitimate." They argue that taking the Scholia program as part of his overall intellectual research was "in the intellectual climate of his century...a legitimate task" [McGuire and Rattansi 1966, p. 126]; not "correct," not "admirable," nor even "reasonable," just "legitimate." What, one might wonder, is not legitimate? They offer no answer. Who legitimizes? The standards of the day? In the Royal Society of London which published the book, the standards were inductive, anti-speculative, anti-theological. On such standards, the program was quite illegitimate. Moreover, we do not need to read Frank E. Manuel's *Portrait of Newton* [1968] to know that Newton flagrantly violated standards and took great liberties. Thus, the word "legitimate" is both non-committal and a suggestion that authorizations exist which do not and cannot materialize.

One might suppose that the legitimation or authorization of Newton's wilder researches comes from McGuire and Rattansi themselves: in some sense, *they* find Newton's Scholia research "legitimate" given the contemporary climate of opinion. But this does not clear up the vagueness in the label: for, do they mean legitimate on modern standards? Again, the answer must be no. Yet the authors quite consistently and carefully refrain from expressing any negative view of their own with respect to Newton's researches; neither do they admit deep conflicts in his program. The impression that they themselves have a positive view of his philosophy and program is conveyed quite strongly throughout the paper. Yet they do not state a positive view. Why? Again, one might suppose simple historical detachment is the answer: their opinions, they feel, are not relevant. But, given the untempered implicit suggestion that their view is positive, this will not do; it will particularly not do if their opinion is in fact negative. And their opinion must be negative: on their own evidence, Newton's researches cannot meet their standards [McGuire and Rattansi 1966, p. 108]. Let me illustrate.

Rattansi and McGuire themselves recognize that thinkers have often taken seriously philosophies which we cannot, in principle, find comprehensible. As they point out, Newton's theory of a

> harmony of the spheres rests on the belief that the true system of the world was known to the ancients but has been turned into "a great mystery" which only initiates could penetrate [McGuire and Rattansi 1966, p. 136].
>
> The true meaning of Old Testament prophecies would only become clear in retrospect, in the light of historical experience . . . . Newton's textual analysis of ancient natural philosophy . . . for him . . . represented a deep penetration into *prisca sapientia* possible only when the preliminary work had been accomplished through experience [McGuire and Rattansi, p. 137].

To paraphrase: we cannot agree with Newton's view of antiquity. Yet, even here the authors avoid any suggestion of the need for criticism of the program and rather stress how unproblematically integrated Newton's philosophy was – in Newton's mind now! – with his scientific program. Their reluctance to express any negative opinion suggests that it is similarly unproblematic to them, a position they would not – I claim – support outright. Their reticence represents not a theory of the irrelevance of their own standards of clarity and reasonability, but a conflict between their standards and Newton's – a conflict they do not know how to handle and which they therefore do not want to report outright.

(c) Holton on Kepler's Odd Metaphysics

In a welcome article, Gerald Holton has made a study of Kepler's very unusual and active metaphysical side [1973]. It is a conscious attempt to give a more accurate and full account of Kepler's worldview and its importance in his scientific work than traditional commentators have managed, Holton aims to resurrect Kepler's metaphysics as a significant factor in his research.

Since Holton acknowledges the oddity of Kepler's metaphysics and champions its rightful claim to be part of his science, hopeful readers might expect a view of the scientist not idealized. And indeed, Holton's study does show that Kepler's metaphysics cannot be relegated to a pre-scientific limbo without relegating his entire scientific achievement to the same limbo. For all this, however, the Holton article is rather disappointing. Although the evidence he presents seems clearly incompatible with the assumption that Kepler's views were clear and coherent, and although he never declares the assumption of clarity outright, he tries to uphold it by presenting Kepler's metaphysics as unconfused despite the great difficulties and awkwardness this forces upon him.

Holton dislikes the portrayal of Kepler as a weird schizophrenic who happened to have been a modern physicist part of the time but was a medieval astrologer at others [Holton 1973, p. 69]. Holton prefers to present Kepler's view as a unified and integrated melding of metaphysics, physics, geometry and theology, not an arbitrary and insulated juxtaposition; the science and the philosophy were part of the same enterprise – that of comprehending the world. But were they unified only in Kepler's thought or also on standards we would share? Given the facts and the context, would we share Kepler's unified view, or would we find it confused? Unfortunately, there is no discussion of such a question. And, Holton's presentation suggests that he not only wants to avoid calling Kepler's worldview modern, regressive, schizophrenic, or superstitious; he also wants to avoid calling it confused, even though he knows it was. Thus, he tries consistently to give the impression that by our standards Kepler's worldview was only apparently confused. This conflicts, however, with his other major message: that this worldview was unified and integrated with a science which our standards allow us to accept with only minor revisions.

Holton gives several excellent summary views of Kepler's genuine passion and ecstasy over his investigations and discoveries; he describes Kepler's openness in narrating his mistakes; he reports with some vividness how closely Kepler associated his science with his love of God in passages that are not only enlightening but moving [Holton 1973, pp. 70, 85–86]. The points which are most important for the discussion here are these: Kepler's outlook differs from ours in radical ways; it cannot be separated into modern categories, and it contributed to his scientific feats in fundamental ways.

First, Holton does recognize that Kepler's intellectual outlook is hard to understand, not merely because he had different problems and concerns than we do, but also because he had a worldview that included different intellectual standards. This, indeed, separates him, not merely from us but even from Galileo and Descartes. In Holton's own words:

Even in comparison with Galileo and Newton, Kepler's writings are strikingly different in the *quality* of preoccupation. He is more evidently rooted in a time when animism, alchemy, astrology, numerology, and witchcraft presented problems to be seriously argued. His mode of presentation is equally uninviting to modern readers . . . . *Nor is this impression merely the result of the inevitable astigmatism of our historical hindsight.* [My emphasis] We are trained on the ascetic standard of presentation originating in Euclid as reestablished, for example, in Books I and II of Newton's *Principia*, and are taught to hide behind a rigorous structure the actual steps of discovery . . . . [Holton 1973, p. 69].

Holton is directing attention to the fact that Kepler's standards concerning what is sensible – and how to present it – differ from ours, even from those of some of his contemporaries. The differences, then, cannot be resolved merely by recovering details of the context: Kepler's view will not be ours no matter how much we understand of his context.

It is significant, however, that Holton emphasizes the difference in style of presentation rather than the deeper conflict of standards, although he reveals both [Holton 1973, p. 70]. That is, nearly as soon as Holton takes the enlightened step featured above, he qualifies and soft-pedals it. Although he has just said that historical insight will not reconcile us to Kepler, he sums up his description of Kepler's unusual standards of presentation by saying, "They mirror the many-sided struggle attending the rise of modern science in the early seventeenth century" [Holton 1973, p. 70]. This summary is true but very partial; it suggests that the divergence between our expectations and Kepler's manner of *presentation* do not extend to deeper standards. Yet from his account, it is clear that we do not merely hide the steps to our discoveries whereas Kepler did not; our view of rationality conflicts with his [Holton 1973, pp. 69–70].

Kepler's science and his metaphysics are thoroughly integrated. The following passage reveals not only the unusual nature of the metaphysics, but also this high level of integration:

We shall see that when his physics fails, his metaphysics come to the rescue; when a mechanical model breaks down as a tool of explanation a mathematical model takes over; and at its boundary in turn there stands a theological axiom . . . .

Kepler set out to unify the classical picture of the world, one which was split into celestial and terrestrial regions, through the concept of a universal physical *force;* but when this problem did not yield to physical analysis, he readily returned to the devices of a unifying *image*, namely, the central sun ruling the world, and of a unifying *principle*, that of all-pervading mathematical harmonies.

In the end, he failed in his initial project of providing the mechanical explanation for the observed motions of the planets, but he succeeded at least in throwing a bridge from the old view of the world as unchangeable *cosmos* to the new view of the world as the playground of dynamic and mathematical laws. And in the process he turned up, as if it were by accident, those clues which Newton needed for the eventual establishment of the new view [Holton 1973, pp. 70–71].

It seems clear that here is a metaphysics which is strange – in fact, confused – but which allowed Kepler to keep on his agenda the important aim of forging a unified worldview. Holton's account emphasizes that Kepler's metaphysics "comes to the rescue" both by keeping important problems from being abandoned prematurely and by allowing scope for the passion to unify disparate visions of the world.

Yet, in spite of his attempts to appreciate Kepler's strange vision without modernisms and idealizations, Holton avoids saying that it is confused or inconsistent. His caution makes him inconsistent even on the point that it is incompatible with what *we* would consider acceptable in these areas; he remains vague on the question whether he – with his modern standards – thinks Kepler's views were as sensible as Kepler thought they were. In introducing the worldview in the passage cited above, Holton reveals his hesitance:

Conceptions which we might now regard as mutually exclusive are found to operate side by side in his intellectual makeup. A primary aim of this essay is to identify those disparate elements and to show that in fact much of Kepler's strength stems from their juxtaposition [Holton 1973, p. 70].

In this passage, too much depends on the interpretation of the words "might" and "juxtaposition": the "might" is evasive euphemism; "juxtaposition" suggests distinct and clear categories in the face of Holton's insistence at other times that Kepler's vision was "unified." Also, consider the concluding and perhaps the best of Holton's descriptions of Kepler's worldview. Holton claims both that Kepler's vision was a unified one in which physical causes, mathematical harmonies, and metaphysical priorities were on a level and interchangeable and that this unified vision was not confused on modern criteria:

> So intense was Kepler's vision that the abstract and concrete merged. Here we find the key to the enigma of Kepler, the explanation for the *apparent* [my emphasis] complexity and disorder in his writings and commitments
>
> In one brilliant image, Kepler saw the three basic themes or cosmological models superposed: *the universe as a physical machine, the universe as a mathematical harmony, and the universe as a central theological order*. And this was the setting in which harmonies were interchangeable with forces, in which a theocentric conception of the universe led to specific results of crucial importance for the rise of modern science [Holton 1973, pp. 86–87].

To paraphrase, there is no problem in thinking harmonies are interchangeable with forces; the disorders and complexity of such a view are only "apparent." Yet, Holton's own account is a vivid one of Kepler's struggles with three basically incompatible themes. Clearly, for Kepler they were melded, not merely juxtaposed in distinct compartmentalized versions; they were melded into what Kepler deemed a clear and consistent view in which the "confusion of incongruous elements – physics and metaphysics, astonomy and astrology, geometry and theology" was only "apparent" [Holton 1973, p. 69].

Holton wants to say both that Kepler's views were unified and that they were not confused; his failure here is unavoidable.

The process Holton summarizes is one of an intellectual odyssey – of a great adventure and a great struggle. His hesitance to present Kepler's worldview as unclear and confused partly obscures the difficulty and extent of that adventure and of Kepler's accomplishment. Holton cannot show as clearly as he otherwise would, that in Kepler's effort to understand the world, his metaphysics was both an invaluable *and* a crude, problematic tool; it was not a finely finished instrument.

SECTION THREE: SUMMARY AND EXPLORATION OF THE POLICY OF AVOIDING HINDSIGHT

There is an important reason why the assumption of clarity – that all metaphysics connected with science is clear and coherent – remains uncriticized even by historians who genuinely wish to avoid idealizations, such as I.B. Cohen, or by scholars who deal with *prima facie* obscure and confused metaphysical ideas such as Yates, or Rattansi and McGuire. The historiographic policy of avoiding hindsight – seeing a theory in its own terms, not being wise after the fact – seems to make the assumption of clarity inescapable. Why? How do these two ideas connect? I wish to explore one answer to these questions a bit more. That answer is that the assumption of clarity is part of an independent test of an historical interpretation, and that as such, it is extremely difficult to drop. Moreover, in the absence of a theory of rationality which allows a thinker to be confused and reasonable simultaneously, the difficulty must remain chronic. Any historian as allergic to history as caricature as he is to history as idealization will need to work with some such assumption.

To see why this is so, one must reconstruct the situation of the critical historian with an eye to his philosophic position. He is faced, let us imagine, with a theory or problem which appears alien and unrelated to what are now taken to be successful efforts in a particular field. He knows that many advances in the field have occurred since the theory in question was being debated. These advances may have been made with the help of the theory, but – as Agassi points out[16] – they obscure the problems which formerly were alive, the opponents and difficulties which loomed large, the advantages of the view at the time. Knowing this to be the case in general, the historian does not credit an initial sense that a theory is odd, inconsequential, uninformative, or confused. He tries to reconstruct the problem situation of the intelligent thinkers who invented or advocated the theory. When he does this, the critical historian must use some standard by which to judge whether or not he has been successful – successful, that is, at reconstructing the contemporary problem situation accurately. That is, he must decide how he will tell if his efforts to overcome the parochial and unfavorable modern bias have been sufficient to allow proper credit to the view in question. A ready standard in this endeavor is the attainment of clarity and consistency: when the historical event or theory is seen to be clear, informative, a coherent part of the thinker's worldview, and consistent with his other beliefs or with popular views, one has certainly done one's best for it.

By the above route, the need for an independent standard transforms the aim of "seeing a theory in its own time" into "seeing a theory as clear and coherent in its own time." To repeat, an historian sensitive to the liability of being wise after the event can try to find background theories, problems, or traditions which make a seemingly unclear theory clear. But, to succeed in the reconstruction of the view and make it as clear and coherent as possible, this easily becomes the test of success in overcoming one's modern point of view. When this happens, the assumption of clarity becomes part of policy and is beyond empirical criticism.

The logic of this historiographic approach, therefore, leads to uncritical policies—as I hope to have shown earlier – in spite of the original intentions. To restate: one must assume that any incoherence in a theory is merely apparent and will vanish when the historian does his homework fully and adequately. Whenever a seemingly unclear or inconsistent view presents itself, the critical historian is led almost irresistibly to the view that the weakness is one in his reconstruction, rather than in historical fact. If every case of obscurity or contradiction is considered the result of the historian's inadequacy, however, then it becomes impossible to attribute a confused or incoherent view to any thinker.

Of course, one might argue that no historian takes the extreme literal position that the commentator is always at fault for not recovering a proper context whenever a view seems unclear; and further, that the moderate position – to take seriously the possibility that the view is clear, to assume it *prima facie* and look for the context – is very sound policy. This moderate position may be superior, but only when combined with clear standards as to when one reverses the *prima facie* assumption. Unless a standard is provided, the moderate position becomes an extreme one in spite of the historian's better intentions.

One can see that the additional standard is needed by considering the particularly difficult cases of reconstruction when the traditions involved are incompletely documented or especially alien – for example, Hermetic-Cabalistic view of the Renaissance. In such cases, presenting that part of the theory which is clear, and considering the rest as "in need of further study," seems both conscientious and fair; for, it may well be true that the part of the theory which appears unclear or incoherent is in historical fact clear and coherent, and this fact will be revealed once more information or familiarity is achieved [Agassi 1963, pp. 33–40].

The "further study," which seems so reasonable here, may be prolonged indefinitely, however, especially if the need for the "deep and lengthy" search, to which Yates and Cohen refer, is considered appropriate. During this time, one is presenting an image of a view that is completely clear – since the public parts of it *are* clear and the rest is merely alluded to as the object of research; whereas, what one has on hand *in fact* is quite the opposite: a *prima facie confused* and incomplete view which may never reach any different status. Thus, the moderate position is likely to produce the greatest distortion in those cases of odd traditions concerning which it is most tempting to adopt it. The works of Yates, McGuire and Rattansi are cases in point. Hence, at the very least, authors should not conceal the situation as it looks at the time they write.

To go more deeply into the problem, it should be noted that the rule of thumb for good historical explanations – to find the contemporary context which makes a view clear and consistent – is particularly hard to resist when the thinker who propounded it was eminently successful in some other way. Newton, Galileo and Harvey are examples. Were this not the case – that their other achievements were famous – their less current beliefs would not very likely be the subject of historical inquiry. This also means, however, that one is dealing with thinkers who have left evidence of ability to reach fruitful, valuable, powerful theories, which are also clear and coherent – even possibily clear and coherent on the best and highest modern standards. It makes some sense that the less famous views of these thinkers would live up to the same standards. If they seem *prima facie* not to do so, it is easy – indeed, reasonable – to conclude that further investigation will reveal the context which in turn reveals their clarity and coherence. For example, if Newton's *Scholium Generale* appears obscure and silly, since his *Principia* shows him to have been a powerful and rigorous thinker, it is quite reasonable to believe that the other views he took seriously and expressed there were judged by those same rigorous standards [McGuire and Rattansi 1966, p. 138].

The point which must be emphasized is that this assumption is no mere hero worship. One does not assume the clarity of Newton's less familiar views simply because he is a renowned thinker and had other respectable views; rather, one tends to make that assumption in order to avoid appealing to the *ad hoc* theory that Newton was an intellectual schizophrenic – that he used, quite unaccountably, two conflicting standards in judging his intellectual production. That is, one tries to explain why Newton cultivates certain *prima facie* confused or obscure views by showing that they can be explained as clear on further investigation, that they do not constitute an inexplicable lapse. Yet, one can explain his obscure passages by other means, e.g., political means, or even personal ones (as done by Frank E. Manuel). We should allow the possibility of clarifying obscure passages in Newton, but we should not rule out the impossibility of so doing [Agassi 1963, pp. 44–45].

Similarly, one should admit that a person whose views were coherent and clear had a better chance to act and think rationally than a confused one. But we need a theory as to how a confused thinker evolved while clearing up some of his confusion as well as some of his errors – perhaps also how he used his confused views in helpful ways without clearing them up – and then used these improvements to effect others. This is the bootstrap theory. Without it we cannot explain the growth of rationality and the growth of clarity as evident even in the works of Kepler and Galileo [Agassi 1973a, 1973b; Bartley 1964, 1968; Wettersten 1978].

To conclude, historians need a philosophic theory which allows the possible rationality of cultivating confused ideas – of being confused as well as mistaken or out-of-date. Then, there would be a way to reconstruct the history so as to see a thinker's views as confused *and* valuable to his science *and* valuable to *science* at the same time. Right now we can only recommend that Yates, McGuire and Rattansi, or Holton not idealize their subjects for the sake of a theory of science that doesn't fit the facts. But can we describe what alternative they have? Lakatos' aphorism cuts deeply here as I hope I have shown: historians have an obligation not to present meaningless caricatures just as they have an obligation not to present idealizations. Regardless of the historical information available or the historical information recounted, if the record includes evidence against the rationality of the thinker, the record may simply be discounted – taken as a mere caricature.

University of New Hampshire

NOTES

* This essay is part of a Ph.D. dissertation which was summarily rejected twice by examiners at the University of London; it was later revised, then resubmitted at City University of New York. I want to thank Joseph Agassi, Robert Cohen, and my parents for help that allowed the work to survive. Joseph Agassi, John Wettersten and Stephan K. Cutting made extensive criticisms and corrections to later drafts of the essay; I am very grateful for their help.

[1] A modified version: Lakatos [1971], pp. 105, 121, 122. For a summary of the oral tradition: Agassi [1977], Ch. XII, n. 27, pp. 216–217.

[2] Burtt [1932], p. 126; Koyré [1935–37], p. 11; [1968c], p. 6. Also, Agassi [1958–59], pp. 234–236; also my 'Burtt and Koyré as Revolutionaries'.

[3] For Burtt and Koyré as opponents of Duhem [1954], see Agassi [1958–59]; Agassi [1963], pp. 48–54, 57, 64; also Keynes [1921].

[4] For some of this see Agassi [1964]; also my dissertation *The Assumption of Clarity* [1977], Ch. 3 and Ch. 4, (C.U.N.Y.).

[5] Burtt protests against the consequences of the metaphysics on the human self-image and the fraudulent pretense that it is proven theory, but not its power as a vehicle for science; e.g., [1932], p. 137.

[6] For Koyré's reticence on these matters, see Agassi [1974], pp. 415–416.

[7] Besides the works mentioned below, see Pagel [1967]; also his other works.
[8] See H. A. Wolfson's classic study of Talmudic textual analysis [1929], pp. 24–27.
[9] For historiographic remarks: Yates [1964], pp. 447–449, 451–452, 455; [1966], pp. 352, 364–365, 371–374; [1967], pp. 261, 263, 265, 269–271; [1972], pp. 128, 130.
[10] See my review essay of Yates' *Bruno* for elaboration.
[11] Yates [1964], pp. 105–106, sp. 156; also humble or utopian aspects of the urge to "operate" as discussed 147ff.
[12] References to further work: Yates [1964], pp. x–xi, 106, 142, 155, 157, 162; [1966], pp. 13–14, 362, 366, 371–372; [1967], pp. 258, 269–270; [1972], pp. 128, 130.
[13] Rosen, E., 'Was Copernicus a Hermeticist?' in Stuewer [1970], pp. 163–171.
[14] McGuire and Rattansi [1966], pp. 109, 125, 138, perhaps also pp. 119, 121 are the places at which one would expect such discussions; only one or two phrases in qualification occur.
[15] McGuire and Rattansi [1966], pp. 124–125; and text to note 62 below on the misleading use of "apparent' in these contexts.
[16] Agassi, see n. 1 above; also his [1968].

REFERENCES

Agassi, J.: 1958–59, 'Koyré on the History of Cosmology', *British Journal for the Philosophy of Science* **9**, 234–236.

Agassi, J.: 1963, *Towards an Historiography of Science, History and Theory*, Beiheft 2, Mouton, The Hague. Reprinted Wesleyan U.P., 1967.

Agassi, J.: 1964, 'The Nature of Scientific Problems and Their Roots in Metaphysics', in Mario Bunge, (ed.), *The Critical Approach to Science and Philosophy*, Free Press, New York, 189–211; reprinted in [1975].

Agassi, J.: 1968, 'The Novelty of Popper's Philosophy', *International Philosophical Quarterly* **8** 442–64; reprinted in [1975].

Agassi, J.: 1974, 'Criteria for Plausible Arguments', *Mind* **83**, 406–416; reprinted in [1975].

Agassi, J.: 1973a, 'Testing as a Bootstrap Operation', *Zeitschrift für allgemeine Wissenschaftstheorie* **4**, 1–25; reprinted in [1975].

Agassi, J.: 1973b, 'Rationality and the *Tu Quoque* Argument', *Inquiry* **16**, 55–67.

Agassi, J.: 1975, *Science in Flux*, D. Reidel, Boston and Dordrecht.

Agassi, J.: 1977, *Towards a Rational Philosophical Anthropology*, Nijhoff, The Hague.

Bartley, W. W., III: 1964, 'Rationality versus the Theory of Rationality', in Mario Bunge (ed.), *The Critical Approach to Science and Philosophy*, Free Press, New York, pp. 3–31.

Bartley, W. W. III: 1968, 'Theories of Demarcation between Science and Metaphysics', in Lakatos and Musgrave (eds.), *Problems in the Philosophy of Science*, North-Holland, Amsterdam, pp. 40–64.

Burtt, E. A.: 1925/1932, *The Metaphysical Foundations of Modern Science*, 1925, Routledge and Kegan Paul, London. (Rev. ed. 1932).

Cohen, I. B.: 1966, 'Alexandre Koyré (1892–1964) Commemoration', *Isis* **57**, Part 2, 157–165.

Duhem, P.: 1954, *The Aim and Structure of Physical Theory*, Princeton University Press, Princeton.

Holton, G.: 1973, 'Johannes Kepler's Universe: Its Physics and Metaphysics', Ch. 2 in *Thematic Origins of Scientific Thought*, Harvard University Press, Cambridge, Mass.; Orig. publ. *American Journal of Physics* **24**, (May 1956), 340–351

Keynes, J. M.: 1921, *A Treatise on Probability*, Macmillan, London.

Koyré, A.: 1935–37, *Études galiléennes*, Hermann, Paris. (*Galileo Studies*, Harvester Press, Brighton, 1978).

Koyré, A.: 1968a, *The Astronomical Revolution*, Cornell University Press, Ithaca, N.Y., orig. publ. 1961.

Koyré, A.: 1968b, *From a Closed World to the Infinite Universe*, Johns Hopkins University Press, Baltimore, orig. publ. 1957.

Koyré, A.: 1968c, *Metaphysics and Measurement*, Chapman and Hall, London.

Koyré, A.: 1968d, *Newtonian Studies*, Chicago University Press, Chicago.

Kristeller, P. O.: 1961, *Renaissance Thought*, Harper and Row, New York.

Kristeller, P. O.: 1972, *Renaissance Concepts of Man and Other Essays*, Harper and Row, New York.

Kuhn, T.: 1968, 'History of Science', in David L. Sills (ed.), *International Encyclopedia of the Social Sciences*, vol. 14, Macmillan, New York, 74–82.

Lakatos, I.: 1971, 'History of Science and Its Rational Reconstructions', in R. Buck and R.S. Cohen (eds.), *PSA 1970. In Memory of Rudolf Carnap*, [*Boston Studies in the Philosophy of Science*, vol. 8], Reidel, Boston, 91–136.

Lindholm, L. M.: 'Burtt and Koyré as Revolutionaries', manuscript in circulation.

Lindholm, L. M.: 'The Influence of Caballa vs. the Clarity of Science in the Renaissance – a Review Essay on Yates' *Giordano Bruno*', manuscript in circulation.

Manuel, F.: 1968, *A Portrait of Newton*, Harvard University Press, Cambridge.

McGuire, J. E. and P. Rattansi: 1966, 'Newton and the "Pipes of Pan" ', *Notes and Records of the Royal Society of London*, pp. 108–144.

Pagel, W.: 1967, *William Harvey's Biological Ideas*, S. Karger, The Hague.

Stuewer, R. (ed.): 1970, *Historical and Philosophical Perspectives on Science*, [*Minnesota Studies in the Philosophy of Science*, vol. 5] University of Minnesota Press, Minneapolis.

Walker, D. P.: 1958a, 'The Astral Body in Renaissance Medicine', *Journal of the Warburg and Cortauld Institutes* **21**, 119–133.

Walker, D. P.: 1958b, *Spiritual and Demonic Magic from Ficino to Campanella*, Notre Dame University Press, London.

Wettersten, J. R. (with J. Agassi): 1978, 'Rationality, Problems, Choice', *Philosophica* **22**, 5–22.

Wolfson, H.A.: 1929, *Crescas' Critique of Aristotle*, Harvard University Press, Cambridge, Mass.

Yates, F. A.: 1964, *Giordano Bruno and the Hermetic Tradition*, Routledge and Kegan Paul, London.

Yates, F. A.: 1966, *The Art of Memory*, Routledge and Kegan Paul; also Penguin Books, 1969; (references taken from Penguin edition).

Yates, F. A.: 1967, 'The Hermetic Tradition in Renaissance Science', in Charles S. Singleton (ed.), *Art, Science and History in the Renaissance*, Johns Hopkins Press, Baltimore, pp. 255–274.

Yates, F. A.: 1972, *The Rosicrucian Enlightenment*, Routledge and Kegan Paul, London.

HENRY MARGENAU

PHYSICS AND THE DOCTRINE OF REDUCTIONISM

The most careful, critical and illuminating logical philosophical analysis of the doctrine of reductionism has been given by Mario Bunge.[1] Analysis of several sciences, primarily biology, leads him to accept what he calls moderate reductionism or "the strategy consisting of reducing whatever can be reduced without however either ignoring emergence or persisting in reducing the irreducible."

The present paper addresses the problems of physics from the reductionist point of view, reaches a conclusion similar to that of Bunge but adds further provisos and proposes a terminology more descriptive of present-day procedures than terms like reduction, level, and hierarchy.

Since the process of reduction involves important features of the method of science, indeed is often claimed to combine them all into an overarching principle, thought has to be given to the precise manner in which science operates, constructs "physical reality" in its entire complexity and cohesiveness. I have attempted to do this[2] in terms which have aroused few contradictions and have found their way into the teaching of physical science. I therefore, ask the reader's indulgence for using the technical jargon there proposed, including such phrases of protocol (P) plane, rules of correspondence, constructs, construct (C)-field and metaphysical principles. As will be seen at the conclusion of this paper, the problem of reduction has a very simple graphical representation in terms of them.

Two features of the scientific method need to be recalled and emphasized in the context of the present theme of reductionism or reducibility. One is inherent in the entire scientific enterprise and is so clearly understood by its workers that it often escapes discussion. It is the temporary character of scientific truth, its continual need of refinement, extension, and at times rejection followed by new research. Elsewhere scientific truth was called asymptotic truth, the light at the end of an infinite road of discovery, an ideal probably never within human grasp. It is part of the scientist's creed that his endeavor, as it alters or enlarges physical reality, is not a random search, but an approach to an ideal. We use the word creed, for science too has articles of faith.

A strange terminology has spread among some philosophers. They

J. Agassi and R.S. Cohen (eds.), Scientific Philosophy Today, 187–199.
Copyright © 1981 *by D. Reidel Publishing Company.*

speak of the history of science as a series of revolutions, as though the abandonment of one theory in favor of another were a unique, uncommon and upsetting occurrence. The inappropriateness of this usage is clearly evident. The "revolutions" are continual, they are zigzag parts of the asymptotic movement which never subsides. At any given time there is a prevalent scientific view (some call it a paradigm, which literally means example, and is normally used to designate a typical grammatical sequence of forms such as AMO, AMAS, AMAT, etc.). But unless it is dogmatized the prevalent view is viscous but fluid (i.e., *not* given up without resistance).

The second feature of interest in connection with the reducibility problem, in fact a feature located in the center of interest, is a metaphysical principle earlier called "extensibility of constructs." It is usually employed as a criterion for eliminating useless theories; it is an elaboration of Occam's razor and demands that of two theories, both of which explain a certain range of observations (P-facts), the one with the wider range is to be accepted. This may mean that, with a certain reinterpretation of constructs two theories covering different fields merge into a single, larger one, or it may mean that one will be rejected, or, less frequently, that deeper understanding reveals two theories to be one. An example of the latter change is the mathematical proof that Heisenberg's matrix mechanics is isomorphic with Schrödinger's use of his differential equation.

Ultimate reductionism is equivalent to unlimited extensibility of a single theory. It usually takes the form of a belief that a theory of simple phenomena (whatever that may mean!) when properly understood and refined, will ultimately explain all experience. And since physics is conventionally regarded as the simplest, or at least the most concrete science (even though the modern physicist will probably deny this belief), the most common form of reductionism is therefore physicalism, the view that everything can ultimately be reduced to physical terms. A very naive form of physicalism is materialistic monism.

The simple notions concerning reducibility or extensibility have given rise to a wide-spread terminology. Biologists and philosophers, experts who are particularly concerned with the problem of reducibility, are prone to speak of hierarchies and of levels, either term referring to domains of existence and to theories explaining them. The word "hierarchy" denotes a pyramid of power, authority or control.[3] It has some relevance for the description of biological organisms, where specific parts of a living system monitor its actions and its genetic processes with ascending degrees of power. Evolution, too, seems to lead from chaos to increasingly greater order, resembling a temporal bureaucracy, and if it is regarded as a theory attempting to explain what has happened in the past the name "hierarchy of explanation" is tolerable in reference to it. But there is mounting evidence that old-style theories of evolution, e.g., Darwinism, illuminated by Mendel's Laws, Neo-Darwinism, and the more recent astounding discoveries in genetics still lack essential explanatory elements and are at any rate better described by systems and information theory.[4] In the physical sciences the term hierarchy is rarely used, is indeed useless, and we shall ultimately dispense with it.

The term 'level', on the other hand, may refer to differences in complexity of existing, observable things or aggregates of things, or it may designate more-or-less complex theories explaining them. We should therefore at once distinguish between levels of existence and levels of explanation. We discard the former as useless or redundant for, as shown elsewhere,[5] physical reality is constructed from elements involving human perceptions and aspects of reasoning, and if the latter are more complex or more embracive, i.e., form a higher level, so does the reality that corresponds to them.

In the sequel we present examples drawn from sciences that are reasonably complete and most widely accepted, examples that allow classification in terms of levels of explanation. Later the problem of reductionism will force us to adopt a more specific language and an extension of our scope.

EXAMPLES ILLUSTRATING "LEVELS" OF EXPLANATION:

1. There are some trivial examples, cited here because they exhibit very clearly some features of more complex ones. One-dimensional analysis of space is a most primitive kind of science. Its constructs are points and lines, and its only observable is distance. The next higher level of geometry is represented by two-dimensional space, where we encounter lines as well as two-dimensional figures like polygons, circles, and so on. Length and distance are still valid observables but a new one appears: area. A one-dimensional being (a creature knowing only fore and aft and moving along a line) would have no conception of an area. That concept would be strange for it[6]. The reverse, however, is not true. A two-dimensional being can operate in a meaningful way with distances. If two-dimensional geometry is a higher level, say of complexity, then we find what might be termed continuity of explanation from above, but not from below.

A similar rise from two to three dimensions will occasion the appearance of further observables, like volume, and new constructs like solid figures, with which solid geometry operates. Here again we witness continuity of explanation from above, not from below.

One-way continuity of this type (not always from above) is typical of some of the less trivial examples which follow. More important and more general, however, is the fact that the observables and the explanatory laws at the "higher" level (e.g., solid geometry) could not be formulated, *nor even conceived* in terms of the lower level observables; additional ones are needed.

2. More interesting is the relation between Newtonian dynamics and thermodynamics (or its counterpart, statistical mechanics). Here some of the features of example 1 repeat themselves, but in a more complex way.

Newton's theory describes the motion of individual particles. Observables are position, velocity, acceleration, and force. His laws establish connections among these quantities for each individual particle. Their success is so impressive that physicists hesitated to abandon them, or rather modify them, when new discoveries in relativity theory and quantum mechanics limited their extensibility. For the present we ignore these innovations.

If we pass from consideration of individual, separately recognizable particles to a large assembly of them, such as a stationary gas or a liquid, knowledge of the Newtonian observables is no longer useful or indeed obtainable. New ones enter the scene, such as pressure, volume, temperature and entropy. In terms of them we discover the laws of thermodynamics. Note that there is, in theory at least, what was called one-way reducibility, for if position and velocity of every molecule were known precisely at a given instant, the thermodynamic variables could be calculated – in principle, though not by any human mind. The customary view follows Laplace, whose principle of causality affirms that if position and velocity of every molecule in a gas are known at a given instant, the laws of nature determine, and therefore in principle allow prediction, of these two observables for every molecule at any later time.[7]

In spite of the validity of this reasoning, however, what we said about one-way reducibility remains true so long as the laws of conservation of energy and of momentum hold during molecular collisions. Hence there is continuity of explanation from below, but not from above. It should also be noted that, as in Example 1, the "higher-level" observables are meaningless at the lower level; entropy and temperature mean no more for a single molecule than volume does with respect to a plane figure.

The upward extensibility of the new theory, thermodynamics, however, is immense. Temperature and entropy are not observables confined to finite fluids; their meaning stretches to infinite domains, in fact to the entire universe. The second law of thermodynamics implies that the entropy of the universe continually increases. Nor are ideas like entropy and temperature always material observables: we assign temperature to the radiation filling empty space, volume to the entire universe. Evidently the extent of the thermodynamic "level" is tremendous.

A word might be said about certain constructs associated with the two "levels". Those encountered in Newtonian dynamics, like mass and momentum, are all simple, direct, and easily visualized or apprehended. Their relation to everyday experience is close. But the useful ideas which enter into the constructional explanation of the laws of thermodynamics are often abstract and defy visualization. One of the most successful theories of thermodynamics is Gibbs' statistical mechanics. It operates with ideas like phase space, which is no longer three-dimensional but his six times as many dimensions as the gas has particles. It refers to an ensemble, which is a large number of replicas of the gas under discussion, but with each replica containing the individual molecules in different Newtonian states. The theory becomes highly abstract, and the philosophically minded physicist is prone to ask: Is phase space real? Does an ensemble exist? The answer, we feel, must be affirmative if these constructs play a necessary role in our explanation of thermodynamic phenomena. But do phase space ensembles "reduce" to the constructs of thermodynamics, do they "evolve" from them or do they in the strictest sense transcend them?

3. It is possible to climb a little higher on the echelon of "levels" into theories describing phenomena closely related to thermodynamics, theories in which we encounter relatively few and less surprising new variables. Thermodynamics, as described, is concerned with *stationary* fluids, although its laws permit infinitely slow expansions. The study of fluids which move with finite speeds may be regarded as a slightly higher level of analysis. To be sure, here the use of the adjective high becomes questionable. It is called elementary fluid dynamics.

In its simplest form, which we consider here, it ignores thermodynamic variables and focusses upon observables characteristic of fluid motion.

Chief among them are ρ, the fluid's density, and **I**, its current (volume of fluid crossing unit area perpendicular to its flow in unit time). The fundamental law regulating the flow involves these observables; it is the equation of continuity, which reads

$$\partial\rho/\partial t + \nabla \cdot \mathbf{I} = 0$$

It is not obtainable from the laws of thermodynamics. Here the new observables, while not derivable from temperature, entropy, pressure, etc., show a certain intuitive affinity with them, so that their appearance and their need are not surprising. It is nevertheless true, however, that they are new, and additional to the observables of thermodynamics. It is hardly proper, therefore, to speak in this instance of reducibility, not even one-way reducibility. The situation is better described by saying that the laws of thermodynamics and hydrodynamics can be brought together without conflict, that they merge into a larger theory or, to use a term we prefer, that the laws of the two "levels" are not reducible but *compatible*. Compatibility was clearly also characteristic of our previous example, the rise from Newtonian mechanics to thermodynamics.

4. Another methodically rather similar ascent from steady-state thermodynamics to a "higher level" is possible. It takes us to situations where a fluid shows differences in temperature so that a new phenomenon, heat flow, occurs. Again, we encounter new observables appearing in the fundamental law of the heat-flow equation, which reads

$$\partial T/\partial t = (k/\rho s)\, \nabla^2 T + A/\rho s$$

Here T is the old concept of temperature which need no longer be uniform; $T = T(x, y, z)$. The new observables are:

Rate of heat generation per unit volume, $A(x, y, z)$
Specific heat, be s
Thermal conductivity, k

Some of these have no meaning for a steady state, are not reducible to its "level," and it is a moot question whether their need in the analysis – and perhaps many others – could have been foreseen with certainty. Again, while there is no reducibility in any normal sense, there is compatibility: two theories, outwardly distinct, merge into one.

5. In the last two instances the meaning of the term "level" has become diffuse, though it remained faintly meaningful. We now present an example

from another field where, perhaps, its meaning is restored, and where again we witness what we called one-way reducibility. It is similar to the transition from Newtonian mechanics to thermodynamics or its explanatory counterpart, statistical mechanics. And like that transition, it opens enormous vistas into unexplored realms. We refer to electromagnetism.

Its earliest and simplest description involved the observables electric charge, magnetic pole strength, together, of course, with the ubiquitous Newtonian concepts of time, space (distance), speed, and acceleration. Charge and pole strength were new; the basic law was Coulomb's. In essence, it is an extension of Newton's physics resulting from the introduction of new sources of force. After Coulomb the theory became enriched, not only by new constructs like the ether and Faraday's lines of force, but also by the addition of new observables like electric and magnetic field strength, which are alien to Coulomb's Law. Nor could anyone have foreseen the connection between moving charges and magnetic fields which was discovered by Oersted.

The rise from the level of Coulomb's laws for electric charges and magnetic poles to its highest level of complexity occurred in the discovery of Maxwell's equations, which are laws in terms of wholly new observables known as electric field strength, displacement, magnetic field strength, and induction. In terms of them the laws attain an unexpected measure of elegance and symmetry and open vast perspectives previously closed. The concept of what we now call a radiation field evolved in time from the law of force between charges and poles. Here is a case where elaboration, if that is the correct term for the opposite of reduction, opened ranges of physical reality which, like entropy and temperature, defied adherence to matter, filled space, and extended to infinity. Even the medium which was thought to be their carrier, Faraday's ether, was shown to be nonexistent by experiments leading to the theory of relativity.

Clearly, the observables needed in Maxwell's equation could not have been discovered from an analysis of Coulomb's law, even though Coulomb's law is logically implied by them. One feature, however, is common to the present example and a former one, the ascent from Newtonian mechanics to thermodynamics: if the observables of the constituents of a fluid were known at every instant, the observables of the fluid itself (temperature, entropy, etc.) could in principle (though not actually) be deduced; the reverse is not true. If the observables entering Coulomb's law were known for every one of many charges producing an electromagnetic field, that field could be predicted. But if a complicated field were given, the locations and velocities of many charges producing it are not inferrable, for there are different charge distributions that can give rise to the same electromagnetic field (at least within any finite region of space). Again, there is continuity of explanation from below, but not from above. This, as already pointed out, is a very special form of reducibility, indeed one which hardly deserves the name.

Before going on to more complicated examples let us review some of the preceding ones from a slightly different point of view. This may provide further evidence for the inadequacy of terms like level, hierarchy and continuity of reduction, evidence which casts further doubt on the meaning which the term reductionism claims to possess.

Consider once more the first example. Level *A* shall designate two-dimensional, *B* three-dimensional geometry. *AO* stands for *A*-observable, *BO* for *B*-observable. We then find that

(a) the *BO*'s mean nothing in *A*, and

(b) the *AO*'s are meaningful, can be "foreseen" from the *BO*'s.

If, as would generally be supposed, the "*B*-level" is higher than the "*A*-level," this state of affairs would be described by saying that the higher level observables determine the lower ones! There is continuity of explanation from *above*.

In our second example, level *A* shall designate Newton's theory of motion as it applies to individual molecules. Observables are position, velocity and force. Level *B* is a gas at rest enclosed in a vessel; observables are temperature, pressure, etc. To be sure these have meaning only if certain restrictions are placed upon the *AO*'s (Maxwellian distribution); but in a short time these restrictions realize themselves automatically. Reductionists would speak of *B* as the higher level. Here we find

(a) the *BO*'s mean nothing in *A*, and

(b) the *BO*'s depend on, can be "predicted" from *AO*'s.

This state of affairs might be described by saying that the lower level observables determine the higher ones: There is continuity of explanation from *below*.

We conclude from these two examples, which can be amplified by numerous others, that continuity of explanation is not a one-way matter. This suggests the lack of meaning of the terms above and below, indeed of the term levels.

We now examine the relation between the pre-Maxwellian theory of electrodynamics which operates with observables like magnitude, sign, positions, velocities, acceleration, etc. of electric charges, calling it "level"

A, and the Maxwell-Lorentz theory of radiation, *B*. Here the *AO*'s are meaningful in *B*, are in fact among the *BO*'s. But some of the *BO*'s are meaningless in *A*. We encounter what one might be forced to call partial continuity of explanation from *A* to *B*, complete continuity from *B* to A. The concept of level becomes confused. On the other hand, were we to identify *A* and *B* we should have no room for the radiation field which belongs only to *B*. This example shows perhaps more clearly than the others that the concept of reduction must be replaced by *transcendent elaboration with continuity*, without reference to levels and hierarchies.

This example is interesting for a special reason. Having transferred attention from *A* to *B*, one concept, namely the radiation field (whose observables are electric and magnetic field strength) takes on overwhelming importance, detaching itself as it were from its originators, the moving charges. First, because of its finite velocity, its present state depends on a previous condition of the electric charges (retarded potentials); it can even exist when the charges are annihilated. To complicate matters still further, the destruction of the charges (e.g., electron – positron annihilation) creates its own radiation field which is superposed upon the former, then pervades space in the absence of the material charges but retains its identity – though its observables will change their values in time.

A most important, cosmic example of this has occupied the attention of physicists and astronomers quite recently. The "big bang," presumed to be the origin of the present universe, must have occurred some 15 billion years ago. Precisely what happened could be inferred only from considerations of certain likely astrophysical processes which included a cataclysm among charged particles; most probably the destruction of many of them. Whatever its details, whatever the precise nature and number of its no longer existing originators, the radiation field remained, has been identified and measured. The achievement has been honored by the award of Nobel prizes. A person cherishing a belief in immortality might be tempted to regard this as a physical example of the survival of a material entity upon the death of matter.

In the present context, however, it shows the irrelevance of every kind of reductionism which does not involve transcendence, does not allow for the possible role of unpredictable entities in any given mode of explanation.

Our examples have shown the term hierarchy to be generally meaningless. Certainly this concept does not characterize physical theories. Quantum mechanics is an elaboration of classical physics. Its concepts are different from those of classical mechanics, indeed more elaborate, less simple. They reduce, however, to the latter in the molar realm. If physics harbored hierarchies in the simplest sense which refers to size of objects, the class of atoms and elementary particles ought to lie below that of ordinary objects; yet quantum theory is far more complex, and it uses in elementary form some of the ideas, properly modified, of classical physics. But both are compatible. Viewed from above, the observables of quantum mechanics cannot be inferred; from quantum mechanics, however, the observables in the molar world can be understood and visualized. In the language of hierarchical relations the two cases are clearly opposite. Similar aberrations appear in some of the other examples we have discussed. We therefore dismiss the term hierarchy from our discourse altogether.

The same difficulties can be seen to attach to the word level, which is also common in previous discussions of reductionism. But as we have seen, it is often difficult to say which of two levels is higher and which is lower. There is greater or lesser complexity in different modes of explanation, which often means that the constructs involved in one mode are more difficult or less familiar. Perhaps the only sciences for which the terms hierarchy and levels of organization remain meaningful are descriptive botany and zoology. In general, science is not a bureaucratic system, not a hierarchy of levels. Even such terms as higher and lower, inner and outer, may be meaningless.

Henceforth we shall mean by the word level a *mode* or domain *of explanation*, and by reduction a reversal of *transcendent but compatible elaboration.* As mentioned at the outset, in these deliberations I have profited from the analysis of Bunge, who has subjected reductionism to logical scrutiny. He, too, criticizes the unprincipled use of the term but retains it in the limited region where it does in fact apply: in a set of domains characterized by the size of their constituents, i.e., the levels recognized chiefly by biologists which can be arranged in the rather obvious sequence of cells, organs, organisms, populations, ecosystems and biosphere.

Here we have been interested, not so much in size of constituents as in the theories which regulate their behavior, in the observables called into being as we pass from one domain to another, in their complexity and their coherence. What matters in our approach is the interconnectedness between fundamental theories or, as we have occasionally expressed it, between modes of elaboration, between old and new observables which appear in them. Hence there is this difference between Bunge's (entirely correct) analysis and the present: He emphasizes parts and sizes, while we consider the nature of observables and the detailed nexus of their interrelation, the epistemological point of view being that presented in an earlier more detailed account.

The appearance of new observables Bunge chooses to call emergence, a term whose use he shares with others, among them Popper[8] though not always with precisely the same meaning.

For reasons already stated we replace it by the stronger word *transcendence*. Strictly speaking, what emerges was already there, invisible and often unexpected. We, however, wish to emphasize the uniqueness, the novelty of the newly appearing observables, their creation by a new theoretical approach, the scientist's inability of even conceiving them from a domain of explanation in which they have no meaning.

In a study of the literature on reductionism one finds extreme confusion arising from the rather arbitrary use of innumerable words that are nearly synonymous. Since our view is close to Bunge's, which preceded it, we thought it appropriate to state explicitly how our terminology is related to his.

Conclusion[9]

The meaning of reduction, which we replaced by the term transcendent elaboration with continuity, is most easily explained in terms of our P-plane-C-field figure. For what it implies is simply an advance to the left in the C-field, a further departure from P, usually coupled with an extension of the range of constructs parallel to the P-plane. In this sense reduction means nothing more than systematic increase in understanding the world.

From P, where we observe masses, positions, velocities, and accelerations, observables which, after being defined operationally, we assign to systems called bodies, we proceed into C. The first law which combined these observables satisfactorily was Galileo's: all bodies fall with an acceleration of about 32 ft per second per second. It related to that part of the P-plane which concerned observations on bodies falling near the earth's surface.

This law was greatly amplified, with an accompanying enlargement of its range of application, when Newton formulated his three laws of motion and gave a new and more adequate operational definition of the concept force. Prior to Newton, the meaning of force had been controversial. In this one, simple respect Newton transcended previous theories, in which his idea of force played no role. This addition, however, increased the range of application of mechanics vastly, for it allowed the explanation of Kepler's laws, thus becoming applicable to the crucial problems of astronomy of the day.

Further elaboration occurred in the centuries following Kepler, but no essential elements transcending Newton's theory were introduced until Einstein advanced his theory of universal gravitation. This brought into line several phenomena contradictory to certain P-experiences (e.g., the precession of the perihelion of mercury), and the crucial construct which transcended previous theories was Riemannian space. It introduced the coefficients of a metric as new observables. To be sure, non-Euclidean geometries were already known to mathematicians, but physicists regarded them as interesting curiosities not suggested by, and hence transcending, their known laws of motion.

The abstractness of the general theory of relativity places it far to the left in the C-field. In fact, it marks its present boundary and acts as a basic postulate, beyond which we cannot go. Increasing abstractness accompanies the passage to greater distances from the P-plane, and an enlargement of the domain of observable phenomena is established.

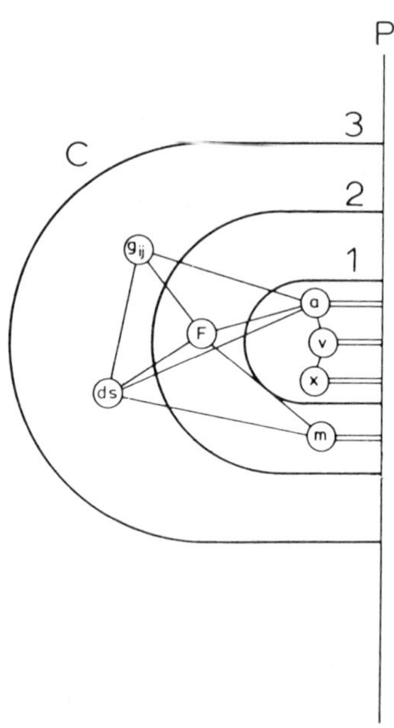

Fig. 1.

All this is represented simply in our diagram, Figure 1. Enclosure 1 maps Galileo's theory; enclosure 2 maps Newton's, enclosure 3 Einstein's theory. In ordinary, inadequate reductionist parlance the figure represents 2 basic reductions in physics. Notice the area outside of enclosure 3. Nobody can say with any assurance that it may not some day contain constructs like mind, or consciousness and others of interest to psychologists.

In writing this paper the author has profited from discussions with L. LeShan, to whom he wishes to express his acknowledgments.

Yale University

NOTES

[1] The topic appears in numerous of Bunge's books. Special reference is here made to his paper: 'Levels and Reduction', *Am. J. Physiol.* **233** (3), 1977, and to other articles there cited.

[2] See *The Nature of Physical Reality* (Ox Bow Press, New Haven, 1978).

[3] Specifically ecclesiastical control, for the word 'hierarchy' contains *hieros* (sacred) and *archein* (to rule).

[4] Cf. E. Laszlo, *Introduction to Systems Philosophy* (Gordon and Breach, New York, 1972).

[5] Similar arguments will be found in M. Bunge, 'The Metaphysics, Epistemology and Methodology of Levels', in L. L. Whyte, A.G. Wilson and D. Wilson, eds., *Hierarchical Structures*, American Elsevier, New York, 1969.

Approaches to the view we are about to present appear in *Main Principles of Modern Thought*, ch. 2, H. Margenau, ed. (Gordon and Breach, New York, 1974), and in Laszlo, op. cit.

[6] If it lived in a 3-dimensional world and had the power of reason, it could arrive by systematic study at the concepts of the higher level.

[7] Max Born (private conversation) has questioned this view. He pointed to the human element that enters the argument. To assume *exact* knowledge is a divine prerogative. Man, who must obtain knowledge of position and velocity by empirical means, must allow for errors in their values. In other words, he must also assume an error, say Δx_i, in the determination of the position of the *i*th molecule, and an error Δv_i in its velocity. Thus, as a basic epistemological principle, errors must be included in every empirical magnitude. There is no true value in the absolute sense. When a scientist uses this term, he refers to the arithmetical mean of a large number of observations. These errors, Born pointed out, increase in every collision between two molecules, and one can easily show that infinitesimal Δx_i's and Δv_i's will grow into extremely large errors in a very short time, so that within a microsecond the original knowledge of positions and velocities of the molecules composing the atmosphere of a room of ordinary size would be completely wiped out.

[8] Popper, K. R. and J. C. Eccles: 1977, *The Self and Its Brain*, New York, Springer.

[9] I am indebted for this material to Professor Harold Morowitz.

JOHN MAYNARD SMITH

SYMBOLISM AND CHANCE

Perhaps the hardest concept in science for most of us to think about is that of randomness. The theme of this essay is that the difficulty arises because of a conflict between thinking in terms of probabilities and another kind of thinking which I will call symbolic, but which could as well be called analogical.

In his book, *Rethinking Symbolism*, Sperber describes how, when he started anthropological field work among the Dorze, he selected topics for further study. He writes,

> Someone explains to me how to cultivate fields. I listen with a distracted ear. Someone tells me that if the head of the family does not himself sow the first seeds, the harvest will be bad. This I note immediately.

And again,

> When a Dorze friend says to me that pregnancy lasts nine months, I think 'Good, they know that.' When he adds 'but in some clans it lasts eight or ten months', I think 'That's symbolic'. Why? Because it is false.

How do people come to believe things which are in some sense "false," but which nevertheless convey important truths?

The answer seems to be that if you tell a child (or, often enough, an adult) a story which is clearly not true in a literal sense, the child does not react by saying to itself "that is an obvious lie". Instead, the child seeks for some symbolic meaning in the story, and does so by seeking a formal analogy between the structure of the story and other structures already familiar to it. This, I think, is the main insight which redeems the structuralist approach to anthropology from total incomprehensibility.

To illustrate this point, let me take an example from Sperber. The Dorze assert that leopards are Christians, and therefore that it is unnecessary to guard one's flocks against them on fast days of the church. (Of course, in practice, the Dorze do not relax their guard on fast days.) Sperber suggests the following interpretation. The Dorze make sense of the story by forming the analogy, "leopards are to hyenas as the Dorze are to their neighbours." They are led to the analogy by the following facts: the leopard and the hyena are the two common large carnivores in the area; the former, unlike

J. Agassi and R.S. Cohen (eds.), Scientific Philosophy Today, 201–206.
Copyright © 1981 *by D. Reidel Publishing Company.*

the hyena, kills all that it eats, and eats only fresh meat; the Dorze differ from their neighbours in that they only eat freshly-killed meat; the Dorze, unlike their neighbours, are Christians. Hence, by saying that leopards are Christians, something is being said about the relation between the Dorze and their neighbours.

I do not know how correct his interpretation is, and it does not much matter. The assumptions behind the interpretation, however, are important, and I think they are correct. They are as follows:

(i) Human beings attempt to make sense of "stories" which they know to be untrue in a literal sense.

(ii) The main method adopted is to seek for formal analogies between the new story and things already known (which may be things accepted as literally true, or other things accepted as only symbolically true).

(iii) The "truth" so acquired is moral or aesthetic rather than literal.

(iv) The force of a moral belief is greater if it has been acquired by an individual through an active attempt to perceive meanings in symbolic inputs, rather than (or as well as) by a set of explicit rules such as the ten commandments.

In case it should be thought that this process is peculiar to societies less technological than our own, remember that many of the stories we read aloud to small children are obviously untrue, even to the child, because they are about talking animals, and even (in a series of stories I read to my own children) about talking steam engines.

What is happening in symbolic thinking is that we are making sense out of nonsense; in Sperber's words, "Symbolic thought is capable, precisely, of transforming noise into information." The same thing happens when we interpret a Rorschach ink blot. There is nothing surprising about this. It is an inevitable consequence of the way we perceive reality. It is now generally appreciated that in order to perceive things we form hypotheses about what those things might be.

The recognition of structural similarities between things is a primary method of understanding not only myths but also reality. It is a method which reached its highest level of sophistication in China. The Chinese theory of five elements – wood, earth, metal, fire, water – in some ways seems a more promising basis for science than the four-element theory of the Greeks, because it was a theory of transformations, in which each element in the cycle overcomes the next. However, the five elements became the basis of a complex system of analogies; to each element there corresponded a planet, a colour, a taste, a family relationship, a point of the compass (the centre counted as a fifth), and so on. A still more complex system of analogies was developed in the Book of Changes, the I Ching. Here, every conceivable object or event is allocated to one of the sixty-four hexagrams.

One can see the I Ching as a method of classification (Needham suggests that it is a reflection of the bureaucratic mind), and of ascribing symbolic meanings. However, it is based on and helps to reinforce a view of the world so different from that of modern science that it is hard for someone with my background to grasp it. Some idea of that view can be given by two quotations from Needham's *History of Chinese Science*. He writes,

> It was like the spontaneous yet ordered movement of dancers in a country dance of figures, none of whom are bound by law to do what they do, nor yet pushed by others coming behind, but cooperate in a voluntary harmony of wills.

Also,

> If the moon stood in the mansion of a certain constellation at a certain time, it did so not because anyone had ever ordered it to do so, even metaphorically, nor yet because it was obeying some mathematically expressible regularity depending upon such and such an isolable cause – it did so because it was part of the pattern of the universal organism that it should do so, and for no other reason whatsoever.

Needham sees this organic view of the world as one which scientists will have to acquire as they come to grips with biology. This may be so, although it would require a major shift in outlook; at present, biology seems set on a very different course. However, I did not mention the Chinese world outlook because I wanted to discuss this possibility. Rather, I mention it to illustrate one way in which thinking based on a recognition of analogies can develop. In Western science, the road has been different. We recognize the analogy between, say, mechanical and electrical phenomena by recognising that they are described by the same differential equations. Mathematics has become the language of analogy in science; it makes possible the recognition of much more complex structural similarities than the classical structuralist $A : B :: C : D$. Of course, much analogical thinking in science, particularly in the early stages of hypothesis formation, is verbal or pictorial rather than mathematical. The aim, however, is usually to replace such loose analogies by mathematical description; then we can kick away the analogical ladder by which we first reached our hypothesis.

It is time that I passed from a general description of symbolic and analogical thinking to the problem of randomness. As it happens, the I

Ching provides a bridge. Although it was not its primary purpose, the I Ching has been extensively used, in China and more recently in the West, as a device for foretelling the future. A hexagram was selected by a process comparable to the drawing of lots, and its correlations used for divination. This use of what most scientists would regard as a randomising device in divination is, of course, widespread. The petals of a daisy, the tealeaves in a cup, the livers of birds, the cracks in a scapula thrown in the fire, the fall of dice or cards – all have been pressed into service. The logic is unassailable. If indeed nothing happens by chance, if every event, like part of a holograph, reflects the whole, there is nothing unreasonable about the procedure. It is only necessary to learn the nature of the correlations, and divination becomes a possibility.

It should now be clear that the concepts of randomness and of symbolic thinking are incompatible, at least in their more extreme forms. As a definition of randomness, I suggest that to a scientist a random event is an event into whose causes it is not at present efficient to enquire. This leaves open the question of whether the causes are thought to be in principle unknowable, as in quantum theory, or whether, as in theories of mutation in evolution, the causes of mutation can be and are studied in molecular genetics, but are held to be irrelevant to the evolution of adaptations. It also leaves open the question of how a random sequence of events is to be defined mathematically. For the present purpose it is sufficient that a scientist, by treating some set of events as random, is asserting that there is no structure in these events which is relevant to the phenomena he is attempting to understand. Most of us would assert that the distribution of tealeaves in a cup, or the fall of the milfoil sticks in divining by the I Ching, is irrelevant to the future.

If the function of symbolic thinking is to transform noise into information, most scientific theories rest on the assumption that, for any particular purpose, some events (and usually most events) are indeed noise, from which no information can be derived. There is no *a priori* reason for regarding the scientific approach as more reasonable; if anything, the opposite is the case. Further, an obstinate refusal to accept some input as random has sometimes been scientifically fruitful; where would we be today if our predecessors had accepted that the wandering of the planets was random?

Our justification, in so far as we have one, for adopting randomness as a component of scientific theories is that such theories have proved to be fruitful. Darwin is interesting here. In his book, *Darwin on Man*,

Gruber shows how, when Darwin first sought for a mechanism of evolution, he saw the problem as one of explaining the nature and causes of mutation (i.e., of the *origin* of new variation). In his final theory of evolution by natural selection, the *existence* of variation has in effect become a postulate, not a problem. Darwin's theory does indeed treat mutations as "random" in the sense defined above, as events into whose causes it was not efficient to enquire. Darwin himself was well aware that an understanding of evolution did ultimately require an understanding of mutation; this may have been one of the reasons for his long delay in publishing. With hindsight we can see that there was no way in which Darwin at that time could have made much progress in a study of mutation, so his achievement did depend on his treating them as random.

Given the universality of symbolic thinking, and its incompatibility with concepts of randomness and probability, it is worth asking whether the relatively slow development of the mathematical theory of probability and its applications arose from this incompatibility. The basic idea of the "fundamental probability set" emerged *circa* 1550 in Italy in the work of Cardano and Ferrari. A hundred years later Huygens, basing himself on the work of Fermat and Pascal, published the first methodical treatment of the calculation of probabilities. It was not until the middle of the nineteenth century that scientific theories were formulated in which the concept of probability was central.

I find it hard to account for this slow development either by the intrinsic mathematical difficulties, or by the absence of any external demand for the theory. When the theory did develop, it did so in response to questions about games of chance. There was plenty of gambling for money in the ancient world, and a surprisingly accurate knowledge of the probabilities of various throws. However, this knowledge appears to have been empirical; if so, people must have been prepared to spend a long time throwing dice and astragali to get it. The mathematical calculations required to get the information, given a clear idea of how to combine probabilities, would not have been very difficult. The block seems to have been conceptual. However, my remarks are based not on serious research, but on superficial reading. I want only to suggest that the problem would repay further study.

I cannot end without commenting on one other aspect of the conflict between the symbolic and the random. One cannot spend a lifetime working on evolution theory without becoming aware that most people who do not work in the field, and some who do, have a strong wish to believe that the Darwinian theory is false. This was most recently brought home to me when my friend Stephen Gould, who is as convinced a Darwinist as I am, found himself the occasion of an editorial in the *Guardian* announcing the death of Darwinism, followed by an extensive correspondence on the same theme, merely because he had pointed out some difficulties the theory still faces.

Why should this be so? It happens because people expect evolution theory to carry a symbolic meaning which Darwinism manifestly fails to do. All previous societies have had myths about origins. These myths, like the Dorze belief that leopards are Christians, were not intended to be taken literally. The truth they conveyed was a symbolic one, contributing to an understanding of the purpose of the universe and of man's role in it. Darwinism is also an account of origins, but it is intended to be taken literally and not as a myth. As Bernard Shaw said so eloquently in the preface to *Back to Methuselah*, Darwinism is a rotten myth.

The University of Sussex

REFERENCES

David, F. N.: 1962, *Games, Gods and Gambling*, Cambridge University Press, Cambridge.
Gruber, H. E.: 1974, *Darwin on Man*, Wildwood House, London.
Levi-Strauss, C.: 1966, *The Savage Mind*, Weldenfeld and Nicholson, London.
Needham, J.: 1956, *Science and Civilization in China*, Vol. 2, Cambridge University Press, Cambridge.
Sperber, D.: 1975, *Rethinking Symbolism*, Cambridge University Press, Cambridge.

CARLOS-ULISES MOULINES

A STUDY IN PROTOPHYSICS

In his *Foundations of Physics*, Mario Bunge for the first time introduced his notion of protophysics as a special field located on the borderline between physics and the philosophy of physics – a field of interest both to working scientists and philosophers. I think it is not exaggerated to evaluate Bunge's inception of protophysics as a crucial step opening new perspectives in present-day philosophy of science, and not only in the narrower field of philosophy of *physics* proper: We could adapt the prefix "proto-" in Bunge's sense to any other scientific discipline.

Bunge's treatment of protophysics in the above-mentioned work was more programmatic than systematic. He has refined and further developed his ideas on the subject in subsequent writings, especially in the third and fourth volumes of his recent *Treatise on Basic Philosophy*. There he expounds his "general systems theory" (actually an outgrowth of a *part* of his former protophysics) with more detail. However, I think his basic programmatic statements in *Foundations of Physics* still provide the best clue to get a synoptic view of what the content of protophysics is or should be. Consequently, for the first section of this essay, where I discuss protophysics in general, I shall mainly rest on *Foundations of Physics* for my proposed interpretation of what the content and method of protophysics in general should look like. It is hoped that this reinterpretation will not appear to be incompatible with Bunge's original intentions. In the second section I shall try to make a contribution of my own to the protophysical program in Bunge's sense by setting up an axiomatic basis for proto-*thermodynamics*. This case study is intended to give support to the contention that a wide and stimulating field is opened to philosophy of science by the very idea of protophysics. By applying the general idea to concrete cases of scientific disciplines like thermodynamics it is hoped that the fruitfulness of the general program will appear more conspicuously. As far as I know, Bunge himself has never attempted to engage in a systematic study of the protophysical foundations of thermodynamics—perhaps because he thinks that this particular case cannot be dealt with in a "clean" way. I would feel very satisfied if the second part of the present study would

J. Agassi and R.S. Cohen (eds.), Scientific Philosophy Today, 207–224.
Copyright © 1981 *by D. Reidel Publishing Company.*

persuade Bunge, among others, that a fruitful and rigorous treatment of protothermodynamics is indeed possible.

I. PROTOPHYSICS IN GENERAL

Much of present-day philosophy of science is devoted to the task of constructing a general concept of a scientific theory. This is the general tool needed to attain the goal of many philosophers of science, viz. to provide adequate logical reconstructions of single scientific theories. By applying a suited general theory-concept to existing expositions of theories we would be able to explicate the implicit structures of particular theories. Theories are thereby viewed as the basic units of a philosophical analysis of science. The crucial question, then, appears to be this: What metatheoretical approach do we have to take in order to get at a formally and intuitively satisfactory theory-concept which is suited to our task of identifying and reconstructing single scientific theories? We have got a number of alternative approaches. The formal axiomatics of Carnap and his followers, the informal set-theoretic approach of Suppes and his collaborators, and more recently the rather sophisticated metatheories of such authors as Sneed, Ludwig, or the Polish group of methodologists – they are all different (but not necessarily incompatible) ways to respond to that question.

I take this kind of metatheoretical undertaking in the philosophy of science (and especially in the philosophy of physics) certainly to be worthwhile. They have led us to ask some fruitful questions and to gain some deep insights into the (implicit) structure of theories. On the other hand, we should not forget that this kind of approach can help us only in the analysis of the *structures* of theories – not of their *content*. More precisely, those approaches do not help us in clarifying what I would like to call questions of the *physical semantics* and *ontology* of scientific theories. By this I mean the following. When we undertake a structural reconstruction of a theory, we are only interested in the way the theory's basic concepts and principles all hang together – what the form of the building is. But we cannot make any systematic statements about the piece of "outside world" the theory is supposed to be applied to or to be related to. This is primarily a question of the ontology presupposed in the given theory, and secondarily a question of its semantics – since in order to answer questions about the semantics of a physical theory, we have to know something about the reference of its concepts, and this can, in turn, only be done by clarifying the theory's ontological presuppositions.

Now, this kind of ontologico-semantical aspects of physical theories are not (intended to be) covered by the usual theory-concepts of present-day formal philosophy of science. The reason is that they escape any approach which restricts itself to the level of theories as units of analysis. To treat those aspects adequately we need go farther in two opposite directions: to the level above as well as to the level below theories.

To be a bit more precise, a thorough discussion of the semantics and ontology of physics necessarily leads us to leave the restricted frame of single physical theories and to attempt an explication of concepts and principles of a more general nature than those belonging to single theories, such as the notions of space, time, system, process, stochastic process, cause, combination, etc., and the principles governing them. Such concepts and principles are not the "property" of particular physical theories; rather, they either "underlie" or "overarch" theories – "underlie" them in the sense of constituting their implicit presuppositions or "overarch" them by unifying different theories into a more general picture of reality. Because of their general nature they are essential to the constitution of those families of theories we call scientific disciplines. But, for this very reason, they are only implicit in most standard expositions of physics. In fact, these all-pervading concepts and principles do not properly belong to physics, at least in the sense that most of them do not appear as belonging to the extant theories of physics. In a sense, they are prior to physical theories, since they systematize their ontological and semantical presuppositions. Therefore, it might seem appropriate to label their study as "protophysics." Protophysics in this sense is a (much neglected) branch of the philosophy of science that we should distinguish from the philosophy of (existing) physical theories.

I think this characterization agrees with Bunge's own notion of protophysics. In *Foundations of Physics*, Bunge defines the object of protophysics as "a quaint collection of nonformal yet generic and important principles and theories" [1967, p. 85]. More concretely, protophysics studies general presuppositions common to a whole range of physical theories – principles not belonging specifically to any particular theory. They constitute *necessary*, but by no means *sufficient* conditions for setting up full-blown physical theories: "... protophysics leads nowhere in particular though without it we get nowhere" [1967, p. 85]. Furthermore, we should not think of protophysics as a systematic theory of *physics-in-general*: "...protophysics is fragmentary rather than unitary: it is not a single theory or even a set of contiguous theories but an heterogeneous assortment of principles and theories" [1967, p. 85].

Nevertheless, this heterogeneity of the content of protophysics does not mean conceptual confusion. Indeed, we can distinguish several "branches" within protophysics and treat them in a more or less systematic, sometimes even axiomatic way. This is what Bunge tries to do, at least programmatically, in his work. It will be helpful to give a more detailed picture of the content of protophysics as Bunge views it.

The most difficult portion to characterize is what Bunge calls "Zerological Principles," which apparently represent the most "philosophical" area of protophysics. In *Foundations of Physics* Bunge offers two sorts of such principles: a first group is of an ontological kind and could be considered as the fundamentals of a general theory of causality; the second group is metatheoretical in kind: They are regulative principles of theory construction.

If we leave aside these "zerological principles" (whose status is not clear to me – nor, as far as I can see, to Bunge himself), the three main areas of protophysics according to Bunge are:

(1) Probability theory
(2) Chronology, physical geometry, and generally any theory of spacetime
(3) Protophysical systematizations proper, which in turn subdivide into:
    (a) Analytical dynamics (it would be more adequate to say: abstract frameworks for "dynamical" theories)
    (b) General systems theory (mereology).

With respect to the first area, I have serious doubts that it is appropriate to include it within protophysics. Certainly, probability theory is very fundamental for a wide range of physical theories in the sense that physics essentially uses probabilistic constructs in very different contexts. But the same goes for many other abstract theories of pure mathematics, like group theory, topology, and others, that go into a great deal of different physical theories just because of the reason that, due to their abstract nature, they are susceptible of many different physical interpretations. I think the inclusion of probability theory into protophysics is misleading and does not correspond to the intended content of this new field.

On the other hand, it is undoubtedly meaningful to ascribe any general theory of space and time to protophysics, since spatial and/or temporal constructs are in some way presupposed by practically any physical theory. It is just for historical grounds that the study of space and time has not usually been subsumed under the general label of protophysics: Theories of space and time are much older than the idea of protophysics as such. Indeed, this is the only branch of protophysics that has been really developed by philosophers and scientists. Since there is already a vast and well-known literature on this subject, I think it is more interesting to concentrate now on the still rather unexplored parts of protophysics to which Bunge has directed our attention. Lacking a better label for this latter area let us call it "protophysics proper."

In *Foundations of Physics* two different areas are distinguished within protophysics proper: one dealing with qualitative, elementary concepts and operations, another introducing highly abstract mathematical constructs which systematize common features of a wide range of physical theories. The first area is mainly represented by what is usually known as "mereology" in the philosophical literature.[1] The second area is mainly devoted to the construction of what we might call "abstract analytical theory-frames."

One example of work done within mereology, i.e., "the theory of the most general relations among objects of any kind" [1967, p. 112] is Bunge's clarification of the notion of a *system* in general, which he started in *Foundations of Physics* and has culminated in his recent *Treatise on Basic Philosophy*. Another example, dealing with a more concrete case, is the study of protothermodynamic systems and operations offered in the second section of the present essay. An example of work done within the second area is a generalization of the Hamiltonian "formalism" due to J.L. Martin, which Bunge discusses *in extenso* in *Foundations of Physics* [pp. 113–127]. This is apparently the most general abstract frame for physical theories to date. The Hamiltonian, Lagrangean, and Jacobian formalisms are specializations of it, though they by themselves are still *frames* only, not full *theories*. We get real theories out of them only when we provide a full interpretation of the basic notions, e.g., a mechanical interpretation leading to a mechanical theory. The explication of a thermodynamic theory-frame as expounded in Moulines [1980] could also be understood as an attempt along these lines.

Both mereology and the construction of analytical theory-frames are of much relevance to a discussion of ontological and semantical issues in the philosophy of physics. The construction of theory-frames clearly helps in the logical reconstruction of inter theoretic relations (theory-frames are by their very nature inter-theoretical entities), and this in turn should be regarded as an essential part of physical semantics – as has been argued at length in Moulines and Sneed [1979].

As for mereology, since it is especially devoted to physical systems and physical operations, it is obvious that it should have a semantical and ontological import for physics. Certainly, mereology has been usually neglected by working scientists and left to philosophers with a strong tendency towards operationalism or radical empiricism. In fact, an operationalist could be described as someone tending to substitute mereological protophysics for theoretical physics – because of the philosophical prejudice that qualitative notions of systems and operations on systems are safer and more fundamental than quantitative functions of full-blown theories However, Bunge's example clearly shows that this impression is misleading, i.e. that mereology has a right of its own quite independently of whether or not one subscribes to an operationalistic credo.

At this point, it may be appropriate to add a word of terminological caution to avoid misunderstandings. The prefix "proto-" in "protophysics" could be interpreted as an attempt to provide an epistemological or methodological foundation for existing physical theories very much in the sense of the operationalistic view that has just been criticized. This would be a misunderstanding. The aim of protophysics as propounded here should not be confused with any sort of "fundamentalist" reconstruction of science in the sense of trying to give absolutely firm and unmodifiable foundations to physics. On the contrary, the development of protophysics depends at least as much on the development of physical theories as these depend on the former. In this respect, it is important not to confuse Bunge's (and my own) use of the term "protophysics" with the notion of a "Protophysik" as introduced by the German group of philosophers of science known as the "Erlangen school." The "Protophysik" of Lorenzen and his followers is a kind of sophisticated version of operationalism in that it tries to provide an absolute foundation of the theoretical concepts and principles of physics by means of a systematic study of operational notions which are supposed to be epistemologically "safe." The tenor of the "Protophysik" of the Erlangen school could be summarized in the slogan: "Physics is suspect unless it becomes founded on the prior and safer basis of 'Protophysik.'" Nothing could be farther removed from Bunge's intention than this sort of fundamentalist attitude. Bunge is well aware of the hypothetical, tentative, and thoroughly theoretical nature of protophysics as well as of science in general. Protophysics in his sense can only be a supplement to, not a substitute for, the reconstruction of full-blown physical theories.

II. PROTOTHERMODYNAMICS

Let us now attempt a protophysical study of the foundations of thermodynamics within an exact conceptual framework. This kind of study is what we might call "protothermodynamics" (as a special branch of protophysics in general). Actually, to be more precise the present study should be labelled: "mereological protothermodynamics," since protothermodynamics as such should not only encompass the mereological investigation of the basic thermodynamic concepts but also the building of a general frame of a high-level degree of abstraction analogous to "analytic dynamics" in mechanics. It seems possible to build such a highly abstract frame for thermodynamic theories as well. Tisza has offered a frame of this sort in his *Generalized Thermodynamics* (cf. Tisza [1966]). More recently, I have tried to do a similar sort of thing along somewhat different lines (cf. Moulines [1980]). But it is not to this part of protothermodynamics that the present study is devoted. Our concern now is with the mereology of thermodynamics; more specifically: with the "qualitative" notion of a thermodynamic system and the "qualitative" relations among thermodynamic systems. ("Qualitative" here means: "non-metric and directly related to physical operations.") Surprisingly enough, the mereology of thermodynamics has been a much neglected field – still more neglected than protophysics in general. This is surprising because, by its very essence, thermodynamics itself is one of the scientific disciplines where the importance of mereological considerations is most apparent. The neglect perhaps is just due to the fact that people interested in the foundations of science tend, for whatever reasons, to forget, ignore, or dismiss thermodynamics. The only systematic attempts to study the mereology of thermodynamics I know are due to authors with an operationalistic bias. Bridgman, Falk, Jung, Giles, Duistermaat, Buchdahl, Hornix have made substantial contributions to the field and gained important insights – although probably for the wrong reasons: trying to reduce thermodynamic metrical parameters to laboratory operations.

In fact, historically, the first real axiomatization of thermodynamics, namely Carathéodory's, was at the same time an attempt to make use of *proto* thermodynamic notions in our sense for the purpose of reconstructing thermodynamics along operationalistic lines. In his epoch-making study *Untersuchungen über die Grundlagen der Thermodynamik* [1909], Carathéodory tried to build up thermodynamics by assuming only observational parameters. By applying sophisticated constructs borrowed from differential geometry, he mainly aimed at a definition of entropy in terms of a fundamental "laboratory operation": accessibility through adiabatic isolation. Carathéodory was certainly influenced by the positivistic *Zeitgeist* of his times, according to which the conceptual basis of a theory should contain only macroscopic observational notions. Though we can still admire the ingenuity of Carathéodory's construction, his program has the characteristic shortcomings of any (too ambitious) operationalistic program: In order to obtain the desired reduction of entropy to "laboratory notions," Carathéodory had to make some implicit or explicit assumptions of an *ad hoc* character and much to restrict the scope of the theory and of its central concept, entropy.[2]

Nothing of this sort is attempted in the following pages. The program is avowedly much more modest in spirit. No definitory reduction of full-blown metrical functions of thermodynamics like entropy or energy to laboratory operations like adiabatic transition is attempted. Instead, we restrict ourselves to the axiomatization of the mereology of thermodynamics, without considering metrical functions and without any operationalistic commitment. What we want to reconstruct through protothermodynamics is not the whole of thermodynamics (an impossible task) but only one (neglected) aspect of thermodynamics, namely, its purely mereological part.

As Bunge rightly emphasizes, the first thing we ought to do when starting protophysics is to introduce the adequate notion of a (physical) *system*. In the present case, we need the notion of a *thermodynamic system*. When dealing with the mereological foundations of any discipline the crucial issue to get at is the specificity of the systems the discipline is about – in the present case: thermodynamic systems. This involves two sorts of things: first, to be clear about the specific physical interpretation of the systems (not every physical system will do as a thermodynamic system); secondly, to find out what the basic relations defined on the systems are, i.e., how the systems are structured. Actually, one task is dependent on the other. We cannot give a complete formal characterization of a thermodynamic system without introducing the basic relations holding in it. This is a well-known story in every axiomatic approach. The individuals the theory deals with are formally defined as the elements of the domain where the primitive relations of the theory hold, and these in turn are "implicitly defined" by the axioms. In principle, this is what we shall do here with the concept of a thermodynamic system. However, in protophysics at least, some intuitive pre-axiomatic characterization is always to the point. Let us say two things in this line.

First, following Bunge's general characterization of a system, we shall understand a thermodynamic system as "a concrete object ... acting as a unit in some respect" (cf. Bunge [1968], p. 299). This means, in particular, that we should not view a thermodynamic system as an abstract entity, e.g., a class of states (as I myself have done before in Moulines [1975] following Giles' proposal in his foundational work). There is nothing formally wrong in defining a thermodynamic system as a class of states; it does not lead to theoretical inadequacies. But it is intuitively unsatisfactory, and this seems to be reason enough for trying to avoid this kind of definition if nothing is lost in the way of a precise reconstruction.

Secondly, the basic units of our axiomatization of protothermodynamics should not be viewed as entities developing in time (as processes) but actually as *system-slices*. Or, slightly more formally, we could say that our systems (as the basic individuals the theory is about) will be ordered couples consisting of a spatial region and a state: If r is a given region in space and z is the state in which this region is, then $\langle r, z \rangle = s$ will be the corresponding system. Admittedly, this is a more restrictive notion than the common intuition of a system that appears in textbooks. But I think it is also more suitable for an axiomatic approach. In this way, we can forget about regions and states, and only talk about systems and define all relations on them.

Thus, our "universe of discourse" will only consist of a set S of systems s_i. What else do we need for the axiomatic presentation? We need some relations structuring S. In any systems theory, "combination" of systems is an operation which immediately comes to our mind. The study of thermodynamic systems is not an exception to this. Combination or coupling of systems is especially important in thermodynamics – as everyone knows. So, we could introduce a combination operation on systems as a primitive relation. However, an important insight I have gained from Bunge is that it is convenient to distinguish two sorts of combination operations in any systems theory: additive and multiplicative operations. They should be kept apart – at least in protophysics.

Following Duistermaat [1968], my previous papers on the foundations of thermodynamics [1975] and [1980] rested on the assumption that one can formalize combination of systems by means of just one operation – so to say, the logical sum of Bunge's addition and product. Again, this is formally possible, and one can do a lot of foundational work without worrying about further differentiations. But when one comes to the mereology of thermodynamic systems, a "tidy" treatment of the subject requires the two sorts of combination to be distinguished. Let us then adopt the standard notation and write $s \dotplus s'$ for the sum of two systems and $s \dot{\times} s'$ for their product. $s \dotplus s'$ is that system which results from the physical (spatial) connection of systems s and s' *without* any thermodynamically relevant interaction. On the other hand, $s \dot{\times} s'$ is the system resulting from an *interpenetration* or real thermodynamic interaction of s and s'. Now, a serious objection could be raised against this interpretation of $\dot{\times}$. One could object that since there are many possible sorts of thermodynamic interactions between systems, the operation $\dot{\times}$ is not univoque (thus actually no operation at all). However, this objection can be circumvented by considering the way we have previously interpreted our concept of a system. Remember that our systems are actually system-*slices*, having the concept of a state implicitly contained. Therefore, in different kinds of interactions the states will be different – therefore also, in our terminology, the systems will differ, though they might be in the same spatial region. This means that $s \dot{\times} s'$ represents just *one* interaction. In a different kind of interaction we would have different systems interacting, say, $s_1 \dot{\times} s_1'$ — although from the physicist-common-sense point of view s_1 might be considered "the same system" as s and s_1' the "same" as s'. But we need not adopt the physicist-common-sense point of view here; therefore, $\dot{\times}$ can be interpreted as a univoque relation on (appropriately interpreted) systems.

Having introduced addition and product of thermodynamic systems, it seems also plausible to have a "null-system" and a "one-system" in S. Let us call them 0_S and 1_S, respectively. 0_S is to be interpreted as a thermodynamically inert system. We assume that all thermodynamically inert systems are equivalent with respect to their interactions with actual thermodynamic systems. Thus, we are entitled to speak of *the* thermodynamically inert system. As for 1_S, it will have a natural interpretation as the thermodynamic universe viewed as a single system. (Remember that we should not confuse S, which is an abstract entity, with 1_S, which is a concrete system.)

The complement $\bar{s}$ of a system s also has a natural interpretation in this context: It is what in thermodynamics is sometimes called the "surroundings" of a given system – or also the "universe," though this last denomination is misleading. It is plausible to admit that for any thermodynamic system s there is another system $\bar{s}$ such that $s \dotplus \bar{s} = 1_S$ and $s \dot{\times} \bar{s} = 0_S$.

At this point it might appear as the next natural move to assume that S together with addition, product, complement, 0_S, and 1_S (with the proposed interpretations) constitutes a Boolean algebra, and that with this step we shall have covered all fundamental notions of protothermodynamics. Actually, the general systems theory presented in Bunge [1967] just postulates that systems will always constitute Boolean algebras. However, Bunge himself has opened the way to other possible structures of systems later on. For example, in Bunge [1973] he works with rings and envisages the possibility of using lattices instead of Boolean algebras (cf. Bunge [1973], p. 147–151). I think these other alternatives are relevant to the present case study. A Boolean algebra is too strong a structure for thermodynamic systems. It is dubious that the law of distributivity with respect to $\dot{+}$ and $\dot{\times}$ should be satisfied. For one thing, according to our interpretation of the complement of a system as the "surroundings" and of product as interaction of systems, it seems natural to admit that one and the same system might have a number of different complements. Consider an isolated system s. Its surroundings $\bar{s}_i$ may be very different systems but all of them fulfill the conditions $s \dot{+} \bar{s}_i = 1_S$ and $s \dot{\times} \bar{s}_i = 0_S$.

Now, we know there is an algebraic structure which is very similar to a Boolean algebra but different from it in that it admits many complements to a given element by just dropping the distributivity law. This is a non-distributive, complemented lattice. Therefore, I propose, at least as a working hypothesis for the time being, to define the basic structure $\langle S, \dot{+}, \dot{\times}, ^{-}, 0_S, 1_S \rangle$ of protothermodynamics as a (non-distributive, complemented) lattice.

We still need something more to get a complete picture of the mereology of thermodynamics. We need two further dyadic relations called "*transition*" and "*equilibrium*." Both are needed to understand what is specifically *thermodynamic* in protothermodynamics – as we shall see. Roughly speaking, a transition $s \Rightarrow s'$ is a relation between systems s and s' that may be read "s goes into s'" or "s is transformed into s'." As this reading suggests, a non-zero, finite lapse of time is needed for a transition, but since we do not introduce a time parameter into our set of primitives, this temporal explication should be taken only as an intuitive visualization. A given transition $s \Rightarrow s'$ should not always be understood as actually realized; all we need in general is to express that system *s may go into* system s' – given the required amount of time and the necessary conditions. Moreover, a further intuition we may associate with transition is that it is a "spontaneous" process undergone by a system s in the direction of s'. "Spontaneous" should not be understood too literally here, since the process is usually triggered by an exterior agent. In any case, "spontaneous" is not meant as a synonym for "in isolation" – as Giles seems to imply in his own explication of transition.[3]

Equilibrium (actually: *partial* equilibrium is what we mean here) is a dyadic relation between systems (let us represent it by: "$s \sim s'$"), where s and s', though thermodynamically interacting, are quiescent with respect to each other. And, of course, if s and s' are relatively quiescent, they are also absolutely quiescent, i.e., each system will be *in* equilibrium. (The last assertion will come up as a theorem in the following axiomatization.)

Transition and equilibrium are left rather unspecified at this stage. This is all right. There are many sorts of transitions and equilibriums in thermodynamics, according to the particular sorts of systems considered and to the particular thermodynamic theories which consider them. Protothermodynamics cannot take all these possibilities into account. It can only deal with their most general, common features. Also, it is a conceptual trap to think that one can give a full characterization of transition and equilibrium (as well as of all other protothermodynamic notions, for that matter) in purely qualitative terms – without introducing thermodynamic metrical functions. It is only through the introduction of the metrical functions as related to the operational concepts that we can fully characterize the last. If this were not so, operationalists would be right. But they are wrong. Here, I shall not say anything about the connection between the qualitative, operational concepts of protothermodynamics and the metrical functions of a full-blown thermodynamic theory. I have tackled this problem elsewhere (cf. Moulines [1980]). Here, we are only considering the mereology of thermodynamics; metrical functions have no place at this stage. But this means, of course, that our protothermodynamic concepts are left underdetermined to a certain extent.

To set up our axiomatic system of protothermodynamics it is convenient to introduce some special notation. The logical symbolism we shall use is well known: "$\neg$" for negation; "$\wedge$" for conjunction; "$\vee$" for disjunction; "$\rightarrow$" for the conditional; "$\leftrightarrow$" for the biconditional; "$\forall$" for the universal quantifier, and "$\exists$" for the existential quantifier. We shall make use of the convention that, when free variables appear in a systematic sentence, they are supposed to be universally quantified. A sequence of n elements $a_1, \ldots, a_n$ shall be written as an ordered n-tuple $\langle a_1, \ldots, a_n\rangle$. We shall use a so-called "projection function" π_i on tuples such that, if $x = \langle a_1, \ldots, a_n\rangle$, then $\pi_i x = a_i$. (We follow the convention that $\pi_i\, x = a_n$ for all $i \geq n$). If $\langle a_1, \ldots, a_n\rangle$ and $\langle a_n, \ldots, a_p\rangle$ are two ordered tuples, then their set-theoretic composition is $\langle a_1, \ldots, a_n\rangle \circ \langle a_n, \ldots, a_p\rangle =$

$\langle a_1, \ldots, a_p \rangle$. If $\Rightarrow$ is the transition relation, we shall write "$s \Leftrightarrow s'$" instead of "$s \Rightarrow s' \wedge s' \Rightarrow s$."

The first set of axioms we postulate is intended to systematize the most general framework to protothermodynamics. Their content is intuitively quite straightforward. They should be presupposed in any thermodynamic theory. Some simple and intuitively very plausible consequences will follow from them. After this first step, and after the appropriate definitions are set, a somewhat stronger (and more problematic) axiom will be added, which is intended for the protophysical foundations of classical thermostatics.

A 1: $\langle S, \dot{+}, \dot{\times}, ^{-}, 0_S, 1_S \rangle$ is a non-distributive, complemented lattice with a null-element 0_S and a one-element 1_S.

S shall be called a "set of systems" s, s'', s_i, and so on. $\dot{+}$ is called "sum of systems" and $\dot{\times}$ "product of systems." $\bar{s}$ is a complement of system s.

A 2: $\Rightarrow$ is a dyadic, transitive relation on S. ($\Rightarrow$ is called "transition.")

A 3: $\sim$ is a dyadic, symmetric, and transitive relation on S. ($\sim$ is called "equilibrium").

A 4: $s \dot{+} s'' \Rightarrow s' + s'' \rightarrow s \Rightarrow s'$.

A 5: a) $s \dot{+} s' \sim s'' \leftrightarrow (s \sim s'' \wedge s' \sim s'')$.
b) $s \dot{\times} s' \sim s'' \leftrightarrow (s \sim s'' \wedge s' \sim s'')$.

A 6: $s \sim s \leftrightarrow s \Rightarrow s$.

REMARKS: The first group of three axioms sets up the formal structure of each of the primitive notions, whereas the second set of three axioms establishes pairwise connections among the primitive relations. By the very notion of a lattice, *A1* implies that the combination operations $\dot{+}$ and $\dot{\times}$ are commutative and associative, which is plausible for all we know in at least classical thermodynamics. The lattice being non-distributive the possibility is left open that a system have more than one complement (for the reasons discussed above). Note that $\sim$ is not required to be reflexive; therefore, strictly speaking, it is not an equivalence relation. Some systems may not be in equilibrium with themselves: They are precisely those systems which are not in equilibrium. But, of course, some other systems are in equilibrium with themselves: They are equilibrium systems. *A4* expresses a sort of "cancellation law" with respect to $\dot{+}$ and

$\Rightarrow$, which is very plausible under the proposed interpretation of $\dot{+}$. But it would not be plausible to make an analogous requirement for $\dot{\times}$, since interaction of systems may change the conditions for a possible transition. The last axiom relating equilibrium to transition might presumably appear as the most problematic of the whole set: That reflexivity of equilibrium is equivalent to reflexivity of transition might seem to be too strong a requirement. But this is not so if we remind ourselves of the proposed interpretation of transition not as "always-actual-transition" but as a *possibility* of spontaneous transition of one system into itself after a nonzero, finite amount of time: This last seems to be possible only when the system is in equilibrium.

Let us derive some interesting consequences from the axioms just started. First, we define an equilibrium system formally.

D1: $\mathrm{Eq}(s) \leftrightarrow_{\mathrm{df}} s \sim s.$

T1: $s \sim s' \rightarrow \mathrm{Eq}(s) \wedge \mathrm{Eq}(s')$

That is, if two systems are in equilibrium with each other, then they are themselves equilibrium systems. This immediately follows from *D 1*, *A 3*.

T2: a) $s \dot{+} s' \sim s'' \rightarrow s \sim s'.$
b) $s \dot{\times} s' \sim s'' \rightarrow s \sim s'.$

Roughly, if a combined system is in equilibrium with a third system, its component parts are also in mutual equilibrium.

Proof: From $s \dot{+} s' \sim s''$ it follows by *A 5a*) that $s \sim s''$ and $s' \sim s''$. By the symmetry and transitivity of $\sim$ (*A 3*) we then get $s \sim s'$. An analogous argument holds for *T 2b*).

T3: a) $s \sim s' \leftrightarrow \mathrm{Eq}(s \dot{+} s').$
b) $s \sim s' \leftrightarrow \mathrm{Eq}(s \dot{\times} s').$

Roughly, two systems are in mutual equilibrium when their combination is also in equilibrium.

Proof: If $s \sim s'$, by *T1* we also have $s' \sim s'$. Applying *A 5a*), respectively *A 5b*), to $s \sim s' \wedge s' \sim s'$ we get $s \dot{+} s' \sim s'$, respectively $s \dot{\times} s' \sim s'$. And by *T1* again, $\mathrm{Eq}(s \dot{+} s')$, respectively $\mathrm{Eq}(s \dot{\times} s')$. Conversely, if $s \dot{+} s' \sim s \dot{+} s'$, by applying *A 5a*) (taking $s \dot{+} s' = s''$) we obtain $s \sim s \dot{+} s' \wedge s' \sim s \dot{+} s'$. And from this, by the symmetry and transitivity of $\sim$ (*A 3*), we derive $s \sim s'$. The same argument holds for $\mathrm{Eq}(s \dot{\times} s')$.

T4: a) $s \sim s' \leftrightarrow s \dotplus s' \sim s' \dotplus s$.
b) $s \sim s' \leftrightarrow s \dot{\times} s' \sim s' \dot{\times} s$.

This follows directly from *T3* and *A 1* (commutativity of $\dotplus$, respectively of $\dot{\times}$).

Now, in thermostatics the notions of quasistatic, reversible, and isolated processes play a very important role. These notions can be defined within the present framework in a very natural way. Also, some very general connections between these notions are sometimes stated in textbooks. For example, we would like to have that every reversible process is a quasistatic process but not conversely (see, e.g., Callen [1960], p. 63) and that any isolated process terminates in equilibrium (within the range of application of standard thermostatics – cf. Buchdahl [1966], p. 8). People often think that these notions and the corresponding assertions can *only* be stated once we have introduced metrical functions like entropy, energy, etc. into thermodynamics. I think the following shows that this need not be the case.

The first thing we have got to do is to define the general notion of a *process*. Of course, it will rest on the notion of a transition.

D2: *Process*

$$\mathrm{Pr}(p) \leftrightarrow_{\mathrm{df}} \exists s_0, s_1, \ldots, s_f(p = \langle s_0, s_1, \ldots, s_f \rangle \wedge s_0 \Rightarrow s_1 \Rightarrow \cdots \Rightarrow s_f).$$

If s_0 and s_f are, respectively, the initial and the terminal systems of a process p, we shall express p also as $\langle s_0/s_f \rangle$.

A corollary follows immediately from this definition and *A 2*.

T5: $\mathrm{Pr}(p) \wedge \mathrm{Pr}(p') \wedge p = \langle s_0/s_f \rangle \wedge p' = \langle s_f/s_h \rangle \rightarrow \mathrm{Pr}(p \circ p')$.

D3: *Continuating process*

$$\mathrm{CP}(p, p') \leftrightarrow_{\mathrm{df}} \mathrm{Pr}(p) \wedge \exists p_1, \ldots, p_n(\mathrm{Pr}(p_1) \wedge \cdots \wedge \mathrm{Pr}(p_n) \wedge p' = p \circ p_1 \circ \cdots \circ p_n).$$

(read: "p' is a continuating process of p").

For the following we need a precise notion of an isolated process and for this, in turn, we have to define the basic notion of an *isolated system*. Different possibilities can be offered within the present framework. The most natural seems to be in terms of the interpenetration operation $\dot{\times}$. The intuitive idea behind it is that an isolated system is one that does not interpenetrate with any other system which is not a part of it or of which it is not a part. I have not explored the consequences of this definition any further, and the results we obtain afterwards are quite independent of this specific notion of an isolated system. Whether a better explication could be found is still an open question. But, in any case, the present definition seems intuitively appealing and also very simple to me.

D4: *Isolated system*

$$\mathrm{Is}(s) \leftrightarrow_{\mathrm{df}} \forall s'(s \neq s \dotplus s' \neq s' \rightarrow s \dot{\times} s' = 0_S)$$

The antecedent of the defining conditional expresses the idea that s is not a (mereological) part of any given s' nor is s' a (mereological) part of s. If this condition is fulfilled, then we require of an isolated system s that its interaction with any such given s' be null.

With the notion of an isolated system we can state the stronger axiom we need for thermostatics.

A7: $\quad \mathrm{Is}(s) \rightarrow \exists s'(s \Rightarrow s' \wedge \mathrm{Eq}(s'))$.

Now we define three important kinds of processes.

D5: *Isolated process*

$$\mathrm{IP}(p) \leftrightarrow_{\mathrm{df}} \mathrm{Pr}(p) \wedge \forall i(\mathrm{Is}(\pi_i p)).$$

D6: *Quasistatic process*

$$\mathrm{QP}(p) \leftrightarrow_{\mathrm{df}} \mathrm{Pr}(p) \wedge \forall i(\mathrm{Eq}(\pi_i(p)).$$

D7: *Reversible process*

$$\mathrm{RP}(\langle s_0/s_f\rangle) \leftrightarrow_{\mathrm{df}} \mathrm{Pr}(\langle s_0/s_f\rangle) \wedge \mathrm{Pr}(\langle s_f/s_0\rangle).$$

T6: $\quad \mathrm{RP}(\langle s_0/s_f\rangle) \rightarrow s_0 \Leftrightarrow s_f$.

This immediately follows from *D2*, *D7*, *A2*.

Now we want to prove that, according to the previously defined concepts, every reversible process is also a quasistatic process.

T7: $\quad \mathrm{RP}(\langle s_0/s_f\rangle) \rightarrow \mathrm{QP}(\langle s_0/s_f\rangle)$.

Proof: From $\mathrm{RP}(\langle s_0/s_f\rangle)$ it follows by *T6* $s_0 \Leftrightarrow s_f$; by *A2*, $s_0 \Rightarrow s_0 \wedge s_f \Rightarrow s_f$. By *A 6*, this implies $\mathrm{Eq}(s_0) \wedge \mathrm{Eq}(s_f)$. Furthermore, take any s_i in the process $\langle s_0/s_f\rangle$, i.e., $s_0 \Rightarrow s_i \Rightarrow s_f$. Since also $s_f \Rightarrow s_0$, we have $s_i \Rightarrow s_f \Rightarrow s_0 \Rightarrow s_i$; therefore, by *A 2*, $s_i \Rightarrow s_i$, and by *A 6*, $\mathrm{Eq}(s_i)$. Thus, $\forall i(\mathrm{Eq}(\pi_i \langle s_0/s_f\rangle))$, therefore, $\mathrm{QP}(\langle s_0/s_f\rangle)$.

It is easy to see that the converse does not hold, as should be expected.

Another interesting result we obtain within the present axiomatization is that any isolated process terminates in a quasistatic process.

T8: $\text{IP}(p) \rightarrow \exists p'(\text{CP}(p, p') \wedge \text{QP}(p'))$.

Proof: Let be IP(p) for $p = \langle s_0/s_f \rangle$. By *D5*, we have in particular Is(s_f). By *A 7*, there is an s_n such that $s_f \Rightarrow s_n \wedge \text{Eq}(s_n)$. Now take $p' = \langle s_n/s_n \rangle$. By Eq($s_n$) and *A 6*, $s_n \Rightarrow s_n$. Therefore, p' is a process (though a "trivial" one): Pr($\langle s_n/s_n \rangle$). Moreover, it satisfies the conditions of *D3*: CP(p, p'). And trivially, $\forall i(\text{Eq}(\pi_i p'))$, thus QP($p'$).

We could certainly derive some further consequences from the above axioms and definitions, but I think the theorems derived so far already have sufficiently illustrated the potentialities of the present axiomatic system for protothermodynamics. This shows, among other things, that some simple but interesting and intuitively appealing results can be achieved in protophysics without resorting to metrical functions and "real" theories. Of course, this protophysical approach has, by its very nature, essential limitations. It certainly cannot be regarded as a substitute for the full-blown quantitative theories characteristic of physics. *Protophysics* is not *physics*. What I have tried to do here is not to replace physics by protophysics (as the operationalist would like to do), but rather to make out a case for the (dormant) potentialities of protophysics understood as a systematic study of the elementary notions of physics.

Instituto de Investigaciones Filosóficas, U.N.A.M., Mexico

NOTES

1 Bunge also calls it "general systems theory", but since the label "systems theory" is used in a variety of contexts not all of which correspond to Bunge's notion of protophysics, I still prefer the traditional term "mereology".

2 A critical appraisal of Carathéodory's attempt is offered by Tisza in his brief history 'Evolution of the Concepts of Thermodynamics' (included in *Generalized Thermodynamics*). More technical criticisms of Carathéodory are to be found in the last section of Falk and Jung's *Axiomatik der Thermodynamik*.

3 Giles [1964] also includes transition among the primitive notions of the foundations of thermodynamics. But his notion is too restrictive for our purposes. For one thing, he considers adiabatic transitions only. For another, he interprets transition in straightforward operationalistic terms, which makes the whole approach too problematic epistemologically.

REFERENCES

Buchdahl, H.A.: 1966, *The Concepts of Classical Thermodynamics*, Cambridge University Press, Cambridge.

Bunge, M.: 1967, *Foundations of Physics*, Springer, New York.

Bunge, M.: 1968, 'Conjunction, Succession, Determination, and Causation', *International Journal of Theoretical Physics* **1**, 299–315.

Bunge, M.: 1973, *Method, Model, and Matter*, D. Reidel, Dordrecht.

Bunge, M.: 1977, *Treatise on Basic Philosophy*, vol. 3, D. Reidel, Dordrecht.

Callen, H.B.: 1960, *Thermodynamics*, Wiley, New York.

Carathéodory, C.: 1909, *Untersuchungen über die Grundlagen der Thermodynamik*, *Math. Ann.* **67**, 355–386.

Duistermaat, J.J.: 1968, 'Energy and Entropy as Real Morphisms for Addition and Order', *Synthese* **18**.

Falk, G. and H. Jung: 1959, *Axiomatik der Thermodynamik*, in S.Flügge (ed.), *Handbuch der Physik*, III/2, Springer, Berlin.

Giles, R.: 1964, *Mathematical Foundations of Thermodynamics*, Macmillan, New York.

Moulines, C.-U.: 1975, 'A Logical Reconstruction of Simple Equilibrium Thermodynamics', *Erkenntnis* **9**/1.

Moulines, C.-U.: 1980, 'An Example of a Theory-Frame: Equilibrium Thermodynamics', in J. Hintikka, D. Gruender and E. Agazzi (eds.), *Pisa Conference Proceedings*, vol. II, D. Reidel, Dordrecht.

Moulines, C.-U. and J.D. Sneed: 1979, 'Suppes' Philosophy of Physics', in R.J. Bogdan (ed.), *Patrick Suppes*, D. Reidel Dordrecht.

Tisza, L. 1966. *Generalized Thermodynamics*, M.I.T. Press, Cambridge, Mass.

MIGUEL A. QUINTANILLA

MATERIALIST FOUNDATIONS OF CRITICAL RATIONALISM

The main argument Popper puts forward for making his metaphysical theory of the three worlds plausible is based on the idea that World 2 of the consciousness or human subjectivity and World 3 of ideas, problems, myths and theories invented by man are essential in order to understand the meaning of rationality, in other words, to understand the possibility and value of rational criticism. On the other hand Popper considers that the theory of the three worlds is incompatible with a materialist ontology. In fact a good deal of his contribution in Popper and Eccles [1977] is devoted to the criticism of materialism and the defense of the metaphysical hypothesis of the immaterial reality of the mind and its products. Obviously this means that for Popper rationalism and materialism are incompatible.

Such a conclusion poses an interesting challenge to the philosophy of Mario Bunge. On the one hand Bunge (editor of the first Festschrift for Popper, the title of which is significant, *The Critical Approach to Science and Philosophy*) is a philosopher who may be justifiably included in the ranks of critical rationalists. On the other hand he has created and defends a metaphysics which may be considered, and again justifiably so, as an important contribution to materialist ontology [Bunge 1977 and 1979]. One must come to the conclusion, therefore, that either Popper is wrong or that Bunge is inconsistent.

I believe that the topic is of special interest in relation to a particularly appealing aspect of Bunge's philosophy: we seem to be confronted with, roughly speaking, a "pre-Kantian" style of philosophy. That is to say, we are dealing with a philosophy which goes straight to things and does not linger for a moment on the "transcendental" meditation on the possibility conditions of the task in hand. Thus, those who associate the critical nature of a philosophical theory with this self-reflecting trend of thought might view Bunge as an uncritical philosopher.

This is in no way my opinion. On the contrary I will endeavor to demonstrate that Popper's theory of the three worlds is unacceptable, that Popper's arguments against materialism do not affect Bunge's ontology,

J. Agassi and R.S. Cohen (eds.), Scientific Philosophy Today, 225–237.
Copyright © 1981 *by D. Reidel Publishing Company.*

and that starting from this ontology the foundations of rationality can be framed in a more consistent and more "critical" manner.

1. THE ARGUMENT AGAINST MATERIALISM

Popper's line of argument [in Popper and Eccles 1977] against materialism may be reconstructed schematically in these terms: The materialist must admit that both World 2 of the psychical processes or the consciousness and the world of the products of this consciousness, and especially the world of theories, logic, conceptual objects (World 3) are a result of biological evolution. To accept this implies accepting that World 2 and World 3 consist of entities and processes endowed with an autonomous reality and a specific adaptive efficiency in the evolutionary process (by means of the Darwinist mechanism of natural selection).

The adaptive efficiency of World 2 and World 3 lies in the fact that with these appears rational criticism and, therefore, the possibility of transforming World 1 (the physical or material world) in accordance with theories and scientific method.

So, according to Popper, this is equivalent to saying that World 2 and World 3 are biologically functional to the extent that with them appear things such as the objective logical standards which allow us to evaluate an inference or to compare the contents of two scientific theories. It is true that the materialist could reduce the operations of the consciousness to material operations of an organism, and the meaning of conceptual properties like the consistency of a theory to properties definable in terms of World 1 (as, for example, to the maximum practical utility of the theory or to the degree of its social acceptability, etc.) But when this reduction is made all possible reference to objective standards of rationality is lost (the absolute notion of logical validity of an inference, for example disappears), and in this sense one can also say that the idea of the specific biological functionality of human rationality disappears too.

Consequently, materialism (the reduction of Worlds 2 and 3 to World 1) and evolutionism (the emergence and specific biological functionality of rationality) are incompatible.

So far, in brief, we have Popper's line of argument.

The structure of this argument does not fail to be striking. At first one might expect that, if one adopts what Popper calls the Darwinist point of view, the result would be a greater plausibility for the materialist hypothesis of the reduction of World 2 and World 3 to World 1. Instead Popper turns the whole situation around: if we take seriously enough both Darwinism (that is to say the idea that the mechanism of natural selection is responsible for the appearance of emergent properties or entities, endowed with their own adaptive efficiency) and the idea that human reason has a specific biological value in the evolutionary process, then one must be consistent and admit that biological evolution transcends itself giving rise to the emergence of immaterial entities.

In Popper's argumentation, however, several problems and theories come together which it is convenient to distinguish:

(a) On the one hand there is the problem of the scientific nature of logical relations of the type "proposition *p* implies proposition *q*." The objectivity of this type of relations between constructs (and other similar properties) is an indispensable requisite for rational thought.

(b) On the other hand Popper assumes that the objectivity of logical relations and properties alluded to in (a) together with the fact that we can talk of an influence of World 3 on World 1 through World 2, compels us to consider these logical entities, properties and relations as *real*.

(c) Finally, Popper conceives the reality of World 1 (material) in such a way that it is impossible to reduce the reality (b) of Worlds 2 and 3 to that of World 1 without the former losing their fundamental properties.

We consider assumption (a) as acceptable, but assumptions (b) and (c) taken together as unacceptable. Before entering upon the discussion of these in the light of Bunge's ontology, it is desirable that we should stop to analyze the difficulties that the ontological theory of Popper present.

2. THE DISADVANTAGES OF IMMATERIALISM

One may argue against Popper's theory from several different points of view. Firstly it is an imprecise theory. On the other hand, if one tries to pinpoint, for example, the strict content of World 3, one falls into contradictions [Keuth 1974]. Finally it may be argued that it is an ontology not coherent in relation to the rest of the ontology which serves as a foundation for science [Bunge 1980]. In spite of everything, one might defend the idea that it is the best theory available as long as no other alternative exists which obeys the same methodological demands. However, there is a basic methodological reason which makes Popper's theory unacceptable: a good theory ought not to limit itself to being a mere verbal reformulation of the problem which it is trying to solve. But it may be demonstrated that Popper's line of argument in favor of the real nature of immaterial entities of World 2 and World 3 are condemned to being, at best, purely verbal.

In fact, the criterion which Popper employs in order to be able to consider an entity as real may be formulated thus: "If x is a real entity and y exerts an influence over x, then y is a real entity." So, argues Popper, since the entities of World 1 may be considered without difficulty as real and since World 2 influences World 1 (the mind has influence over the body) and World 3 influences World 2 and, through it, World 1 (ideas are the driving forces of nations, theories impose their logical consequences on human reason) one might conclude that both World 2 and World 3 are real.

Now, this conclusion is unacceptable, because Popper's criterion of reality makes no sense when what we are trying to justify is the reality of immaterial entities for which the notion of causal influence or interaction remains undefined. In short, what Popper is doing is equivalent to seeking to found the concept of immaterial reality on a concept, that of *immaterial causal influence*, which is as confusing and problematic as the former (if not more so). Nothing is gained, save having changed the terms in which the problem was posed: we have moved from the problem of justifying and understanding the autonomous existence of immaterial entities to that of justifying and understanding the existence of immaterial causal influences.

It would be possible to apply in a more fruitful way Popper's criterion of reality. But to do so it is necessary to define the notion of causal influence between two entities. One way of characterizing this notion is the one which Bunge proposes in the framework of his approach to the description of the state of a system and to the notion of event or process [Bunge 1977, 119–140].

Every thing or concrete object of some kind K has a set of properties, known or hypothetical. Relative to some reference frame, each property of the thing may be represented by a certain function F_i which, in the simplest case, is a time dependent real valued function, i.e., $F_i: T \rightarrow R$, where R is the set of real numbers. The list $F = \langle F_1, F_2, \ldots, F_n \rangle$ of the n functions representing the known properties of the thing constitutes the state function of it relative to the given reference frame. The value $F(t)$ of F at time t represents the state of the thing at the given time. And the set of all such values, or $S(K)$, is the state space of things of kind K. But not all the mathematically possible values of F (i.e., points in $S(K)$) are really possible. One must take into account the restrictions imposed by the laws of nature which "govern" the things of kind K. The outcome is the lawful state space for entities of kind K, i.e., $S_L(K)$. Each lawfully possible state of a thing of kind K at time t can be represented by a point in $S_L(K)$, and each event or change of state by a vector in $S_L(K)$.

Let us assume then two systems x and y, a common reference frame M and both state functions F and G respectively. Let us also assume a function H dependent on F and G ($H = g(F, G)$). H will constitute a state function of the system consisting of x and y. So if $H \neq G$, then there exists a causal influence of x over y; if $H \neq F$ there exists a causal influence of y over x: and if H is not equal to either F or G then there exists a causal interaction between x and y. (In other terms: x and y behave in the compound system in a different manner to that in which they behave in isolation.)

According to this, in order to define the notion of causal influence between two things we need a common reference frame, the existence of state functions for each of the systems, and the existence of a state function for the compound system. But no one has said so far how a state function of an immaterial system (the mind, for example) may be defined, nor even is it known what the notion of state or event in such a system might mean.

In conclusion, Popper's criterion of reality is not defined for immaterial realities, and therefore cannot be used to justify the existence of the same. We will see, however, that it is perfectly applicable in the framework of a materialist ontology.

3. THE BASIS OF PLURALIST MATERIALISM

It has been pointed out [Moulines 1977] that the general formulations of materialism of the type "everything is matter" are empty statements, since a set of defined properties which characterize what is understood by matter are not specified. Generally, it is understood, in this type of affirmation, that the properties of matter are physical properties, or characterizable in terms of physical theories. But this is equivalent to identifying materialism with physicalism and, therefore, closing the doors on any ontology which tries to be at one and the same time materialist and pluralist. This is why the theory of Bunge is not formulated in these terms. Instead he takes as basic the concept of thing or concrete system and as the criterion of reality the existence of interactions with concrete things.

I believe, nevertheless, that there is no danger in formulating the general postulate of materialist ontology in traditional terms such as "only material entities are real" since a definition of the predicate "x is material" may be given which explicitly states what is assumed in the characterization of "thing" or concrete system in Bunge's ontology, avoiding physicalism at the same time. On the other hand, the characterization which we propose takes account of the intuitive concept of material entity which obviously refers to macroscopic physical bodies, and includes the notion of natural class (which is defined in Bunge [1977, p. 140ff]) and the idea of causal influence expounded in §3.

In short the following recursive characterization of the predicate "x is material" may be given:

(a) If x is a macroscopic physical body then x is material.

(b) If x is material and y is influenced by, or has an influence over x then y is material.

(c) If x is material and x is a member of the natural class X, then any member of that class is material (X is a class of material things).

This characterization of the predicate "being material" is not empty, since it is based on a not very problematic basic concept, that of the class of macroscopic physical bodies. But neither is it physicalist, since in order for a thing to be material it is not necessary that it have identical properties to those of macroscopic physical bodies, but only that it may have influence over, or be influenced by, any one of them. Consequently, materialism is compatible with pluralism and with the emergence of new properties of matter. Materialism is especially compatible with the following *pluralist postulate*: If X is a class of things formed by evolution [Darwinian] from things of class y, then the members of X have emergent properties which the members of Y do not possess.

This postulate, together with that of materialism, is sufficient to question Popper's argument: materialism and emergentism are not in principle incompatible. What is left to be done is to construct a theory of the mind and of the products of the mind in this framework of pluralist ontology. Before moving on to this point we will stop, however, to discuss the gaps and the problems which may be posed with regard to the materialist postulate.

Popper himself points out that present-day physics has overcome the conceptual framework of materialism: "The universe now appears to be not a collection of things, but an interesting set of events or processes" [1977, p. 7]. In other words, to the extent that materialist ontology depends on a certain substantialist conception of reality (as in the case of

Bunge and the characterization which we have just outlined for the concept of matter) it could be said that it is an antiquated ontology in relation to present knowledge in physics. I think, however, that what is under discussion here is not so much an ontological question as a question of semantics, a problem of interpretation of the theories of present day physics (and in particular of quantum mechanics). Bunge himself has made clear that the non-substantialist interpretation of the theories of physics is neither the only possible nor the most adequate one.

In fact, the interpretation of a theory is not complete if the referent, that is to say the class of things the theory is trying to represent, is not indicated [Bunge 1974a and b]. Thus, a theory that only deals with processes without subjects, or events without things in which the events take place, is not a theory which reduces reality to processes and events, but just a theory which is not sufficiently (or coherently) interpreted. It is true that it is in quantum mechanics where the most difficulties seem to exist in order to carry out this program of semantic analysis which Bunge proposes. The reason, however, is that the problem of finding a concrete referent for quantum theory is confused with that of finding a referent which may be described in terms of classical mechanics. Instead of this Bunge proposes a literal interpretation of quantum theory [Bunge, 1967a, 1967b, 1973] in which the referent (the quantons) are concrete entities endowed with quantum properties, not classical ones, and in which the metaphors taken from the language of classical physics, such as wave-particle duality, play no part.

From what I have seen so far, nobody has demonstrated that Bunge's interpretation is inconsistent or unacceptable for any other reason. Therefore, and while the contrary is not proven, the conception of matter in terms of the theory of state space of things or concrete systems may be viewed as applicable at all levels of reality and makes the materialist postulate compatible with present-day physics.

4. A MATERIALIST THEORY OF THE MIND

In Bunge's most recent work [1980], the general program of semantic analysis and foundations of scientific theories which we have just made reference to is applied to the field of psychology. The result is the development of a materialist ontology of the mind which was already outlined in Bunge [1979]. Here again the starting point is the search for the referent of a theory: since psychological theories refer to processes, events and psychic properties, it is a question of determining in which concrete system (or class of concrete systems) such processes and events occur, or which thing (or class of things) is the one that is considered endowed with such properties.

Bunge proposes that we consider the psychic functions as bio-functions of a class of organisms endowed with a plastic central nervous system. A central nervous system is understood to be *plastic* if over a period of time the synaptic connections vary as a result both of its internal activity and of experience. The variation of the synaptic connections allows subsystems of neurons specialized in psychic functions to form (Bunge gives them the name *psychons*); consequently psychic events and processes such as learning, thinking (conceiving concepts or propositions), reasoning, evaluating, deciding, etc., are produced.

We are not going to expound in detail the several postulates and definitions of Bunge's theory. We will limit ourselves to commenting on the conclusions which can be obtained from this theory pertinent to the argumentation of Popper against materialism. We must remember that the challenge Popper sets up for a theory of the mind is double: on the one hand it must account for the specific nature of mental activities of a rational kind, and on the other it must account for the objective nature of the logical properties of our thoughts.

With regard to the first point, Bunge's theory offers an adequate framework for solving the problem. Since mental activities are activities of specific systems of neurons (psychons), there is no reason to suppose that the materialist theory leads to an elimination of the peculiar nature of such processes.

As far as the intellectual processes are concerned, the approach proposed by Bunge is, in short, the following. We take as the elementary operation that of conceiving the concept which corresponds to a natural class. This operation may be defined as the activity of a determinate plastic neural system (a psychon) which reacts in the same way with regard to each member of a natural class. To conceive or think of a proposition may be understood as the successive activation of two psychons that conceive the two constructs (subject and predicate) that take part in the proposition; and the propositional functions may be interpreted in terms of the corresponding chains or processes of propositional thoughts. The specific properties of some of these processes, as for instance the deductive inferences, would be explained then by assuming that there are some thought processes endowed with a particular structure. To put it in the words of Bunge, "An animal possesses *logical* intelligence if it is capable of thinking, and some of its thoughts have the structure of a uniquely complemented lattice" (Bunge 1980, ch. 7).

In short, Bunge's approach in principle allows us to understand the specific nature of intellectual operations (even those which consist of reasonings guided by logical principles) without needing to recognize an immaterial mind. On the other hand we are confronted with an approach coherent with scientific research in the field of biological psychology [Hebb 1949; Bindra 1976; Piaget 1967].

However, the second requirement of Popper still stands; the problem of understanding the objective nature of logical properties, that is, the meaning of expressions such as "theory T is consistent," "the inference X is logically valid" or "there is an infinity of even numbers." Both Bunge and Popper accept that these types of expression make perfect sense and are indispensable for the development of scientific knowledge. The differences lie in the interpretations which are given to them. In order to understand the position of Bunge it is necessary to take into account his theory about logic and his treatment of the concept of existence.

Bunge's semantics [1974] starts from a basic assumption: it is possible to distinguish two types of objects, conceptual (or constructs) and real (material objects). The constructs are fictitious objects, their only reality lies in the mental (brain) processes by which such constructs are conceived; however, we may consider them fictitiously as autonomous objects endowed with logical, mathematical, properties etc. This is the fiction that allows us to construct our logical and mathematical theories. There appear in these theories existential statements (for example, "there is an infinity of even numbers") which may lead to error. In order to avoid confusion we must analyze the concept of existence as a relative predicate in a fixed context, and at the same time interpret the particular quantifier of logic so that it has no existential content. As far as the predicate of existence is concerned, relative or contextual, Bunge distinguishes two main types: conceptual existence and real existence. In both cases the existence may be understood as the characteristic function of the corresponding set of constructs, or of material things. So that to say "x exists conceptually" is the same as saying that there is a set of constructs to which x belongs. And to say "x exists in reality" is the same as saying that there is a set of real things to which x belongs. Consequently the analysis of logical or mathematical expressions such as "there is at least one prime number greater than 2" leads us to the following result:

"some x has the property of being a prime number greater than 2 and that of existing conceptually."

In this way we can assume the whole language of logic and mathematics without needing to confuse logico-mathematic objects with real objects. We can even employ known properties of conceptual objects to judge the logical value of an inference or of a theory. But this does not imply that such fictitious objects exert a causal influence on our mind or on our real world. Only our thoughts (psychic processes) are real.

I believe that this theory of Bunge is compatible both with scientific research in the psychology of reasoning and the demands of a nondogmatic rationalist philosophical theory. We shall look at the latter in the next section. As regards the psychology of reasoning, I believe that Bunge's theory is in perfect harmony with Piaget's theory on the genesis of logical operations. Piaget [in Piaget and Beth 1961] refers to two types of "abstraction" in order to explain cognitive development, which he calls empirical abstraction and reflexive abstraction. The latter is based on the experience of the very actions and operations of the individual and constitutes the basis for the formation of logical concepts. In terms of Bunge's theory we may say that reflexive abstraction is equivalent to the mental activity which comprises thinking about one's own thoughts, the result of which is the construction of these fictions we call concepts, propositions or theories.

5. THE FOUNDATIONS OF CRITICAL RATIONALISM

It is to be hoped that by replacing Popper's ontology of the three worlds with the materialist ontology of Bunge a displacement of the problem of the foundations of rationality is brought about. In fact, Popper's conception of rationality is basically normative, it defines rational behavior and thought by means of a system of norms which must be employed in order to act rationally. It does not tell us what rational thought is like but what the thought must be like for it to be considered rational. On the other hand, Bunge characterizes rationality from a descriptive point of view: rational thought is a special type of thought of some superior animals (which otherwise do not always act or think rationally). In other words: whilst Popper's theory of critical rationality is a proposal put forward against the sceptic, the dogmatist and the irrationalist, the conception of reason on the part of Bunge is directed against the nonmaterialist philosopher. That of Bunge is an ontological theory of reason. That of Popper a gnoseological or methodological theory of rationality. The problem is to know whether starting from Bunge's ontology a basis for the founding of the characteristic normative proposals of Popper's conception of critical rationalism may be found.

I do not believe that a simple answer can be given to this question. But it may be illuminating to analyze a problematic aspect of the Popperian conception of critical rationalism and see its possible reinterpretation starting from Bunge's philosophy.

In Popper's theory a certain paradox is presented related to the problem of the justification of his own philosophy: the basis of the activity of rational criticism lies in an irrational decision. In other words, rationalism cannot be justified rationally. Of course this statement need be in no way contradictory. But even if it is not, it does not seem a satisfactory solution to leave things like that. Firstly, because in so doing a door is left open to uncontrolled metaphysical speculation: in a way this is equal to reproducing the same intellectual process by which Kant, after having placed the basis of knowledge in human reason, went on to consider this as a last unanalyzable fact whose value was a *factum* not submitted to criticism. Faced with a theory of this type the irrationalist will always be able to argue *ad hominem* as Bartley [1964] has pointed out. Nevertheless, it is possible to frame the problem of the founding of critical rationalism not as a problem of logical justification, but as a problem of factual explanation.

Indeed, this would be the question of applying Bunge's general criterion of the fictitious nature of the entities of reason. When we say, for instance, that human reason is conjectural or critical, that there does not exist a royal road to discover the truth, or that rational criticism is based on the moral decision of exercising this criticism, we are talking of reason, of the rational individual, of true knowledge *as if* they were entities, we are pretending that they are entities, we are building entities of reason. But this does not authorize us to confuse them with real entities. For this very reason the questions that can be meaningfully framed do not refer to the rational justification of rationality, but rather to the factual explanation of the characteristics of rational thought. One might think that in this way the descent into dogmatism is inevitable. Rather the opposite happens; if to the question about the grounds for the rational critical attitude we reply by referring to an unanalyzable and irreducible decision adopted by some free and autonomous individual all we are doing again is to introduce ghosts into our ontology, to raise up barriers against the development of our understanding of the phenomenon of rational criticism.

On the contrary, from a materialist perspective, a possible answer would be the following: in fact, we might say, we exercise rational criticism (when we do so) because we have no alternative. Our neural systems function in such a way that at times we invent new theories and generally we are continually learning from experience. We may, of course, conceive of the possibility of denying ourselves the opportunity of thinking rationally or of accepting rational criticism. But it is even more plausible to think that our minds will continue working in favor of what we call reason. There exist no other foundations beyond these beliefs for rational thought. Nor are they needed. A materialist may defend rational criticism in this way (in imitation of the erroneous conception that everything real is rational): let us be critical because matter (not all matter, but our brain, and also social reality) is dynamic, changing and creative. That is to say, we might be permitted the metaphor–the materialist's slogan is "let us be critical because matter is critical!"

University of Salamanca.

REFERENCES

Bartley, W. W., III: 1964, 'Rationality versus Theory of Rationality', in M. Bunge (ed.) [1964], pp. 189–211.

Bindra, D.: 1976, *A Theory of Intelligent Behavior*, New York, Wiley.

Bunge, M. (ed.): 1964, *The Critical Approach to Science and Philosophy*, Free Press, New York.

Bunge, M.: 1967a, 'A Ghost-Free Axiomatization of Quantum Mechanics', in M. Bunge (ed.), *Quantum Theory and Reality*, Springer, New York, pp. 105–117.

Bunge, M.: 1967b, *Foundations of Physics*, Springer, New York.

Bunge, M.: 1974a, *Sense and Reference* [*Treatise on Basic Philosophy*, vol. I], D. Reidel, Dordrecht.

Bunge, M.: 1974b, *Interpretation and Truth* [*Treatise on Basic Philosophy*, vol. II], D. Reidel, Dordrecht.

Bunge, M.: 1977, *Ontology I: The Furniture of the World* [*Treatise on Basic Philosophy*, vol. III], D. Reidel, Dordrecht.

Bunge, M.: 1979, *Ontology II: A World of Systems* [*Treatise on Basic Philosophy*, vol. IV], D. Reidel, Dordrecht.

Bunge, M.: 1980, *The Mind Body Problem. A Psychobiological Approach*, Pergamon, Oxford.

Hebb, D.: 1949, *The Organization of Behavior*, Wiley, New York.

Keuth, H.: 1974, 'Objective Knowledge out of Ignorance: Popper on Body, Mind and the Third World', *Theory and Decision* **5**, 391–412

Moulines, C.-U.: 1977, '¿Por qué no soy materialista?', *Crítica* **9**, 26, 25–39.

Piaget, J.: 1967, *Biologie et connaissance. Essai sur les relations entre les processus cognitifs*, Gallimard, Paris.

Piaget, J. and E. W. Beth: 1961, *Épistémologie mathématique et psychologie*, P.U.F., Paris.

Popper, K. R. and J. C. Eccles: 1977, *The Self and Its Brain*, Springer, New York.

GERARD RADNITZKY

ANALYTIC PHILOSOPHY AS THE CONFRONTATION BETWEEN WITTGENSTEINIANS AND POPPER

ABSTRACT. The current discussion in philosophy of science is interpreted as a dispute between the Wittgensteinian and the Popperian positions. Logical empiricism, the successor of the Vienna Circle, has been shaped by the early philosophy of Wittgenstein. Its favored tool is the formal language. The significance of this tool is investigated, and in the light of Tarski's definition of truth, is posed as an exemplar the question of what contribution formal semantics makes to the philosophical problem of truth and to Popper's explication of "nearness to the truth." The chief critic of the Vienna Circle and its successors is Popper. He criticizes not only their results, but in particular the ideal of science lying at the basis of logical empiricism. A new ideal of science and new tasks for methodology arise from the critique. The most important critics today of Popperian methodology have been influenced by the later philosophy of Wittgenstein. In light of the tenets of the so-called ordinary-language philosophy it is shown that such a philosophy (as exemplified by T.S. Kuhn and P. Feyerabend) loses the normative problem. The loss of the idea of objective, although fallible, evaluative yardsticks would have serious consequences not only for methodology but also finally for the political sphere.

0. HISTORICAL SURVEY, PREVIEW AND PRELIMINARY STANCE

So-called analytic philosophy has its thematic focus on language. When someone says "so-called," he intends, with P.F. Strawson, to reserve analysis in the narrower sense for the "ordinary language approach" and to contrast it with the style of thought which works with ideal languages, which constructs formal languages and reconstructs theories with their help. He wants to highlight differences rather than common ground. From an etymological point of view, "analytic" is a justifiable label: analysis was originally clarification. In Plato "clear concepts drive away phantastic ideas" – those of the alchemists, mystics and ideologues of all sorts. At the time of the Vienna Circle such analysis was as pressingly necessary as it has ever been. *The activity of the Vienna Circle was clarification.* Its central problem, that of demarcating science from non-science, or more precisely from a sub-class of non-science, pseudo-science, is not such an important problem for methodology. A demarcation criterion is primarily a means for combating the pollution of the intellectual environment by certain ideological doctrines which for propagandistic reasons want to drape their dogmas in the cloak of scientificality – from National

J. Agassi and R.S. Cohen (eds.), Scientific Philosophy Today, 239–286.
Copyright © 1981 *by D. Reidel Publishing Company.*

Socialism to "scientific" socialism. To unveil such fraud was at that time the most important contribution a philosopher of science – *qua* philosopher of science – could make to the political debate. And so for Popper a critique of the initially offered solution to the demarcation problem, a critique of verificationism, was also crucial, since he had seen that this attempt was bound to run aground on Hume's critique of inductivism. Ayer's "testability criterion" of 1937 seems to be a response to Popper's critique (though to my knowledge this has never been acknowledged).

If both analytic approaches coincide in that they see themselves as clarification and in fact can contribute to clarification, and in that they give language a central position, their *differences* are similarly evident and well known. *Their attitudes towards ordinary language are diametrically opposed.* "Ideal language philosophy" ("*IL* philosophy" for short) fundamentally mistrusts natural language and claims to be able and obligated to counter the "bewitchment of our intelligence through language" by working with artificial, formalized, language-like systems. "Ordinary language philosophy" ("*OL* philosophy" for short) trusts natural language so much that it regards ordinary language as our secure foothold on the world.

As with every style of thought, there lies the *danger of a possible totalization.* An indication that at least sometimes some people have succumbed to this attempt is the discussion of problems which Mario Bunge has labelled "non-problems." The *IL* style of thought, which in any case can only treat a narrow field, runs the risks that the intellectual tools themselves will induce many problems and that real problems will simply be postponed. In caricature, one could compare at least some analytic philosophers of science with craftsmen who receive an order from their customers (customers as analogous to those interested in methodological problems), and who tell the customers, "We have developed a fascinating new tool; therefore we only handle problems that this tool can deal with; so we will not handle your problem, but rather the problem which results when we apply our tool to your problem – for the main thing for us is that we work exactly and precisely." With the *OL* style of thought, totalization leads to holding natural language, because it "functions," to be sacrosanct, even to making it the arbiter of philosophical disputes, so that finally "the world as language" becomes language as the world.

Although one can maintain that the *OL* style of thought is as old as Socrates, it did not exist as a philosophical approach at the time of the Vienna Circle. However, it is false to designate it as the "last phase" of analytic philosophy, for it is not a successor to and also not really a competitor of the *IL* approach, since it sets itself different tasks. What there was at approximately the same time was the Prague Circle. The Vienna Circle wanted to reconstrue physical theories and to articulate the ideal of a universal unified science. It mistrusted ordinary language and wanted to replace it in philosophical reflection with the language of the "ideal unified science." The Prague Circle was concerned with the scientific construction of linguistics and saw in teleological nexuses, which the Vienna Circle wanted to reduce to causal relations, the real key to uncovering the specific structure of language. Despite all their differences, it may well have more affinities with the Vienna Circle than it does with the *OL* approach.

Philosophical approaches can, in a somewhat idealized way, be profiled by articulating *their philosophical presuppositions and their methods*. Presuppositions and procedures are of course connected; but they can be represented separately and sometimes can even be separated. Here we address ourselves to the following issues: (A) The philosophical framework of the Vienna Circle and of its successor, *logical empiricism*, is fundamentally influenced by *Wittgenstein's early philosophy*. In what follows this influence will be briefly characterized. (B) Working with formal languages can in my opinion be considered and evaluated as a method apart from an evaluation of the philosophical framework of logical empiricism. The question should be posed, "Why philosophize with ideal languages?" This is a question of research policy, which can only be answered with justificatory arguments. (C) In this connection, the relations between the *IL* approach and currently important problems of Popperian methodology (e.g., that of the formal definition of verisimilitude) should be touched upon. (D) The philosophical framework of the *OL approach* has likewise been influenced by Wittgenstein: *by the later philosophy of Wittgenstein*. Since *OL* philosophy's method of proceeding seems to be more dependent on its philosophical presuppositions than *IL* philosophy's is, it appears to be appropriate particularly with reference to *OL* philosophy to investigate these assumptions or presuppositions. The primary concerns of this *essay are issues (A) to (D)*.

In this connection *Popper* appears above all as *critic of the Vienna Circle* and of its successor, logical empiricism. Both attempt to articulate a particular ideal of science. Popper criticizes this ideal of science and sets up an opposed ideal. The problem of the circumstances under which a scientific hypothesis can be accepted, a problem resulting directly from logical empiricism's ideal of science, becomes for Popper the problem of justifying a rule of preference for theories. This is only possible in that *from the critique of logical empiricism's ideal of science a new ideal of science grew and was legitimated.* In my view, *this development of Popper* represents *a Copernican revolution in methodology.*[1]

The most important critics of Popperian methodology have been decisively influenced by Wittgenstein's later philosophy. This is true of Kuhn, Toulmin and even Feyerabend. The critics do not necessarily have to be aware of this influence. And of course they don't for this reason have to be proponents of the *OL* approach. (Thus Paul Feyerabend, for example, has only devastating criticism for the *OL* approach.) In my view, the above-mentioned critics of Popperian methodology have indeed stimulated the discussion greatly, but have lost hold of the normative problem, which is a central problem for the logical empiricist philosophy and for Popperian methodology. A similar claim can be made about the *OL* approach. The *maître à penser* of *IL* philosophy, of *OL* philosophy and even of Popper's most important critics is thus Wittgenstein, or are the "two" Wittgensteins. The name is almost a philosophical Rorschach test: whether it awakens associations with the earlier or the later Wittgenstein says a great deal about the thinker who reacts to the cue word 'Wittgenstein'. It is not unusual for an adherent of the one to distance himself from the other. (Thus, for example, Gustav Bergmann and his school of "ontology after the linguistic turn" are much indebted to the Wittgenstein of the *Tractatus*, but would distance themselves from Wittgenstein's later philosophy.) Now in any comparison one can emphasize either similarities or differences. Wittgenstein himself, as is well known, gave up on the attempt to systematize the "sketches of landscapes" of the *Investigations*, and did not even try to fit them together with the *Tractatus's* blueprint of a transcendental logic; and he was openly a sharp critic of his early philosophy.

It is not an exaggeration to say that *philosophy of science*, general philosophy of science or at least philosophy of the natural science, *has been shaped in our century by two outstanding personalities: by Wittgenstein and by Popper.* The much-discussed "Kuhn-Popper debate" is in the final analysis a result of the opposition between these two.

1. *IL* PHILOSOPHY: METHODS AND STYLE OF THOUGHT

1.0. A pronounced preference for a particular kind of method is always based on particular suppositions about reality. These presuppositions come to expression upon occasion in a programmatic definition of the philosophy. Of course, all this is perfectly legitimate. However, as soon as a method is totalized, that is to say, as soon as claims to exclusivity are advanced for it, the philosophic tradition's presuppositions which this method has declared to be the only adequate ones become so over-burdened that it can be clearly recognized that at least in this form they must be untenable. The *IL* style of thought is usually associated with logical empiricism. It is correct that logical empiricism has made use of this style of thought, but that in no way makes the two identical.[2] Logical empiricism is a philosophical position, and its presuppositions do not necessarily have to be accepted by those who make use of the *IL* style of thought. In principle many of the research enterprises of logical empiricism could also be carried out without formal languages. Such an undertaking would merely be not much more practical than using Roman numerals for calculating. The *IL* approach is primarily an attempt to apply mathematical methods in work in philosophy and philosophy of science. Without doubt it has brought in certain areas of philosophy a precision and clarity never before achieved. One cannot condemn a method for being also liable to misuse, for allowing precision to become apparent precision and formalization to become pretentiousness. In §1 the question will be posed of what good reasons there are for working with formal languages; afterwards it will be shown in an example – Tarski's semantic definition of truth – what this approach can achieve for philosophical problems and what significance Tarski's explication has for Popperian methodology. Our example has been chosen for the following reasons. The discipline concerned with formal languages is an area of mathematics. Among its accomplishments, two in particular are celebrated as epoch-making achievements: Gödel's incompleteness theorem for the language of *Principia Mathematica* (henceforth abbreviated as '*PM*') and number theory as well as Tarski's explication of the concept of truth for formalized languages. Gödel's theorem shows, as it were, limits in principle for mathematics. It has significant consequences for the picture of humankind and thus for philosophy.[3]

1.1. *Basic Characteristics of the IL Approach as a Way of Working*

In this section we want briefly to recall the familiar. Keystones of this way of working are, as is well known, axiomatization and formalization. As soon as a formal language is interpreted, it contains certain presuppositions about the sector of reality to which it is supposed to be applicable.

But since one wants to make explicit assertions with its help, it contains or rather expresses a formalized theory as an effective calculus. A formal language as a system of structures, sign designs, functions as something similar to a language when it has been interpreted. Since this interpretation must in the final analysis be carried out by means of a natural language, one will make an effort beforehand to clarify as much as possible that portion of the natural language that one wants to call upon for the interpretation, and to improve it for this quite special purpose, because the interpreted formal language, as far as its meaning is concerned, cannot be clearer than the part of the natural language which has been called upon for interpreting it. The semantic rules which achieve this interpretation must be formal; that is, only sign designs can be assigned a meaning. For the meaning of a concrete token of a particular sign design is unambiguously determined by the meaning of this sign design. For this reason, through a formal interpretation a formal language as a mathematical structure becomes not a language in the true sense but instead an entity resembling a language. It is neither intended for nor useable for purposes of communication. In it the meaning of a concrete token expression is wholly independent of context, while in communication it is first made clear through the speech act how a linguistic expression is to be taken, and function and meaning vary with the speech act. Moreover, a formal language normally contains symbols which function as pure labels, i.e., which say nothing about the object designated and whose correct reapplication therefore cannot be controlled.[4] From these differences it follows that the tasks which pose themselves when a formal language, as one of the structures belonging to the objects of pure mathematics, is investigated, are entirely different from the hermeneutic task of understanding texts or actions, a task which conceives itself as an entering into dialogue with the text and which therefore includes within itself the structure of question and answer.

1.2. *Why Does One Concern Oneself with Formal Languages?*

1.20. For the "*IL* philosopher," formal languages are *ex definitione* his most important tool. At least the *significance of this tool* is doubted by advocates of other tendencies. In order to be able to reach a judgment in this controversial issue, we will first have to ask to what end this instrument is introduced. But the question of justification poses itself also for the *IL* philosopher: whether it is justified for him himself to invest in this particular kind of research, whereby he must bring into consideration as opportunity costs the results he presumbably could have achieved with this expenditure of time and effort if he had used other methods. In the final analysis this problem has an aspect in research policy: what good reasons can be advanced for spending public funds for research about formal languages or research using them, i.e., for a particular kind of mathematical research? In answering this question, two main types of justificatory arguments may be used.[5] (1) "*Arguments of utility*": Research as a means for realizing a concrete practical goal, usually for raising technological potential ("technological" being understood in the widest sense, including social technologies, etc.). The line of thought in this justification is simple: if one wants the goal and if the means are seen as effective and efficient (including "costs" of all kinds), then using the means is rational in the sense of goal-directed rationality (Max Weber's *Zweckrationalität*). (2) *Justificatory arguments from intrinsic value or from cultural value*: In such arguments one does not appeal to utility of any sort, neither with a direct nor with an indirect and unspecified appeal (as in the so-called "overhead" arguments). The possibility of such utility is in no way ruled out, but it is irrelevant to this sort of justificatory argument.

1.21. For research with formal languages *as its object*, as previously stated, a particular kind of mathematical research, an argument of utility is easy to advance. Formal languages are namely the appropriate intellectual instrument for constructing and working with a particular kind of instrument which is becoming ever more important for our future: somewhat crudely put, formal languages are "the language" of the computer.

1.22. Research using formal languages *as a means* is to be approached as a particular kind of mathematical research which is pursued in order to reach particular philosophical goals.

(a) The method of axiomatization, one of the keystones of the *IL* approach, is often practiced for the sake of a particular *ideal of knowledge*: the ideal of knowledge in axiomatized form with its great tradition from Euclid to Peano, Frege, Carnap, etc. A particular quantity of information put into axiomatic form is more transparent; and one can manipulate it better and more securely. This has evident advantages. For example, if the correctness of the (few) postulates can be established, then there is no need to examine the individual theorems (whose number is in principle infinite). In the philosophy of science the axiomatic method is used especially in the so-called foundational studies: theories, considered as completed items, are cast into axiomatic form in order to make basic assumptions obvious and to clarify basic concepts. These investigations join seamlessly onto the corresponding discipline, usually mathematical physics. As a magisterial example, the work of Mario Bunge can be pointed to. The value of these investigations for a better understanding of the theories thus investigated is evident, and for this reason justification is also here unproblematic.

(b) Work on and with formal languages is done with the goal of eventually being able to produce an *improved universal language*, which one hopes to be able to use as an important means for perfecting "the discovery of truth," decision procedures, etc. For example, to stylize a particular decision theory by means of an interpreted calculus fit into a formal language and thereby to get a better grip on it. Behind this intellectual motive also stands a great philosophical tradition, from Leibniz to the logical empiricists. One aspect here is that of economy of thought: if it can be decided mechanically in a formal language whether a particular sequence of symbols represents a proof, etc., so that such decisions and similar ones can in principle be left to a computer, then labor in thinking can be spared and quarrels can be simply and definitively settled. To this tradition belongs Leibniz's famous contention that with the help of a suitable *characteristica universalis*, instead of fighting, people would say, *Calculemus*!

The notion of such a universal language has *utopian characteristics*. In Paul Lorenzen's "ortholanguage" the attraction of this utopia is combined with the promise of improving the world. For those speaking with each other in the "orthotongue", the "so-called scientific distinction between is and ought"[6] is overcome and with it pluralism. The entire affair is carried out in the style of thought of foundationalism – which places this trend in sharp contrast to Popper.

(c) Besides introducing formal language to reach very general philosophical goals, formal languages are also worked with to reach particular goals in specialized areas of philosophy. The intellectual motive is the hope that the formal language will yield a better "key" to reality than any possible natural language. This motive is quite evident in Wittgenstein's utopia of a logically complete language, of a "transcendental logic." The *Tractatus* is intended to develop such a complete language system, with which one can "depict" the important aspects of the material world.[7] The *IL* functions similarly for Gustav Bergmann, as a theoretically not indispensable but in practice necessary instrument for making his "ontology after the linguistic turn" possible. His enriched version of logical empiricism's *IL* ("exemplification tie," "meaning nexus", etc.) serves to produce diagrams mapping the world. That is, he really uses only the formation rules of the *IL*. For the logical empiricists, a formal language like *PM*-ese (one similar to that developed in *Principia Mathematica*) is the linguistic garb of the "ideal unified science." More or less: *PM* with an empiricist interpretation.

(d) From a historical point of view there are *epistemological and in particular mathematical-philosophical motives* for concerning oneself with formal languages: the endeavor to deduce mathematics from logic, i.e., to make mathematics a part of logic, in particular to deduce arithmetic from (axiomatized) logic with the help of formal languages. This endeavor to "reduce" mathematics to logic, as it were, inspired the philosophy of mathematics from Frege to the *Principia Mathematica*. In the *PM* this program even seemed to have been realized. Today Russell's reduction thesis, according to which mathematics is a part of logic, has been almost universally abandoned. The new era in the philosophy of mathematics can be considered to have been inaugurated with Gödel's incompleteness proof for *PM* and even for axiomatized number theory [1931]. Popper writes, "Mathematics is not only not derivable from logic, but has even led to an essential refinement of logic and, one may well say, to a critical correction of logic: ... and to the critical insight that our logical intuitions do not reach all that far" [1974, p. 5]. According to Popper, there are various ways of constructing the various branches of mathematics, but no possibility of "grounding" them – for, according to Gödel, we can never prove that any system as rich as a language capable of expressing arithmetic is both consistent and complete. The opposed position, the antipodal point to the position of Mario Bunge and Sir Karl Popper with respect both to the reduction thesis and to the "grounding" is again the so-called Erlangen School. Thus Paul Lorenzen writes, "Namely, while logic has prevailed anew since about 1900 as the foundational discipline particularly for mathematics and natural sciences (it suffices to recall Frege, Russell and their logical empiricist followers from Carnap to Popper) ...." And the Erlangen School not only grounds mathematics on logic but also grounds logic in turn.[8]

Closely related to the above-mentioned motive from epistemology and the philosophy of mathematics is the special challenge which has resulted at the turn of the century from *the appearance of antinomies*, particularly in connection with the genesis of number theory. Russell's Theory of Types of 1908 presents a first attempt at resolution, which essentially consists of a prohibition. As a response to this special challenge in the realm of number theory, metamathematics arises – really a special area of mathematics. Antinomies are made impossible from the very beginning by producing mathematical calculi which are guaranteed to be consistent. What can be achieved with metamathematics is, of course, first of all an improvement of knowledge within the realm of pure mathematics.

(e) It is not an exaggeration to say that metamathematics was then taken over as a model by the group of philosophers here called *IL* philosophers. They attempted to extend the range of application of the approach from mathematics to other areas as well: to use formal languages and the knowledge about such languages, results of mathematical research, for stylizing particular aspects of natural languages or texts formulated in such languages. Logical empiricism then also took *metamathematics as the model* for its theory of science. A scientific theory is stylized as a system of postulates possessing in addition to a logical-mathematical interpretation an empirical (physical, biological, etc.) interpretation. The intellectual motive of this endeavor is the idea that propositions, arguments, etc., can be improved if, instead of leaving them in natural language, in which the sense of words and sentences is vague and often ambiguous and indefinite, one recasts them into a formal language. If such a stylization is available, one can let oneself be guided in critically judging the validity of an argumentation by unambiguous and explicit formal rules, and the logical structure of everyday-language sentences can be made more transparent and clearer by formalization. In short, formal languages serve for making logical *stylizations* – models – *of texts, systems of sentences, etc., formulated in natural languages*.[9]

Whether in the interpretation of the non-logical constants of an as yet uninterpreted calculus the derived theorems yield true propositions with this interpretation depends, as is well known, on the truth of the assumption that in relevant aspects the abstract system, the calculus or model corresponds sufficiently with the real system to which the model is applied. In other words, it depends on the constitution of reality. The case of the logistic stylizations as models of, e.g., sentences of a natural language, is analogous. The criticism that many aspects of the sentences formulated in natural languages are not captured by such models is accurate but not compelling, because the stylization only serves to get a better grip on a few, entirely specific aspects or partial functions of natural languages, such as deduction. Nevertheless, this criticism brought forward by ordinary language philosophers has brought a useful warning against the threatening over-evaluation of the possibilities of the methods of formalization.

As is well known, formal semantics investigates certain relations between a formal language as a structure built out of sign designs and certain other structures; and since sign designs can be equated with numerals, a formal language can be regarded as a kind of "numeral" structure. Accordingly, formal languages belong among the structures investigated by pure mathematics. This becomes particularly evident if one considers the theory of proof. The proof-theoretic investigation of a mathematical theory is indeed described as metamathematics, but metamathematics is a special branch of mathematics. That the investigation of the proof theory of a calculus with a logical interpretation is often called "metalogic" and that the proof theory for a theory T is often denoted as "the metatheory of T" and designated as MT do not alter the fact that investigations of this sort belong to mathematics. The same thing applies to various kinds of applied logic: e.g., logical formalizations of theories [e.g., Woodger 1937, 1959] non-logical interpretations of logic (e.g., Boolean algebra interpreted as a theory of nerve networks [e.g., McCulloch and Pitts 1943; Shannon 1938; Hohn 1960]). Applied logic is a branch of applied mathematics, while proof theory (metamathematics) belongs to pure mathematics. A metamathematical proof must, of course, itself be formulated within the framework of a mathematical theory.

A milestone in the history of proof theory, with enormous repercussions on our notions of human capacities for knowledge, is Gödel's famous incompleteness proof. If one may attempt to summarize his result in a few sentences, one could try the following. (Here I follow A. Wedberg's presentation [1966, p. 294].) Any formalization of a theory which is rich enough for adding and multiplying natural numbers and which includes the concept of a prime number, is, if consistent, incomplete. Moreover, the consistency of a formalization T cannot be proven within T itself, or within any MT identical with T, unless T is inconsistent. If one can prove in MT the consistency of theory T, the "metatheory" MT must include more than T does and accordingly must allow methods of argumentation which do not occur in T and on which we are nevertheless willing to rely more than on those which occur in T. But this could almost look like a so-called bootstrap operation: a conjecture which is questionable, or at least not assuredly true, is used to criticize another conjecture and perhaps to improve it. This also seems to be a strong argument against any sort of "foundationalist" thought and an argument for Lakatos' hypothesis – endorsed by Popper – that most mathematical theories are, like scientific theories, hypothetico-deductive, and thus pure mathematics stands closer to natural sciences than has widely been supposed until quite recently.[10]

What is the balance of this inventory of possible justificatory arguments? In all, it is not hard to admit that the *IL* approach as a strain of mathematical research in philosophy has become an ineradicable component of philosophy as a discipline. A danger lies, as everywhere, in a possible monopolization, if one declares it to be the only legitimate approach for philosophizing. There are also examples of this,[11] but one cannot use this as criticism of the approach, but only of some of its "practitioners."

1.3. *To What Extent Can the Results of Formal Semantics, a Branch of Mathematics, Contribute to the Solution of Philosophical Problems?*

1.30. This question is to be made concrete here with an example. For the above-mentioned reasons, Tarski's definition of truth has been chosen. What does Tarski's result contribute to the philosophical problem of truth, and what significance does it have for Popper's theory of science?

Tarski's classic work proposed a precisified concept for use in connection with formal languages, designed to replace our intuitive idea of accurate representation. It can hardly be correct to say (as Tarski himself seems to do) that he has explained one theory of truth among others, namely the so-called correspondence theory. The idea of correct representation (*zutreffende Darstellung*) belongs to the semantic structure of every functioning ordinary language; not until a communications system has, in addition to the expressive and signalling functions, the function of more or less accurately representing states of affairs, does it become a language. This assertion implies the further assertion that every ordinary language is based on a realistic ontology. Without the concept of truth and theory realism, which maintains that one of the functions of a theory is to represent certain aspects of reality as accurately as possible, the idea of empirical testing loses its meaning and with it the ordinary procedure of empirical science and thus it is certainly impossible to get a grip on the problematic of progress in knowledge. To be sure, scientific theories are also, among other things, instruments for producing explanations and predictions. The error of the instrumentalist conception of theories lies again in totalization. Even the vulgarly pragmatic conception of theories, which sees scientific theories as nothing but instruments for solving practical problems (a position often advocated in Marxist writings), does not get around bringing realistic points of view into play. For predictions are needed for technological efficiency, and a prediction is useable only if it to a certain degree accurately represents future events. Sir Karl Popper and Mario Bunge have always been defenders of a realist view of scientific theories. Foundationalism (whether in the form of "intellectualism"/ "classical rationalism" or in the form of "empiricism") systematically confounds concept and method of ascertainment, in that it defines the concept of truth through a method ascertainment.[12] Here foundationalist epistemology retained the fundamental error of operationalism, which in large measure thanks to Mario Bunge's criticism has long since been overcome in methodology. Plainly epistemology can first begin when operating with expressions like "... is a more accurate representation than —," 'true' and 'false' already functions in ordinary language, just as methodology can first make an onset – clarifyingly, critically, constructively – where research findings, hypotheses, theories are on hand claiming to represent to some degree accurately certain aspects of reality.

1.31. *Clarification of the explicandum as preparatory stage in Tarski's work.* The explicandum is the assertion that something is actually the case. This assertion is made by means of a sentence of the object language, but which can be made *explicit* by means of a sentence of the meta-language, i.e., by claiming that the sentence of the object language is true, that it accurately represents, describes the facts. Tarski implies here certain assumptions about natural languages which, although not relevant for his explicatum, can nevertheless be of interest. They illuminate in an exemplary way the attitude of *IL* philosophers and mathematicians towards natural languages. They suggest the presumption that Tarski may suppose his explicatum to be applicable to natural languages. This even though the realm of explication foreseen by Tarski for his explicatum is indicated quite unmistakably: his explicatum is to be predicated of sentence forms of a particular (class of) formal language(s). Tarski holds that natural languages, because of their many-sided application, cannot be consistent. This seems indeed to presuppose that "the normal laws of logic" also obtain in everyday language. But this assumption appears to be equivalent to the supposition that ordinary language can also be regarded as a kind of calculus, that it can be treated as if it had postulates and derivation rules fixed for once and for all. But that is certainly not the case. One ground for Tarski's presumption of the, so to sepak, endemic inconsistency of ordinary language is the ancient "liar" paradox.

1.32. *Tarski's explicatum.* His result can, as is well known, be summarized as follows: Tarski defined a particular concept of truth for a particular formal language and also indicated a method for defining this concept for a particular class of formal languages. Of course, his explicatum relates only to sentential forms, since only such forms are brought into consideration in formal languages. Since every formal language has or needs to have a formal interpretation, and since thus every sentential form in such language must be either true or false, he can define his explicatum in a categorical form: "The sentence (sign design) *S* is true if and only if ..." If a formal interpretation, fixed for all time, is not assumed, then it seems imperative explicitly to relativize: "The sentence *S* is true in formal language *L* (and that means, relative to a particular interpretation *I* of *L*, fixed for all time) if and only if ..."

The formal language in which Tarski formulates his definition of the concept of truth contains as a part the formal language (the "object language") of whose sentential forms the defined predicate "truth" is predicated. If we permit ourselves – here I follow A. Wedberg's presentation [1966, p. 299][13] – to act as if the ordinary-language sentence "Snow is white" were a sentence of a formal object language and this object language were a part of our language, in which we now speak, then we could say, "The sentence 'Snow is white' says that snow is white," and thus also "The sentence 'Snow is white' is true if and only if snow is white." Of course, we have not yet specified Tarski's definition at all, and we cannot do so. For in order to do that we would first have to specify the formal language *in* which we formulate our definition – only then can the definition in this formal metalanguage be given. This could scarcely be carried out in the format of an article. For our purpose it suffices to see that Tarski has set up a particular criterion of adequacy for his explication, and indeed a "material" criterion which does not have to do with the form of the definition. It is the well-known Carnapian *metacriterion of "sufficient" similarity between explicandum and explicatum*: roughly the paradigmatic cases of the positive (negative) application of the expression for the explicandum must also belong to the assured positive (negative) extension of the expression for the explicatum. This metacriterion is plainly a *necessary* but not a sufficient condition for the adequacy of an explicatum, for if it is not fulfilled then one has plainly not made a proposal for replacing an "old" concept already on hand with an improved concept in particular contexts, but rather has proposed another concept which has too little or nothing to do with the original one. Such a procedure would simply not be an explication. This similarity criterion could be roughly formulated thus: an explicatum of the correspondence notion of truth can be adequate only if, by means of the definition which introduces the expression for the explicatum into the metalanguage, every sentential form of the object language is derivable with the following form: The sentence '(*mention* a sentence, i.e., write down here a sentence of your choosing)' is true if and only if (*use* this sentence, i.e., write down here exactly the same string of symbols). Thus the first time the sentence is mentioned, designated, with the help of the metalanguage; the second time it is used, appears as a sentence of the metalanguage. If the definition of "true in language *L*" has been accepted for the metalanguage *ML* of *L*, sentences of the form in question are analytic in *ML*. The sharp distinction between object language and metalanguage is necessary in order to avoid paradoxes of the Liar type: the price that must be paid.

The type of entity that Tarski considers is the *type* sentence of a *formalized* language which is *interpreted*, not the meaning of expressions. (The meaning could after all be expressed in various languages, at least in the form of near-synomyms.) The achievements of his explication are the following: He shows (i) that the concept of truth is not inconsistent, (ii) that the concept of truth is not redundant, and (iii) he gives a striking example of the importance of the *distinction* between the *concept of truth* and a *method* for ascertaining whether the concept of truth applies in a concrete case, whether a particular type sentence or token sentence is true. *He restricts himself specifically to an explication of the concept of truth* – making no assertions about methods for ascertaining the truth of particular sentences, about criteria of truth. The explicandum is the descriptive content (*Sachgehalt*) of sentences which have a descriptive function (*Darstellungsfunktion*), the assertion of a sentence in the object-language to the effect that a certain state of affairs obtains. To this assertion corresponds a metalinguistic statement which says that the sentence in the object-language is *true* and contains an *accurate description* of the actual state of affairs. In the metalinguistic sentence, the sentence in the object-language is placed in relationship to the facts; the *truth* of the *description* of possible facts which the sentence provides is recognized as accurate (*zutreffend*). The provision that such a thing is at all possible is that the reference to reality is present even in the meaning asserted in the object-language sentence. It is not possible to speak about this object-language sentence except in a meta-language. In doing so we make explicit the meaning asserting in the sentence, i.e., claim that the description in the object-language is accurate. After having tested a descriptive utterance we can conjecture that this sentence is true (i.e., if the results of the test are positive). Only if we view the sentence as a means of describing a real state of affairs does it make *sense* to attempt to carry out an *empirical test*. For the *concept of "empirical testing" to be at all meaningful*, one must presuppose the concept of correct description, i.e., the concept of truth and therefore REALISM. Some would like to discard the concept of truth as a regulative idea and, along with it, realism as a hypothesis, i.e., as an assumption about reality and the human capacity to know. Realism assumes that we possess by means of our perceptual apparatus some access to certain aspects of reality. However, to discard the idea of truth and along with it realism, would deprive the usual methodology – the procedure employed in the empirical sciences – of all its meaning.

As mentioned, Tarski restricts himself specifically to an explication of the concept of truth with respect to type sentences of an interpreted, formalized language. In situations involving ordinary language the designation "true" is only then relevant when truth claims are explicitly made a topic of discussions: "Is it true that ... ?". or "I claim that ... is true," etc. In the sentences (there is no distinction between object-language and meta-language in natural languages) the "pragmatic meta-function," as we might call it, also plays a role in addition to the descriptive function of language. In such sentences a communicative act is carried out and simultaneously reflectively designated. This pragmatic meta-function has been the object of study since the Stoics, and has continued up to the Ordinary Language philosophers. The aforementioned communicative act involves self-consciousness, or self-reflection; it includes what we see from the other's point of view, as well as a prediction from our point of view, of the expected, future perceptual experience which the other will have. Hence, it also demands self-reflection on the other person's part (that he should believe us, etc.). The element of self-reflection is essential to this performative meta-function. Even when so-called "animal languages" are supposed to also have a descriptive function, for example, as the content of statements which basically function as calls, this self-reflection appears to be missing as far as we know. For instance, animals do not lie. In this way Ordinary Language studies complement Tarski's explication.

1.33. *What significance does Tarski's explicatum have for Popper's methodology and what application has it found there?* Popper's claim that he *uses* Tarski's definition of truth [1972, pp. 45, 326, *et passim*], taken exactly, or perhaps really somewhat pedantically considered, is not entirely correct. For he does not use the definition; rather, he uses the above-mentioned adequacy criterion of the "sufficient" similarity between explicatum and explicandum. His assertion that Tarski's definition is a *rehabilitation* of the concept of the truth of a judgment [e.g., 1972, pp. 44, 60, 308, 314, 323, 328; 1934/1971, p. 328] seems to us easily capable of giving rise to misunderstanding. The similarity criterion already presupposes the correspondence concept of truth – if this presupposition were not made, the entire project of explication would not have much point. It looks as if Popper's contention had a procataleptic function: to counter the contrary assertion, frequently advanced by opponents of the *IL* approach, that Tarski's definition re-applied to ordinary language is trivial. But one wouldn't really even need to counter this anticipated objection, for literally taken it is nonsense, since the Tarskian definition, formulated in the formal metalanguage, only applies to formal languages and thus cannot itself be applied at all to ordinary language. But in case the similarity criterion under discussion is meant with the anticipated accusation of triviality, the accusation does not thereby win in cogency, because in demanding similarity one gives no more information about the explicandum by means of the explicatum or by means of the information about the explicatum than was already put into it in advance. In brief, the often repeated accusation of triviality rests on a misunderstanding.

Thus Popper uses the similarity criterion, which holds for every explication of the correspondence notion of truth, and moreover Tarski's definition serves him as a kind of model and as an auxiliary concept for the attempt to give a formal definition for the concept of verisimilitude, i.e., as a "logical underpinning" for his theory of nearness to the truth.[14] In order to be able to judge what is achieved with such a "logical underpinning," one must "play through" his examples of application in detail with him and test out the formalism. David Miller and Pavel Tichy have showed that Popper's comparative concept of verisimilitude cannot be an adequate explication of "nearness to the truth," since it only enables one to compare those genuinely axiomatizable theories which are true, but not to compare false theories, whether they are axiomatizable or not. G. Andersson [1978] has now shown that the result about true axiomatizable theories can also be derived from Popper's theorem on truth content. He also gives an elegant demonstration of why Popper's quantitative explicatum of the notion of nearness to the truth does not enable one to discriminate between any false theories of the same logical probability. If one comes to the conclusion that the previous attempts to introduce the concept of verisimilitude by means of a definition which makes use of a partially formalized language have not led to any satisfactory result, one cannot infer from that that such a definition is impossible in principle. First, there is no compelling reason for such pessimism. Second, in case someone should (*per impossibile*) succeed in producing a kind of impossibility proof for the production of an adequate explicatum for the concept "as a representation of *z*, *x* is more accurate than *y*" by means of a formal definition, i.e., by means of a partially formalized language, even this proof would only hold for a particular formal language or a particular class of formal languages. Third, it would not follow from that that the correspondence notion of truth was not suitable for serving as an explicandum or that is was dispensable. Incidentally, in my opinion nothing is to be expected from the attempts to introduce the concept of 'verisimilitude' by means of a definition formulated in a partially formalized language, except that these attempts will lead to difficulties entirely analogous to the well-known difficulties that emerged from the attempts to introduce an explicatum of the concept of 'empirical significance' by means of a formal language. These difficulties are certainly already imbedded in the means, in the *IL* [Cf., e.g., Radnitzky 1968/70, vol. I, pp. 112–45].

1.34. *Why is a justification for the efforts to give an explication for the concept of comparative nearness to the truth by means of a partially formalized language not as unproblematic as a justification for Tarski's endeavor to explicate the concept of truth for formal languages?* As was already said, the point of an explication is to refine a concept by specializing it. It is proposed within a particular research project to replace a particular concept by an improved version of the concept. This proposal is legitimated by the assertion that the improved concept represents a better conceptual instrument than the explicandum for the research project concerned: that it is more "fruitful." As necessary conditions for this fruitfulness, "sufficient" similarity between explicandum and explicatum must be demanded (to insure that an entirely different concept does not smuggle its way in), and a degree of precision which seems appropriate for the research project concerned must be demanded. Accordingly, increased precision is only a means to an end, and not an end in itself.

If one formulates a theory (physical, biological, etc.) as a calculus, one does this in order to *use* the calculus. To this end one must establish what is to be meant by the assertion that a particular kind of real system is a "model" for the calculus, i.e., by the assertion that sentences of the calculus become true sentences with this interpretation (related to this kind of real system). As we argued in §1.2, good reasons can be advanced for working with theories in the form of calculi. If one opts for this way of working *then the task of explicating the concept of truth for formal languages is an indispensable component*. Since in the case of Tarski's explication the issue is one of theories with a mathematical interpretation, the question of how far and with what means empirical theories can be cast into the form of an interpreted calculus is irrelevant. For the research projects for which Tarski's explication is envisioned, the issue concerns *mathematics* and not the methodology of the empirical sciences.

The context in which the need for an explicatum of the concept of nearness to the truth arises is an entirely different one, namely methodology. Methodology is here conceived as a discipline which makes proposals about how one is to conduct oneself in particular kinds of research situations (e.g., in evaluating theories), and which attempts to legitimate these proposals, to legitimate them by hypothesizing that if one follows them, the chances of succeeding in bringing about scientific progress are better than would be the case if one followed the proposals of a rival methodology. In connection with the task of evaluation, as soon as the fallibility in principle of the results of empirical science has been recognized, *evaluating the comparative achievements of rival problem solutions* becomes the chief task of methodology, and thereby also the task of giving good reasons for a preference rule for theories. To this end one needs to make precise, or to improve, the intuitive idea, "... is a more accurate representation than (*zutreffendere Darstellung als*)—." At the risk of being repetitious, we must here emphasize the *difference* between this *task of explication* and the *task of giving* (*fallible*) *methods of ascer-*

tainment by means of which good reasons can be advanced for the conjecture that one of the theories gives a more accurate representation than the rival.

The explicandum, the comparative notion of nearness to the truth, is again in turn built into ordinary language and is obviously also applicable to pairs of sentences when both components are assumed false. To explain this intuitive notion, the comparative concept "... has a greater content of empirical information than—" is introduced, where one thinks of "empirical content" as the composite of "truth content" and "falsity content" – roughly, the true and false consequences respectively of the sentence. From here it is not far to build an approach to an explication on the ideal that "*r* is closer to the truth than *s*" claims either that *r* has more truth content and not more falsity content than *s*, or that *r* has at least as much truth content and less falsity content. In the attempt to introduce the explicatum of nearness to the truth by means of a definition formulated in a partially formalized language, difficulties arose, as is well known.

In order to be able to judge the adequacy of a proposed explicatum of the concept of nearness to the truth, one must consider the question, "What is an explicatum of 'nearness to the truth' supposed to achieve for *methodology*?" It should be better suited than the intuitive, ordinary language notion of nearness to the truth (more accurate representation than) to the task of giving good reasons for a preference rule which is to guide us in theory appraisal. In accordance with the rules of the game of explication, this "better suited than" must be more than mere increase in precision. Moreover, the explicatum must also be applicable in appraising pairs of theories in which both components are falsified.

The attempts at explication up to now presuppose the explication of the auxiliary notion "content of empirical information" and attempt to clarify and improve the auxiliary notion by means of formal language. Two strategies have been tried for this: (a) To clarify the auxiliary notion "*x* has a greater information content than *y*" by means of auxiliary concepts from mathematics (logic and set theory): as consequence set or subset. If one adopts this approach, the Tarskian explication is a necessary preliminary. (b) Alternatively, one attempts to explicate the auxiliary notion by means of auxiliary concepts from the formalized theory of logical probability. Content of information is conceptualized as the complementary concept to (absolute) logical probability, or the measure of the former as inversely proportional to the measure of the latter. The difficulties of this approach are well known: the probability of a universal proposition relative to a finite number of accepted singular propositions equals zero; the notion of the "absolute logical probability" of a single theory may be a notion without cash value (and really be relative to implicitly presupposed background knowledge). These technical problems, especially the problem of probability, cannot be entered into here. Here we should only mention that in the abovementioned work of G. Andersson no claim is made to have achieved more than an explication; criteria are not promised. Incidentally, methods of ascertainment are only used in practice to compare competing hypotheses anyway, and not to "measure" verisimilitude. In every explication primitive concepts are unavoidable; it is impossible to define everything. It may be a matter of taste whether one wants to define "nearness to the truth" by means of "probability" (or "content of information") or vice versa. In any case, the concepts of probability and content of information are indivisibly bound to each other as a "complementary" or "inverse" pair of concepts. G. Andersson takes both possibilities into consideration: to choose "probability" or "content of information" as the primitive concept. He opts for probability because at least there is a well-developed theory for the formal properties of this concept. To be sure, absolute logical probability is not (as is so often maintained) defined by the concept's fulfilling the axioms of the probability calculus. That can be no more than a necessary condition for the adequacy of a concept of probability. It is often attempted to give the concept content by, say, the idea of "number of models (or state description *à la* Carnap) divided by the total number of models" (whereby both numbers are infinite). But that hardly says anything more than the idea of "relation of 'favorable' to possible cases." When the issue is one of methods of ascertainment, one is left to making conjectures and estimates. The same may hold for methods of establishing content of empirical information. He who contends that "nearness to the truth" is intuitively clearer than "probability" or "content of information" would have to reject all such attempts at definition as pointless. On the other hand, it could nevertheless disquiet him if attempts to achieve a definition by means of standard logic miscarry. If all attempts to give a "logical underpinning" of a particular concept miscarry, there are in principle two options: (a) to conclude that the explicandum is hopelessly unclear, or (b) to conclude that the price of such a "logical underpinning" is too high. But the case at hand seems to be special, for the first option does not offer a real alternative: we cannot communicate and argue at all if we do not already presuppose the idea that one proposition, one hypothesis can be a more accurate representation of certain aspects of reality than another hypothesis. With "too high a cost" one would then mean that the problem of the "logical underpinning," which is supposed to be solved by means of the system of standard logic, would not exist at all without this system, that at least an essential part of the difficulties is induced by the system itself. At any rate, that would be a debatable standpoint.

And so we come to the question of allegiance for all offers of explicata for "nearness to the truth" which make use of formal languages: *why* should one expect that an explication of "nearness to the truth" which is made for a particular *formal* or partially formalized language and is introduced by means of such a language will be more *fruitful* for the *methodological* problem of giving a preference rule for theories than an explicatum which doesn't relate to formal languages or doesn't appeal to such languages? In the case of Tarskian explication an analogous question could not be posed at all. It might be that for methodologists a discussion with logicians who do not want to answer this question, but rather reject it as spoiling the game, will not be very fruitful. For, taking this question seriously differentiates the investigator who wants to employ formal languages here as a means from the "chess player" for whom the means has become an end in itself. How far Tarski's semantic explication can contribute to solving the problem relevant to the methodology of empirical sciences of explicating "nearness to the truth" thus depends on answering the question under discussion. In any case the distinction between Tarski's project of giving an explication of the concept of truth for formal languages and the project of giving an explication of "nearness to the truth" which can help us legitimate a methodological preference rule is very striking.

2. LOGICAL EMPIRICISM: A PHILOSOPHICAL POSITION AND APPROACH.

2.0. The *IL* is certainly the main instrument of logical empiricism. But it is, as was already mentioned, wholly consistent to recommend the *IL* approach for particular areas of research in philosophy and in philosophy of science and simultaneously to regard the philosophical position, i.e., the presuppositions, of logical empiricism to be untenable. The distinguishing mark of logical empiricism can perhaps be most clearly depicted by placing its source of inspiration at the center: Wittgenstein's *Tractatus*. Let us for a moment consider the *Tractatus*, with Ernan McMullin [1974], as a thought experiment which tells us what theory of language and, in consequence, what theory of physics would be adequate if certain presuppositions were fulfilled. Then the following would be the most important. (1) When the issue is one of developing a theory of science, then all *relevant* nonextensional contexts can be reduced to *extensional* ones, i.e., can be expressed in the *IL*. (2) The semantics (formal interpretation) of the *IL* possesses correspondence rules which are theory-independent so that atomic propositions can "depict" facts or possible states of affairs – "depict" in the same sense in which one mathematical function can be depicted by another. (3) There is a class of propositions which are epistemologically unproblematic, i.e., which build a *secure "basis,"* and we are in a position to discover such propositions. From these assumptions would follow an ideal of science according to which research would essentially consist in continually expanding the stock of secure basic propositions through truth-functional combination or through logical operations in accordance with inductive-logical or probabilistic rules. If this is the case, *then the only task belonging to philosophy of science is that of articulating the various aspects of this ideal of science and making them more precise*: it then construes, e.g., its models of patterns of explanation in the "Ideal Unified Science," which are formulated in the *IL*, gives an explication of the notion of "empirical significance," in which either basic propositions or basic predicates are brought into play, develops models of a scientific theory as an interpreted calculus, etc. A scientific theory (explanation, etc.) represents progress over its predecessor insofar as it offers a better approximation of the model of the theory (explanation, etc.) of the Ideal Unified Science than its predecessor does. For our present purposes it suffices to indicate more closely in what follows the *ideal of science*, so that we can then give the Popperian critique and the Popperian alternative.

2.1. *Logical Empiricism's Ideal of Science*

At the basis of logical empiricism's philosophy of science there lies a particular ideal of science. In what follows, a highly stylized portrait of this ideal is to be given. The ideal of science may be regarded as an explanation of our intuitive ideal of science, an ideal whose attractiveness, if we disregard for the moment its realizability, could scarcely be doubted.

In the view of this intuitive, naive ideal of science, the aim of research is to produce *genuine* knowledge which is as *comprehensive* and *deep* as possible: i.e., it should explain a lot and these explanations should help us understand the world and humankind better and better. A successor theory represents progress over the predecessor theory if it achieves more in at least one of these aspects. This in no way represents an essentialist definition of "science." Rather, if an explication is to be a meaningful undertaking at all, there must be an explicatum there, an (at least) intuitively sufficiently determinate concept, a concept which through a sufficiently stable use of language has a more or less stable "semantic spectrum".

The explicatum which logical empiricism works out is introduced through the basic idea that scientific knowledge should be so reliable that it *idealiter* offers certainty. Of the three desiderata, "genuineness" in the sense of certainty, universality and "depth," *certainty* is granted the *highest priority*. Methodology accordingly gets the task of formulating and legitimating a *rule of acceptance* which accords with the conviction that a proposition of the sciences is in the final end acceptable only if it is true. In order to make such a rule operative, one needs in addition to an explication of the concept of truth also a method of ascertainment by means of which it can be decided in concrete cases whether a particular sentence is true or not, i.e., a method of ascertainment with a guarantee of truth, an infallible criterion. A sentence is acceptable – in the sense given by the ideal – just in case it is not only true but recognized to be true, "proven": this is the core of *verificationism*.

The *role of experience* in such a demonstration is *positive*: it serves *as an establishing arbiter*. Verificationism in the strict sense seems practicable only if one is willing to assume a secure basis, e.g., a particular kind of sensory experience as a sure source of knowledge about reality as a foundation on which the edifice of science can be erected. One could almost call it "proof empiricism." The intuition lying at the basis of this approach is the ideal that it is possible to "cover" the informative content of a sentence with the informative content of the set of sentences introduced as evidential support for it. If one acknowledges – entirely disregarding the problems of the basis – that this demand cannot be fulfilled for universal sentences, then one lowers his sights and demands merely a partial informational covering, in order to "probabilify" a universal sentence by means of its degree of confirmation so much that its adoption seems justified. It seems that at the basis of this *probabilistic version of verificationism* lies the hope that if more and more evidential support is produced, it would be possible "in the long run" to realize or at least asymptotically to approximate this ideal condition of complete informational covering.

Logical empiricism's philosophy of science can be regarded as an effort to articulate the crucial aspects of this ideal of science.[15] The *IL* is here the major tool. The models of the various aspects are then articulated by means of at least partially formalized languages. Thus the explicatum of the concept of "empirical significance" as a necessary precondition for a proposition's or a theory's qualifying as part of a science connected with experience; the deductive-nomological model of explanation, which articulates the criteria for the adequacy of the formal aspects of a pattern of explanation in the ideal science etc.

If one were to consider the ideal to be utopian in the negative sense, one would plainly not enjoy elaborating and making precise its individual aspects so much. If one accepts this ideal, one then also holds a particular *descriptive picture of science* to be not too unrealistic. This ideal of science projects an ideal end-state: evidential support accumulates and the "covering" of universal hypotheses with these pieces of evidence improves until finally certainty is reached. To this there corresponds a representation of the course of actual, historically given science according to which science conserves what knowledge it has achieved and, since it is *cumulative*, will someday reach a complete end-state, so to speak. This representation sees science in analogy with the exploration of the earth's surface since the Middle Ages: as an ongoing "exhaustive search" of an essentially finite domain [Rescher 1978]. A descriptive picture of science subscribed to by many renowned historians of science and also by many contemporary physicists.[16]

3. POPPERIAN CRITIQUE OF LOGICAL EMPIRICISM'S IDEAL OF SCIENCE

3.0. The following critique of the ideal of science results from the standpoint of Popperian methodology; it is implicitly present there, even if Popper does not explicitly speak of a critique of this ideal of science. Every critique of a methodology has to come to terms with two kinds of questions: are this particular methodology's results adequate solutions to the problems it has set for itself; are these ways of posing the problems really the appropriate ones?

3.1. In regard to the critique of solutions of the problem offered, one would have to say from the Popperian standpoint that the various models (of "empirical significance," of the scientific theory as a calculus interpreted with an "(enlarged) observation language," etc.) have not been able to help the researcher, even if he wanted to, to bring his results – even if these are highly stylized – closer to realizing or approximating the desideratum mentioned. For, first, the presupposition of a secure "basis" is not fulfilled, if only because of the dependence in principle of linguistically formulated observations on universals or theories – an insight which has become a commonplace. Second, even if that were not so, we know since Hume that strict verificationism with respect to universal propositions is a logical impossibility; and the so-called logical probability of a universal proposition on the basis of a finite set of singular propositions equals zero according to probability calculus. To be sure, it must at once be admitted here that the possibility of an "inductive logic" is today the object of a lively controversy. Thus Adolf Grünbaum has argued that if "background knowledge" is interpreted as those theories which are not currently problematized, the explication of the key concept of falsificationism, "degree of severity of tests," encounters difficulties analogous to those of so-called eliminative inductivism (which cannot avoid relativizing with respect to the current competitor theory) or of Bayesian inductivism, which also takes "background knowledge" into consideration [Grünbaum 1978]. Here it should be remarked that if the concept of "background knowledge" is used in evaluating the relative "severity" of tests, this use is not inductivistic. Be that as it may, if Grünbaum's thesis about equal difficulties is correct, this circumstance does not improve the chances of probabilism. Probabilism as a proposed problem solution for fulfilling the task of approximating the desideratum of certainty as fully as possible must lead to an inability to fulfill the desideratum of the *deductive* structure of the ideal science. For *ex definitione* not only deductive (non-amplificatory) methods of transformation are admitted.

3.2. A Popperian critique of logical empiricism's *ideal of science* could be formulated as follows. Considered as an explicatum of our above-mentioned intuitive idea of an ideal picture of scientific knowledge, the ideal of science of logical empiricism would have to be charged with not at all satisfying the similarity criterion, not to mention the overriding metracriterion of "fruitfulness." That it has brought an essential increase in precision cannot garner many points for it, since precision, as stated, serves only as a means in the service of "fruitfulness." (1) The desideratum of "certainty" is in principle unattainable. In case we had an absolutely true empirical sentence, we would not be able to say with certainty that this was so. (2) Since certainty and content of empirical information are inversely proportional and since logical empiricism has given top priority to the demand for certainty, this must necessarily be at the cost of the status of the desideratum of "high informative content" or "increase in content." (3) The desideratum of "depth" is not only lost but anathematized: the fundamentally anti-metaphysical stance is a further justification for speaking of "positivism" here. The positive lies on the surface; for this reason positivists are against "deep" knowledge or deny the existence and possibility of such knowledge. Postulating a class of secure basic sentences formulated in a theory-independent observation language has led to the pseudo-problem of the relation between theoretical languages and observation language and to attempts at reduction, which Mario Bunge has called "theory-demolishing techniques."

The descriptive picture of the course of science fitting this ideal of science is refuted not only by the history of science as soon as the lesson of fallibilism has been learned; but it appears untenable already on epistemological grounds.

4. POPPER'S ALTERNATIVE

After the fundamental stance of logical empiricism has proved to be foundationalism, with the realization of the untenability of all such foundationalisms, scepticism would be a possible rational reaction. But it is not the only possibility. Popper offers an alternative: the striving for certainty, the mirage of foundationalism, as well as the utopia of an end-state, is given up in favor of a *fallibilism in principle* at least for empirical science. But the fallibilism is *united with holding fast to the ideal of absolute truth as a regulative principle, in particular to the comparative notion of "more accurate representation than."*

The question of the acceptability of a theory has its locus primarily in technological questions, when the issue is one of applying theories as software instruments. If in the context of a methodological discussion a hypothesis or theory is designated as accepted, this is an abbreviated way of saying that it has not (yet) been falsified and has been corroborated to such an extent that we propose to continue working with it and improving it, i.e., among other things to test it further. In basic research the concern is to develop theories further and upon occasion to replace one theory by a better one, i.e., the issue is progress. *A central task for methodology is thus not formulating an acceptance rule, but formulating and legitimating preference rules*: the issue is one of evaluating the comparative achievements of competing problem solutions. The goal of research could be outlined in one phrase thus: more and more accurate representation of those aspects of reality whose comprehension leads to new, fruitful perspectives and thereby to new and deeper problems. This concept is the core of the Popperian explication of the above-mentioned intuitive ideal of science. Accordingly, methodology is allotted the task of making the ideal of the goal just stated more precise, i.e., articulating the ideal of science implicit in it and proposing procedures with whose help the chances for research to approximate this goal can be improved. In brief, out of this ideal of the goal the individual components of the Popperian ideal of science develop and with them the corresponding tasks of methodology.

(1) From the *desideratum of "more and more accurate representation"* there arise two interconnected *tasks: explicating the comparative concept of nearness to the truth* and *developing methods of ascertainment, fallible indicators*, which help us adduce good reasons for the conjecture that one particular theory in fact comes closer to the truth than its rival. *The role of experience here is exclusively that of a critical arbiter*. As is well known, the degree of corroboration expresses, as a fallible indicator, the balance of the tests to date, the attempts at falsification, and assumes a key position in the reasons for preferring or not preferring a particular theory with respect to its representational function, which, to be sure, can only be one aspect of evaluating theories.

(2) Since the achievable degree of corroboration is essentially dependent on a theory's degree of testability in principle (an information-theoretic concept) and the degree of testability is equivalent to the content of empirical information, the theory with more content has better possibilities for being tested: that is, more risk of being falsified and thus more chance, upon occasion, of achieving a higher degree of confirmation. For this reason, from the first desideratum follows the desideratum of *increase in content*. From this desideratum in turn follows the recommendation, *ceteris paribus*, to prefer the theory richer in content, and thereby for methodology the task of developing the corresponding instruments of evaluation: i.e., explication of the comparative information-theoretic

concept "... has more empirical content than—" as well as development of corresponding methods of ascertainment. If content is defined as the consequence class of a theory, it is denumerably infinite and the class of consequences deduced to date can be used as a fallible indicator in order to advance good reasons for the conjecture that one particular theory has more content – or, after tests have been carried out, more corroborated content – than a particular rival theory. (Of course, strictly speaking, contents are comparable only if subset relations obtain.)

(3) A dramatic increase in informative content or, upon occasion, in new knowledge is possible only in cases where the successor theory contradicts the predecessor theory. Only then does the successor theory open really new perspectives, which can lead to deeper explanations and deeper problems. What is here meant by "depth" can most succinctly be illustrated with an example. The explanation of a particular solar eclipse with the help of Kepler's laws reaches a certain level of depth. The explanation of the Keplerian laws with the help of Newtonian theory reaches a greater depth. For not only is Newtonian theory more general, but while explaining the Keplerian laws it also corrects them. That, says Popper, is a sure sign that a new theory is deeper than the old [cf. 1963, p. 202]. The deeper explanation becomes possible through the successor theory's introducing new concepts, new categories: in the present example causal concepts. Another example of increase in depth would be the relation between Newton's and Einstein's theories. Einstein's general theory, which contradicts Newtonian theory, has only a very few corroborators over and above Newton's: (a) the precession of the perihelion of Mercury, (b) the bending of light rays (both are results supporting Einsteinian theory, while (a) discorroborates Newtonian theory) and (c) red shift. But it brought a new perspective and enabled an improvement of our world-picture, e.g., with regard to causality, made new questions "askable" and also posed them. In consideration of the above, it seems that *the third desideratum, "greater depth" is the distinguishing mark for essential progress.* A key position seems to attach to this desideratum, because without the introduction of new concepts/categories, new perspectives, neither an essential improvement of the world-picture nor a gain in corroborated content to a very large extent seems possible.

The desideratum of the *deductive* construction of science, or the use of exclusively non-amplificatory methods of transformation, does not need to be specified as a desideratum within the Popperian ideal of science;

for it follows from the desideratum of "higher degree of confirmation" together with the method of falsification that all methods of transformation must be deductive. The striving for certainty is retained in the only place where it is legitimate: in the transmission of truth from the premisses to the conclusion as well as in the transference of falsehood from a falsified conclusion back to the premisses. According to the following schema: $T \& A \rightarrow (I \rightarrow P)$. A theory T together with auxiliary hypotheses A implies logically that if initial conditions I prevail, the event or condition P occurs. If instead of P, P' occurs and P' implies not-P (and P' is accepted as authentic), and I is not placed in question, and the validity of the argument is also not placed in question, then the premisses ($T \& A$) are falsified: the falsehood is transferred back to the set of premisses. This is plainly logically correct; for which reason the transference of truth from conclusion *back* to premisses which contain universal sentences – something demanded by all versions of verificationism – is not permitted.

The picture of historically given science which fits the Popperian ideal of science is, as would be expected, contrary to the descriptive picture of science which fits the logical empiricist ideal of science with its hope for a final state, some day to be reached, which again presupposes that the number of basic problems is finite.[17] It is primarily supported by methodological reflections, but – entirely apart from that – it is also illustrated by the history of science. The corpus of scientific knowledge, the set of hypotheses and their implications which are accepted by the scientific community at a given time and represent the state of knowledge in the current discipline, does not grow cumulatively but organically, that is, certain parts are retained, to be sure normally in a revised version, but other parts are replaced by new ones. Research is an unbounded process and there is no reason to assume that at some time a final state will be reached. Science has its limitations, i.e., there are problems which are not scientific problems, but it has no limits. Kant said this very clearly in 1783: "In the... natural sciences, human reason does indeed recognize limitations, but no limits, that is, that something does indeed lie outside it, into which it can never reach, but not that it will itself be completed somewhere in its inner progress. The enlargement of insights ... proceeds into the infinite ..." [1783 §57].[18] The key position in the Kant-Popper picture of science – as a descriptive picture of historically given science for Kant and also for Popper the core of his ideal of science – is that *progress is conceived as a progressing from problems to deeper problems.* The *Kant-Popper thesis of problem propogation says* that every solved problem gives birth to several new ones. The measure of progress, of the gain in knowledge, is the degree of the deepening of problems: how much deeper are the new problems which first become formulable/askable through the new concepts and presuppositions of the new theory? Here the new theory as an answer to an older problem may introduce new presuppositions which replace the presuppositions lying at the basis of the predecessor theory, so that certain questions are no longer askable. (For example, the problem of the behavior of phlogiston in various experimental situations is a question which can no longer be asked once the phlogiston theory has been abandoned.) Kant underlines this problem propagation: "... we may begin as we like, every answer given according to basic propositions of experience always gives birth to a new question ..."[19] Popper stands here entirely in the line of succession from Kant: "... Science should be visualized as *progressing from problems to problems* – to problems of ever increasing depth." And he emphasizes, as we have already mentioned, that if the progress is significant, reaching the order of magnitude of a "scientific revolution," the new problems will be essentially deeper ("the new problems will be on a radically different level of depth"). If one considers the development of the most recent history of philosophy and of philosophy of science, it is indeed a rough approximation, but correct, to say that *Popper* is *the chief critic of that philosophy of science* (and the ideal of science on which it is based) *which owes its decisive impetus to Wittgenstein's* early work. The *Tractatus's* attempt to create a transcendental logic invites, so to speak, conceiving an ideal of science which places a premium on certainty and projecting a description of science which envisions a final state. The debate between Wittgenstein and Popper, which has not in fact been carried out directly by them, but rather implicitly between the logical empiricists as the continuers of the ideas of the Wittgensteinian *Tractatus* on the one side and Popper on the other, even has a further chapter: namely, in the debate between Popper and his adherents on the one side and those *critics of Popperian methodology* who in turn are *decisively indebted to Wittgenstein's later philosophy* on the other. Indebted entirely independently of the degree to which they are aware of this. This is true of T.S. Kuhn, S. Toulmin, Paul Feyerabend, to name only a few of the most prominent. In order to make this tie with Wittgenstein's later philosophy clear, one must indicate the distinguishing marks of this later philosophy.

5. THE *OL* APPROACH: WITTGENSTEIN'S LATER PHILOSOPHY AND THE SCHOOLS AND POSITIONS SPRINGING FROM IT

5.0. As is well known, the most important source of inspiration for the so-called analytic philosophy of language is Wittgenstein's later philosophy. Analytic philosophy of language – "ordinary language philosophy" (henceforth abbreviated "*OL*") – is not only a style of thought, but a philosophical position. At any rate, it is much easier to regard the *IL* approach's systems of theses as thought experiments and to let work in formal semantics begin, so to speak, with the words "We assume the following:..." The *OL* approach implies its presuppositions in an entirely different way.

5.1. Naturally, the activity of analysis and of clarifying concepts requires no justification, for every philosophical activity must involve consideration of such things. But if the analysis of ordinary language is declared to be the core or even the sole legitimate manner of philosophizing, there has been a totalization which clearly betrays an anti-philosophical attitude. What would be excluded with such a narrow, programmatic definition of philosophy? Above all two philosophical approaches: *IL* philosophizing as well as philosophizing by means of a conceptual apparatus produced for this goal, and that means the entire specialized philosophy of the philosophical tradition.

The objection to the *IL* approach seems to rest on a misunderstanding: in philosophizing by means of formal languages, all aspects of communication, the so-called pragmatic aspects of language, are abstracted from. But that only means that a division of labor is proposed, which is economical and even indispensable. Insofar as it is remembered that an abstraction has been made so that one's own approach does not become totalized, there is nothing to object to in such a division of labor. Only in a totalization of the *IL* approach would the accusation of the so-called "abstractive fallacy" (K.O. Apel) be justified. Disregarding that, studies such as those of P.F. Strawson were useful for making the limits of the possibilities of formalization clear. On the other hand, excluding that *IL* approach from "philosophy" would mean quite unnecessarily renouncing methods which can be very useful in certain areas. Raising to the status of a method the renunciation of special concepts of professional philosophy would be much more serious. This is not only the expression of an anti-philosophical attitude, but is also wholly impracticable. It is difficult, to put it mildly, to be a specialist in non-specializing. The demand for such an asceticism in concepts seems paradoxical, for every research is forced to develop a suitable technical language, since ordinary language, because of its enormous versatility, plainly cannot be a suitable instrument of thought and mutual understanding for the purposes of a specialized investigation. The rejection of every technical jargon by *OL* philosophers therefore remains declamatory. Thus J.L. Austin, for example, develops a terminological flora which departs widely from ordinary language.

5.2. *Programmatic definition of philosophy: thesis of the inconsequentiality in principle of all philosophizing*

At the basis of the programmatic definition there lies a presupposition: philosophy can contribute nothing to the content of our knowledge about world and human, because all problems about content are a matter for special sciences and practical problems are a matter for mastering in everyday life. Accordingly, the philosopher can do nothing more than analyze various usages in ordinary language, illuminate them, etc.: "Philosophy leaves everything as it is" [Wittgenstein 1953 §§599, 124, 126; Brand 1975, p. 202f]; "in the end it can only describe" the world, [Wittgenstein 1953, §§599, 124; 1969 p. 39; Brand 1975, pp. 202, 204, 207] "it spreads before us ... does not explain, does not deduce" [Wittgenstein 1953, §126]. These are classic passages from Wittgenstein's later philosophy. Thus a totalized version is conceivable here, according to which ordinary language would always have the last word, so to speak. In Wittgenstein it sometimes appears as if that were the case: "Philosophy may in no way interfere with the actual use of language ..." [Wittgenstein 1953, §124]. J.L. Austin clearly postulates a weaker assumption: that ordinary language should have the first word, but not (necessarily) the last [e.g., 1961, p. 133; cf. Copleston 1976, p. 113]. But that would hardly be anything more than a heuristic recommendation for the initial phase of working on a philosophical problem. Thoroughly plausible for certain questions in philosophy. But what comes after ordinary language or after the analysis of ordinary language which analytic philosophy of language gives? Again the specialized disciplines. Such a position would have to be designated as linguistic or language-analytic positivism, or as scientism. The same scientistic basic attitude even gave rise to the idea of a "scientific conduct of life" in the Vienna Circle in the wake of Wittgenstein's *Tractatus* – an idea whose difficulties Musil then brought out in his novel *The Man without Qualities* (*Der Mann ohne Eigenschaften*).

The scientistic basic attitude is also a distinctive feature of the linguistic-phenomenological anthropology of G. Ryle. If the *OL* philosophers are sometimes accused of reducing the task of philosophy to assisting with an improved edition of the Oxford English Dictionary, this accusation definitely does not hold for Ryle's *Concept of Mind.* Ryle's anthropology is based on a materialist metaphysics, a position characteristic not only of Ryle but also of the earlier and later Wittgenstein. It is obvious that this materialist metaphysics is not compatible with the metaphysics on which ordinary language is based. Certainly the fact that a functioning ordinary language is based on it is no criterion for the worth of a particular metaphysics. But how is it compatible with the project of only elucidating ordinary language to develop a particular metaphysics contradicting it, as Ryle has done? And how is the entire plan compatible with the declared methodological precept of avoiding specialized philosophical expressions and conceptual apparatus? The materialist ontology on which metaphysical behaviorism builds is already laid out in the *Tractatus.*[20]

The historical currents within contemporary philosophy of science obviously lie in the stream of Wittgenstein's later philosophy: they proclaim the fully analogous thesis that methodology has no effects: it leaves everything as it is; it can really only elucidate descriptively what successful researchers have done or do. Prescriptive methodology (in the sense of a quasi-technology) is not possible.

5.3. *"The limits of my language are the limits of my world" – would it therefore be impossible to overstep the limits or to fix the boundaries?*

Against linguistic positivism the following can be made to stick: it is correct that philosophy is in a certain sense secondary to science and to mastery of the everyday world, somewhat in the sense in which methodology is secondary to scientific research, and epistemology can first make an onset when successful performance with concepts like "true," "more accurate representation than (*zutreffendere Darstellung als*)," etc., already occurs. On the other hand, philosophy transcends science in that it is one of the traditional main tasks of philosophy to use the results of specialized disciplines as raw material in philosophical reflection, by means of which philosophical cosmology and philosophical anthropology have to develop and constantly to improve the image of the world and the image of humankind. In just this way one may hope to overstep or to enlarge "the limits of his language," exactly by enlarging "his" language. Otherwise there would be the danger of going from the world as language to language as the world.

Proponents of the style of philosophizing which consists in analyzing ordinary language sometimes explain that they do not investigate words or uses of linguistic expressions; rather, they analyze the key concepts of ordinary language. What does one hope to discover by analyzing such key concepts? These key concepts are in the end nothing but sedimentations of older philosophical doctrines which have not only left traces behind them in ordinary language, but have actually molded it. This claim is not controversial, but it has undergone very diverse interpretations. For example, Bertrand Russell thought that in ordinary language the metaphysics of the Stone Age was preserved. Austin, on the contrary, was of the opinion that in it was stored the human experience of millenia, which according to him provided a kind of guarantee of its "wisdom."

The more specialized an area of activity is, the more Russell's view may hold for it. So, for example, in the area of economics there are practitioners who think they can get along without theory and can operate on the basis of their practical experience, as a rule blind followers of a long-dead economic theoretician.[21] The same holds for the realm of research: a researcher who thinks he can get along without methodology as a the rule a blind follower of an outdated methodology which he can't even subject to criticism at all, since it is not articulated and he is probably unaware that he has opted for it.

On the other hand, Russell's interpretation may be much less or not at all true of the everyday world. In fact, Wittgenstein's later philosophy attempts to probe how things stand in this area and how one acts: "Both *show themselves*, are included *in acting* itself, are in the end given to us to be accepted in *forms of life*." Philosophy is then, as Gerd Brand writes further, "a description laying open ... how people understand the world and how they act in it – thus a phenemonology of the life world (*Lebenswelt*)" [Cf. Brand 1975, p. 16]. The problem of the normative may be incapable of being treated in any other way for the realm of the everyday. A substantive ethics may only be binding for us when it can base itself on tradition, religion or myth. But the situation is quite different in the realm of goal-directed action; as in the examples of economic activity and scientific research mentioned above.

5.4. *Can the limits of the monadic language game as forms of life be spanned by "descriptive metaphysics" and "descriptive ethics"?*

The analysis of ordinary language is sometimes carried out with the intention of pushing forward to "our" everyday world or to a particular language community's image of man and of the world by investigating the key concepts, ideas and themes with whose help "we" understand ourselves and the world or the language community investigated understands itself. This "phenomenology of the life world," has been called "descriptive metaphysics" by P.F. Strawson. "Descriptive ethics" would be a subdivision or a complement. In contrast to normative ethics, it would not at all aim at changing our basic attitudes. In such investigations one would naturally have to concentrate on concepts which play a central role in dealing with our existence and with whose help elemental experiences like "birth," "love," "death," and "happiness" are conceptualized. This has been emphasized particularly by Peter Winch. Then one would really have to investigate not only "our" form of life, but also the various roles and spectra which these concepts central for dealing with one's existence play in various forms of life, "language-game communities." But this may only be possible in interdisciplinary research, in co-operation with researchers in comparative linguistics, ethnopsychiatry, cultural anthropology, etc. But such interdisciplinary investigations have been up to now rarely, if ever, led or executed by analytic philosophers of language.

Can such investigations lead beyond projecting alternative possibilities for living, a kind of *musée imaginaire*? Does one intend thus to work out the "human in general"? Stuart Hampshire has hinted at this. What about the problem of the normative? From the beginning it was explicitly set aside. Will it always remain thus set aside? Are the various forms of life – such as African sorcery, Russian communism, leftist dictatorships and rightist dictatorships, liberal democracy – of equal value? Or are they even incommensurable? A moral and aesthetic relativism, as the thesis that it is not possible to establish a rankordering of various forms of life in these dimensions, which could claim universal validity, or for whose validity "compelling arguments" could be advanced, seems in fact plausible. It becomes most implausible when the thesis of relativity is also extended to the realm of goal-directed activity, particularly to that of scientific research. But it seems that many advocates of analytic philosophy of language conceive relativism so broadly; above all this is done by the historians and philosophers of science who, often without being fully aware of it, move in the shadow of Wittgenstein's later philosophy.

5.5. *The loss of the problem of the normative*

Analytic philosophers of language have been compared with people who clean their spectacles in order to be able to see better with them, but then don't look through these spectacles. They have been accused of wanting to be nothing but "specialists in eliminating confusion" (H. Lübbe) who attempt by means of language-cleansers to lead the "battle against the bewitchment which forms of expression exercise on us" in order "to eliminate misunderstandings, untangle knots" [Wittgenstein 1969, p. 51; 1932, p. 115; Brand 1975, pp. 196, 197]. The resignation in the thesis of the inconsequentiality in principle of all philosophizing shows the skepticism of these philosophers about philosophy. "To show the fly the way out of the fly-bottle" is then also an attempt doomed to failure from the start to "emancipate" philosophy – "emancipation" in the Marxist sense, i.e., "freeing humanity *from*..." – this time not from Jewry but from philosophy. For "philosophizing" may belong to life, and our freedom of choice consists in being able to do it consciously or implicitly. Even if an ordinary language has functioned satisfactorily in mastering the everyday for as long as you like and the universally human is grasped in the analysis of "key concepts," an evaluation of the "metaphysical" fundamental concepts lying at the basis of this ordinary language and of other key concepts could not be derived from such descriptive hypotheses, no matter how well confirmed, neither in the epistemological nor in any other dimension. From descriptive sentences about the functioning of a particular language or of "language" the evaluative statement can just as little be derived that a moral validity attaches to the norms lying at the basis of the language game or presupposed in it; a validity, that is, which is constituted differently from the political and social bindingness of norms which is based on "correct" procedures for establishing norms [Lübbe 1976]. If one emphasizes the (so to speak) built-in "wisdom" of a smoothly functioning ordinary language, there is a danger of confusing the social bindingness of norms with the validity of conceptions which can only be supported on objective (although fallible) methods of ascertainment, i.e., methods where not a human decision but an arbiter outside the human sphere of influence settles the issue: hypotheses which have to prove true in empirical testing. Confounding claims to truth and social bindingness would, as H. Lübbe has shown [Lübbe 1976; Radnitzky 1977], eventually lead to the destruction of the specifically civil rights of freedom of creed and of opinion, and to the imputation that the conceptions whose bindingness the victorious majority was convinced of are *moreover* to be recognized as true. (Obviously, socio-political bindingness also says nothing about *moral* value of such a conception.) Indeed, that which seems plausible in moral relativism would also be repudiated thereby: identifying the political bindingness of norms and truth-claims of statements would have totalitarian consequences.

What would the analytic philosopher of language advance against such objections? Presumably he would say that in his opinion the question of the philosophical quality of metaphysical hypotheses and that of the bindingness of substantive moral rules are meaningless questions. In that case, Ernest Gellner's well-known accusation that analytic philosophy of language does not at all attempt to broach normative questions would seem in fact to be justified. As is well known, in analytic philosophy of language ethics is reduced to descriptive ethics, which basically must be an empirical discipline, or it shrivels into metaethics. Here, to be sure, the question of the ethical presuppositions and implications of the various kinds of metaethics then arises again.[22]

6. ON THE CRITICS OF THE POPPERIAN ALTERNATIVE WHO ARE INFLUENCED BY WITTGENSTEIN'S LATER PHILOSOPHY

T.S. Kuhn and Paul Feyerabend are today presumably the most influential of these critics. *Kuhn's* ideas were largely anticipated by Ludwick Fleck in 1935; but these ideas first attained a wide influence with Kuhn. As is well known, with the aid of a collection of examples from the history of science, Kuhn accuses Popperian methodology of being unrealistic or at least oversimplified as a basic model for stylizing the history of science. In his commentary on Kuhn's contribution in the Schilpp volume, Popper admitted that he has previously paid little attention to what lies between the important developments, namely "normal science." That a methodological model must make oversimplifications lies in the nature of the affair. But Kuhn completely misunderstands the character of Popperian methodology: it is prescriptive, i.e., it formulates recommendations for how to proceed in particular types of research situations if the researcher wants to improve his chances of achieving progress and it attempts to legitimate such recommendations. Therefore it can only be criticized argumentatively: by showing that there are no good reasons on hand for regarding these recommendations as helpful. But for this the normative, the evaluating concept of "scientific progress" is indispensable. The Kuhnian approach, with its monadic research groups held together by paradigms and its systems of theories which are incommensurable if divided by a paradigm change, cannot in principle carry out such a critique.

The Kuhnian picture of science is descriptive. Its central concept, "paradigm" has indeed many meanings, but certainly the meaning corresponding with the *Wittgensteinian concept of the language-game as form of life* is central. The complementary concept to normal science, the concept of the so-called scientific revolution, corresponds with the *Wittgensteinian use of Gestalt switch.* Kuhn can speak of progress *only* in connection with a succession of theories developed under one and the same paradigm, and there progress is spoken of in terms of increased potential for puzzle-solving. As soon as he claims that theories within one discipline are incommensurable, he has already given up on the task of explaining or argumentatively showing why it is rational to prefer one theory to another one. He is also criticized for this, e.g., by S. Toulmin, another critic of Popperian methodology influenced by Wittgenstein's later philosophy. In short, the problem of evaluation is completely lost exactly where it is most pressing. To the question, "When should a paradigm be accepted, when is it better than a rival paradigm?" Kuhn has no answer except that sometimes there exists a consensus of scientists with regard to a particular paradigm. The search for objective (although fallible) criteria of evaluation is abandoned in favor of a subjective criterion of consensus: the strong tendency of the Kuhnian position towards subjectivism and relativism is obvious. Something similar can be said of Stephen Toulmin. In a survey article in which he summarizes the development of the philosophy and history of science since the 1950's he maintains the following: during a phase "oriented towards specialization," the acceptable focal points of academic studies were those aligned with an "abstract, specialized idea of the truth" and they remained without direct relation to or interest in the capacity for doing good. In contrast, during the ensuing "problem-oriented" phase this point of reference shifted to a "more concrete, generalized standpoint", and accordingly "the capacity for Good is a necessary element in the Truth itself."[23] As Imre Lakatos has pointed out, historiography of science is methodology-impregnated. Kuhn's history of science is as "paradigm"-dependent as the rest of scientific knowledge. Peter Munz rightly emphasizes that if Kuhn's theory is correct, the theory that scientific revolutions are not governed by effort to come closer to the truth, then he must be wrong in his belief that he has proved his case by reconstruction of the history of science.

Paul Feyerabend, who has sharpened the Kuhnian theses, can only speak of progress obliquely. He recognizes this explicitly when he writes, "... whereby the word 'progress' is to be understood as the defender of a particular rule understands it, thus differently for different social and professional groups" [Feyerabend 1972, pp. 168f]. Thus there can be no objective indicators of scientific progress for Feyerabend. Further, not only do we have no method for ascertaining whether a particular theory represents progress over another, but also it is impossible to give an adequate explication of our intuitive idea of cognitive progress. A presupposition of this thesis is the descriptive hypothesis that up to now there has not been any recognizable scientific progress, i.e., in the sense of a concept of progress which Feyerabend would accept. A provocative thesis in fact, especially since scientific progress represents the paradigm of progress in general and is perhaps the only area in which progress can incontestably be spoken of. Intentionally provocative, for Feyerabend explicitly says that an "epistemological anarchist" is "like the Dadaist, whom he resembles much more than he resembles the political anarchist" and that "his favorite pastime is to confuse rationalists by inventing compelling reasons for unreasonable doctrines" [Feyerabend 1975, p. 189]. So then, Feyerabend is also always amusing and his theory of science is simultaneously an anti-theory of science. (He must develop both, for he himself says, "to be a true Dadaist, one must also be an anti-Dadaist" [Feyerabend 1975, p. 189].) A useful challenge for critical rationalism. If one followed the proposals of these two critics, then one would *give up methodology – but in favor of what*? Perhaps in favor of a *sociologistic reduction*? We would designate as a relapse into sociologism the train of thought which runs: since there are no objective measures for evaluating the products of research, let us attempt to develop better measures for evaluating the producers! This would be a prime example of what Imre Lakatos has called a "degenerating problem-shift": if one possessed objective indicators for the scientific quality of products, it would be senseless to choose such a problem-shift. If one has no objective indicators, then it is a vain hope to want to solve the problem by retreating to evaluating the producers. For a non-circular evaluation of a producer as a "good" producer of X is possible only if we have independent methods of ascertainment (which must also be objective, but may be fallible) for deciding when a product of type X is "good" or "better." To reach back here to empirical consensus, e.g., of a particular group of producers, would be a relapse into the naive form of sociologism.

More dangerous than reductive sociologism – because more misleading – seems to be the trend in current philosophy of science which one could call 'historiographism': namely, using the history of science as arbiter in evaluating prescriptive (quasi-technological) methodologies. But the normative problem of progress cannot be avoided in this way either. The attempt to criticize or to legitimate methodological rules on the basis of descriptive (methodology-independent) history of science cannot succeed. If it is maintained that with the aid of an example one can show that a particular methodological rule led to successs in a particular historical research situation, or would have hindered success if it had been applied, the one must also be able to show why it was successful (or would have been a hindrance) and why one would have reason for trusting that in the future it would also lead to success in similar situations. But above all the problem of evaluating theories would arise anew if the judgment should be justified that the development in question represents "scientific progress." History of science without methodology would be like a ship without a compass. One would not know at all what to describe and explain, for in order to know which developments were the important, the crucial ones, one needs methodological measures for evaluation. This claim is not at all to be equated with declaring methodology to be independent. Methodology needs the history of science not only in order to get material for its thought experiments; it cannot evaluate a historically given theory at all without becoming involved itself in research in the history of science. For a theory is something which has grown historically. For example, the designation "the Newtonian theory" is ambiguous: there is a historical succession and later even various versions alongside each other, various formulations of the "same" version, and one particular version must first be extracted from the historically given texts and perhaps be reconstrued. Thereby a formalizing reconstruction of course represents a kind of cross-section: a kind of experimental preparation which is laid under the microscope and arranged. Neither methodology nor the history of science can get along without the other.

Our polemic is here directed against a successor to Wittgenstein's later philosophy in methodology, according to which methodology "leaves everything as it is" and any theoretical developments which are divided from each other by a "paradigm change" – since they are conceived in analogy to Wittgensteinian monadic forms of life – can only be criticized "from within", i.e., in the framework of the currently accredited paradigm. A toothless critique in fact. In the realms of goal-directed action, and in particular in the realm of scientific research, relativism (which may be justified in the realm of morals) is untenable. A relativism in philosophy of science *à la* Kuhn or an anti-theory of science *à la* Feyerabend would moreover have dangerous political implications[24] if it found sufficient dissemination, sufficiently many "intellectuals" who believed in it. That, e.g., the authors mentioned in no way want the political consequences does not help much, because those who want them will know how to exploit relativism in methodology for their goals. An intellectually free society can only arise by codifying a concept of truth which makes truth objective. The idea of empirical investigation, of experimenting, of empirical science itself, expresses the insight that a decision about what is true lies *outside* the human sphere of influence. That the same insight cannot be disregarded in practical action is taken care of if necessary by foundering upon reality. But one can always veil this in rhetoric and negate it in proclamations. That exactly today many attempts are made to blur the distinction between "is" and "ought" is a phenomenon of our time, of a time in which ideological trail-blazers, in order to be able to bring their secular doctrines of salvation to the people, have to attempt to clear this distinction out of the way. For only in this way can an epistemological basis be laid for a totalitarian party's demands to be the sole competent authority for all realms of life. From this viewpoint, Popper's social philosophy, the philosophy of the Open Society and the politics of small steps – the politics which wants to test the consequences of political decisions empirically, instead of using promises of a Millennium to support a program of totally reshaping society – seems to be more important than everything else. Its topicality could be no greater. But this social philosophy is indivisibly linked with a methodology which sees the development of objective, although fallible, measures for appraising the comparative achievements of competing problem solutions as one of its primary tasks. This methodology is the core of its epistemological basis, whereby research serves as the paradigm for rational problem-solving behavior in general, i.e., also in the public-political sphere.

Criticism first makes the open society possible, not vice versa. But in this criticism the problem of the normative plays an essential role. A philosophy of science which understands itself in analogy with the later Wittgenstein's dictum for philosophy – "Philosophy is really 'purely descriptive'" [Wittgenstein 1969, p. 39, Brand 1975, pp. 202, 204, 207] – seems dangerous, because it contributes in practice to relinquishing the terrain to those who have no self-criticism and make exorbitant ideological demands or give promises of salvation. There may be much which is in need of improvement in Popperian "critical rationalism," a position including not only methodology and social philosophy based on it but also a general theory of goal-directed activity – in any case, the position is in the highest degree worthy of improvement, for without it a form of life which respects individual liberties could not be maintained. Whether one wants that is, to be sure, a personal, existential decision, that is to say, a decision for which there are no universally binding arguments.

Universität Trier

NOTES

[1] This is spelled out, e.g., in Radnitzky [1979b, 1981c].

[2] Cf., e.g., Radnitzky [1968/70] I, pp. 40ff.

[3] Among his successors, questions are also treated today such as that of the possible 'empirical' boundaries of mathematics: a statement decidable in principle could perhaps require such a copious proof that it cannot even be accomplished by computers, not only by technically possible but also by empirically possible computers (Michael Rabin, Hebrew University; Larry Stockheimer, IBM).

[4] Cf. Bergmann [1964], pp. 137, 179, 93, 238. Wedberg [1966] gives a very clear survey of the nature of formal languages; a work which unfortunately is so far available only in Swedish.

[5] The various types of justificatory arguments are treated in Radnitzky [1976a].

[6] The political implications of denying or blurring the distinction between 'is' and 'ought' are treated in the author's contribution to the Ernst Topitsch Festschrift (K. Salamun, ed. [1979]).

[7] Thus Gustav Bergmann calls the "world" of the *Tractatus* "a world without mind."

[8] Cited from Lorenzen's declaration of his program ("Erlanger Schule der konstruktiven Wissenschaftstheorie," *UniKurier Erlangen* **3** [1977] 16–18) in which, because it is directed to a broad public, the basic position of the school is set out particularly clearly.

[9] A classic example would be the analysis of the sentence, 'The present king of France is bald'; cf. e.g., Radnitzky [1968/70], p. 40, fn. 34.

[10] Stated explicitly in Popper [1974 – Alpbach lecture]. Cf. also Note 9 above concerning 'grounding' mathematics and logic. The work of Hans Lenk and of Hans Albert are highly relevant in this connection.

[11] This style of thinking has dominated philosophy in Sweden, for example, since the end of the Second World War.

[12] As is well known, attempts at 'grounding' founder on Hans Albert's "Münchhausen trilemma" (*circulus vitiosus, regressus infinitus* or breaking off the attempt dogmatically); cf. Albert [1968, 1975], also Albert [1978].
[13] Scheibe [1973] also regards the schema given as an adequacy condition.
[14] In Popper [1963] App., Popper [1972] ch. 9. As is well known, Popper first developed this theory in the 1960's.
[15] Radnitzky [1968/70]. Vol. I attempts to show this.
[16] Thus, for example, the well-known historian of science G. Sarton [1931], esp. pp. 10f., and [1936], p. 5. D. Bromley in Bromley, D., et al. [1976], p. 26, to which Nicholas Rescher refers.
[17] This presupposition also lies at the basis of the so-called finalization theory, whose status is now wavering in West Germany. It is criticized in, e.g., Andersson [1976], Radnitzky [1976a], pp. 398ff., Radnitzky [1976b], §3.1, pp. 28–31. For a comprehensive presentation of the debate, see Andersson [1977].
[18] p. 352 in the Kant [1911] edition.
[19] *Ibid.*, The title of the Italian edition of Popper's intellectual biography is particularly appropriate: "La ricerca no ha fine."
[20] The materialist ontology (on which metaphysical behaviorism builds) is laid out in the *Tractatus*. A key passage is 5.542: "But it is clear that '*A* believes that *p*', '*A* thinks *p*', '*A* says *p*' are of the form ' "*p*" says *p*'..." In this attempt to reduce intentional contexts to extensional ones, the species of the act is lost. 'Thinking', 'believing', etc., are acts, 'depicting' is not. " '*p'says p*" can mean two things: either, the proposition-that-*p* *depicts* the state of affairs, or the thought-that-*p* (in the objective Fregian, Popperian sense) *intends* this state of affairs. If "*A* believes-that-*p*" were of the same form as "*A* depicts *p*", then the actually psychic (or a statement about something mental) would be replaced by something non-psychic (or a statement about a sentence *form*), therefore something physical. Here is the crucial error. If it is not recognized, it must be concluded: if only extensional contexts are allowed, then "*A* believes that *p*" cannot be said at all. Soul/mind are ineffable, inexpressible. And thus the thesis of ineffability rests on the above error. In Wittgenstein there is 'ineffability' for yet another reason. Ineffability concerns, for example, that which Gustav Bergmann calls "the tie of exemplification" (e.g., an individual exemplifies a property – the reverse relation would be "ontologically impossible"). This quasirelation ('quasi' because one of the relata without anything more can only have the status of "possibility") in any case shows itself, but if one had not already always understood it, it could not be understood at all and it would thus be pointless to elucidate it verbally, to want to explain it. This thesis of ineffability is convincing. See Bergmann [1964] for this. Incidentally, it has great similarity with the so-called hermeneutic circle.
[21] Cf. Keynes [1936], p. 383; Albert [1976], esp. p. 150 – the reference to Keynes.
[22] Analytic philosophy of language is not alone in its reticence about the normative problem. Something similar could also, *mutatis mutandis*, be maintained of Gadamer's hermeneutics – e.g., [1960/1965], pp. 343f., 314f. In the German-speaking world several of the tendencies which have occupied themselves intensively with analytic philosophy of language have attempted to derive a substantive ethics from "language" or from "preconditions for communication."
[23] Cf. Toulmin [1977], p. 160 – the loss of the correspondence concept of truth, the real concept of truth, could not be more obvious. Regarding this, cf. also Radnitzky [1976b].

REFERENCES

Agassi, J.: 1975, *Science in Flux*, Reidel, Dordrecht.

Agassi, J.: 1977, *Towards a Rational Philosophical Anthropology*, M. Nijhoff, The Hague.

Albert, H.: 1963, *Traktat über kritische Vernunft*, J.C.B. Mohr, Tübingen. 3rd rev. ed. 1975.

Albert H.: 1978, 'Science and the Search for Truth: Critical Rationalism and the Methodology of Science', in Radnitzky and Andersson (eds.) [1978], pp. 203–220.

Andersson, G.: 1976, 'Freiheit oder Finalisierung der Forschung', in Hübner et al., [1976], pp. 66–76.

Andersson, G.: 1978, 'Truth and Scepticism: The Problem of Verisimilitude', in Radnitzky and Andersson [1978], pp. 291–310.

Austin, J.L.: 1961, *Philosophical Papers*, Clarendon Press, Oxford.

Bartley, W.W.: 1962, *The Retreat to Commitment*, Knopf, New York.

Bartley, W.W.: 1974, *Wittgenstein*, Quartet, London.

Becker, W. and K. Hübner (eds.): 1976, *Objektivität in den Natur- und Geisteswissenschaften*, Hoffmann und Campe, Hamburg.

Bergmann, G.: 1964, *Logic and Reality*, University of Wisconsin Press, Madison.

Brand, G.: 1975, *Die grundlegenden Texte von Ludwig Wittgenstein*, Suhrkamp, Frankfurt.

Bromley, D., et al.: 1976, *Physics in Perspective*, student ed., National Research Council, National Academy of Science Publ.

Bunge, M.: 1967, *Scientific Research*, Springer, New York. Vol. I: *The Search for System*. Vol. II: *The Search for Truth*.

Bunge, M.: 1972, 'A Program for the Semantics of Science', *Journal of Philosophical Logic* **1**, 317–328.

Bunge, M.: 1973a, *Philosophy of Physics*, D. Reidel, Dordrecht.

Bunge, M.: 1973b, *The Methodological Unity of Science*, D. Reidel, Dordrecht.

Bunge, M.: 1973c, *Method, Model and Matter*, D. Reidel, Dordrecht.

Bunge, M.: 1974, *Treatise on Basic Philosophy*, Reidel, Dordrecht. Vol. I: *Semantics I. Sense and Reference*. Vol. II: *Semantics II. Interpretation and Truth*.

Bunge, M.: 1977a, *Treatise on Basic Philosophy:* Vol. III: *Ontology I: The Furniture of the World*, D. Reidel, Dordrecht.

Bunge, M.: 1977b, *Foundations of Physics*, Springer, New York.

Bunge, M.: 1977c, 'The Interpretation of Heisenberg's Inequalities', in Pfeiffer [1977], pp. 146–155.

Bunge, M. (ed.): 1973, *Exact Philosophy. Problems, Tools and Goals*. D. Reidel, Dordrecht.

Cohen, R.S., P. Feyerabend and M. W. Wartofsky (eds.): 1976, *Essays in Memory of Imre Lakatos*, [*Boston Studies in the Philosophy of Science*, vol. 39], D. Reidel, Dordrecht.

Copleston, F. 1976. 'Sprache und Realität. Gedanken zur analytischen Philosophie', in Becker and Hübner [1976], pp. 109–122.

Feyerabend, P.: 1975, *Against Method. Outline of an Anarchistic Theory of Knowledge*, New Left Books, London.

Feyerabend, P.: 1977, 'Marxist Fairytales from Australia', *Inquiry* **20**, 372–397.

Feyerabend, P.: 1978a, *Science in a Free Society* (sequel to *Against Method*), New Left Books, London.

Feyerabend, P.: 1978b, 'In Defence of Aristotle. Comments on the Condition of Content Increase', in Radnitzky and Andersson [1978], pp. 143–180.

Feyerabend, P.: 1979, 'Dialogue on Method', in Radnitzky and Andersson [1979], pp. 63–132.

Fleck, L.: 1935, *Entstehung und Entwicklung einer wissenschaftlichen Tatsache. Einführung in die Lehre vom Denkstil und Denkkollektiv*, Schwabe, Basel.

Gadamer, H.G.: 1960/1965, *Wahrheit und Methode*, J.C.B. Mohr, Tübingen.

Gödel, K.: 1931, 'Über formal unentscheidbare Sätze der *Principia Mathematica* und verwandter Systeme', *Monatshefte für Mathematik und Physik* **38**.

Grmek, M., R.S. Cohen and G. Cimino (eds.).: 1981 *On Scientific Discovery. The Erice Lectures 1977*. (*Boston Studies in the Philosophy of Science* **34**). D. Reidel, Dordrecht.

Grünbaum. A.: 1976, 'Can a Theory Answer More Questions than One of Its Rivals?', *British Journal for the Philosophy of Science* **27**, 1–23.

Grünbaum. A.: 1978. 'Popper vs. Inductivism', in Radnitzky and Andersson, [1978], pp. 117–142.

Harré, R., (ed.): 1975, *Problems of Scientific Revolution. Progress and Obstacles to Progress in the Sciences*, Clarendon Press, Oxford.

Hohn, F.: 1960, *Applied Boolean Algebra*, Macmillan, New York.

Hübner, K., N. Lobkowic, H. Lübbe and G. Radnitzky (eds.): 1976, *Die politische Herausforderung der Wissenschaft. Gegen eine ideologisch verplante Forschung*, Hoffmann und Campe, Hamburg.

Kant, I.: 1783, *Prolegomena zu einer jeden künftigen Metaphysik, die als Wissenschaft wird auftreten können*, in *Kants Gesammelte Schriften*, vol. 4, Reimer, Berlin, 1911.

Keynes, J.: 1936, *The General Theory of Employment, Interest, and Money*, Macmillan, London.

Kuhn, T.: 1962, *The Structure of Scientific Revolutions*, Chicago University Press, Chicago; 2nd enl. and rev. ed. 1970.

Lübbe, H.: 1976, 'Dezisionismus. Eine kompromittierende politische Theorie', *Schweizer Monatschefte* **55**, 949–960.

McCulloch, W. and W. Pitts: 1943, 'A Logical Calculus of the Ideas Immanent in Nervous Activity', *Bulletin of Mathematical Biophysics* **5**.

McMullin, E.: 1974, 'Two Faces of Science', *Review of Metaphysics* **27**, 655–676.

Pfeiffer, H. (ed.): 1977, *Denken und Umdenken. Zu Werk und Wirkung von Werner Heisenberg*, Piper, Munich.

Popper, K.: 1935, *Logik der Forschung*, Springer, Vienna; 4th rev. ed. J.C.B. Mohr, Tübingen, 1971. Engl. tr. *The Logic of Scientific Discovery*, Basic Books, New York, 1959; rev. eds. 1960 and 1968.

Popper, K.: 1963, *Conjectures and Refutations*, Routledge and Kegan Paul, London.

Popper, K.: 1972, *Objective Knowledge. An Evolutionary Approach*, Oxford University Press, London.

Popper, K.: 1974, *Wissenschaft und Kritik* (Lecture delivered in "Alpbach Seminar"), mimeo. MS.

Popper, K.: 1975, 'The Rationality of Scientific Revolutions', in Harré [1975], pp. 72–101.

Popper, K.: 1976, *Unended Quest. An Intellectual Autobiography*, Fontana Library, London. (Repr. from Schilpp, [1974]).

Popper, K.: 1976, 'A Note on Verisimilitude', *The British Journal for the Philosophy of Science* **27**, 147–160.

Radnitzky, G.: 1968, *Contemporary Schools of Metascience*, Esselte Studien, Göteborg, and Humanities, New York; 2nd ed. 1970; enl. pbk ed. Regnery, Chicago, 1973; from Gateway Editions, Chicago, 1977.

Radnitzky, G.: 1974, 'From Logic of Science to Theory of Research', *Communication and Cognition* **7**, 61–124.

Radnitzky, G.: 1976a, 'Popperian Philosophy of Science as an Antidote against Relativism', in Cohen, et al. [1976], pp. 505–546.

Radnitzky, G.: 1976b, 'Prinzipielle Problemstellungen der Forschungspolitik', *Zeitschrift für allgemeine Wissenschaftstheorie* **7**, 367–403.

Radnitzky, G.: 1976c, 'Dogmatik und Skepsis. Folgen der Aufgabe der Wahrheitsidee für Wissenschaft und Politik', in Hübner, *et al.* [1976], pp. 24–54.

Radnitzky, G.: 1976d, 'Science and the Search for Values. Motives and Dangers', *Proceedings of the IVth International Conference on the Unity of the Sciences* (New York 1975), International Cultural Foundation, pp. 221–234.

Radnitzky, G.: 1977, 'Science and Values. The Cultural Importance of the Is/Ought Distinction', *Proceedings of the Vth International Conference on the Unity of the Sciences* (Washington, D.C. 1976), International Cultural Foundation, pp. 789–805.

Radnitzky, G.: 1978, 'The Boundaries of Science and Technology', in *The Search for Absolute Values in a Changing World. Proceedings of the VIth International Conference on the Unity of the Sciences* (San Francisco, Nov. 1977), International Cultural Foundation, New York, pp. 1007–1036.

Radnitzky, G.: 1979a, 'Die Sein-Sollen-Unterscheidung als Voraussetzung der liberalen Demokratie' in Salamun [1979], pp. 459–493.

Radnitzky, G.: 1979b, 'Justifying a Theory vs. Giving Good Reasons for Preferring a Theory. On the Big Divide in the Philosophy of Science', in Radnitzky and Andersson [1979], pp. 213–256.

Radnitzky, G.: 1981a, 'From Logical Empiricism to Critical Rationalism by Way of Critical Theory', in A. Mercier and M. Svilar (eds.), *Philosophers on Their Own Work*, P. Lange, Bern, pp. 145–178.

Radnitzky, G.: 1981b, 'Progress and Rationality in Research', in Grmek *et al.* [1981], pp. 43–102.

Radnitzky, G.: 1981c, 'Wertfreiheitsthese: Wissenschaft, Ethik und Politik' in Radnitzky and Andersson [1981], pp. 47–126.

Radnitzky, G. and G. Andersson (eds.): 1978, *Progress and Rationality in Science*, *Boston Studies in the Philosophy of Science*, vol. 58, D. Reidel, Dordrecht.

Radnitzky, G. and G. Andersson (eds.): 1979, *The Structure and Development of Science*, *Boston Studies in the Philosophy of Science*, vol. 59, D. Reidel, Dordrecht.

Radnitzky, G. and G. Andersson (eds.): 1981, *Voraussetzungen und Grenzen der Wissenschaft*, Mohr, Tübingen.

Rescher, N.: 1979, 'Some Issues Regarding the Completeness of Science and the Limits of Scientific Knowledge', in Radnitzky and Andersson [1979], pp. 19–40.

Russell, B.: 1903, *The Principles of Mathematics*, Cambridge; 2nd ed. New York and London, 1938.

Salamun, K. (ed.): 1979, *Sozialphilosophie als Aufklärung: Festschrift für Ernst Topitsch*, J.C.B. Mohr, Tübingen.

Sarton, G.: 1931, *History of Science and the New Humanism*, Harvard University Press, Cambridge, Mass.

Sarton, G.: 1936, *The Study of the History of Science*, Harvard University Press, Cambridge, Mass.

Schilpp, P. (ed.): 1974, *The Philosophy of Karl Popper*, 2 vols., *Library of Living Philosophers*, vol. XIV, Open Court, La Salle, Ill.

Schmidt, A.: 1950, 'Mathematatische Grundlagenforschung', in *Enzyklopädie der mathematischen Wissenschaften* **1** 1/2, Leipzig.

Scholz, H.: 1938–39, 'Natürliche Sprachen und Kunstsprachen', *Blätter für deutsche Philosophie* **12**.

Shannon, C.: 1938, 'A Symbolic Analysis of Relay and Switching Circuits', *Transactions of the American Institute of Electrical Engineers* **57**; repr. Newtonville, Mass., 1952.

Tarski, A.: 1935, 'Der Wahrheitsbegriff in den formalisierten Sprachen', *Studia Philosophica* **1**, 261–405; Polish original 1933.

Toulmin, S.: 1972, *Human Understanding*, vol. I., Princeton University Press, Princeton.

Toulmin, S.: 1977, 'From Form to Function. Philosophy and History of Science in the 1950s and Now', *Daedalus* **106**, 143–162.

Wang, H.: 1955, 'On Formalization', *Mind* **64**.

Watkins, J. W. N.: 1978, 'The Popperian Approach', in Radnitzky and Andersson [1978], pp. 23–44.

Wedberg, A.: 1958–, *Filosofiens historia*, Bonniers, Stockholm, Vols. I and II, 1958–59; vol. III, 1966.

Whitehead, A., and B. Russell: 1910–1913, *Principia Mathematics*, 3 vols., Cambridge University Press, Cambridge; 2nd ed. 1925–1947.

Wittgenstein, L.: 1921, 'Logisch-philosophische Abhandlung', *Annalen der Naturphilosophie* **14** 185–262; bilingual ed. with Engl. *Tractatus Logico-Philosophicus*, D. Pears and B. McGuinness, tr., Routledge and Kegan Paul, London, 1961.

Wittgenstein, L.: 1932, *Philosophische Grammatik*, Rush Rhees (ed.), Blackwell, Oxford. (Also in Ludwig Wittgenstein, *Schriften* **4**, Frankfurt, 1969).

Wittgenstein, L.: 1953, *Philosophische Untersuchungen*, G. Anscombe and R. Rhees (eds.), Blackwell, Oxford. (Also in *Schriften* **1**, pp. 279–544).

Wittgenstein, L.: 1960–, *Schriften*, Suhrkamp, Frankfurt.

Wittgenstein, L.: 1969. *The Blue and Brown Books* (1933–35), Rush Rhees, ed., Blackwell, Oxford. (Also in *Schriften* **5**, pp. 15–116, 117–282; references to this edition).

Woodger, J.: 1937, *Axiomatic Method in Biology*, Cambridge University Press, Cambridge.

Woodger, J.: 1959, 'Studies in the Foundations of Genetics', in L. Henkin, P. Suppes and A. Tarski (eds.), *The Axiomatic Method*, part III, North-Holland, Amsterdam.

FRIEDRICH RAPP

DISTRUST OF REASON

> Ich habe schon lange gedacht, die Philosophie wird sich noch selbst fressen. – Die Metaphysik hat sich zum Teil schon selbst gefressen.
>
> G.C. Lichtenberg, *Aphorismen – Schriften – Briefe*, (ed. by W. Promies), Hanser, Munich, 1974, p. 176.*

COMPLAINTS

Various types of objections to the modern, rationally shaped world are put forward today. Together with the desired civilizational patterns based on both science and technology, criticism of these same patterns is spreading all over the globe. This is not surprising, as the worldwide trend towards uniformity of ideas constitutes a natural counterpart to the unification of the physical environment. In this respect the critique of the modern, rational understanding corresponds to other intellectual movements like the questioning of the traditional authority of the older generation by young people, which is also apparent on a universal scale. The broad diffusion of such ideas parallels the spread of the science-based "superstructure" of "airports, throughways, skyscrapers, hybrid corn and artificial fertilizers, birth control, and universities" to every part of the world [Boulding 1969, p. 347]. The following notes are intended to point to the philosophical background relevant here and to the interplay between the rational and the irrational elements involved.

Due to the ever-increasing pace of change induced by the efficiency-oriented industrial civilization, a latent uneasiness and even a deep feeling of anxiety about the future are arising on the emotional level. When formulated as arguments, the objections in question amount to a fundamental criticism of the very intellectual pattern on which our modern world is built. Strictly speaking a discussion of them would only be possible to the degree in which they are cast in the form of consistent theses. But not all of the issues concerned are actually treated on the level of philosophical reasoning. Hence, even the following brief outline must partly rely on an interpretation and reconstruction of the atmosphere prevailing. Two different aspects can be distinguished here.

J. Agassi and R.S. Cohen (eds.), Scientific Philosophy Today, 287–298.
Copyright © 1981 *by D. Reidel Publishing Company.*

On the level of *individual existence* doubt is cast on the meaning of life. In the conflict between dignified ideas and crude reality the rising generation, in particular, tends to be disappointed by the value system that is actually dominant. In a world apparently deprived by science of all its mysteries and in a social environment thoroughly organized in terms of effectively prefabricated civilizational patterns, no real challenge seems to be left. Hence all kinds of alternative approaches are put forward in order to arrive at an 'authentic' style of life. Their range is broad, stretching from alternative food and living in family communities to Zen Buddhism and transcendental meditation.

The complaint is not about the lack of physical facilities or material conveniences but rather their abundance. The mind and not the body is in distress, since there are no challenges left necessitating adventure and meaningful creative activity. Indeed, the world which yields an affluence of material goods inconceivable to former generations is organized by perfect division of labor, technological efficiency and ruthless competition. The feeling of being lonely and powerless in the thoroughly organized anonymous society results in the desire to flee into the safety and comforting contact of a small group or even into a state where the mind is completely given up in favor of an intuitive 'suprarational' feeling.

On the *collective level* of understanding, doubt is cast on the very idea of progress. A growing consciousness of the limited character of resources, the apocalyptic prospects of a never-ending arms race, and the problematic possibilities emerging from biochemical research are shaking confidence in the belief that the future holds nothing but bright prospects and that technical ingenuity will solve all problems. The doubt affects the very core of the modern, secular understanding of history, which claims that by means of scientific method and enlightened ideas higher and higher levels of perfection will be achieved. To use a famous metaphor: mankind as a whole is in the position of the sorcerer's apprentice. Man was able to put the desired forces of science and technology into action, but now these forces follow their own momentum and can hardly be guided any more. The awareness of this situation results in a pervading uncertainty about the hitherto unquestioned ideal of continuous progress. This is the general background from which the individual search for meaning arises.

THEORY-BASED PATTERNS

Strangely enough, the given state of affairs does not result from a lack of rational thinking. Just the contrary. Any observer of the outward appearance of our time will arive at the conclusion that our world is shaped to a higher degree by deliberately considered, theoretically conceived ideas than any other age in history. It would hardly be an exaggeration to call our epoch the theory-shaped age. What we are experiencing around us is, as it were, materialized theories. To prove this it may be sufficient to call attention to a few instances.

Perhaps in the field of *science and technology* the relevance of theory becomes most evident. All branches of modern scientific activity are based on highly abstract conceptual frameworks which can be appropriately understood only by the insiders. Outside the group of specialists in the scientific community concerned hardly anybody will be able to evaluate the judgment of the scientific experts. Since scientific ideas penetrate into common knowledge in a simplified form, every understanding is also more or less permeated by (partially distorted) scientific theories. The role of theories is also evident on the level of sensory experience. In the industrial nations everybody finds himself surrounded by sophisticated artifacts, which incorporate scientific findings and technological knowledge. The production of television sets, atomic power plants and supersonic aircraft would in fact be impossible without systematic method and abstract physical theories. Even the larger part of our empirical knowledge of nature is highly theory-laden, since it is not only formulated in terms of theoretical concepts but also gained by means of complicated, theoretically conceived experiments, in which the 'behavior' of the physical world is revealed.

The various *types of information* which constitute the elements of our understanding of the world are also permeated to a high degree by theoretical patterns. This is an immediate consequence of the sophisticated technological devices which instantaneously spread any knowledge from the farthest place all over the globe. As a result, an immense potential of news of all types is offered to us, which far exceeds our capacity to digest it. In order to arrive at manageable portions, some sort of selection and structuring is needed, and this necessarily implies the possibility of many ways of manipulation. Thus the theory based physical means of communication have to be supplemented by some intellectual framework – and hence by another theoretical construct – which allows the abundance of information to be cast into a meaningful pattern. The theoretical bias is especially evident in political affairs, where the very same event is often judged in completely different terms according to the conceptual background applied. An example of the second-hand character of the information on which modern man relies is a person sitting in an armchair and watching a television set. Without having any real first-hand experience of the points in question he receives 'authentic' information about nuclear engineering or a political upheaval on the other side of the world.

Ideology is a further area clearly determined by theoretical reasoning. In tradition-oriented societies, as in developing nations, people are prone to take the basic ideas about man and the world for granted, simply because they are part of the heritage of their ancestors. In contrast to this, in modern society theoretical reasoning and not tradition constitutes the basic point of reference. In the event of a dispute only those claims will be accepted which can be justified by reference to theoretically conceived rational understanding. Points in question are the concept of the autonomy of the individual, the idea of the equality of all men or the notion of the sovereignty of the people. The fundamental political and social ideas on which our life is based are indeed of a theoretical nature. Thus even overtly totalitarian governments eagerly pretend to comply with the ideals of democracy, and clearly imperialistic rivalries are rationalized in terms of ideological divergencies. In this connection it may be noted that it is by reference to the theory of K. Marx that the social and political structures of one third of the world's population are fashioned.

PHILOSOPHICAL CRITICISM

The foregoing remarks have made it evident that modern life cannot be criticized for a *lack* of theoretical reasoning but only for being based on a *wrong* type of rational understanding. Now the ideas mentioned as shaping our life are the outgrowth of the history of Western philosophy. Hence the only way open consists in taking the historical development as the target of criticism, the point being that some sort of false and distorted rational understanding and not the right and genuine one has come to dominate our thought, and all types of criticism put forward do in fact follow this scheme more or less explicitly.

Clearly, in the question under consideration mere reference to the factual state of affairs cannot yield any solution. Ideals are by their very nature of a normative character. For this reason, in the last analysis theoretical arguments and not empirical evidence will constitute the court of appeal. In fact the quarrel is about diverging ideas of rational understanding, which lead to different normative statements about such basic issues as the good life, the structure of society, scientific progress, or the meaning of history. Within the philosophical arguments raised against the modern intellectual development there are, roughly speaking, three major trends to be distinguished.

The *Neo-Marxist critique*, as formulated especially by T. W. Adorno, M. Horkheimer and H. Marcuse,[1] concentrates on a shift of rational understanding during the history of philosophy. Their claim is that in modern times the idea of acting in terms of rational principles has been narrowed down, rational understanding being no longer taken in its entirety as a guide for establishing moral values, but rather being reduced to merely formal, instrumental means of effective procedure. To their mind a perversion of the very idea of reason has taken place, since the real task of reasoning consists in revealing the *ends* to be realized and not in the scientific and technological optimization of the *means* on hand. The trend prevailing amounts to giving up the goal-setting function of reason and submitting to the alleged constraints imposed by the given technological and organizational patterns. On the theoretical level their remedy consists in re-establishing the rationalist tradition. This would allow normative statements to be deduced from rationally established philosophical principles and thus open the way to 'emancipatory' practice.

Another point of criticism arises from the *development of science*. The Vienna circle movement and Logical Empiricism tended to take continuous progress in science for granted. But the work of T. S. Kuhn [1970] on the changing conceptual frameworks in scientific research and the ensuing discussion made it clear [Lakatos and Musgrave 1970] that the various stages of science are based on different theoretical approaches and do not fit into the scheme of a continuous process of accumulation of knowledge. In pushing forward the argument, P. Feyerabend advocates an anarchistic epistemology, since "the only principle that does not inhibit progress is: *anything goes*" [1975]. By means of a review of the historical process he arrives at the conclusion that there is no guarantee of reason in the development of science, since no safe criterion can be named on which one could rely when judging about the rational or irrational character of competing theoretical systems. Thus various alternative approaches that were given up during the history of science and are neglected even nowadays might very well turn out worthwhile, as all types of theoretical ideas are in principle equally justified.

The most fundamental argument concerns the *historical status of reason* itself. As the rationalist tradition has it, the capacity for intellectual understanding and deductive reasoning is by its very nature not bound to historical change. Yet, for example, E. Topitsch [1958] has been able to show to what extent hidden mythological patterns are still alive even in the most enlightened scientific and philosophical reasoning. And in the discussion about historical relativity (historicism), it was especially W. Dilthey who pointed to the historically contingent systems of thought, which differ from epoch to epoch and determine understanding in all fields of intellectual activity, i.e., in everyday understanding as well as in science, philosophy, art, and religion. Recently ideas of this type have been put forward by K. Hübner [1978]. He regards the relevant conceptual frameworks as *a priori* stipulations, beyond which no metatheoretical judgment, and hence no evaluation, will be possible, since it is always a specific system which constitutes the ultimate point of reference. For this reason, along with scientific understanding non-scientific approaches, like thinking in terms of mythology, could also be qualified to function as the *a priori* system needed for the structuring of our experience.

NON-RATIONAL ELEMENTS

In dealing with the problem of rational understanding those components of the mind also have to be considered which are not reducible to intellectual activity and consistent reasoning. Even a superficial glance at the history of Western philosophy will show that at any time men of thought have been aware of the non-rational elements which, as it were, disturb intellectual understanding as well as human conduct. This is not astonishing at all, since, to put it in general terms, the aim of philosophy is to extend the realm of rational understanding as far as possible. In this process some sort of clash with the non-rational elements must obtain. Thus along with the *formal* capacity for discursive reasoning additional *material* ingredients always come into play, since any argumentation needs a definite point to start with. This is to say that – save for a *regressus in infinitum* – one cannot avoid accepting certain material elements as incapable of being decomposed into some conceptual structure. For this reason, the relation between the rational and the non-rational could readily be chosen as a clue for analyzing the structure of any philosophical approach whatever. Since it is not possible here to exhaust all the relevant problems, some short remarks must suffice.

In *conceptual analysis* the non-rational elements are those which cannot be described exhaustively, however appropriately the categories may be chosen. Since reasoning is about universal and necessary relations, the singular and contingent traits of the world will ultimately escape rational understanding. Following Aristotle's teaching, this idea was generally accepted in scholastic philosophy, the principle in question being *individuum est ineffabile*, which is to say that the specific characteristics (accidents) of an individual cannot be expressed by means of universal categories (substantial forms).[2] In a different perspective Leibniz considers propositions instead of ideas and assumes a basic distinction between the contingent *vérités de fait* and the necessary *vérités de raison* [Leibniz]. In his analysis of the structure of experience Kant takes up the traditional dichotomy between matter and form; to him the formal, structuring element is supplied by the *a priori* given categories which impose their patterns on the material yielded by the contingent *a posteriori* given sense-impressions [Kant].

Practical philosophy too, escapes perfect rational understanding. The rationalist attempts at a certain and positive foundation of ethics, starting from Spinoza's application of the deductive method to Kant's formalist approach did not turn out to be very satisfactory, since they neglect the complexity of the real situations of life and of the emotional element involved in human conduct. As regards the non-rational element of the individual moral decision, Aristotle's approach is more sophisticated, since he takes into account the difference between the intellectual virtue of understanding (sophia) and the virtue of prudence (phronesis) which leads to the right and reasonable choice [Aristotle]. At this point one could also mention the ironic French saying "Appuyez-vous fortement sur les principes, ils finissent toujours par céder" or even Schiller's joke about the incapacity of philosophy: "Einstweilen bis den Bau der Welt Philosophie zusammenhält, erhält sie das Getriebe durch Hunger und durch Liebe" [1954]. Other famous examples of criticism of a rationally stylized ethics are Erasmus' *In Praise of Folly* [1946], Voltaire's *Candide* [1961] or Mandeville's *The Fable of the Bees* [1924].

A further trait not reducible to consistent, rational patterns is *creative activity*. Descartes' classical methodological advice for effective scientific procedure is only able to give general hints, not clear-cut prescriptions for successful scientific procedure. Notwithstanding this, it would be false to underestimate the importance of his rules for the development of modern science. The combination of mathematical description and empirical method has turned out to be an extremely effective means for unravelling well-defined processes of nature and putting them to use by means of technology. One could also regard the paradigm cases and the conceptual framework of "normal science" [Kuhn 1970; Lakatos and Musgrave 1970] as an additional help for specifying the general methodological principles of scientific research. Yet the point is that in the intellectual field merely algorithmic procedures will only result in combinations of given elements but never in really new findings. Of course, after a creative process is completed, ex post facto reconstructions are possible, and even necessary, in order to eliminate useless ideas and to arrive at a consistent presentation of argument. But the endeavor to reveal the new always demands creative intuition. Since discursive reasoning refers to given elements it cannot cope adequately with the hitherto unknown. The reverence for men of genius is perfectly justified, since their creative capacity surpasses the standards of rational understanding common to all men.

In *irrationalist positions* the value of rational understanding is questioned altogether. Whereas the points considered hitherto have concerned *particular* traits of the world that resist rational understanding, now doubt is cast on the capacity of reason *in principle*. Of course, varying degrees have to be distinguished here. But they all have the idea in common that some sort of intuition – and not conceptualization and discursive reasoning – is the real source of insight. In contrast to those who along with the rational elements of our knowledge accept some *non*-rational ones, here the stress is on the decisive role of the *ir*rational traits, i.e., on the failure of rational understanding as far as basic issues are concerned. An example is the romantic movement, since it considers poetry, dreams and subconscious awareness to be superior to deductive reasoning and scientific method. The most extreme case is obviously that of mystical experiences, usually cast in religious terms, which, in one form or another, are to be found in all civilizations. Experiences of this kind are by their very nature highly subjective and defy all attempts at consistent systematization. Certain epistemological notions, like those of Shaftesbury or Rousseau, could also be mentioned here, as they rely more on sentiment than on rational understanding. Furthermore, irrational elements are present in the German "Lebensphilosophie"[3] in Heidegger's "non-concealment of Being" [Richardson 1974, p. xii] and in the existentialist idea of the specific human type of existence deemed to escape conceptual analysis.

A REALISTIC CLAIM

From the short excursion undertaken one may conclude that rational understanding has indeed reached a critical stage. The ambitious idea of continuous progress in the use of nature for human purposes and in the perfection of moral standards has been shaken in many respects, and this in spite of the fact that the world we live in is constantly being reshaped in terms of rational ideas. From the various philosophical criticisms put forward it becomes evident that the rationally conceived ideas that are dominant have been shaped in a contingent historical process and might very well look different. Furthermore, diverse types of non-rational elements must be noticed, from which certain restrictions of the realm of rational understanding arise. Provided this analysis is correct in its main points, the following conclusions may be drawn from it.

(1) It is impossible to renounce reference to reason. Theoretical criticism and practical changes are certainly needed. But if they are to be of a rational type, they have to be shaped in terms of the idea of rational understanding. Thus on a *metatheoretical* level the discerning capacity of reason must come into play again. As long as rationally conceived understanding – and not just intuitive feeling – is in question, reason is the last resort. The philosophical criticisms mentioned constitute an example of this, since they all make an appeal to the capacity for rational understanding. In the field of rational discourse only reason can fight against reason [Bunge1959].

(2) But reason is not only necessary in terms of the theory of normative, philosophical discourse. It is also indispensable in terms of practical considerations. Since this world of ours is shaped by reason, it is only by making use of rational ideas that its functioning can be maintained. Not only the instrumental type of effective, rational thinking, applied in science, technology, and management is in question here. As mentioned above, the basic ideas of our social and political life also result from rational understanding. For the domination of the two possibly hostile forces of nature and of our neighbors we have to rely on rational thinking.

(3) Yet, it is a common-place, that the rational methods for the control of nature and of social life not only solve problems but also create new ones, as is exemplified by the arms race, the ecology problem, and the various restrictions on personal freedom resulting from increasing state activity. It would be all too simple to ascribe the problems arising here only to ignoring the common truth that one cannot have one's cake and eat it (although in the question of future energy supplies some people would apparently like to do just this).

In actual fact, the disparity prevailing between the theoretically conceived ideas and what happens in practice is due to a great extent to an over-estimation of the capacity of reason. When pointing out the merits and the indispensable function of rational understanding philosophers were inclined, as it were, to impose on reason too heavy a burden. As a result, the non-rational elements were unduly neglected. This is of special relevance in the field of ethics. Here Spinoza's naturalistic theory seems to be more appropriate. To him the basic drive in human conduct arises from passions and not ideas. These passions are passive emotions, which can be changed by the mind into active emotions, and in this way it is possible for intellectual advance to lead to moral advance [Spinoza]. A similar approach is that of Freud, who holds that it is through sublimation, i.e., through the channelling of instrumental impulses into socially acceptable behavior, that culture is generated [Freud 1953]. M. Scheler's idea of the 'blind' spontaneous drive, which is to be guided by mind and rational understanding, could also be mentioned here.[4] This is to say that in a realistic perspective one should not claim the omnipotence of reason, but rather face the unquestionable emotional element in human nature.[5] This approach offers a better chance of guiding the passions by means of reason than the simple way out of ignoring their importance altogether. And even the emotional drive in the desire of young people for comforting contact and meaningful tasks can be dealt with in terms of this analysis.

(4) When considering rational ideas, it may also be worth-while taking into account the state of affairs in the real world. Ideals, by their very nature, are conceived in theoretical terms; but they are to be realized under empirically given circumstances. The two elements may clash and unfortunately the ideals will not win in every case, however perfectly they may be shaped.[6] For this reason, especially on the level of society as a whole, a pragmatic approach may in fact be better than insisting on ultimate ideals. Yet, in our rationally shaped world, when a failure occurs there is a tendency to blame the circumstances and not the theory. Instances of this can be observed all over the world, but perhaps in communist countries the divergency between theory and practice is most explicit. Here again it would be worthwhile to reduce the claim of rationally justified ideas and the activity derived from rational theory to a realistic level.

(5) This by no means amounts to a plea for irrationalism. Just the contrary. Since the aim is to shape the world to as high a degree as possible in rational terms, a modest – but feasible – procedure will lead much further than a theoretically stylized, unrealistic approach. Here M. Weber's basic dichotomy between ethics of conviction (*Gesinnungsethik*)

and ethics of responsibility (*Verantwortungsethik*) applies. Whereas the former pays heed only to the rational principles, the latter takes into account the actual results for men and hence proves to be the really responsible way of action.[7]

The outcome is that rational understanding cannot be renounced as an epistemological principle and as a guide for action. But in order to apply rational principles effectively, the emotional drive in human conduct and the practical consequences of political actions have to be considered.

Technische Universität Berlin

NOTES

* "I've thought for a long time that Philosophy will yet devour herself. Metaphysics has already partially devoured herself" (*The Lichtenberg Reader. Selected Writings of Georg Christoph Lichtenberg*, trans., ed. and introduced by F.H. Mautner and H. Hatfield, Beacon Press, Boston, 1959).

[1] The objections in question are put forward most explicitly in Horkheimer [1947]; the need to consider possibly negative traits of enlightenment and progress as well is underlined in Horkheimer and Adorno [1947]. Of Marcuse's writings especially *One-Dimensional Man* [1964] is relevant here.

[2] Cf. Lalande [1962], p. 497 for further references, especially on Leibniz' divergent idealistic notion that "omne individuum est species infima."

[3] A sympathetic account is Bollnow [1958].

[4] Scheler [1975], pp. 61–64. Evidently Bacon's idea of conquering nature by obeying her laws does not only apply to the physical world but also to human conduct.

[5] Kant insists that only the motive of duty and never the motive of inclination, can give moral value to an action. But he still finds it necessary to pay tribute to human nature. He explicitly states that nevertheless it is an *indirect* duty to seek our own happiness, since lack of satisfaction might easily constitute a great temptation to act against real duty (*Grundlegung zur Metaphysik der Sitten*, B 11 f.).

[6] In terms of psychology the reason for the dispropriation between theoretical expectations and actual practice consists in an 'explosion of needs', since the level of aspiration tends to increase faster than the means available. It is an important feature that this psychological fact is based on the philosophical idea of continuous progress in the domination of nature and a corresponding perfection in human affairs. For a discussion of the role of scarcity on the individual and social level see Balla [1978].

[7] Weber, [1956], p. 175–82. From a systematic point of view Weber's idea is akin to the distinction between *deontological* and *teleological* theories, as described, for example, in Frankena [1963], pp. 13–16.

REFERENCES

Aristotle: *Nicomachean Ethics*, VI, 1144b30–1145a14.

Balla, B.: 1978, *Soziologie der Knappheit. Zum Verständnis individueller und gesellschaftlicher Mangelzustände*, Enke, Stuttgart.

Bollnow, O.F.: 1958, *Die Lebensphilosophie*, Springer, Berlin.

Boulding, K.E.: 1969, 'The Emerging Superculture', in *Values and the Future*, K. Baier and N. Rescher (eds.), Free Press, New York, London.

Bunge, M.: 1959, *Metascientific Queries*, Thomas, Springfield, Ill.

Erasmus, D.: 1946, *In Praise of Folly*, L.F. Dean (ed.), Packard, Chicago.

Feyerabend, P.: 1975, *Against Method*, NLB, London.

Frankena, W.: 1963, *Ethics*, Prentice-Hall, Englewood Cliffs, N. J.

Freud, S.: 1953, *Abriss der Psychoanalyse – Das Unbehagen an der Kultur*, Fischer, Frankfurt.

Horkheimer, M.: 1947, *Eclipse of Reason*, Oxford University Press, New York.

Horkheimer, M., and T.W. Adorno: 1947, *Dialektik der Aufklärung*, Querido, Amsterdam.

Hübner, K.: 1978, *Kritik der wissenschaftlichen Vernunft*, Alber, Freiburg i. Br.

Kant, I.: *Kritik der reinen Vernunft*, Transzendentale Analytik.

Kuhn, T.S.: 1970, *The Structure of Scientific Revolutions*, 2nd ed., University of Chicago Press, Chicago, London.

Lakatos, I., and A. Musgrave (eds.): 1970, *Criticism and the Growth of Knowledge*, Cambridge University Press, Cambridge.

Lalande, A.: 1962, *Vocabulaire technique et critique de la philosophie*, PUF, Paris.

Leibniz, G.W.: *Nouveaux essais sur l'entendement humain*, Book 4, ch. 2, §1.

Mandeville, B.: 1924, *The Fable of the Bees*, 2 vols., F.B. Kaye (ed.), Clarendon Press, Oxford.

Marcuse, H.: 1964, *One-Dimensional Man*, Beacon Press, Boston.

Richardson, W. J.: 1974, *Heidegger. Through Phenomenology to Thought*, Nijhoff, The Hague.

Scheler, M.: 1975, *Die Stellung des Menschen im Kosmos*, Francke, Bern.

Schiller, F.: 1954, 'Die Weltweisen', in *Werke*, vol. II, P. Stapf (ed.), Deutsche Buchgemeinschaft, Berlin, Darmstadt.

Spinoza, B.: *Ethics*, part III, prop. 59.

Topitsch, E.: 1958, *Ursprung und Ende der Metaphysik*, Springer, Vienna.

Voltaire: 1961, *Candide ou l'optimisme*, Engl. tr. by D. M. Frame in *Candide, Zadig and Selected Stories*, Indiana University Press, Bloomington.

Weber, M.: 1956, *Soziologie, Weltgeschichtliche Analysen, Politik*, J. Winckelmann (ed.), Kröner, Stuttgart.

MICHAEL RUSE

TELEOLOGY REDUX

Although Mario Bunge is best known as a philosopher of physics – and deservedly so – he has in fact turned his keen attention to just about every other area of science. Truly he might be called the twentieth-century's answer to William Whewell![1] And like Whewell, Bunge has considered the biological sciences, and has things of interest to say about them. I think it is fair to conclude that, generally speaking, Bunge and I agree about biology. We are both suspicious of attempts to put biology apart from the physical sciences. In particular we both think that some sort of axiomatic or hypothetico-deductive ideal is appropriate in the biological sciences as well as in physics and chemistry [Bunge 1967; Ruse 1973].

But Bunge is too young and vigorous to bury with praise or agreement. I want therefore to turn to another aspect of biology, hoping thereby to provoke disagreement and to stir Bunge into action yet again. My concern in this paper will be with the problem of teleology. I think it is probably true to say that, inasmuch as the philosophy of biology can be said to have a "hot" topic, the nature of teleological explanation is that topic.[2] But what is meant here by "teleology"? Simply that in biology we find all kinds of funny language apparently making reference to, and indeed explaining in terms of, *ends*. Let me give just one example, and then I will explain why teleology has attracted philosophical attention.

The stegosaurus was a large herbivourous dinosaur of the Jurassic period. What made it very distinctive was the existence of a set of bony plates running along its back (see Figure 1). The question which paleontologists have long asked is, what "end" or "purpose" or "function" did they serve? What "problem" were they supposed to "solve?" A number of answers were put forward. Some said that the plates existed "in order to" make the stegosaurus seem bigger and more fearsome, thus frightening off predators. Some said that the plates existed "in order to" facilitate courtship recognition – with plates like that down one's backside it would be pretty difficult to make a mistake and spend one's time making romantic overtures to a member of the wrong species. And some said that the plates existed "in order to" aid heat regulation – the plates would act as sorts

J. Agassi and R.S. Cohen (eds.), Scientific Philosophy Today, 299–309.
Copyright © 1981 *by D. Reidel Publishing Company.*

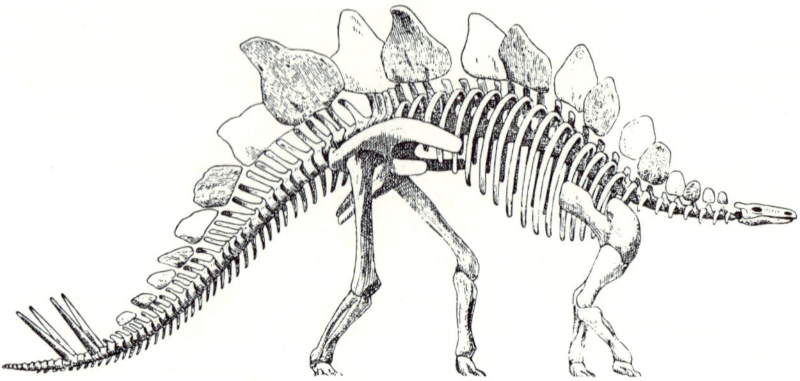

Fig.1. The STEGOSAURUS. This skeleton in the American Museum of Natural History is 18 feet long. (From Lewontin, 1978)

of cooling devices, radiating heat, thus enabling the non-sweating dinosaur to move about in the heat of the sun and get on with the business of living.

We see therefore the frankly teleological language of the biologist. And let us make no mistake about its value. By thinking in this way, biologists make great progress. Indeed, there is good reason to think that paleontologists have now solved the stegosaurus problem. The plates most probably existed for heat regulation – a conclusion based on their "design" for such regulation:

> The hypothesis that the dorsal plates of *Stegosaurus* were a heat-regulation device is based on the fact that the plates were porous and probably had a large supply of blood vessels, on their alternate placement to the left and right of the midline (suggesting cooling fins), on their large size over the most massive part of the body and on the constriction near their base, where they are closest to the heat source and would be inefficient heat radiators [Lewontin 1978, p. 218].

And yet teleology is a problem. Not for biologists. They use it and they are happy to use it. But teleology is a problem for philosophers. If one looks at the physical sciences, rightly or wrongly still taken as the paradigm, one does not find the overt use of teleological language. It is true indeed that in the last century when Sir David Brewster – Scottish Man of Science – was asked what function the moon serves, he was happy to reply that it serves the end of lighting the way of nocturnal travellers [Brewster 1854]. But the fact that we today (at least in Ontario) find such an answer amusing, points to the incongruity of using teleology in physics. The moon does not exist "in order to" do anything. The very suggestion is not false, but absurd. It is like asking whether Tuesday is more tired than Monday. Tuesdays are not the sorts of things that can be tired, and moons are not the sorts of things that can have ends.

We see therefore that, philosophically speaking, teleology (or rather, the teleological way of thinking) is an interesting and important question. Physicists do not use it. Biologists do. What we need are answers to questions like, why biologists use a teleological way of thinking, how they can use it and get away with it, what they mean when they do use it, and whether or not biologists could stop using teleological language and still get as far in their biology? This last question might suggest a somewhat arrogant attitude – why on earth should biologists stop using teleological language? But I think it is true that biologists generally feel an urge to make their science as much like physics as possible. It is after all the paradigm science (not necessarily the nicest science), even though biologists will never admit to the position of physics. (As is well known, biologists tend to suffer from physics envy.)

In recent years there has been, as I have suggested, a great deal written about teleology, although non-philosophers should not be deceived by this productivity. Much that has been written stems not from the fact that the topic of teleology is important, which it is, or from the fact that the topic of teleology is difficult, which it also is. Rather the popularity of the topic of teleology stems from the undeniable fact that the average philosopher of science wants as little to do with science as is humanly possible. Trained in one of the neo-scholastic education mills of North America, he has no contact with science, and, searching desperately for a thesis-topic, he has little inclination or pressure to make such contact.[3] Teleology therefore seems tailor-made. All one needs to know is the (false) fact that the heart beats in order to pump blood and that it does not beat in order to create heart sounds, and one is off and running. From then on one can drop all reference to things organic, and one can deal exclusively in such delightful philosophical phenomena as door knobs serving as paper-weights, sewing machines with knobs to be pressed when you want them to self-destruct, multi-purposed vibrators, or what have you.

But let me stop carping about my fellows and state my own position. This is that I think one can make some progress with the problem of teleology, but that this simply cannot be done in isolation from real science. One must stay close to science, both as it is now and perhaps even more importantly, as it was in the past. (I say this notwithstanding the fact that references to the history of science are about as popular amongst philosophers of science as are references to contemporary science.) I want to go further and doubt that an all-encompassing analysis of teleology can be offered. Or rather I want to suggest that if an all-encompassing analysis of teleology is possible, the way to get at it is not by trying to produce it immediately, but rather by producing analyses in the various branches of science, and then seeing if they can be put together. Following my own prescription therefore, what I want to do is look at biology and its history, and see how to unpack its teleological content.

If we look at the history of biology, we see that 1859 was a watershed. That was the year of the publication of Charles Darwin's *On the Origin of Species*. Literally, biology was revolutionized [Ruse 1979]. Before 1859 people believed that the organic world had been created miraculously by God. After 1859 people believed that the organic world was caused naturally by the action of unbroken law, and those who followed Darwin believed that the chief causal agent was natural selection working on random variation. From the point of view of teleology, 1859 and the publication of Darwin's *Origin* were crucial. And yet the *Origin's* importance is not straightforward. I am reminded of Conan Doyle's marvelous Sherlock Holmes story about the theft of the race-horse Silver Blaize. Holmes turned to Watson and said: "The most important fact my dear Watson, is the dog that barked in the night." "But the dog didn't bark in the night." "Precisely." And the most important fact about the arrival of the *Origin*, is that from the point of view of the teleology in biology, *it did not make the slightest bit of difference*. Before Darwin people cheerfully said that the eye existed in order to see. After Darwin people cheerfully said that the eye existed in order to see. The causes were different perhaps, but the teleology was not!

This being so, let us ask first about the teleology of pre-*Origin* biology. Perhaps then with this answered we can carry forward to modern biology. Now, the answer to questions about pre-*Origin* biology comes fairly readily. As every student of introductory philosophy knows well, people thought that they could speak teleologically about the organic world because they thought the organic world was teleological – it shows the ends put there by God for the benefit of man and other organisms. (Whether everything ultimately reduced to man's benefit was a nice point of natural theology.[4]) Thus, as Paley [1819] was happy to argue, we can speak of the eye having the function of seeing, because the eye does have that function – it was designed (literally) that way by God. The eye, if you please, is one of God's *artifacts.* And let it not be thought that only professional natural theologians thought that way. In a discussion in 1834 in Britain's leading science journal, the *Philosophical Transactions* of the Royal Society, of the adaptations of the kangaroo for feeding its young, we find Britain's leading zoologist, Richard Owen, calmly stating that such adaptations show "irrefragable evidence of creative forethought."

Of course, in an important sense, pre-Darwinians were using a metaphor or analogy. Literally they thought the eye was one of God's artifacts; but they were modelling it on human artifacts. We make the telescope, having our ends in mind. For this reason we can speak and think teleologically about the telescope. The eye in many respects seems very similar to the telescope. Therefore it must have been made by God. ("Things which are so design-like just don't happen by chance or blind law.") So for this reason we can speak and think teleologically about the eye.

The key to pre-Darwinian biology therefore is the *artifact model.* Because the organic world seems *as if* it were designed, it was felt permissible and appropriate to speak of it as actually being designed. And coming straight across the *Origin* to post-Darwinian biology, the same point holds. Because the organic world seems *as if* it is designed, it is felt permissible and appropriate to speak of it as being designed. A molecule or the moon does not seem very much like an artifact, so we do not think such language appropriate in physics. The fins on the stegosaurus however are another matter. They look like turbo blades or the Heath Robinson contraptions that solar energy buffs are into. So why not talk and think that way? For Charles Darwin himself, incidentally, the continued use of teleological language came very naturally, because he had been brought up on Paley's *Natural Theology* [Darwin 1969].

None of this is to deny that pre- and post-*Origin* there is a difference. Pre-*Origin* the eye was caused by God. Post-*Origin* the eye was caused by natural selection. The teleology however remains the same.

> Much of evolutionary biology is the working out of an adaptationist program. Evolutionary biologists assume that each aspect of an organism's morphology, physiology and behavior has been molded by natural selection as a solution to a problem posed by the environment. The role of the evolutionary biologist is then to construct a plausible argument about how each part functions as an adaptive device. For example, functional anatomists study the structure of animal limbs and analyze their motions by time-lapse photography, comparing the action and the structure of the locomotor apparatus in different animals. Their interest is not, however, merely descriptive. Their work is informed by the adaptationist program, and their aim is to explain particular anatomical

features by showing that they are well suited to the function they perform. Evolutionary ethologists and sociobiologists carry the adaptationist program into the realm of animal behavior, providing an adaptive explanation for differences among species in courting pattern, group size, aggressiveness, feeding behavior and so on. In each case they assume, like the functional anatomist, that the behavior is adaptive and that the goal of their analysis is to reveal the particular adaptation [Lewontin 1978, pp. 216–17].

Let me sum up now what I have tried to say or hint at so far. I argue that the teleology in modern biology is analogical. The organic world seems as if it is designed; therefore we treat it as designed. The *artifact model* is the key to biological teleology. The phenomena of physics and chemistry do not seem as if designed. Therefore we do not think teleological thought and language appropriate in those sciences. Of course, we do not today think biological phenomena really are designed – at least, not by the direct intervening agency of a miracle-working God. Rather, we think that natural selection working on random variation is the causal key. Therefore, if one wants to cash in a functional or teleological statement, "The *function* of the fins on the stegosaurus is to effect heat regulation," "The fins of the stegosaurus exist *in order that* its heat might be efficiently regulated," "The fins *solve the problem* (*serve the end*) of heat regulation," one must cash it out in terms of natural selection. The stegosaurus has fins because those of its ancestors which had fins survived and reproduced, and those that did not did not. In short, the fins are an *adaption*, and to talk in functional language in biology is to refer to an *adaption*. If x has the function y, then x is an adaptation, and y is adaptive – it helps survival and reproduction.

This is the crux of what I want to say about biological teleology. In a sense it is all rather simple. But that does not mean it is not true. It makes sense of biology and of its history. Moreover, for once as a philosopher I am not prescribing what biologists ought to do. Rather, I am describing what they do do. For instance, in one of the most highly praised books of the past two decades, the evolutionist G.C. Williams commits himself to precisely the position on teleology I have just unpacked and endorsed.

Whenever I believe that an effect is produced as the function of an adaptation perfected by natural selection to serve that function, I will use terms appropriate to human artifice and conscious design. The designation of something as the *means* or *mechanism* for a certain *goal* or *function* or *purpose* will imply that the machinery involved was fashioned by selection for the goal attributed to it. When I do not believe that such a relationship exists I will avoid such terms and use words appropriate to fortuitous relationships such as *cause* and *effect*.

Thus I would say that reproduction and dispersal are the goals or functions or purposes of apples and that the apple is a means or mechanism by which such goals are realized by apple trees. By contrast, the apple's contributions to Newtonian inspiration and the economy of Kalamazoo County are merely fortuitous effects and of no biological interest [Williams 1966, p. 9, his italics].

Enough by way of defence. Let me now conclude with four corollaries which follow from my main theorem.

First, one might ask why teleology is possible in biology? In one sense I have already answered this question; in another sense I do not have to answer it. My answer is that teleology is possible because the organic world is design-like. My answer does not depend on a correct answer to the further question: Why is the organic world design-like, or rather, why does natural selection make the organic world design-like? But having exempted myself from the responsibility of having to give a correct answer (!), what I would suggest is that the answer lies in a suggestion to be found in David Hume's *Dialogues Concerning Natural Religion*, namely that irrespective of whether there is any similarity in the production of human artifacts and organic characteristics, the latter have to be design-like because if they were not, they simply would not work. The eye is like a telescope or some other human artifact for seeing, because if it were not like a telescope or related artifact, one would not see at all.

Second, is biological teleology real or hard-line teleology in the sense of explaining or trying to understand the past in terms of the future? And if it is, how do biologists avoid classic problems like that of the missing goal-object? (If one explains x in terms of future y, that is fine if y occurs; but what happens if x occurs, and then y does not?[5]) My position, which I shall more state than argue for here, is that the teleology is fairly hard-line. I would argue that when the biologist says x exists in order to y, or the function of x is y or (y ing), then the earlier x is being explained in terms of the later y. Consider (for support) the following two passages by G. G. Simpson, one of the world's leading evolutionists. He writes:

In order to realize the new functions of a changed environment, an organism must, at the moment when the change or the occupation of a new environment begins, have at least some functions prospective with regard to the new environmental functions. This is merely a more technical way of saying that organisms can continue to live only under conditions to which they are already at least minimally adapted [Simpson 1953, p. 189].

If a mutation, whether adaptive, non-adaptive, or inadaptive with respect to the adaptation of an ancestral population, does become fixed and spread by selection it is adaptive from the start with respect to the descending populations [Simpson 1953, pp. 194–5].

This, it seems to me, makes it all pretty clear that x is being explained in terms of the later (future) y.

But what then about the missing goal-object problem (and related problems)? I think these do not really arise (or only very rarely arise), because by the time that the biologist gets around to working out the function of x's, the y's (which were future to the x's) are now past too – in short, the biologist knows that they occurred and were not missing! Assuming that the stegosaurus explanation is correct, one can as a paleontologist explain the plates in terms of (later) heat regulation, because the heat regulation is in our past also, and cannot therefore go missing. As far as today's organic world is concerned, the biologist is still working on organisms essentially as they have gone (or are going) past. If the biologist wants to state categorically that at this moment the function of x is to do y and y really is still future, then frankly I think he/she is taking a gamble. She/he might come up against the missing goal object problem. But of course, practically speaking, the biologist is on safe ground. The future, at least for this year's generation, is usually like the past – eyes go on seeing, and legs go on walking – and so by the time that anyone gets around to deciding that the future is not like the past, the future is not future any more.

Third, the above analysis raises the question of whether all of the teleology of biology has been located, and perhaps incidentally throws some light on a concept that keeps recurring in the philosophical literature on teleology, namely that of a *goal-directed system*.[6] Some philosophers have argued that all of the teleology of biology is analyzable in terms of goal-directedness, namely the ability to get back on target despite disruptions. This is clearly untrue – the assumption that the plates of the stegosaurus exist in order to effect heat-regulation says nothing about how the stegosaurus would react were something untoward to happen. But I think that goal-directedness is nevertheless important in biological teleology, and the above analysis shows how.

Consider: We get the teleology of functions in biology by analogy from human teleology, namely the teleology of human artifacts. Is there any other kind of human teleology? Clearly there is, namely our own teleology. We strive to achieve certain ends, just as we try to impregnate ends into our artifacts. I see no reason why we should not, and indeed would rather expect that we would, read this other kind of teleology into the organic world also. But how would we recognize this teleology, or rather what would lead us to impute such teleology to the non-human world? I suppose the existence of brains in other organisms would go part way, but I suspect that identifying goal-directed behavior is the really crucial factor. Suppose I am teleologically oriented towards something, say getting a Ph.D. The real mark that I have that Ph.D as a goal is the fact that I strive for it, despite all obstacles like unpleasant professors, irrelevant language exams, and so forth. Similarly, we see a teleology in the organic world, over and above the teleology of functions, when the wolf-pack shows a flexible strategy to bring down the much larger moose, and when the insect colony regroups and moves in the face of attack, and so forth. In other words, by analogy we see goal-directed behavior as teleological, of a kind different from that of function.

I am not sure however that we see all goal-directed behavior as teleological (and this is why supposed problem examples of goal-directedness drawn from the inorganic world do not worry me). As we get farther from human intelligence and its ability to conceive of and act towards goals, we see less teleology (of this second kind). Take something like sweating and shivering, which is generally conceded to fill the conditions for a goal-directed system. Is this teleological? A body capable of sweating and shivering is certainly teleological inasmuch as such processes are adaptations (which they are!). But is there more teleology (of the kind I find in the coordinating wolf pack)? I rather doubt it – no one thinks that any intelligence is involved in keeping the body at constant temperature through sweating and shivering.

I suspect however that there is a grey area about where this second form of teleology operates. We find it in dogs. We do not find it in trees. Do we find it in earth worms? In part, this is an empirical question. How goal-directed are earth worms capable of being – particularly, how goal-directed in terms of *behavior* are they capable of being? After that, the imputation of such teleology is all a bit arbitrary. But then, of course, this element of arbitrariness is characteristic of metaphor or analogy. Some people are prepared to push metaphors further than others. A grey area where there are disagreements confirms my position, rather than refutes it.

Fourth and finally let me mention reduction and eliminability (a point at which I suspect that Bunge and I might start to part company). Can we get rid of the teleology of biology? Would we want to? I suspect that we could, but that we would not want to. We could certainly talk always in terms of past causes (efficient causes) and drop all talk of function and so forth. But we could not then say as much. We might, for example, talk about the embryological development of the heart, talking in terms of cell-division and so forth, but we could not ask what the heart was for. Moreover, simply to switch to talk of "adaptation" would be no solution, because the forward-looking teleological aspect of biology still then remains. The whole way of looking at organisms in terms of design and artifacts has to be dropped. But why would one want to do this? Biology is not physics, nor should it have to be. If there were no good reason for biology's being different, then one would expect them to be the same. This is why Bunge and I feel justified in neglecting neo-vitalistic attempts to separate biology and physics on such pseudo-grounds as the purportedly peculiarly complex nature of biology. In pertinent respects, biology is no more complex, or unique, or what have you, than physics [Ruse 1977].

But, perhaps putting me apart from Bunge, when we come to teleology, I believe that we do have good reason for separating physics and biology, as they stand now. And the reason for this is that biology refers to phenomena that are design-like and physical phenomena are not, and biology as a science tries to capture and utilize this distinctive facet of the biological world. Moreover, interestingly, at the borderline if anything it is physics that gives. Where biology and physics meet, we do not get a total rejection of teleology. Rather, conversely teleology seeps in. At least, we find molecular biologists talking about such things as the need "to crack the genetic code." And if that is not to talk in terms of design, I do not know what is. In short, the teleology of biology is here to stay. Given its power in the hands of the evolutionist, for instance in finding out the purpose of the plates on the back of the stegosaurus, let us be thankful for this fact.

University of Guelph, Ontario, Canada

NOTES

[1] I intend this as high praise.

[2] Useful recent reviews include Woodfield [1976]; Wright [1976]; Nagel [1977].

[3] I say "he" in this context deliberately. My experience is that female philosophers of science are far more sensitive to science than are their male counterparts.

[4] See, for example, Buckland [1836].

[5] This and related problems are discussed in Scheffler [1963].

[6] Woodfield [1976] admirably digests and criticizes the literature on goal-directedness.

REFERENCES

Brewster, D.: 1854, *More Worlds than One: The Creed of the Philosopher and the Hope of the Christian*, Murray, London.

Buckland, W.: 1836, *Geology and Mineralogy: Bridgewater Treatise* 6, Pickering, London.

Bunge, M.: 1967, *Scientific Research*, Springer, New York.

Darwin, C.: 1859, *On the Origin of Species*, Murray, London.

Darwin, C.: 1969, *Autobiography*, N. Barlow (ed.), Norton, New York.

Lewontin, R. C.: 1978, 'Adaptation', *Scientific American* **239**, 212–30.

Nagel, E.: 1977, 'Teleology Revisited', *J. Phil.* **74**, 261–301.

Owen, R.: 1834, 'On the Generation of the Marsupial Animals, with a Description of the Impregnated Uterus of the Kangaroo', *Phil. Trans.*, 333–64.

Paley, W.: 1819, *Natural Theology*, in *Collected Works*, Rivington, London, 1st published 1802.

Ruse, M.: 1973, *Philosophy of Biology*, Hutchinson, London.

Ruse, M.: 1977, 'Is Biology Different from Physics?', in R. Colodny (ed.), *Logic, Laws, and Life*, University of Pittsburgh Press, Pittsburgh, pp. 89–127.

Ruse, M.: 1979, *The Darwinian Revolution: Science Red in Tooth and Claw*, University of Chicago Press, Chicago.

Scheffler, I.: 1963, *The Anatomy of Inquiry*, Knopf, New York.

Simpson, G. G.: 1953, *The Major Features of Evolution*, Columbia University Press. New York.

Williams, G. C.: 1966, *Adaptation and Natural Selection*, Princeton University Press, Princeton.

Woodfield, A.: 1976, *Teleology*, Cambridge University Press, Cambridge.

Wright, L.: 1976, *Teleological Explanations*, University of California Press, Berkeley.

ERHARD SCHEIBE

INVARIANCE AND COVARIANCE*

According to its title the present paper has a twofold aim. As far as invariance is concerned I want to argue that the numerous invariances of physical laws can essentially be reduced to one concept of invariance that is already part of the concept of kinds of mathematical structures that are used to formulate physical theories. The second aim of the paper is to present the concept of covariance not as any kind of invariance but rather as a concept of equivalence between two formulations of a physical theory that are already invariant, one of which, however, has a higher 'degree' of invariance than the other. As regards the actual use of the terms 'invariance' and 'covariance' in physics and the occasional discussions of it in the philosophy of physics I do not want to entertain any criticism of the literature although the somewhat deplorable state of affairs concerning the two concepts would justify this. The present paper will thus be self-contained, and I will pay attention to what usually is meant by invariance and covariance only to the extent that makes it sufficiently clear what I am talking about.

To give a brief outline of my argumentation I begin with calling attention to the fact (part I) that one of our fundamental metamathematical concepts, the concept of a species of structures, is essentially defined by a condition of invariance. On the other hand, the mathematical part of every physical theory (the 'formalism' that is peculiar to it) presumably can be reformulated as a species of structures. Consequently, to the extent to which this is true, an invariance condition would be built into a physical theory from the very outset. This naturally leads to the question what this kind of invariance, canonical invariance as it will be called, has to do with the well-known invariances appearing in physics as characteristic properties of its fundamental laws. One proper context in which this question can be discussed is given by those physical theories whose mathematical formulation is based on species of structures called manifolds (part II). Most important examples are theories, e.g., Einstein's theory of gravitation, that are founded on a theory of spacetime as the basic manifold. But also the configuration spaces and phase spaces of classical mechanics are cases in point. For a large class of these theories (based on manifolds) it can be

J. Agassi and R.S. Cohen (eds.), Scientific Philosophy Today, 311–331.
Copyright © 1981 *by D. Reidel Publishing Company.*

proved that canonical invariance is essentially equivalent to the common invariances of physical laws, and it seems plausible that complete generality can be achieved. As a consequence, no invariance requirements beyond the one that is already part of the definition of a species of structures would have to be imposed on a physical theory. This result can be viewed as a substantiation and justification of what is nowadays sometimes recommended as the 'coordinatefree method' in physics.

Transcending the concept of invariance the concept of covariance (part III) points to the fact that two species of structures can be equivalent and as such can become the formalisms of one and the same physical theory although their 'degrees' of invariance are different. They are, however, supposed to be comparable in the sense that the groups underlying the invariances are related by inclusion. Covariance as equivalence takes up the old issue that came up with the general theory of relativity to increase at will the invariance of a physical law. Maxwell's equations in their Lorentz-invariant form, for instance, can be *made invariant* under arbitrary differentiable transformations by explicitly introducing the Minkowski metric as a tensor field. If this is done two new facts occur: the invariance of the new equations and their equivalence to the equations from which one started. But the new equations are invariant under the larger group in exactly the same sense as were the original equations with respect to the smaller group. Therefore, if there is any need for a new concept then it is with respect not to invariance but to equivalence.

This can be shown even more convincingly in the context of the so-called 'principle of general covariance'. If this principle is taken to mean that physical laws should be invariant under arbitrary differentiable coordinate transformations then, as is well known and will here be confirmed anew, it can be trivially satisfied unless further requirements are included. With respect to the converse relation of covariance, however, the situation is altogether different: The reduction of the group of coordinate transformations is not always possible, and if it is possible this may very well be a nontrivial result. In applying this relation of reduction I shall argue that the general theory of relativity has to be estimated on account of its irreducibility and *not* as is frequently done with regard to the possibility of finding covariant versions of other physical theories on the invariance level of general relativity.

The three concepts of species of structures, of the invariance inherent in them and of the equivalence between them are due to Bourbaki [1968, Ch. IV]. They are, however, founded on a particular formulation of logic and set theory that I do not want to take over. I will rather assume a modern formulation of the system of Zermelo-Fraenkel (ZF). This means that set theory is presented as a standard theory [cf. Shoenfield 1967]. For convenience I will even work with an extension by definitions of ZF that is rich enough to contain all the auxiliary concepts that will constantly be needed (ZFT). The reader may or may not assume that ZFT (and hence ZF) is given an interpretation in the modeltheoretical sense. Such an assumption would at least not be necessary since all metatheoretical concepts that will be introduced will concern only syntactical entities (formulas or terms).

I

Roughly speaking a species of structures is an *axiom system* of a special kind: There is a subdivision of its concepts into basic concepts $X_1, \ldots, X_n$ and typified concepts $s_1, \ldots, s_m$ as well as a corresponding subdivision of the axioms into proper axioms α_i and typifications $\gamma_1, \ldots, \gamma_m$ such that
(1) γ_μ determines the type of s_μ relative to the X_ν, i. e., it determines the 'nature' of the entities falling under s_μ relative to those falling under any of the X_ν, whereas
(2) the γ_μ by themselves and the α_i by *fiat* do *not* determine the 'nature' of the entities falling under the X_ν relative to each other or to any concept 'presupposed' by the axioms.

To give the details in set-theoretical terms [cf. Bourbaki 1968, Ch. IV], we think of an axiom system as being a finite set of formulas of ZFT depending on certain variables indicating the sets ('concepts') that are interrelated by the axioms. We then have their subdivision into the *base sets X* (short for: $X_1, \ldots, X_n$) and the *typified sets s* (short for: $s_1, \ldots, s_m$). The *typifications* γ_μ of (1) are formulas

$$s_\mu \in \sigma_\mu(X) \tag{1}$$

where σ_μ is a scale term, i.e., a term constructed from its arguments by successively applying one of the operations that yield a power set (Pot) or a cartesian product (×). In the case before us the arguments are the X and possibly further constant (!) terms that are available in ZFT. To keep the notation simple the latter will not be made explicit but must be kept in mind. The 'nature' of the s (or – as above – of their elements) is determined by (1) relative to the X and constant sets from ZFT in the sense that the s are elements of the power set of ... the cartesian product of ... the X and the constants. The remarkable thing about (1) is that it provides counterparts of all the predicates and terms of arbitrary arity and order as they appear in the various independent logical calculi. In the present context a scale term σ is called a *type of structures* and a system of sets X and s satisfying (1) a *structure of type* σ.

Having done with the determination of the s with respect to the X and constant sets announced in (1) we have now to prepare the ground for the invariance condition taking charge of (2). To do this we define the *canonical σ-representation* to be the following assignment: Given a scale term σ and bijections f from sets X onto sets X' (i. e., according to our convention bijections f_ν from X_ν to X'_ν) a bijection f^σ from $\sigma(X)$ onto $\sigma(X')$ is uniquely determined by the two recursive conditions that for the power set operation

$$f^{\mathrm{Pot}\sigma}(S) = \{f^\sigma(s) \mid s \in S \wedge S \in \mathrm{Pot}\sigma(X)\} \tag{2a}$$

and for the cartesian product operation

$$f^{\sigma_1 \times \sigma_2}(\langle s_1, s_2 \rangle) = \langle f^{\sigma_1}(s_1), f^{\sigma_2}(s_2) \rangle \tag{2b}$$

We go on to define f to be a *σ-isomorphism* from $\langle X, s\rangle$ onto $\langle X', s'\rangle$ if and only if $\langle X, s\rangle$ and $\langle X', s'\rangle$ are structures of type σ and the f are bijections from the X onto the X' such that $f^\sigma(s) = s'$. It then *follows* immediately for the typifications (1) that

$$\mathrm{iso}_\sigma(X, s; X', s'; f) \dot{\rightarrow} s \in \sigma(X) \leftrightarrow s' \in \sigma(X') \tag{3a}$$

is provable (in ZFT). By analogy we now *require* that the proper axioms are *canonically σ-invariant* in the sense that for their conjunction

$$\mathrm{iso}_\sigma(X, s; X', s'; f) \dot{\rightarrow} \alpha(X; s) \leftrightarrow \alpha(X'; s') \tag{3b}$$

is provable (in ZFT).

An axiom system consisting of typifications (1) and proper axioms satisfying (3b) is called a *species of structures*. In Bourbaki [1968, Ch. IV] the crucial property (3b) of α is called 'transportability'. By using the term 'canonical invariance' instead I want to emphasize that, if anything, then (3) is a condition of invariance. The intuitive idea standing behind any definition of invariance is the idea of something that remains unaltered while something else on which it depends is changed. In the present case what remains the same is, expressed in semantical terms, the truth value of a proposition, and what is changed is that about which the proposition is a proposition: If the structure $\langle X, s\rangle$ is replaced by a structure $\langle X', s'\rangle$

isomorphic to the first, then α is true about the second if it is true about the first and false about the second if false about the first. Extreme cases of invariance of truth values would be given by formulas provable or refutable in ZFT as, for instance, (3a) or (3b) themselves. In a model of ZFT such formulas would be true or false no matter what interpretations their variables are given, and this means that any replacement of the values of these interpretations would leave truth values invariant. This complete freedom of replacement is restricted in (3) in two ways: The base sets X may only be replaced by sets resulting from them by bijections f, and the typified sets s only by their images under the canonical representations of the f. This makes room for many more formulas becoming invariant in the sense of (3). On the other hand, the restrictions are still wide enough to prevent the axioms from saying anything about the 'nature' of the X in the sense that their elements were determined relatively to each other or to any constant set. Without bringing this idea under precise terms let us imagine some clear cases in which one would say that such a determination occurs. Cases in point would be given by all typifications (1) with one of the base sets instead of s_μ and the rest of the base sets instead of X as our axiom α. It is evident that such a relation can be destroyed by isomorphisms and therefore would contradict (3b). The same would be true for axioms α saying that one base set X has an empty (or nonempty) intersection with another base set or constant set. Thus even such negative determinations are ruled out by our invariance condition.

Examples for species of structures abound from mathematics: In point of fact all the well-known concepts of a group, ring, vector space, topological space, manifold, fibre bundle etc. are defined by axioms that can easily be reconstructed as so many typifications (1) and axioms proper satisfying (3b). It is likewise a fact that these mathematical concepts are frequently applied in theoretical physics, and at least in the mathematical treatment of higher level theories such as quantum theory and the theory of general relativity the use of species of structures has been generally acknowledged: "A general lesson to be drawn from the development of the theory of relativity is that it is desirable to analyze in detail the various structures inherent in the mathematical models used to describe physical phenomena" [Trautman 1972, p. 85]. The multifarious application of species of structures in theoretical physics does not mean, however, that the common conceptual basis of this mathematical field is sufficiently understood. By now there exists only one thoroughgoing attempt at a *general* conceptual reconstruction of physics on the basis (as far as mathematics is concerned) of the concept of species of structures [Ludwig 1978]. There is, admittedly, also the set theoretical approach of Suppes, Sneed and others, and it has recently been claimed that this approach could be viewed as an extension of the Bourbaki program to science [Stegmüller 1979]. Apart from several objections that could be made against this claim the set theoretical approach can be left out of consideration here because it never does focus on the defining conditions of a species of structures which, however, make all the difference with regard to the goal of this paper.

As I have tried to argue in a previous paper [Scheibe 1978] species of structures occur in physical theories not only as incidental concomitants. Rather they can be used to characterize a physical theory as a whole as far as this is possible without regard to their empirical interpretation. In particular this characterization includes physical laws in the narrower sense and with them the sort of things in whose invariance properties we are primarily interested. However, the characterization can not be achieved solely by the simple and general species of structures that are the building blocks of modern mathematics. Rather we must have recourse to richer species built up from the simpler ones: "As a rule, rich structures are used in physcis; those of differential manifolds carrying additional geometric objects and of Hilbert spaces with preferred sets of operators" [Trautman 1972, p. 85].

A general concept particularly helpful in making clear what is at issue here is the concept of an *extension* of a species of structures. Given a species $(\sigma_0; \alpha_0)$ it can be extended by adding a (not necessarily) new typification $s \in \sigma(X)$ and axiom $\alpha(X; s_0, s)$, both referring to the old base sets, such that $(\sigma_0 \times \sigma; \alpha_0 \wedge \alpha)$ again is a species of structures. It follows from the assumption that $(\sigma_0; \alpha_0)$ already is a species of structures that the canonical $(\sigma_0 \times \sigma)$-invariance of $\alpha_0 \wedge \alpha$ is equivalent to the provability of

$$\begin{aligned} \mathrm{iso}_{\sigma_0 \times \sigma}(X, s_0, s; X', s_0', s'; f) \wedge \alpha_0(X; s_0) \\ \dot{\rightarrow} \alpha(X; s_0, s) \leftrightarrow \alpha(X'; s_0', s') \end{aligned} \tag{4}$$

In this way the 'invariance part' of (3b) can be isolated for the *added* axiom α. Now, in physical practice one frequently proceeds from axioms that in a systematic account of the species of structures involved would be added only at the very end. It is this methodological inversion that can be elucidated by the concept of extension.

Take ordinary quantum mechanics as an example. The core of it, the physical law that really matters, is Schrödinger's equation (with $\hbar = 1$)

$$i\dot{\psi} = H\psi \tag{5a}$$

Asking for the axiomatic background of this equation it turns out that it is 'only' the last step in a series of extensions. The series would start with an abelian group X with addition A which then would be extended into a complex vector space by adding a scalar multiplication B. The next extension would lead to a Hilbert space by adding a metric C, and this is the point where the structural aspect usually comes to a stop. But in principal the extension could be continued with a self-adjoint linear operator H and an X-valued and $\mathbb{R}$-argumented function ψ as two new structures that are related by (5a). The only requirement would be that the whole axiom that we have obtained is canonically invariant with respect to the total typification of the five structures imposed on X, and this is indeed the case. In this way not only is the crucial equation (5a) submitted to a species of structures. It also has received the 'right' invariance property: The invariance of (5a) under the transformations

$$\psi'(t) = U\psi(t), \qquad H' = UHU^{-1} \tag{5b}$$

with a unitary transformation U of X that is usually considered is but a special case of canonical invariance. It occurs if in the canonical representation of U not only X but also A, B and C are left invariant which is but another way of saying that U is unitary. All this comes out of (4) if α_0 refers to the Hilbert space and α to the properties of H and ψ including (5a) and if, finally, $X' = X$ and $s_0' = s_0$. Then $s' = f(s)$ turns out to be (5b), and the conclusion of (4) is the restricted invariance usually considered.

II

The foregoing example is a particularly favorable case to my point that the invariance properties of physical laws can be reduced to the canonical invariance in species of structures. The reason is that general quantum mechanics is usually presented at a fairly high level of abstraction which sometimes comes very close to its presentation as a species of structures. Actually the situation is not altogether different in the field that I am going to enter now: *Manifolds* of this or that kind are frequently introduced in a manner that makes it not too difficult to reformulate the definitions in terms of species of structures. Sometimes the abstract presentation is even highly recommended as a 'coordinate free'' or 'intrinsically invariant' method. But seldom, if ever, is it precisely stated in what sense we get rid of coordinates and by what features the invariances so characteristic of the old coordinate dependent formulations are replaced.

In order to clear up the matter I start this investigation with a particularly simple class of species of manifolds which in turn will be introduced in two steps. In the first step species of *global coordinate manifolds* (*gcm*) are defined. In a *gcm* the intuitive idea of a set F of preferred coordinate systems used to label all (!) elements of an otherwise arbitrary set M is given a precise formulation. The species of *gcm* are parametrized by two external parameters (i.e., constant terms being available in ZFT): a natural number n and a group G of homeomorphisms of the n-dimensional real number space $\mathbb{R}^n$ with its usual topology. Typification and axiom defining the species *gcm* (n, G) are

$$F \in \mathrm{Pot}^2(M \times \mathbb{R}^n) \wedge \alpha_{gcm}(M; F) \tag{6a}$$

where α_{gcm} says that the $\varphi \in F$ are bijections from M onto $\mathbb{R}^n$, that $\psi\varphi^{-1} \in G$ for $\varphi, \psi \in F$, and that $g\varphi \in G$ for $\varphi \in F$ and $g \in G$. These requirements make F into a complete set of global coordinate systems on the space M with respect to the group G of coordinate transformations. It can easily be proved that α_{gcm} is canonically invariant with respect to the typification in (6a).

The most important examples of *gcm*(n, G) applied in physics are those in which M is interpreted as spacetime (hence $n = 4$) and F is a class of preferred coordinate systems in spacetime with one of the following groups of coordinate transformations [cf. Ehlers 1973]:

$$\begin{array}{ccccc} & G_{\mathrm{gal}} & \subset G_{\mathrm{ros}} & \subset G_{\mathrm{kin}} & \\ G_{\mathrm{new}} \subset & & \subset G_{\mathrm{aff}} & & \subset G_{\mathrm{dif}} \subset G_{\mathrm{top}} \\ & G_{\mathrm{poi}} & & & \end{array} \tag{7}$$

This hierarchy provides us with so many *gcm*$(4, G\ldots)$, as $G\ldots$ is one of the following transformation groups of $\mathbb{R}^4$: the direct product of the Euclidean groups of space and time (G_{new}), the Galileo (G_{gal}) and Poincaré (G_{poi}) group, the affine group (G_{aff}), the enlargements of G_{gal} allowing for accelerative translational (G_{ros} [cf. Rosen 1972]) and completely arbitrary (G_{kin}) rigid motions, the group of diffeomorphisms (G_{dif}) and of homeomorphisms (G_{top}) of $\mathbb{R}^4$.

Besides the direct spacetime applications of the species of *gcm* other applications of interest are the (global) configuration spaces and their cotangent bundles (phase spaces) in classical mechanics. All these cases, however, and in fact the species of *gcm* in general have still to be enriched with further structures that can be used for a more complete description of physical reality. This is achieved in a second step by defining a species of *global manifolds* (*gm*) to be any extension of a species of *gcm*. According to the definition of extensions in Section I this means that (6a) is supplemented by a formula

$$s \in \sigma(M) \wedge \alpha(M; F, s) \tag{6b}$$

such that their conjunction is a species of structures. Depending, as it does, on the parameters n and G, belonging already to $gcm(n, G)$, as well as on σ and α in (6b) a species of *gm* will be designated by $gm(n, G, \sigma, \alpha)$. With the introduction of the new structure s typified and axiomatized in (6b) all sorts of extensions of the previously mentioned physical applications of a *gcm* come within view. Outstanding examples are to be found in Newtonian particle mechanics (based on G_{gal}), nonrelativistic quantum mechanics (usually based on G_{new}, but generalizable to G_{gal} and even G_{ros}, [cf. Rosen 1972]), classical electrodynamics (based on G_{poi}), relativistic quantum mechanics of the Dirac equation (based on G_{poi}) and Einstein's gravitational theory (based on G_{dif}), the latter being restricted to the global case. In all these cases the structure s in (6b) would stand for the physical objects that are under investigation, e. g., orbits of particles, spacetime extensions of fields, probability functions for quantities etc., and α of (6b) would, among other things, state the essential physical laws that the objects have to obey, e.g., a law of motion, a law of propagation etc.

For the moment I do not want to go into any details about the question how the theories just mentioned actually can be reconstructed as so many species of global manifolds. The answer to this question is intimately connected with the main problem to be treated in this section. If, for instance, we were asked to give a formulation of Newton's law of gravitation for a system of mass points as a special case for (6b), then a very common answer would be to write down the well-known gravitational equations in Galilean coordinates, justifying this procedure by the remark that the equations are invariant under any change of this kind of (Galilean) coordinate systems. It is here where we meet the 'common invariance' of physical laws as invariance of certain 'concrete' mathematical equations representing the laws in question under an appropriate representation of a 'concrete' mathematical group. At the same time, the formulation of Newton's law mentioned before would *not* be a formulation in terms of species of *gm although* we feel that it may be equivalent to such a formulation and that the reason for this is just the 'common invariance' of the equations representative for the law.

We may therefore try to solve our main problem by seeking a *characterization* of the species $gm(n, G, \sigma, \alpha)$ by means of appropriate equivalents realizing the idea of a coordinate formulation of (6b). One thing, however, that must be clearly understood about such an attempt is that it has no unique, intrinsic solution but rather depends on a decision as to how an object $s \in \sigma(M)$ is going to be represented in a coordinate system and, accordingly, how the group G of coordinate transformations is going to be represented on the set of representatives of the s. In the first characterization to be given we decide to take canonical representations throughout. Then, given n, G and σ, a canonical coordinate formulation $ccf(n, G, \sigma, \alpha')$ is a formula

$$s \in \sigma(\mathbb{R}^n) \wedge \alpha'(s) \tag{8a}$$

where α' is *invariant* in the sense that

$$g \in G \wedge s' = g^\sigma \cdot s \wedge s \in \sigma(\mathbb{R}^n) \dashrightarrow \alpha'(s) \leftrightarrow \alpha'(s') \tag{8b}$$

is provable. (8b) with g^σ as the canonical representation of g is here offerred as a condition of invariance in the usual sense, and (8a) will have to replace (6b) in the wanted characterization.

This characterization can most conveniently be expressed in terms of a relation between two formulas α and α'. Given n, G and σ the relation is to hold if and only if

(c_1) α extends the data to a species $gm(n, G, \sigma, \alpha)$, or – equivalently – (4) holds for α with σ_0 and α_0 being given by (6a),

(c_2) α' extends the data to a canonical coordinate formulation, or – equivalently – α' satisfies (8b),

(c_3) the formula

$$\varphi \in F \wedge s' = \varphi^\sigma \cdot s \wedge s \in \sigma(M) \dashrightarrow \alpha(M; F, s) \leftrightarrow \alpha'(s') \tag{9a}$$

or – equivalently and slightly more to the point –

$$\varphi \in F \wedge s' = \varphi^\sigma \cdot s \dashrightarrow s \in \sigma(M) \wedge \alpha(M; F, s) \dashrightarrow s' \in \sigma(\mathbb{R}^n) \wedge \alpha'(s') \tag{9b}$$

is provable from (6a).

It turns out that this relation is itself invariant under two equivalence relations for α and α'. For α it is given by the provability of

$$s \in \sigma(M) \dot{\rightarrow} \alpha(M; F, s) \leftrightarrow \alpha_1(M; F, s) \tag{10a}$$

from (6a), and for α' by the provability of

$$s \in \sigma(\mathbb{R}^n) \dot{\rightarrow} \alpha'(s) \leftrightarrow \alpha_1'(s) \tag{10b}$$

(from ZFT alone). The characterization will be possible only *modulo* these equivalence relations. It says (equivalence theorem 1) that the relation defined by (c) is one-to-one *modulo* (10) and that to every α satisfying (c_1) there exists α' such that the relation holds and, conversely, that to every α' satisfying (c_2) an α exists such that the relation holds. An explicit solution of (9) for α, given α', is

$$\bigvee s', \varphi \colon \varphi \in F \wedge s' = \varphi^\sigma \cdot s \wedge s' \in \sigma(\mathbb{R}^n) \wedge \alpha'(s') \tag{11a}$$

and for α', given α,

$$\bigvee M, F, s, \varphi \colon \varphi \in F \wedge s' = \varphi^\sigma \cdot s \wedge s \in \sigma(M) \wedge \alpha(M; F, s) \tag{11b}$$

In the present case the right side of (11b) even reduces to $\alpha(\mathbb{R}^n; G, s')$ and has been given the complicated formulation only for reasons that will become clear in the sequel.

The proof of equivalence theorem 1 is, although a bit tedious, a straightforward matter and need not be given here. The theorem clearly shows in what sense our main problem has been solved (under the present restrictions): On account of the one-to-one correspondence established by the theorem the common invariances (8b) of coordinate formulations are *exactly matched* by the canonical invariances (4) in species of structures. At the same time the theorem provides a *standard method* to produce reformulations of the mathematical part of a physical theory as a species of structures if a canonical coordinate formulation of that theory is known. This is the case, for instance, with many subtheories of classical mechanics, Newton's theory of gravitating mass points being the most prominent example. On the other hand, the usual coordinate formulation of electrodynamics, for instance, is not canonical. The group of coordinate transformations (G_{poi}) gets a tensor representation instead. To include such important cases in our reduction program the equivalence theorem will have to be generalized.

Before this will be done one other remark should be made now. Equivalence theorem 1 is, of course, nothing but a methodologically purified version of the transition from the old way of coordinate geometry to the modern, abstract and coordinate-free approach to geometry. Besides showing the ordinary invariances becoming canonical invariances in species of structures in this transition the theorem also shows in what sense we get rid of coordinates. A coordinate formulation (1) of necessity picks out *one* of the preferred coordinate systems, but (2) leaves it completely arbitrary *which* one is chosen. As can be most clearly seen in (9b) what we get rid of is the arbitrariness of saying what has to be said in terms of one coordinate system. What we *do not* get rid of is the very set of preferred coordinate systems out of which the choice has to be made. Far from becoming eliminated this set rather is introduced explicitly as a structure in a coordinate manifold. This is sometimes blurred by an otherwise very convenient piecewise application of the standard method. In most physical theories the structure s as well as what is said about it in (6b) are both fairly complicated, and one may wish to break them up in order to get simpler units. This method consists of three parts: (1) a decomposition of (8b); (2) the transition of the pieces to (6b) by means of the equivalence theorem; and (3) their recomposition in accordance with their original connection. Since as a rule no reference to the preferred set F of coordinate systems is needed in the third step this may give the impression that F, too, has been eliminated. This, however, is not the case since the second step heavily depends on F.

The result of equivalence theorem 1 may not come as a surprise for the simple reason that (8b) is a special case of canonical invariance. Once we had decided to use the canonical representation of G in the coordinate formulations it was very suggestive to found the relation (c) on the canonical representation of F. But not all representations are canonical. In fact the most important representations used in physics (besides the canonical), namely the tensor and spinor representations of the classical groups, are not. The question therefore arises whether the coordinate formulations of physical theories based on these representations also have species of structures as equivalents and whether the invariances belonging to the former again have the canonical invariances in species of structures as their counterparts.

To answer this question for almost completely general representations in coordinate formulations we introduce a new class of extensions of species $gcm(n, G)$ depending on three additional external parameters. They are two scale terms σ and σ' (each with one principal argument) and an arbitrary representation r of G on $\sigma'(\mathbb{R}^n)$ in the usual sense. (Like G itself

r is meant to be a term, available in ZFT, for which it can be proved that it has the property just required.) A species of *representations in a gcm* – designated by $rp(n, G, \sigma, \sigma', r)$ – is then defined to be an extension of $gcm(n, G)$ with

$$\rho \in \mathrm{Pot}(\mathrm{Pot}(M \times \mathbb{R}^n) \times \mathrm{Pot}(\sigma(M) \times \sigma'(\mathbb{R}^n))) \wedge \alpha_{rp}(M; F, \rho) \tag{12}$$

as the new typification and axiom. The axiom α_{rp} says that ρ is actually a mapping

$$\rho: F \rightarrowtail \mathrm{bij}(S_\rho; \sigma'(\mathbb{R}^n)) \tag{13a}$$

from F into the set of bijections from a set $S_\rho \subseteq \sigma(M)$ (uniquely determined by ρ) onto $\sigma'(\mathbb{R}^n)$ compatible with the representation r in the sense that

$$\rho(\psi) = r(\psi\varphi^{-1})\rho(\varphi) \tag{13b}$$

and with the canonical representation on $\sigma(M)$ in the sense that

$$\rho(\psi) = \rho(\varphi)(\varphi^{-1}\psi)^\sigma \mid S_\rho \tag{13c}$$

for any $\varphi, \psi \in F$. Thus $\rho(\varphi)$ does for the objects $s \in S_\rho$ what a coordinate system does for the points of M: As φ represents the latter in $\mathbb{R}^n$ so $\rho(\varphi)$ represents the former in $\sigma'(\mathbb{R}^n)$.

Species of representations are applied in physics in a peculiar way: A representation being an extension of a *gcm* just as, for instance, some physical field nevertheless does not have the latter's contingency. Rather there are some *distinguished* representations singled out on every *gcm* by means of a deduction. A *deduction* (here restricted to the case where base sets are kept fixed [cf. Bourbaki 1968, Ch. IV]), relates two species of structures $(\sigma; \alpha)$ and $(\tau; \beta)$ by a term P that deduces the latter from the former in the sense that

$$P(X; s) \in \tau(X) \wedge \beta(X; P(X; s)) \tag{14a}$$

together with the invariance condition

$$\mathrm{iso}_{\sigma \times \tau}(X, s, t; X', s', t', f) \rightarrow t = P(X; s) \leftrightarrow t' = P(X'; s') \tag{14b}$$

is provable from $(\sigma; \alpha)$. The following is a deduction of $rp(n, G, \sigma, \sigma', r_{\mathrm{can}})$ from $gcm(n, G)$: $D_{\mathrm{can}}(M; F)$ assigns to every $\varphi \in F$ its canonical image $\varphi^\sigma: \sigma(M) \rightarrowtail \sigma'(\mathbb{R}^n)$ defined by (2). This is the representation on which equivalence theorem 1 was founded. A second kind of distinguished representations deduced from *gcm(n, G)* presupposes that $G \subseteq G_{\text{dif}}$. They concern the tensors and tensor fields of any valence. Whereas in the case of the canonical representations we had $S_{D_{\text{can}}} = \sigma(M)$ and $\sigma' = \sigma$, in the tensor representations $S_{D_{\text{ten}}}$ is restricted to the bundle of tensors of a given valence (k, l) and $\sigma'(\mathbb{R}^n) = \mathbb{R}^n \times \mathbb{R}^{n^{k+l}}$. For tensor fields the corresponding power sets have to be used. The representations themselves are defined in the usual way by giving the tensors and tensor fields their components in a given coordinate system. A third kind of distinguished representations concerns affine connections. As in the tensor case the representation is confined to a special kind of objects, and it is $\sigma'(\mathbb{R}^n) = \text{Pot}(\mathbb{R}^n \times \mathbb{R}^{n^3})$.

Starting now from species *gcm(n, G)* and *rp(n, G, σ, σ', r)* and a deduction D of the latter from the former we can generalize equivalence theorem 1 in the following way: Replace in the formulas (8)–(11)

$$\begin{array}{lll} \sigma'(\mathbb{R}^n) & \text{for} & \sigma(\mathbb{R}^n) \\ r(g) & \text{for} & g^\sigma \\ S_{D(M,F)} & \text{for} & \sigma(M) \\ D(M; F)\cdot\varphi & \text{for} & \varphi^\sigma \end{array} \tag{15}$$

Then the equivalence theorem holds *ceteris paribus* with respect to the new entities, i.e., given n, G, σ, σ', r and D as above, the relation (c) establishes a one-to-one correspondence *modulo* (10) between all species *gm(n, G, σ, α)* and all coordinate formulations *cf(n, G, σ', α')* (equivalence theorem 2). The essential generalization that has been obtained is the step from canonical to (almost) arbitrary representations and corresponding invariances on the side of the coordinate formulations *retaining* thereby canonical invariance on the side of the species of structures.

III

Up to this point the considerations concerning the reduction of the ordinary invariances occurring in coordinate formulations of physical theories to canonical invariance as part of the definition of species of structures were confined to the topologically trivial, global case where the space M is homeomorphic to $\mathbb{R}^n$. In order to gain the full generality in which the theory of manifolds is usually developed in mathematics and even applied in physics we would now have to take a final step by introducing the *local viewpoint*. As will be seen in a moment this causes no problems as far as the concept of a manifold based on a coordinate manifold is concerned. However, as soon as one proceeds from this new generalized basis in order to get at a corresponding generalization of the equivalence theorem considerable difficulties come up. Already the further conceptual apparatus for a new formulation of the theorem would demand an amount of technical detail that would be beyond the scope of the present paper. I will therefore leave the matter at the stage that has been achieved in the previous section and now switch over to the concept of covariance.

The species of global coordinate manifolds *gcm*(n, G) can be generalized by replacing the requirement that G be a group of global homeomorphisms of $\mathbb{R}^n$ by the weaker assumption that G is only a pseudogroup of local homeomorphisms in $\mathbb{R}^n$ [cf. Iyanaga and Kawada 1977, 92 D]. The idea of a pseudogroup of homeomorphisms generalizes that of a group of homeomorphisms of $\mathbb{R}^n$ by allowing variable domains and ranges. The local viewpoint that is opened in this way sometimes is emphasized by the requirement that every restriction of an element of G to an arbitrary open subset of its domain also belongs to G [cf. Iyanaga and Kawada 1977, 108 *Z*; Kobayashi and Nomizu 1963, p. 1]. Since this excludes groups of transformations as special cases of pseudogroups I adopt the more general concept mentioned first. The global case considered so far is recovered by the requirement that all transformations have a common domain (mostly $\mathbb{R}^n$ itself). It occurs if and only if the pseudogroup is a group. The groups $G \ldots$ in (7) all have obvious counterparts $G^0 \ldots$ meeting the requirement of strict locality mentioned above.

Given n and G as a pseudogroup the species of *coordinate manifolds* *cm*(n, G) is then defined in formal analogy to (6a). But the new axiom α_{cm} would only say that F is a maximal set of local coordinate systems compatible with the transformations of G. The species of *gcm* considered so far are those species of *cm* where the transformations of G have a common domain and hence the spaces M are homeomorphic to (an open subset of) $\mathbb{R}^n$. Again on the general level a species of *manifolds* *mf*(n, G, σ, α) can then be introduced to be any extension (6b) of a species *cm*(n, G). This concept is sufficiently general to include all kinds of manifolds that are known from differential geometry such as Riemannian manifolds, manifolds with an affine connection and those that carry additional fields of all sorts that have been studied in theoretical physics.

Being thus prepared to approach covariance on the most general differential geometric level there is even reason to step on the more abstract level of arbitrary species of structures and develop our concept on it. As was already announced in the introduction, our access to covariance will be guided by the idea of equivalent mathematical formulations of one and the same physical theory. Since species of structures lend themselves to such formulations we have to ask for an adequate concept of equivalence for them. This is readily at hand: Given two species $(\sigma; \alpha)$ and $(\sigma_1; \alpha_1)$ as well as deductions P and P_1 of $(\sigma_1; \alpha_1)$ from $(\sigma; \alpha)$ and *vice versa* such that

$$\begin{aligned} P(X; P_1(X; s_1)) &= s_1 \\ P_1(X; P(X; s)) &= s \end{aligned} \tag{16}$$

are provable from $(\sigma; \alpha)$ and $(\sigma_1; \alpha_1)$ respectively $(\sigma; \alpha)$ and $(\sigma_1; \alpha_1)$ are called *equivalent* with respect to P and P_1. That of two species of structures equivalent in this sense, either both or neither of them can be used to formulate a physical theory, hinges on questions of interpretation that cannot be discussed in this paper [cf. Ludwig 1978]. It is, however, safe to say that, given $(\sigma; \alpha)$ together with a physical interpretation, the move to an equivalent $(\sigma_1; \alpha_1)$ by means of equivalence terms P and P_1 cannot affect the content of the theory *if* the interpretation is retained, and it seems plausible that by using P and P_1 the interpretation can always be transferred from $(\sigma; \alpha)$ to $(\sigma_1; \alpha_1)$ to yield an interpretation of $(\sigma_1; \alpha_1)$ in the same sense as the original interpretation of $(\sigma; \alpha)$. In Ludwig [1978] such a transfer is discussed even for the weaker case where the base sets can be different and P_1 is replaced by the assumption that P is conservative in the sense that every structure $\langle X; t\rangle$ of species $(\tau; \beta)$ is isomorphic to a structure $\langle X; P(X; s)\rangle$ with an $\langle X; s\rangle$ of species $(\sigma; \alpha)$. For the treatment of problems of covariance, however, the stronger and simpler concept of equivalence seems to be sufficient.

Let now Σ and Σ_1 be two species of structures and P a deduction of Σ_1 from Σ. Furthermore, let Σ^* and Σ_1^* be two extensions of Σ and Σ_1 respectively and equivalent with respect to the equivalence transformations

$$\begin{aligned} F_1 &= P(M; F), & F &= P_1(M; F_1, s_1) \\ s_1 &= Q(M; F, s), & s &= Q_1(M; F_1, s_1) \end{aligned} \tag{17}$$

where the *given* deduction P appears in the upper left corner. Here the notation is already adapted to species of *mf* but the actual assumptions are meant to be quite general. The following scheme with P^* for (17) may help to apprehend the given data and their mutual relations:

$$\begin{array}{ccc} \Sigma & \xrightarrow{\quad P \quad} & \Sigma_1 \\ \Big\downarrow \text{deduction} & & \Big\downarrow \text{extension} \\ \Sigma^* & \xleftarrow{\text{equivalence}} P^* \longrightarrow & \Sigma_1^* \end{array} \tag{18}$$

We now call Σ_1^* a *covariant version* of Σ^* and – conversely – Σ^* a *reduced version* of Σ_1^*, both with respect to the remaining data Σ, Σ_1, P and P^*. It should be borne in mind that the concepts of covariance and reduction thus defined strictly speaking relate six syntactical entities to each other. In actual practice, however, we can think of the three data in the upper row of (18) as being 'kept fixed'.

This will become clear if we now specialize the general setting in (18) by assuming Σ and Σ_1 to be two species $cm(n, G)$ and $cm(n, G_1)$ with $G \subseteq G_1$. We then have the following natural deduction P of $cm(n, G_1)$ from $cm(n, G)$: Given a $cm\langle M; F\rangle$ of the species $cm(n, G)$ the set $F_1 = P(M; F)$ is the uniquely determined complete set of coordinate systems on M with respect to G_1 satisfying $F \subseteq F_1$. More specifically, since the days when the general theory of relativity was born the cases of prevailing interest became those in which G was one of the classical groups and G_1 the pseudogroup G^0_{dif}. And since then it became a prevailing tendency in theoretical physics to show of as many spacetime founded physical theories as possible that they could be reformulated as certain extensions of the species of differentiable manifolds.

Before extending the exemplification in this direction also to the lower row of (18) two problems shall be formulated that pose themselves in the situation when only the upper row of (18) is fixed. To be as precise as possible their wordings will again be given in general terms. They are:

(co) If in addition to Σ, Σ_1 and P of (18) an extension Σ^* of Σ is given can we then always find Σ_1^* and P^* such that Σ_1^* is a covariant version of Σ^*?

(re) If in addition to Σ, Σ_1 and P of (18) an extension Σ_1^* of Σ_1 is given can we then always find Σ^* and P^* such that Σ^* is a reduced version of Σ_1^*?

The two problems will be called the *problem of covariance* and of *reduction* respectively. There is obviously a certain duality between them. However, on account of the asymmetry of (18) in its upper row the two problems receive entirely different answers.

The answer to the problem of covariance is positive and trivial. Assuming that the axiomatics of the given species Σ, Σ_1 and Σ^* (typification and axiom proper) are

$$\begin{aligned}&\Sigma: \bar{\alpha}(M;F),\ \Sigma_1: \bar{\alpha}_1(M;F_1)\\&\Sigma^*: \bar{\alpha}(M;F) \wedge \bar{\beta}(M;F,\ s)\end{aligned} \tag{19a}$$

then with $P(M;\ F)$ as the given deduction of Σ_1 from Σ the species Σ_1^* that we are looking for can be chosen to be

$$\Sigma_1^*: \bar{\alpha}_1(M;F_1) \wedge \bar{\alpha}(M;F) \wedge \bar{\beta}(M;F,s) \wedge F_1 = P(M;F) \tag{19b}$$

where the variables have already been chosen such that the equations of (17) become identities except, of course, the one that is prescribed by P. Let us exemplify this solution in the context of species of manifolds taking up the settings from the preceding paragraph. Suppose then that we want to have a covariant version of ordinary electrodynamics. In this case $\bar{\alpha}$ would axiomatize Minkowski spacetime, $\bar{\alpha}_1$ the species of differentiable manifolds, and $\bar{\beta}$ (as part of Σ^*) would be a reformulation of Maxwell's equations in terms of species of structures according to equivalence theorem 2. The trick of getting at a covariant version of Σ^* now consists in adding s (the Maxwell field) *as well as* a set F of distinguished coordinate systems (the Lorentz frames) to the set F_1 of 'arbitrary' coordinate systems in Σ_1^*. Precisely this is expressed by $\bar{\alpha} \wedge \bar{\beta}$ and $F_1 = P(M;\ F)$ which in the present case turns out to be equivalent to $F \subseteq F_1$.

In view of solution (19b) in all its triviality there may be two kinds of objections against our version (co) of the problem of covariance. It may, firstly, be objected that the nontrivial problem will be an analogue of (co) which exlcusively deals with *coordinate formulations*. It may, for instance, seem a nontrivial problem to find a coordinate formulation of the ordinary Maxwell equations that is invariant under G_{dif}. However, given the results of section II the objection can easily be met by combining those results with (19b): Equivalence theorem 2 does provide for a coordinate formulation equivalent to (19b). In the Maxwell case this coordinate formulation turns out to say about $\langle F', s'\rangle$ that F' is a right coset of G_{poi} in G_{dif} and s' a solution of the equations that result from the Maxwell equations by transforming them with any $g \in F$ – the former equations being uniquely determined by F' on account of the invariance of Maxwell's equations under G_{poi}.

According to the second objection it would be pointed out that (co) could be made nontrivial if only *additional requirements* were imposed on the species Σ_1^* that is to be found. In the differential geometric context laid down above it could, for instance, be required that the new structure s_1 of $mf(n, G_1)$ extending the coordinate manifold $\langle M; F\rangle$ of $cm(n, G_1)$ has to be a cartesian product of (1) curves in M, (2) tensor fields on M, (3) one affine connection on M, and nothing else. This would be a perfectly clear requirement, and the trivial solution (19b) would certainly *not* satisfy it. Moreover, the new problem (co) would be nontrivial in the twofold sense that it would not any more have a positive answer for every given Σ^*, and if it has, the answer may very well be nontrivial – at least not as trivial as (19b).

To this objection I wholeheartedly agree. At the same time, the existence of the unrestricted formulation (co) will now be justified as the necessary background for it. It is hard to find out whether there is such a thing as *the* prevailing opinion about the concept of covariance and problems connected with it. But I had always the impression that if there is, then part of it is the tendency to view covariance as something which can be had – and in a sense even trivially had – *at all events*. On the other hand, I have never seen any *proof* or even an attempt to prove that this view is correct. In order to explore under what circumstances a proof in the strict sense would be possible the version (co) and its solution (19b) were produced. And it may be repeated with regard to the first objection that together with the results of section II both the problem (co) and its solution cover corresponding questions concerning invariance as they are usually mixed up with covariance.

Now, adherents of the triviality thesis about covariance becoming aware of the solution (19b) might still not feel themselves confirmed in their intuitions. Regarding the additional requirement objection they would perhaps point out that, once differential geometry had become of principal physical interest as a consequence of the empirical success of Einstein's gravitational theory, covariant versions meeting the objection have been found for the formalisms of *all* classical geometrical and physical theories. And since these were theories of widely differing contents the property of covariance shared by all of them cannot be of much physical importance. In other words, the triviality of covariance consists in the lack of any contribution to the *content* of a physical theory. Before commenting on this argument something has to be said about the fact that it alludes to. There could indeed be made a long list of embeddings of geometrical and physical theories in the general differential geometric context mentioned in the additional requirement objection (or something similar to it). But in order to discuss the consequences of this fact with sufficient rigor the preliminary question had to be answered precisely in what sense covariance had been established in all these cases. It seems that the concept of covariance proposed above will cover all the relevant cases known from the literature [see especially the more recent work done in Havas 1964, Trautman 1967, Künzle 1972, Misner *et al.* 1973, Ch. 12]. In all these cases species $mf(n, G^0_{\mathrm{dif}})$ submitted to the restrictions of our second objection are produced as equivalents of species $gm(n, G)$ with $G \subseteq G^0_{\mathrm{dif}}$ that function as the 'original' formulation of a physical theory. The need for a separate concept of covariance, different from that of invariance, becomes particularly clear in a study of the examples: Since the invariance properties of the given species $gm(n, G)$ as a rule are already exhausted by G the covariant version $mf(n, G^0_{\mathrm{dif}})$ cannot be produced without drastic changes that even touch upon the type of the structures involved.

As to the argument that covariance is too weak a property to make any considerable contribution to the content of a physical theory, it has to be admitted that even under the restrictions of the additional requirement objection the class of species $gm(n, G)$ with covariant versions in G^0_{dif} is very comprehensive. But it had to be pointed out that in the light of the utterly trivial solution (19b) the class in question *is* already restricted, the answer to (co) is *not* any more positive throughout, and therefore the historical fact described in the last paragraph is not trivial in the sense that it could not have been foreseen *a priori*. Moreover, particularly with Einstein's postulate of general covariance in view, one should look at the whole matter also in the opposite direction that is provided by the problem of reduction (re). As has been already remarked, contrary to the problem (co) the problem (re) does not have a positive answer for every case of given data. Thus in the differential geometric context, given $mf(n, G_1)$ (for Σ_1^*) and $cm(n, G)$ with $G \subseteq G_1$ (for Σ) it may *not* be possible to find an extension $mf(n, G)$ equivalent to $mf(n, G_1)$. The latter species would then be *irreducible to* G, and it may even be *absolutely irreducible* in the sense that it is irreducible to G for every G included in but different from G_1. With this concept at hand the postulate of 'general covariance' can easily be made nonvacuous by reinterpreting it as the want for species $mf(n, G^0_{\mathrm{dif}})$ that are absolutely irreducible. Questions of real frames of reference putting aside this interpretation comes fairly close to what Einstein had in mind by looking for physical laws in 'arbitrary' coordinate systems: the laws were meant to be such that they would not *prefer* any subset of coordinate systems. Now, in a well-defined sense this is the case for absolutely irreducible species $mf(n, G^0_{\text{dif}})$, and the impressive thing about Einstein's gravitational theory was that it was the first physical theory using such species. With respect to irreducibility 'arbitrary' coordinates really took over exactly the same role in the new gravitational theory (and its general extensions) as had the Lorentz frames (viewed merely as coordinate systems) in physical theories based on special relativity.

Universität Göttingen

NOTE

* The completion of this paper was made possible by a Visiting Fellowship at the Center for Philosophy of Science at the University of Pittsburgh. The author wants to express his gratitude to the chairman and the director of the Center, Professor A. Grünbaum and Professor L. Laudan, for the kind invitation and generous hospitality.

REFERENCES

Bourbaki, N.: 1968, *Elements of Mathematics: Theory of Sets*, Hermann, Paris.

Ehlers, J.: 1973, 'The Nature and Structure of Spacetime', in *The Physicist's Conception of Nature*, J. Mehra, ed., Reidel, Dordrecht, 71–91.

Havas, P.: 1964, 'Four-Dimensional Formulations of Newtonian Mechanics and their Relation to the Special and the General Theory of Relativity', *Rev. Mod. Phys.* **36**, 938–65.

Iyanaga, S., and Y. Kawada (eds.): 1977, *Encyclopedic Dictionary of Mathematics*, 2 vols., M.I.T. Press, Cambridge, Mass.

Kobayashi, S., and K. Nomizu: 1963, *Foundations of Differential Geometry*, vol. I., Interscience, New York.

Künzle, H. P.: 1972, 'Galilei and Lorentz Structures on Space-Time: Comparison of the Corresponding Geometry and Physics', *Ann. Inst. Henri Poincaré XVII*, 337–62.

Ludwig, G.: 1978, *Die Grundstrukturen einer physikalischen Theorie*, Springer, Berlin.

Misner, Ch. W., K. S. Thorne, and J. A. Wheeler: 1973, *Gravitation*, Freeman, San Francisco.

Rosen, G.: 1972, 'Galilean Invariance and the General Covariance of Nonrelativistic Laws', *Amer. J. Phys.* **40**, 683–7.

Scheibe, E.: 1978, 'On the Structure of Physical Theories', in *The Logic and Epistemology of Scientific Change*, I. Niiniluoto and R. Tuomela (eds.), *Acta Philos. Fenn.* **30**, 205–24.

Shoenfield, J. R.: 1967, *Mathematical Logic*, Addison-Wesley, Reading, Mass.

Stegmüller, W.: 1979, *The Structuralist View of Theories*, Springer, Berlin.

Trautman, A.: 1967, 'Comparison of Newtonian and Relativisitic Theories of Space-Time', in *Perspectives in Geometry and Relativity*, B. Hoffmann (ed.), Indiana University Press, Bloomington, Ind., 413–25.

Trautman, A.: 1972, 'Invariance of Lagrangian Systems', in *General Relativity*, L. O Raifeartaigh (ed.), Clarendon Press, Oxford, 85–99.

ABEL SCHEJTER AND JOSEPH AGASSI

MOLECULAR PHYLOGENETICS: BIOLOGICAL PARSIMONY AND METHODOLOGICAL EXTRAVAGANCE

INTRODUCTION

Molecular phylogenetics is a game practiced increasingly by biochemists. It appeared on the scene at the turn of the century, when a few crystallographers and immunologists tried to test evolutionary hypotheses in terms of their specific techniques. In the fifties and sixties the game developed very rapidly, when protein sequences and nucleic acid hybridizations became accessible and culminated in recent years when the techniques for nucleic acid sequencing revealed some intimate details of the genetic material. By now it is so popular, that far from being an esoteric branch of biochemistry it pervades all biological disciplines with the claim that it can, all by itself, describe the past on the sole basis of the molecular structures of the present.

There are different ways of looking at molecular phylogenetics, but from the methodological point of view what we consider here are the rules of the game, and particularly one that is called, and accepted without much argument, the law or criterion of parsimony. We shall come to it in a short while; let us first consider the problem itself.

THE PROBLEM

When we see the similarity between human and gorilla hemoglobins, we naturally take it as a null hypothesis that it is the joint ancestry of both that is responsible for it. But this null hypothesis is much more problematic than seems at first blush. Do we have a theory of the relations between macro and micro phylogenetics? We can say of any similarity that it is due to a common ancestry, due to sheer accident, or due to some convergence or another. In particular, we are certain that some similarity is as accidental as the system permits: we have a finite number of genetic building blocks and a finite length to each message; therefore, some part of some message must overlap with another, and some translation or some execution of instructions contained in some message may likewise overlap with some translation or execution of some totally unrelated message.

J. Agassi and R.S. Cohen (eds.), Scientific Philosophy Today, 333–356.
Copyright © 1981 *by D. Reidel Publishing Company.*

Hence, we must decide *a priori*, what is the information content of a given message in the sense of information theory, i.e., in the sense of deviation from mere chance. Similarly as to convergence, in the evolutionary sense of the word. Convergence is defined as the evolutionary attainment of a similarity imposed by functional requirements in the general ecosystem. Thus, we can have functional similarities due to swimming, or due to flying: but we can also have convergence due to functional similarities of dissimilar functions in dissimilar conditions. For example, the same mimicry may be, in one condition, an attempt of a species to blend into the background, and, in other conditions, an attempt of the species to stand out and resemble a threatening species or to stand out as unpalatable prey. Here, moreover, we cannot as yet speak of statistics since convergence hinges on given functions within given ecosystems. When the ecosystem changes, we cannot know of convergence before reconstructing it. This is a tall order indeed. (Remember the defunct order of ungulates!)

Hence, the problem imposes itself on us in a greater generality. It is of little use to speak of one similarity, whether striking or not, without first developing some general guidelines, so as not to expostulate too much on a mere accident of nature. How, then, can we decide *a priori*, on the difference between similarities due to phylogenesis, and others?

The question, however, should be put the other way around. After all, we are not really interested in similarities, but in phylogenetics, and so the problem should be posed within the concern and intellectual framework of phylogenetics. The problems of phylogenetics are many. In particular, we may just want to know more about a given family tree. More generally, we may want to know whether speciation really is the result of selection pressure, and to what extent. And, we feel, the molecular biologist may help us reconstruct some family trees so as to help us reconstruct the ecosystems within which certain speciations took place and examine the situations more closely. In other words, we want to be able to reconstruct some systems well enough to be able to critically examine the Darwinian tenets.

Since there is very little paleomolecular biology possible, how can the molecular biologist help? We shall now present the research program he recommends and attempts to follow, and some possible criticisms of it.

In what follows we shall discuss the molecular biological approach to evolutionary convergence and divergence, and argue that we have no means of differentiating the two by chemical analysis. We shall then discuss the principle of parsimony which the molecular taxonomists try to employ in order to permit the performance of this impossible job, and we shall try to explain why this ploy must fail. We shall then try to show the fundamental methodological disagreement we have with the molecular taxonomists: they are inductivist and we are hypothetico-deductivists. We conclude by arguing that the molecular taxonomists' inductivism vitiates their statistical studies even from the purely statistical point of view.

I. THE PROTEIN TAXONOMISTS

The paucity of items in the genetic code insures many unsuspected similarities in the messages. Indeed, it was repeatedly claimed that what is so surprising is the enormous similarity between a mold and a man as seen in their genetic codes.[1] But this is not the end of it, and there is a great problem hidden here for all molecular taxonomists. Clearly, similarities in code and in translation are quite different. For example, the code name of a cat in English is much more similar to that of a bat than to that of a leopard; whereas these code names of species in the Linnean code go the other way. Now it is too much to expect similarity traits in codes and in translations to run in parallel, unless the parallel is built in. Is, then, the similarity in the gentic codes and in the species built in or not? That is to say, is a small mutation in a gene responsible for a small morphological difference in offspring? Can we map the history of the changes of the horses' hoof from the eohippo to date on a line of genetic alterations in the code?

This is how we hope to support or reject Darwinism by molecular biological studies; we wish to reconstruct, with the scanty evidence we have, the history of the genetic codes of diverse phyla. Perhaps the hoof is too much, but then we may take hemoglobin or any other protein.

How, then, do we compensate for the paucity of historical data? Here is the program of micro-phylogenetics as seen by leading members of the school [Fitch 1966; Dayhoff and Eck 1967–1968].

We start with today's species. We compare two different proteins from two different species, which, however, have the same function in both. Indeed, often the two proteins go by the same name, e.g., hemoglobin, or cytochrome, or ribonuclease, even though they are chemically distinguishable. Things are even more complex than that. For, at times, we find astonishing similarities between proteins whose functions are quite distinct, even in the same species. For example, lysozymes in tears or in eggs are anti-bacterial, and lactalbumin in mammary tissue helps build lactose, and the two overlap in about 40 per cent of their primary structures [Brew, Vanaman and Hill 1967]. We therefore postulate common ancestry of the two proteins, whether in different species, or in one species with different functions to begin with, on the basis of structure alone.

Now this is not enough; as we have indicated, accidental and convergent similarities are bound to occur. Therefore, we cluster similarities so as not to follow mere accident. How? Let us take an example. We take two proteins, each defined functionally, each from a different species. We find similarities as defined by identical building blocks – amino acids – in identical positions within each protein chain. Of course, this already complicates matters, since positions within a protein chain are not clearly well defined. We may, however, overcome this difficulty by taking two proteins of the same length and defining positions within them strictly ordinally. But for this the chains have to be both of equal lengths. Moreover, we may insist on comparing protein chains with different lengths; which intuitively we may easily see how to do. For example, two protein chains of different lengths but similar in one end may be placed with the similar ends side by side, and positions then defined strictly ordinally as before [Walsh and Neurath 1964]. Another intuitive possibility is this: we slide the short molecule over the long molecule, placing it in different positions over the long one, looking for the position that offers the highest degree of similarity [Fitch 1966]. Intuitively we can see that on the one hand accident can play great tricks here, yet on the other hand if great similarity happens, e.g., when the large molecule can be built by adding two ends to the small one, we may suspect that phylogenetically this is how the large one came into being or, vice versa, we may suspect that the small molecule is the truncated version of the longer one.

Things are still much more complex than that. Deletion is a known genetic phenomenon and if we try to construct the small chain from the larger one by a series of deletions then probability theory tells us that, for some, accidents will enter into the picture quite against our concern. How, then, can we sift proper phylogenetic deletions from mere accidents of similarities?

Again, we take two proteins each from a different species; we compare for as many similarities as we can, and likewise for differences. Now we have a new difficulty; how do we count differences? If we have two chains of equal length then the sum of similarities and differences is the length of the chains. When we have chains of different lengths we compare them so as to find the maximum number of similarities; the number of differences, then, is the length of the longer chains – i.e., the number of its amino acids–minus the number of similarities. The problem is, how do we get the maximum number of similarities? On the one hand, some moves are forbidden, e.g., the rearrangement of the building blocks and of the chain, but, on the other hand, great freedom is permitted in postulating all sorts of delections in both the short and the long chain. The reader can easily ascertain that this way the degree of similarity can at times jump quite unexpectedly.

Now, we take pairs of different proteins with degrees of similarity and difference attached to each pair. We postulate, quite independently both of genetic code and of our knowledge of the phylogenetics of the species concerned, the following: ignoring similarities altogether, we look at the number of differences, and say, the smaller it is the likelier it is that the two proteins are near relatives. In other words, *degrees of difference are monotone functions of degrees of kinship; degrees of kinship are defined by the number of junctions in the evolutionary tree.*

It seems clear that we have here a highly testable hypothesis, since, by measuring differences, we can map existing species on a hypothetical family tree which becomes more and more definite and accurate the more proteins we examine, and we can all too easily compare this tree with the traditional phylogenetic tree offered to us by the traditional taxonomist. If the two tree systems diverge widely enough, the hypothesis will have to be rejected. If, however, there is a close agreement, it will be possible to argue that the molecular biologist is correcting errors of the classical taxonomist, or at least is able to solve problems that the latter left unsolved.

Of course we are very far from being able to make this comparison. Nobody claims that we can derive definite conclusions about phylogenetics from the results obtained when proteins are mapped in any known way. At most, one may – and we shall contest this – use such information together with other relevant data obtained from other sources in efforts to reconstruct some interesting family trees. Thus far, a few proteins were examined over wide ranges of species, and family trees for them were plotted; fortunately, these family trees are obviously reminiscent of the taxonomist's tree [Fitch and Morgoliash 1967; Dayhoff 1972; Brew *et al.* 1967]. But these are only encouraging signs. For, each tree involves a few deviations from classical taxonomy that make each of the trees suspect; but, and we shall soon discuss the methodology of it, the hope is that the aggregate will eliminate the local complications. We shall elaborate on these two points.

Regarding the complications involved in the existing family trees, let us begin by quoting a neo-Darwinian criticism to the earliest efforts of protein phylogeneticists:

"Let us now look a little more closely at some of the values (number of residue replacements). We find, that within a group of related forms these values do not necessarily parallel degree of relationship. Man and horse differ by 12 replacements. Man and Kangaroo – and everybody agrees that the marsupials are much more distantly related to man than are the perissodactyls – differ by only eight.

Pig, sheep and cow seem to have the same cytochrome structure, even though the pig, in spite of being an artiodactyl, is only rather distantly related to sheep and cow.

Thus, the generalization that residue replacement parallels phylogeny is true only in a very loose way. Let us not claim too much for it. Particularly, let us not read this correlation backward and say that 'as proved by the amino-acid replacements', man and horse are more distantly related than man and kangaroo. Single-character classifications never work [Mayr 1965, p. 293].

This, as we have already noted, merely speaks against relying on too few details. The numerical taxonomists' hope, we remember is that the many details converge by reducing the role played by sheer accident and by evolutionary convergence. The hope is based entirely on general suppositions known together as neo-Darwinian.

But, in order to achieve the aggregate effect, we must compare many proteins, each on the whole range of living things. We must remember that in order to compare two proteins, we define them as more-or-less-the-same by the sameness of function. But, strictly speaking, there is hardly ever sameness of function, and even more-or-less-the-same function is something which usually covers only a small band of the spectrum. *A priori*, the loose definition of a function will cover a broader band of the spectrum of living things but will open options of choice of one or of different proteins of a given species. For example, if we define proteins as enzymes, then many species have many kinds of enzymes; whereas if we define them as enzymes that catalyze the conversion of amino adipic acid into lysine, then we shall find them only in fungi and euglenids.

This difficulty is insurmountable, and calls for the alteration of the whole conception considered here. And, indeed, the alteration was offered by Fitch [1966] and adopted by many others. The suggestion is to examine not proteins but nucleotides, and in the genetic code material. That is to say, we wish to compare the building blocks of the DNA molecules of the diverse species.

This is not as easy as it sounds, for the mere technical reason that we do not know which genetic material is responsible for the formation of which protein, except for one or two proteins. Nevertheless, we may assume what we do not know yet. We shall soon see how.

II. THE PRINCIPLE OF SIMPLICITY OR OF PARSIMONY

Let us start as if the situation is unproblematic, align DNA molecules of different species the way we previously have aligned proteins and count similarities and differences in the same way exactly. The big question now is, what do these differences mean?

The question is really many-faced. The phylogenetic answer is, these are different species; the molecular biological answer is, these are different messages generating different proteins. Now, we can assume that the degrees of difference between different species are monotone functions of the differences in genetic code; this will be the assumption of built-in similarities between the protein taxonomist's tree and the classical phylogeneticist's tree. But, for this we need an added hypothesis. We need the hypothesis that the total number of mutations is a monotone function – by and large – of the number of species. Otherwise we cannot correlate the differences between genetic messages with the phylogenetical distances that they engender. For, without such a hypothesis, it is easily conceivable that small differences testify to large distances and vice-versa, just as the difference between the words "bat" and "cat". Such cases can be excluded by hypotheses, and the simplest hypothesis is that of a monotone function. To make the simplest form of a monotone function we can assume that there was a minimum number of mutations in the history of all the species, that is to say, that each single difference in a building block between two DNA molecules should be explained by postulation of no more than one mutation [Fitch 1966]. This is a stronger hypothesis than the one required for our initial purpose, yet it is the one favored by writers on the topic on the basis of the methodological demand for minimum hypotheses, or the principle of paucity of hypotheses, or of parsimony, or the principle of simplicity [Popper 1959; Agassi 1975]. For, we can just as well assume that between two stages of a development of a DNA molecule there was an intermediary step, or two, or a few vascillations between the two, etc. But *the principle of parsimony, as used by the molecular taxonomist, postulates a common ancestry to all genetic codes and the shortest possible paths from the original ancestral code to all the codes found in the present situation.* The present situation plus the principle as posed here, should enable us to construct a unique or almost unique family tree, in order to compare it with the one of the traditional taxonomist.

This, then, is the program of the molecular phylogeneticist. The asset of the final product, if it agrees with the product of the traditional taxonomist, is, as we have stressed, that the similarity between the micro and the macro systems is thus explained as built in.

We now leave this program, and discuss its import to evolutionary theory. Supposing that molecular biologists have succeeded in offering us a family tree sufficiently similar to that of the traditional taxonomist. What does this tell us about evolution?

Before going into this question, we must admit that some scientists will refuse to place it on the agenda and say: when the task will be accomplished we shall see what it can tell us, and not before. If we try now to derive from the outcomes of possible experimental results all sorts of speculations we will simply have too many preconceived notions, whereas one should come to experiment with no preconceived notions so as to accept its verdict whatever this may be.

Holders of this methodology, the inductivists, strongly oppose the hypothetico-deductivists. The hypothetico-deductivists want to explain the classical tree. Their explanation may clash with the tree in some detail and they will have to modify either the explanation or the tree. The inductivists will ignore the classical tree and try to deduce a tree, say, from DNA hybridization (to which we soon come) or protein structure data, or any other. Once the tree is erected, it can be compared with the classical one; if they agree, we can say we have adduced or induced or abducted it from the facts.

Our first objection is that the molecular taxonomists fall between stools because they employ the hypothesis of the shortest route which they call the parsimony rule. Our second objection is that they fall between stools because they neither accept nor ignore the classical tree but use it as a guideline for their own. It is therefore hardly a surprise that they succeed. Let us go into detail.

As we have said before, if the two trees, the molecular and the classical, come up sufficiently similar, and we can never expect the most unlikely case of complete similarity, then we could try to amend the classical tree, as the classical table of elements was amended in the light of further discoveries. For, the alternative is the suggestion, that we do not wish to make, that the molecular tree is parasitic on the classical tree.

Of course, the molecular tree will have to be definite enough, i.e., its minor deviations will have to be smoothed out by sufficiently large and varied samples, before this could be attempted. But criticism is levelled at the very program, that is to say, criticism has either to prove it unfeasible or to criticize it on the assumption of a successful outcome. For the criticism of the successful outcome we should consider how we can modify the classical tree, and what this would amount to; our present criticism is largely of the very feasibility of the program, which, we shall soon explain, is too inductivist for comfort.

The most obvious criticism one could level against the feasibility of the program is against the putative truth of its presuppositions. It is not clear, however, what these are exactly.

The main presupposition which we have already mentioned is the principle of parsimony. We contend that it is unclear and requires spelling out.[2]

The principle comes from scientific methodology, and hence is metascientific, not scientific proper; that is to say, the principle applies to hypotheses, not to trees and DNA molecules. But misnomers are neither here nor there. Let us only concede that a hypothesis which assumes parsimony in nature may, but need not be, a hypothesis commended by the methodological principle of parsimony of hypotheses. But what is the parsimony of nature here postulated?

Suppose we try to imagine all possible animals. Classical taxonomy takes into account a given number of characteristics, such as vertebrae and mammary glands, or number of members, or shape of fruit; we can calculate the number of combinations of these qualities, also in some cases permutations of them, including the case of the backward moving crab. The number would be astronomical. The most rudimentary classical principles of taxonomy, however, will at once exclude a large chunk of these, rightly or not, such as mammals bearing fruits or laying eggs. Now classical evolutionists made much song and dance about the discovery of one measly egg-laying animal – the duck-billed platypus; they never tried to explain the absence of a fruit bearing mammal, though had they found one they might have made an even bigger song and dance about that find.

This is the complaint against the game – heads-I-win-tails-you-do-not-lose – which ensures that sooner or later you lose all and I win all. The rule against this game begins with Galileo and ends with Sir Karl Popper.[3] But many before Popper stated crisply that a hypothesis not refutable by evidence is not confirmable by evidence either. This is not to say that unconfirmable, or more explicitly, untestable, hypotheses are useless. In particular, classical evolutionism has proven useful but it is not testable.

Without going into detail, we believe it is clear that classical taxonomy leaves much to be explicitly stated and explained [Schrödinger 1946]. And, indeed, the discovery of the genetic code is so exciting because it offers a hope for a closer look at evolution. Let us take, then, some of the most obvious facts of classical taxonomy that classical evolutionary theory – Darwin's or otherwise – has left unexplained, such as the sharp division between kingdoms. Can we try to explain it? Can the genetic code help to explain it?

We do not know. We cannot read off of a DNA makeup of any living thing even the simplest fact that it is a plant or an animal. That is to say, if Tom gives Dick the full genetic code of something he has analyzed without telling him anything about it except the genetic code, can Dick say, at the very least, that it is a plant or not? Now, we can say, why should this be possible? It is very conceivable that the part of the genetic code of A which helps A make chlorophyll is totally different from the part of B which helps B make chlorophyll. After all, convergences of all sorts are not excluded in biology, and biochemical convergence on a function is not very different from any other possible convergence. We do have examples of biochemical convergences on functions, such as hemoglobins and hemerythrines which are structurally and phylogenetically unrelated yet converge to the function of oxygen transport, one in vertebrates and others, one exclusively in molluscs, and one exclusively for siphunculides. It is far from being inconceivable that different code structures are responsible for chlorophyll formation in different plants.

But the principle of parsimony as classically understood by all authors, philosophers or scientists, seventeenth century or contemporary, will recommend that in the absence of evidence to the contrary we take as our null hypothesis the one claiming that all chlorophyll, no matter where it is formed, has the same code message.

It should be noted that the proviso is, in the absence of any evidence to the contrary. For, when we know that lysine is formed differently in plants and in fungi, we cannot demand that the formation of lysine in all living things be explained by the same genetic message. But we have no such facts on chlorophyll; moreover, had we found one message for it, we would explain why plants form a kingdom; and, traditionally it is agreed, parsimony is enhanced by high explanatory power in the sense that a hypothesis becomes more parsimonious when new facts that it explains are found.

Now the molecular taxonomists are not even looking for the code name of chlorophyll formation in the genetic code of plants. They are attempting something much more ambitious. They are attempting, as we have already indicated, to provide information that, known together with other pieces of information from different sources, will help us reconstruct family histories. We do not know whether the principle of parsimony recommends more ambitious or less ambitious programs, since parsimony is judged only when the program is executed: a program that ends up with no explanation, or with explanation made up of many ad hoc hypotheses, is not parsimonious.

This is not to say that other techniques of isolating code messages do not exist. Though we do not know yet how to read messages and how to compare them, we can compare them by allowing them to interact by the technique known as DNA hybridization, which we shall discuss below. Suffice it to say, meanwhile, that in this technique we speak of mutual recognition by one message or another, of degrees of this recognition, and of the fact that messages of higher vertebrates recognize each other to a much higher degree than either of them a message of a plant.

III. A MATTER OF METHOD

It is very instructive to compare the researches of the hybridization taxonomists to those of the protein taxonomists; the claims of the former may be straightforward, highly testable, and not global, or they may be involved, untestable, and global – like the protein taxonomy. That is to say, the hybridization taxonomists may also offer a general criterion to place known taxa on a tree: they may say the proximity of any two taxa on the classical tree and the degree of mutual recognition of any two genetic messages of each other are monotone. This may be a schema for highly testable hypotheses, whose value can be examined. The hypotheses will be testable if degrees are uniquely and rather unequivocally determinable by a fairly clear and fairly algorithmic formula.

The exercise we have suggested would take the classical tree *as given*, observe some of its most conspicuous yet unexplained qualities, and ask for explanations. It thus belongs to the hypothetico-deductive school, now made so popular in biology by Schrödinger [1946] and Medawar [1967]. The protein and hybridization taxonomists may take the classical tree as given, and compare distances on it with distances on their newly-fangled scale. We wish to observe that neither the protein taxonomists nor the hybridization taxonomists follow this line. Their exercise begins with given taxa but no given taxonomic tree. They wish to build one with their hypothesis, the rule of parsimony, and then, when it is all done, but not before, compare their tree with the classical one. Yet, they regularly do make up tentative trees and compare them with the classical one; no doubt all they expect to find there is some measure of convergence (in the methodological sense, not in the evolutionary sense, of course), sufficient to encourage them to continue, to indicate to them that they are on the right track. All these temporary trees are, for them, mere hints: that is to say, only if they diverge very grossly from the classical tree will the protein taxonomists give up the game; if the divergence is not too wide, they can still hope that it will iron out. This, to repeat, means that only their final product can seriously be compared with the classical tree; until then, they can and will ignore it.

Why is ignoring the whole traditional tree the right way? Why accept taxa but not phyla? There is a philosophical answer, which is that taxa are more fact-like than phyla, and the protein taxonomists dislike conjecture. This is why they put all their eggs in one basket.

The philosophy in question, inductivism, is not of our liking. We cannot attempt a general criticism of it here. We can only express our strong disagreement with the expression of it in the program of the protein taxonomists; we do not take taxa to be more factual, less hypothetical, than phyla. For example, we think that chlorophyll as a mark of the plant kingdom, is much more obvious than marks that class the German shepherd as a dog rather than a wolf. Nonetheless, we will proceed with the protein taxonomists' program.

Let us take, then, all the possible DNA messages of a given minimum to maximum length —10^2 to 10^4, or thereabouts. They are of an order of magnitude that defies all computers. We have almost no known rules to exclude *a priori* any of the possible messages. This is why the protein taxonomists look at existing messages. But they do consider all possible ones in the very act of measuring distances. For, the distance between any two messages is exactly the smallest (this is paucity) number of possible messages lying in between them.

Still, we are not satisfied. The notion of betweenness here is quite tricky, and there is no shred of proof that the shortest route between two messages is uniquely definable.

Of course, the right thing for us to do is to look over their shoulders: how do they determine betweenness? Let us take slowly one exercise [Fitch 1966; Fitch and Morgoliash 1967]. The genetic message is expressed in the genetic code. We do not comprehend most messages, but the signs are known. They consist of a linear arrangement of four possible simple radicals (in the classical chemical sense of the word) called, AUCG, where A stands for adenine, U for uridine, C for citidine, and G for guanine, all of which are rather simple nitrogenous bases. Each message unit consists of a triplet – three such radicals together – and so we have 64 possible triplets so that we have 64 units of possible message. Of course, like any message, the genetic message must contain redundancy, for example, since otherwise a mouse could by a genetic mistake give birth to an elephant – this by the very meaning of the term "redundancy." In biology, the redundancy is experienced *inter alia* in having 64 units to transmit 20 possible messages, i.e., 20 amino-acids, which is what the following exercise concerns.

We take cytochrome *c* of different taxa. We have as an eleventh amino-acid (since the first ten are identical), isoleucine in one case, valine, in the other.

The genetic codes can be

Ile	Val
AUA	GUA
AUC	GUG
AUU	GUC
	GUU

Suppose the first lines were the ones for Ile and for Val: then, the distance would be 1; suppose the first were true for Ile and the second for Val, then the distance would be 2; and so on. Now we have here twelve options to choose from, offering the distances 1 or 2, though, in general, distances can be anything from between 0 and up to 3, and no more, because of the principle of parsimony. But we do not know the message. So we postulate a new principle of parsimony, which says, the code for each of the two compared messages is such that the distance between them is minimal. Now, as we know, the code contains redundancy, but this new principle excludes it, so that in our example the distance is 1.

But suppose we propose some other way to measure the distance between two codons; suppose, for example, that we average between possible distances, in the case of Ile and Val we would have $2\frac{1}{3}$. Would we get, as a result, a different tree? Suppose we average not on an equidistribution, since we are dealing with mere possibilities, but on a different type of distribution; would it matter? We do not know, but again expect no difference. The purpose of this exercise was to show how many hidden assumptions are there in the very first step and how many alternatives do exist. And to show that the methodological rule of parsimony does not support the protein taxonomists' choice, to say the least. But let us move to his next step.

Again we take cytochrome *c* in diverse animals, on the suspicion that it is homologous in them, i.e., that the cytochromes *c* in the diverse animals are similar thanks to common ancestry. We examine all mutation distances between each couple of animals. Supposing *A* and *B* are more similar than *A* and *C*, then we assume *A* and *B* to be more recent cousins than *A* and *C*, thus:

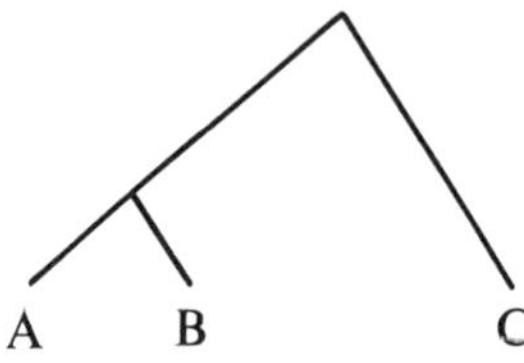

and not thus:

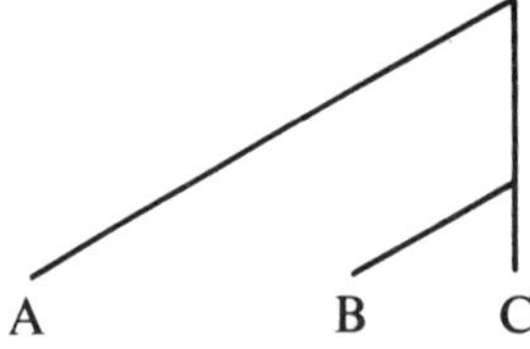

And of course, given sufficiently many species with cytochrome *c*, we *may* easily arrive at a contradiction regarding order, but we need not arrive at one. Of course, the more species, the likelier that data will be inconsistent.

This refutes the theory that protein distances are monotone with phylogenetic distances, if these are taken strictly and literally enough. But the correlation was never intended to be strict: all that was wanted was to lump distances. How should these be lumped?

It seems reasonable, and we think any statistician will concur, to measure distances between any couple of species by the use of a large and varied sample of homologous proteins. These distances will have to be taken more seriously. Trees should be constructed of them, and if these trees lead to an inconsistency, then the whole idea will have to be scrapped: the theory will then be declared refuted.

In other words, increasing the number of proteins first and then slightly increasing the number of species will show the protein taxonomy theory refutable; the other way around will not, because there will always be enough choice for the taxonomist to explain away any possible difficulty by postulating more and more possible proteins to explain the variety of species from which he has started. And we want our theory refutable so that its survival of the attempted refutation will speak in its favor.

Unfortunately, for technical reasons, the protein taxonomists take the road of multiplying the species rather than the proteins. Their test must, therefore, remain inconclusive at best. Yet, we may speculate on the possible outcomes of the more rigorous method. For example, we know that cytochrome *c* of cow, sheep and pig are identical, and that had we compared these three species with respect to dozens of proteins there will be no identity with respect to many of them but most likely their average distances from each other will be smaller than the margin of error. The discussion can be sharpened: not only may we get inconsistencies, we may get no result at all: there is no *a priori* reason, not even in the law of parsimony, to suggest that distances of any three species with respect to any large sample of proteins, will converge into decent averages! The only reason – not *a priori* but conjectural – that we can give is the hypothesis that classical trees should be reflected in protein trees, because of the stability of proteins, because minimal mutations occur in speciation and continue without reversal in further speciation. But since there is much reason to assume, perhaps even facts to support, the claim that further speciation may turn evolutionary blind alleys into broad evolutionary avenues, the conjecture just mentioned cannot be advocated in the name of simplicity, or paucity or hypotheses, or parsimony. And we all understand that the same weapon which may be a burden on a small animal may turn up very useful in a big one, that melanism which fingers a species for destruction may become excellent mimicry in modern industrial environments, etc., etc.

It was E.B. Ford's study of industrial melanism that suggested clearly just this reversal of mutations, and in the following way [1964]. The problem modern evolutionary theory faces is very simple; given one species in a quasi-stable environment, it can mutate only from one relatively stable econiche to another. Most likely, the econiches are not ordered along lines of mutations. How, then, do mutations align themselves to agree with econiches? The reply from randomness is not enough, since the problem is, how do mediating generations survive between econiches? Thus, a part of a complex useful mechanism is a burden; how does the animal live with it?

E.B. Ford conjectured that there are periods of large and small eco-pressures, which permit small and large varieties of mutants to hang about. He claims that the good times allow return to old kinds, developing new kinds, all in an unstable manner; bad times push for stabilization, but not necessarily to the same econiche. In other words, the absence of linearity is of the essence in Ford's theory, but is forbidden by the protein taxonomists.

This does not mean that we must accept one and hence reject the other. It does mean that parsimony does not support the protein taxonomists. What other reason do they offer for us to expect convergence so as to enable us to test their claim of similarity between their protein tree and the classical tree? None.

Back to the way the tree is built by the protein taxonomists. He centers on a few proteins but many species; he observes inconsistencies and the need to remove them. Wishing not to be too arbitrary when deciding on the removal of inconsistencies, he accepts the rule of removing the inconsistencies with the minimum necessary change. But how is this done? It turns out, that the order of choice of species is not irrelevant to the outcome – which is not surprising since the outcome is inconsistent [Fitch and Morgoliash 1967]. Therefore he suggests an algorithm that, after a number of computations, leads him to choose one tree as the best. The tree looks in remarkable agreement with the traditional tree.

We need not go into this algorithm: for others [Dayhoff and Eck 1967–68] have tried somewhat different algorithms and arrived at results not worse than this one. What one might wish to see is how much one can vary the algorithm yet still get good results; for, clearly, it is the essential part of it that signifies and calls for a biological interpretation.

Let us quote Dayhoff and Eck [1967–68]. They say:

> Since there are usually millions of possible topological configurations, it is unreasonable to try them all in a search for an optimum, even on a high-speed computer. The program uses the following scheme:
> (1) Make a first approximation to the topology, determining some unequivocal points of the structure.
> (2) Try a large number of small changes in this topology, finding a best fit.
> This process simultaneously determines the best topology, the dimensions of the tree and also the ancestral sequences.

Now what is at stake here is this. We all expect *a priori* most homology to show more or less common ancestry. The allure of the protein taxonomists is in a much stronger claim: they claim not mere similarity, not frequent similarity, but a systematic correlation of degrees of similarity with phylogenetic distances. All hinges on not confusing the obvious and generally accepted, with the surprising and controversial. The above quoted passage raises the question, is the confusion avoided? Supposing in most cases the approximation mentioned is good enough; what will happen if we begin with a problematic one? Suppose we compare a given group of species on the basis of cytochrome *c*, with birds and turtle included in the group. We will at once find birds and turtle the nearest to each other in the group. Holders of the obvious thesis will mind this anomaly; holders of the precise thesis will have to show that when using more proteins or more species or both, the anomaly vanishes, or that when working out the tree in great detail and with care we still get a surprisingly correct tree, even though the turtle will be wrongly placed in it. We think nothing like this has ever been attempted. Thus, the procedures offered by the protein taxonomists are supposedly precise, but not surprisingly precise if they are precise only part of the time we use them. For, a group of measurements that offer precise results only at times and at other times poor ones, is rather imprecise. In other words, we think the procedures of the molecular taxonomists rely on the imprecise commonly accepted view and seem to support the precise surprising and controversial one, simply by selecting better results from a class of results of varying degrees of precision in accord with the imprecise theory and some obvious distribution of *post hoc* precision.

Let us add in parenthesis, concluding this critique, two variants on the techniques here criticized. First hybridization taxonomy and second taxonomy based on banding patterns of chromosomes. There are other techniques, such as antibody and antigen taxonomy etc. We shall not attempt to complete a list of all possible techniques. To begin with hybridization then.

What seem to us to be the major differences between the hybridization and protein taxonomists are these. First, the former have much less leeway and so produce results that are much less arbitrary. Second, the detail of the material used by the former is much less known or understood than that of the latter.

In hybridization, DNA molecules of one taxon are split both longitudinally into two strands each, and transversally into many rather small messages. They are mixed well with single strands of DNA of another taxon and notice is taken of how well they recognize each other, i.e., what order of magnitude of hybridization obtains. That is the whole experiment of measurement. Clearly, the statistical element is strong as orders of magnitude are very high. Obviously, also, the result is only known in its final stage; there is no analysis of it. The question is, should the method work, and if so why? The factual answer is that it does work. The significance of it is only partially understood and we come to it now, in the conclusion of the present essay.

The second technique is the one based on banding patterns of chromosomes. The technique of banding consists in skimming the chromosomes with fluorescent dyes and is based on the fact that the chromosomes develop patterns of transversal bands of fluorescence that one is unique to each chromosome of each species; the patterns then can be examined for similarity. The hope is that similarities will help reconstruct family trees. The results reported allegedly indicate that the orangutang is closer to the other great apes than is the gibbon, which is keeping with classical taxonomy, but also that man is a closer relative to the gorilla than to the chimpanzee, which is against classical taxonomy. Moreover, the overall results do not agree with local results and much arbitrariness is permitted [Miller 1977].

The hypothesis common to both techniques is that degrees of similarity are monotone functions of degrees of kinship. Now, when we measure, the simplest translation of the hypothesis makes monotony into linearity. Of course, this means very little. For, suppose the function is logarithmic, then by a simple transformation of the co-ordinates we make it linear. Now we can easily declare the mutation rate scale as the time scale. This time co-ordinate should not be confused with physical time or with geological time, or else we may smuggle the hypothesis – be it true or false, it is better stated than smuggled – that all mutation rate is linear with physical or with geological time. For after you smuggle a hypothesis into a system, you may surprise yourself by deducing the hypothesis from the system as if it were new. To make it more explicit. Once we devise a time-scale which is a mutation distance, then, clearly, mutation distance will equal time distance. But this time distance, on the mutation-time scale, need not be monotone with physical time distance or with geological time distance. Yet a number of writers did commit the paralogism.

Nevertheless, similarities do exist, both micro and macro, and they must correlate somehow. We know this *a priori*. We do not mean that we must be right but that the endorsement of neo-Mendelism forces one to correlate them: perhaps even to do away with one of them: it is a problem of overdetermination. That is to say, suppose natural selection alone explains the tree. Then, we can ask, how come we have new mutants to select from every time? The answer should be, and admittedly so, by sheer random process. If so, then there should be no further explanation of any mutation – only of its survival! On the other hand, suppose we ignore natural selection and look at mutations as minimal deviations along minimal courses of events – when this is intuitively understood or somehow well defined, as you will. Then, we do not need any theory of natural selection to explain the tree! Moreover, suppose in fact both selection and minimal mutation are true. Then, as with all overdetermination that works we should suspect a pre-established harmony! We do not want that!

Now, first of all, we do not think it efficient to experiment first and try to solve this difficulty later. Second, we think the resolution-pattern is very simple, even though the details are far from known. What we have here are mutation pressures and selection pressures, both describable as stochastic distributions, and when they both work, we must consider their product by the law of multiplication of probabilities. This means that any new species is doubly improbable, first as a mutant and then as surviving selection pressures. But this is hardly news. The question is, now that the reproduction mechanism is better known, can we say more on the subject? For this we need to select problems, not to work globally and with the aid of computers but without a clear program.

The inductivist program is against the search for grand laws of nature, and for the study of detailed instances of it first. The funny fact is that the molecular taxonomists found a way that leads them by inductivist routes to the search of these grand laws from very minute samples of a few proteins. This is foul play. Yet, we think, we need balance a grand scheme with the study of details, and it is not easy to provide such a balance.

It turns out that hybridization is not only a taxonomical, or other theoretical ploy, but something that claims the status of an observation report. The facts of the matter are as expected: sequences of DNA are now comparatively easy to obtain. In general, when we compare DNA sequences, the results do not substantially differ from the hybridization experiments.

This means that some homology is, indeed, present in nature. However, it has been ascertained that in given sequences, of the codons that are supposedly redundant, some are made ample use of, others not at all [Farabaugh 1978; Fiers and Grosjean 1979]. This means that even the redundant codons are not placed randomly and are not entirely redundant. No explanation is available.

IV. LATERAL TRANSFER

One hypothesis tacit in all studies thus far reported is, of course, the idea that all transfer of information is from generation to generation, that there is no lateral transfer of information, i.e., from coevals, or any other. Can this be tested? Rather, under what assumptions can it be tested?

Taking from classical microbiologists a phylogenetic tree of a given clan of bacteria, let us assume that we have three bacteria, the one nearer to the second than the third, yet the second and third are closer in the structure of their proteins [Ambler *et al.* 1979a; Ambler *et al.* 1979b]. We may wish to explain this as a result of transfer. Once we assume transfer to take place, and the whole idea of phylogenetic tests breaks down. Hence, we have either the phylogenetic transfer pure and simple to work with, as already too complex and too permissive, as we were at pains to explain, or we prefer the more stringent, more highly testable explanatory scheme of transfer, and allow the phylogenetic transfer to remain in the background for generating lateral transfer hypotheses. It is methodologically significant to observe the paradox that the schema of phylogenetic transfer alone is simpler than the schema of both phylogenetic and lateral transfer (and of course, only lateral transfer is absurd), yet the more complex schema includes as special cases some cases of lateral transfer hypotheses that are so highly testable that some such hypotheses are now considered as having the status of factual observations. Under what conditions can this be effected? Clearly, when a homology is observed that is phylogenetically suspect, a lateral transfer can be suggested, and then perhaps replicated in the laboratory. What this shows, then, is that in these cases the phylogenetic transfer is the null hypothesis which is deemed clear enough to be suspected, or even to be refuted. This is not to say that all cases of homology or its absence are equally clear from the purely phylogenetic transfer hypothesis schema.

If the lateral transfer hypothesis is not testable by attempts to replicate, then we are in the soup. If it is testable, and lab reproduction fails, we may deem a specific lateral transfer hypothesis refuted. Since the phylogenetic transfer of the same case was rejected beforehand, in such a case we have a homology neither phylogenetically nor laterally transferred. We may suspect convergence. A convergence hypothesis in this case may be testable. We come to such cases now.

V. THE SUBCELLULAR ORGANELLES

The facts about organelles pertaining to our discussion are these. Mitochondria and chloroplasts have DNA specific to themselves and surprisingly independent of the DNA of their home cells, i.e., of the cells which have them. Even more surprisingly, at least parts of their DNA of these organelles are more akin to bacterial DNA than to that of their home cells [Schwarz and Kossel 1980]. (The surprise results from the enormously large number of combinations of codons possible in principle). This calls for explanations. Obviously, here phylogenetic transfer is impossible. Or is it? Can we not force a phylogenetic transfer hypothesis? One fantastic hypothesis, whose logic, however, is fairly easy, is the view that some eukaryotic cells swallowed some bacteria which since then became symbiotic and integrated into their host cells as organelles.[4] How can such a hypothesis be tested? Bacteria paleontology or histological paleontology is too crude for that.

The amusing fact about this hypothesis is that though concerning a phylogenetic transfer, its origin is supposedly a lateral transfer and as such may perhaps be tried out in a lab. There was, indeed, a successful attempt at creating organelles by accretion [Margulis 1970].

Nevertheless, the accretion hypothesis schema is very problematic. It is not at all clear how a symbiosis can be maintained when the reproduction process takes place. That is to say, not only the organelle has to maintain its individuality through symbiosis, this holds for its code as well. How is this done? Moreover, the code has here to be in charge not only of the reproduction of a part of our organism, but also of its function. Is this not too much to ask of it?

Here we see another hypothesis implicit in all our previous discussion: the codon transfers both structure and function. Or, alternatively, function is a function of structure. (Notice that the term function is used in two different senses in the previous sentence.) Query: how much may structure alter while maintaining function? It was tacit throughout the history of genetics that this query is not pressing. Now it is.

It turns out that many proteins are not coded as integral messages but as sets of fairly unrelated messages [Gilbert 1978]. In such cases, it is easy to conceive of a part of a protein getting lost, or getting suddenly altered so as to change a specific message relating to an important function [Eaton 1980; Goldschmidt 1940]. Since all other functions stay intact, as per hypothesis, the survival value of the altered protein may be high. This will lead, then, to a sharp mutation of one specific function of an otherwise unaltered organism. This amounts, though not by the originally envisaged mechanism, to Goldschmidt's macromutation which he called hopeful monster, much to the opposition of his colleagues. The significance of this result, which now has the status of observation report (i.e., that the monster is hopeful, not that it survives, which is an ecological matter, since we avoid all ecology in the present discussion), is that the sole constraint ever put on the idea of phylogenetic transfer, namely the so-called parsimony law breaks down empirically, since the mutation which by the genetic mechanism is sparse, is ecologically very rich. Parsimony, then, has many different meanings and so breaks down.

Tel Aviv University (A.S. and J.A.) and *Boston University* (J.A.)

NOTES

[1] Sinsheimer [1960, 509]: "The crux of the coding problem is the manner of translation of a series of nucleotides into a sequence of amino acids. For simplicity it is usually assumed that there is, corresponding to each amino acid, a linear sequence of nucleotides in DNA and in the RNA derived therefrom. Recent studies on the heterogeneity of DNA molecules within and among various species have cast doubt upon this coding concept."

Levinthal and Davison [1961, 650]: "Since fewer bases than amino acids have been detected, a one-to-one code is not possible, and each amino acid must be defined by several bases. To this point, agreement is general, but real progress must await further experimental data."

Bennett and Dreyer [1964, 210]: "Although it cannot be said that every organism uses exactly the same coding unit assignment, there is now a considerable amount of evidence that most assignments are universal."

Schwinger [1970, 720–21]: "... the genetic language is identical in all species but ... may be modulated differently in different species, possibly as a function of evolutionary development and differentiation."

Watson [1970, 432–33]: "The genetic code appears to be essentially the same in all organisms. This is not surprising: variations in it from organism to organism would mean that the code had evolved by mutation, and it is almost impossible to imagine a mutation which would change the letters in a code without being lethal."

Biology Today [1972, 283]: "An important and unexpected property of the genetic code is its universality... implies a common ancestor for all biological systems now in existence."

[2] In spite of the central role played by this principle in molecular phylogenetic studies, and in related branches of numerical taxonomy, it has met with little if any comment. Fitch [1966] postulates minimal mutation distances without further discussion: Camin and Sokal [1965, pp. 311] speak of the cladistic possibility of reconstructions "by the principle of evolutionary parsimony" and recognize that their success "depends on the assumption that nature is indeed parsimonious"; Hartigan [1973, 53] finds the minimum mutation model "compelling in its own right" and adds: "It is the assignment which permits representation of the data in a minimum number of symbols"; Cavalli-Sforza and Edwards [1967, 233] come closest to methodological criticism when they say: "The assumptions underlying this method are not too clear" and suggest that "experience with simulated trees should clarify its logical states."

[3] It is no doubt the case that classical classification was problematic. It is no doubt that evolution has introduced a criterion for resolving these. Thus, if we ask why we class a milk-fish as a fish and not as a mammal, but dolphins and whales as mammals and not fish, then we can say, the common ancestry of fish and milk-fish is closer in time than that of milk-fish and whale. But we do not always know common ancestries and so the criterion, though in principle quite satisfactory, is not always operative. Hence, were the molecular geneticists or the hybridization geneticists or anyone else able to speak informatively about common ancestries, the results should prove valuable [Popper 1959]. See also Britten [1977, 286–87].

[4] Sagan [1967]; for a discussion of the simplicity of the accretion hypothesis, see also Uzzell and Spolsky [1974].

REFERENCES

Agassi, J.: 1974, 'Review of Karl Popper, *Objective Knowledge: an Evolutionary Approach*', *Philosophia* **4**, 163–201.

Agassi, J.: 1975, *Science in Flux*, D. Reidel, Dordrecht and Boston.

Ambler, R. P., *et al.*: 1979a, *Nature* **278**, 659–660.

Ambler, R. P., *et al.*: 1979b, *Nature* **278**, 661–662.

Bennett, J. C., and W. J. Dreyer: 1964, *Am. Rev. Biochem.* **33**, 210.

Biology Today 1972: 283 (Communication Research Machines, Inc., Del Mar, CA).

Brew, K., T. C. Vanaman and R. L. Hill: 1967, *J. Biol Chem* **242**, 3747.

Britten, R. J.: 1977, *Science* **198**, 286–87, review of *Molecular Anthropology*, M. Good-

man, R. E. Tashian and J. H. Tashian (eds.), Plenum, N. Y. 1976.
Camin, J. H., and R. R. Sokal: 1965, *Evolution* **19**, 311.
Cavalli-Sforza, L. L., and A. W. F. Edwards: 1967, *Am. J. Human Genetics* **19**, 233.
Dayhoff, M. O.: 1972, *Atlas of Protein Sequence and Structure*, National Biomedical Research Foundation, Silver Spring.
Dayhoff, M. O., and R. V. Eck: 1967–68, *Atlas of Protein Sequence and Structure*, National Biomedical Research Foundation, Silver Spring.
Eaton, W. A.: 1980, *Nature* **284**, 183–185.
Farabaugh, P. J.: 1978, *Nature* **274**, 765–769.
Fiers, W., and H. Grosjean: 1979, *Nature* **277**, 328.
Fitch, W. M.: 1966, *J. Mol. Biol.* **16**, 9.
Fitch, W. M., and E. Morgoliash: 1967, *Science* **155**, 279.
Ford, E. B.: 1964, *Ecological Genetics*, Methuen, London.
Gilbert, W.: 1978, *Nature* **271**, 501.
Goldschmidt, R. B.: 1940, *The Material Basis of Evolution*, Yale University Press, New Haven.
Hartigan, J. A.: 1973, *Biometrics* **29**, 53.
Levinthal, and Davison: 1961, *Am. Rev. Biochem.* **30**, 650.
Margulis, L.: 1970, *Orgin of Eukaryotic Cells*, Yale University Press, New Haven.
Mayr, E.: 1965, In *Evolving Genes and Proteins*, V. Bryson and H. J. Vogal (eds.), Academic Press, New York.
Medawar, P.: 1967, *The Art of the Soluble*, Methuen, London.
Miller, D. A.: 1977, 'Evolutional Primate Chromosomes', *Science* **198**, 1116–1124.
Popper, K. R.: 1959, *The Logic of Scientific Discovery*, Hutchinson and Co., London.
Sagan, L.: 1967, *J. Theoret. Biol.* **14**, 225–274.
Schrödinger, E.: 1946, *What Is Life?*, Cambridge University Press, Cambridge.
Schwarz, Z., and H. Kossel: 1980, *Nature* **283**, 739–742.
Schwinger, A. L.: 1970, *The Molecular Basis of Cell Structure and Function*, World Publ., New York.
Sinsheimer, R.: 1960, *Am. Rev. Biochem.* **29**, 509.
Uzzell, R., and C. Spolsky: 1974, *American Scientist* **62**, 334–343.
Walsh, K. A., and H. Neurath: 1964, *Proc. Nat. Acad. Sci. U. S.* **52**, 884.
Watson, J. D.: 1970, *Molecular Biology of the Gene*, 2nd ed., W. A. Benjamin, Menlo Park, Calif.

TOM SETTLE

LETTER TO MARIO: THE SELF AND ITS MIND

Dear Mario:

I am, of course, delighted to contribute to a Festschrift for you. You know already that I am immensely grateful to you both for your warm personal friendship and for the enormous help you have been to me as a philosopher since I met you in 1968. I rely all the time on the sharp and effective work you have done in axiomatization, in probability, in the interpretation of quantum theory, in how theory meets experience and so on. It is especially refreshing to have you on my side in arguing for the value of metaphysics, though you do not go as far as I would like. But I do not want to use this opportunity simply to rehearse our agreements, which are many, but rather to explore our disagreements. The exercise will take me into a number of pretty fundamental areas, but the main thrust is to redeem an overdue promise to expose as carefully as I can where I differ from you on what you variously call the mind-body or the mind-brain problem. I have come to think that focussing attention on the neural systems involved in deliberate thought, for example the deliberate thought of focussing attention on neural systems involved in deliberate thought, by-passes a main component of the mind-body problem-complex, namely the problem of personal initiative in a world of physical order.

In trying to understand your view as well as I could, I have of course used my recollection of some of your addresses (plus notes I took), principally what you said at your conference in Montreal (February 1978) and your intervention aimed at Sir John Eccles at the Düsseldorf conference (August 1978). These have been supplemented most usefully by the two papers you sent me in response to my request to know where I should look for the best statement of your view on mind, your [1977a] and [1980]. But I have returned to passages of your [1973] and I have spent quite a lot of time with your [1977b].

Some differences between our views have already been advertised in my paper at the Fribourg Conference, September, 1973, (my [1977a]) and my review [1977b] of your [1973] and some of those matters will get raised again in the course of this letter, I don't doubt. But the main issue has only been hinted at in conversation and private correspondence. It

J. Agassi and R.S. Cohen (eds.), Scientific Philosophy Today, 357–379.
Copyright © 1981 *by D. Reidel Publishing Company.*

was not hard to miss that a main target of attack both at Montreal and at Düsseldorf was the interactionist view that has excited public attention recently with the publication of *The Self and Its Brain* by Sir Karl Popper and Sir John Eccles [1977]. Long before the book was out I thought the title was stunning. I want you to distinguish my view from Karl's and both our views from that of John Eccles, so that neither Karl nor I get hit with the strictures you aim at those who believe in a soul and especially one that lives after death and more especially one that lives disembodied. Personally, my preference in my ecclesiastical days was always for St. Paul's view over St. Plato's on the matter of the soul's needing to be embodied. To be fair to Eccles, I do not know whether he believes in the "soul's" survival of a person's death, but his dialogues with Karl in their [1977] certainly show him tempted to.

I fully agree with Karl that the mind is emergent in evolution, but unlike him I identify the self with the person rather than with its mind. And I side with Strawson [1959] in taking the person as primary and its analysis into body and mind as a secondary abstraction (compare Popper [1977], pp. 115–118). I must confess that I am disappointed that in none of your attacks on interactionism do you address either the problem which that view comes to solve or the arguments used, especially by Karl, to pick out interaction as the best solution available. Though I disagree with Karl in the manner just mentioned – at the risk of boring you, I'll present arguments to show why I do so – I do agree with him on central matters. I find his arguments against materialism decisive and I should love to know how you would answer them.

But first I should like to make some fairly detailed criticisms of your [1977a] partly because I think the view I hold did not get a fair representation, but mainly because I should like to show you that the paper did not contain any arguments that could tell against my view, at least not when I had side-stepped your mud-slinging and corrected a few "facts." I'd like to provoke you into finding some such arguments if there are any.

Your certainly slung the mud in that article. But I decided your crack about ideological bias (p. 501) did not refer to me. You said:

> Several doctrines concerning the mind-body problem have some ideological bias or other – and ideologies are not particularly interested in fostering conceptual clarity and empirical investigation.

Rather than me, I thought having an ideological bias might refer to you –

or at any rate to materialists or positivists whose consuming passion for conceptual clarity and empirical investigation may get in the way of solving philosophical problems. (Certainly the problem of induction had 40 years of such passion wasted on it by a quasi-philosophical school that so defined knowledge as to leave no room for philosophy.) And I decided that I was not one of the "rationalist philosophers (who) are radical reductionists and so have claimed that emergence is a myth" nor yet one of the "emergentist philosophers (who) are irrationalists and so have held that there is nothing to be explained: that emergence is as mysterious as it is real" (p. 62). Certainly I agree with you (and with Karl) in thinking emergence real and in need of explanation. But I am perhaps less inclined to believe that much can be done in the way of explanation of emergence. Here I side with Karl in thinking that we cannot give "a full explanation of anything's coming into being in the course of evolution" (Popper [1977], p. 560). I don't think your concept of self-assembly *explains* emergence, though I am sure it elucidates it, and it definitely rejects an alternative-assembly by a presiding Creator. Incidentally, you several times use 'explain' and 'account for' where I would venture only 'partially explain', 'partially account for' or 'elucidate', for instance in Postulate 2 (your [1977a] p. 503): "Every emergent property of a system can be explained in terms of properties of its components and of the couplings amongst these." I strongly suspect that you want to be rid of all mystery, which is, I think, a shame. And possibly hubris. But if I am wrong, one is easily misled by your use of 'explain' without qualification. In any case, I think you merely replace one partially explained mystery (emergence) with another (self-assembly).

You save your juiciest abuse for later. Page 505: dualism is "fuzzy," "not scientifically viable," "double talk," "a disguise for our ignorance." Page 507: dualistic modes of speech "encapsulate our undigested introspective experience and . . . are but metaphorical and vague." Actually, I think you are quite wrong. There has been a lot of digestion of introspective experience, much of the language is quite literal, and it is plainly scientifically viable since it motivates at least some research (Sperry, Eccles). Underlying your abuse is your strong commitment that

> the concepts of mental state, event and process do not fit within the general framework of contemporary science unless they are construed in neural terms, i.e., as, respectively, a state of the brain or an event or process in a brain (p. 505).

Now, I think I can prove you are wrong in this.

First of all, there is absolutely no way that the neural systems, which we all suppose to be connected with mental states, events and processes, can have that connection exhibited except by matching a second party's inspection of the brain with the patient's introspection. That means to say, we could not even make a start at construing a mental state as a neural state until we could connect them. But this raises enormous difficulties if, as you point out, the concept of mental state is fuzzy, informal and so on. How would you know you'd connected the neural state to anything at all unless you *first* knew what a mental state was. Of course, I do not agree with you about mental state, event and process being without sense. Hence I think the program of connecting is possible – indeed I think dualism provokes a very interesting empirical program of finding which neural states, etc. connect with which mental states, etc. and in what way. But this is precisely the program you cannot undertake if you do not think sense has been made of the concepts, mental state, etc. And I must tell you I do not think you have done anything towards that clarification by *declaring* mental states to be states of the brain. There is absolutely no way you would be able to connect those so-called states of the brain with anything else in the brain by merely *inspecting the brain*. You would have to let a person's *introspection* serve as surrogate for second party inspection of your "emergent brain properties" if those properties were ever to be connected with other inspectable states or events in the brain. Your "emergent brain properties" cannot be *identified* except through personal introspection and then only if you had made sense of the mental phenomena that identified them. But in doing so even then, you would have *begged the question* whether the brain states synchronous with certain mental states were their cause, their effect, their twin or their very selves.

In other words, materialism (your version or the Australian) in its zeal for empirical investigation *misses out a stage* in empirical investigation. But the stage missed out is rather crucial in this case. There is no way you can substitute guessing, using law-like hypotheses, for the stage of identifying instrument readings with what is to be measured. In elementary electricity you can invent 'current' as a theoretical construct, investigate its properties and later find out what it is. But with mental activity this cannot be done, if for no other reason because the activity is not completely deterministic nor even indeterministic in the sense of probabilistic. We agree (thank goodness!) that thought *initiates* motion. The neural systems that, in your view, do the thinking are *plastic* ([1980] 91–2, 94).

This is an extremely important characteristic. To an interactionist used to arguing with other materialists that would look like a concession. Certainly, I should not expect Herbert Feigl and those who follow his lead, or the Australian team, to agree with you. Indeterminism is not enough: for human freedom to be genuine there has to be room for behaviour at the mental (or plastic neural system) level that, while breaking no laws, is not lawlike.

Your introduction of a plastic neural system, if it is to work satisfactorily to explain human creativity, has to presuppose a universe that is very orderly but not quite fully orderly; a universe in which there are laws that everything obeys but there are some things that do some acts for which there is no law. It is worth showing how the idea of being above the law is ambiguous and in some senses admirable. A person is wrong to think himself above the law if he means by that that he need not obey it, but not wrong, indeed praiseworthy, if he means he not only does what the law obliges, but goes beyond it and even does things there are no laws about. It is well known that man-made laws are constraints on actions and do not prescribe all actions. By analogy, I would say that the laws we have made in science, clever though we have been at it, do not and should not cover all features of all actions. If they were to, then there would be no novelty, no emergence; worse: no creativity, no individuality. But not only do we admire creativity and individuality, we have enormous evidence for their existence.

I may have got you wrong. You may not mean what I think you mean by the plasticity you endow some neural systems with, though my reading of you is bolstered by your allowing for idiosyncratic properties of things ([1977b] p. 146). I have these misgivings because of your ontological principle:

"M8 Every thing abides by law" ([1977b] p. 17). I do not know how exactly to understand this. I would agree that there is nothing that escapes being constrained by the laws that govern that kind of thing; but I would reject this principle if it meant that there could be no event for which there was no law for it to instantiate, or no property that was idiosyncratic. For the moment let me suppose that my earlier joy at your plasticity was misplaced and you actually hold that mental states, events and processes are either deterministic or indeterministic (probabilistic) rather than (to some extent) idiosyncratic, then you could not know whether your guessed laws for plastic neural systems explained any laws of thinking, etc. until you had made empirical connections between lawlike mental phenomena and lawlike (plastic) neural phenomena.

There seems to me absolutely nothing amiss in hypothesizing plastic neural systems (lawlike or not) and investigating them in their own right. But this can never be claimed to explain mental activity unless the matching can be done. But for the matching to be done, the concepts of mental state, event and process will have to be made to fit within the general framework of science. The introspection reports will have to be treated as *scientific* reports (much as you hinted was possible, on p.509). If this can be done, then you are wrong about the necessity of construing mental activity as neural activity. If it cannot be done, if contemporary science cannot accommodate introspective reports of mental activity as valid reports – so much the worse for contemporary science, it needs mending! – then you are still wrong, because this infirmity will rule out the possibility of connecting specific mental activity with specific neural activity and so rule out your desired construal.

In other words, you cannot *jump* the gap between the mental and the neural, you have to build a bridge. Here I digress to remind you that the first time I saw you in action at a conference in Brockport, N.Y. in 1968, you were arguing that the bridging hypotheses needed to connect events under one kind of description with events under another kind were empirical hypotheses not tautologies nor *a priori* correspondence principles. It is an amusing irony that I am now arguing against you that what are needed to connect phenomena under mentalistic description with phenomena under neurological description will be empirical hypotheses not either tautologies or *a priori* correspondence principles or even definitions. But you should see that my refusal to climb on the monist bandwagon coincides with an aim to foster conceptual clarity and empirical investigation, not to side-step either. Incidentally, I should like to comment that what argues for the identity of the morning star and the evening star (a rather sterile example much abused in the literature) is an empirical hypothesis about Venus coupled with hypotheses about how people observe. Thus the identity is not a logical but an empirical matter. I claim the same in the issue in dispute. If the identity theory is right, which frankly I do *not* believe, it will be an empirical matter. The problem is how to put it to the test – a problem monists and dualists share. This leads us to the matter of what is and what is not a scientific hypothesis.

You keep taking me by surprise both in conversation and in your writing by the keenness you exhibit over drawing the line between science and something else. Why do you use such remarks as "Hence D2 is not a scientific hypothesis" ([1977a], p. 505) as a basis for *rejecting* D2 (psychophysical parallelism)? Do you mean that what is not scientific cannot be true? Sometimes I suspect you of this positivist heresy, despite your having "wandered far away from the Vienna Circle" in not believing "there need or indeed should be any border between science and metaphysics" (your [1973], p. 42). But then your insistence on exactness and generality rather restrict what you can tolerate ("Is scientific metaphysics possible?" [1973], Chapter 8). You continue the attack on dualism by claiming that interactionism "not being a precise hypothesis, ... can hardly be put to empirical tests." So what? The same can be said of emergentist materialism (your favourite). What issue of importance is decided by the empirical testability of an hypothesis? Certainly not its truth. In any case, whatever became of the criterion of agreement with other theories in science – "external consistency" you called it in your [1967] (p. 352)? If we have been offered a theory that does not look testable at present we can at least check its consistency with other theories that have been tested and not yet refuted. This is indeed what Karl does in his discussion of various mind-body theories. Only interactionism stands up to that check. The decisive argument against monism is its bad match with evolution theory (Popper [1977], chapter P3).

Of course what you might have been wanting to say in your [1977a] is not, dualism should be rejected because it's not true and it's not true because it's not testable, but rather, dualism is not much use to neurophysiologists because it does not yield any program of research. But then I should want to challenge that too, since the comparison of reports of introspection with reports of brain inspection, the very program interaction requires, if its workings are to be exposed, is a very fertile program and may even yield tests between dualism and monism. An example of a test that may even now be within sight concerns the clash between your postulate of synchronicity in Definition 12:

> the *mind* of the animal during a given period of time is the union of all the mental processes (functions) that the components of the plastic neural system of the animal engage in during that period ([1980], p. 95).

and Eccles' claim that "there can be a temporal discrepancy between neural events and the experience of the self-conscious mind" (Eccles [1977] p. 362).

Next, the issue of *substance*.

You claim that "the very formulation of the mind-body problem employs . . . (the concept) . . . of substance" ([1977a], p. 501); that dualism acknowledges "one substance's acting upon the other" (p. 505); and that the ideas of state, event and process

> are not transferable to the mental 'substance'. If they are, nobody has shown how. In particular, attention, memory and ideation have not been shown to be properties or changes of properties of a mental substance (p. 505).

On all three counts I disagree. Karl has formulated the mind-body problem deliberately eschewing the (old-fashioned) concept of substance and neither his dualism nor mine acknowledges one substance acting on another. As for the ideas of state, etc., of course I am not wanting to assert that they transfer to a mental *substance*. But I do claim that there are states, events and processes of the mind. I am not quite sure of the force of "nobody has shown how" attention, say, can be a property of the mind. In one sense of the phrase, nobody has shown how attention can be anything else than a property of the mind – or at any rate of a person with a mind (there being some public signals of attention which a behaviourist might want to claim were identical to attention). You certainly could not construe attention as a neural state or a change in attention as a change in a neural state unless you could link attention as it is understood in human self-awareness with some observable neural states.

I think we should try to do without substance. It is not a concept that has received much elucidation and I do not know that it has been anything but a hindrance in the history of science. Correct me if I am wrong. And I see you do not yourself elucidate the concept of substance or "stuff," as such ([1977b], p. 26) despite your very powerful analysis of 'association' and of 'property' and 'thing'. Nor do you show how we would solve the problem of the interaction of things of two different substances, for example a field and a material object. I think you imply on p. 26 with "every real thing consists of some definite stuff endowed with definite properties" that there may be more than one kind of stuff. If you allow interactions between fields and objects, it's hard to see why you should try to rule out interaction between minds and certain (sensitive) parts of brains on the alleged grounds of difference of substance. If I were to go along with your idea of substance for the sake of argument, I would claim that minds are substantial individuals and I should rely for a program to deal with the problem of energy exchange on what Karl says in Dialogues *X̄* and *XII* (his [1977], pp. 543–545, 564, 565). Plainly I cannot prove that minds are substances in a manner that would satisfy you; but nor could you show they were not. Actually I do not care whether minds are substantial though I do care very much that they be considered real. If the only way to have them real in your metaphysics is to have them substantial in some sense, then let them be substantial. It is certainly not a hindrance to have mind a different "stuff" from brains any more than it is to have a gravitational field different from the matter it influences.

In your terms I would say a person is a thing with parts: a mind and a body; of course the body has parts, including a brain. I think the mind emerged from the brain in evolution, probably among lower animals. The person emerged from other organisms with the emergence of self-consciousness in the mind. Of course, the evolutionary development of plastic neural systems is a necessary condition for human autonomy but their property is to *enable* thought rather than, as you would put it, themselves to think.

I have a few further criticisms of your [1977a]. For instance, I do not like your treatment of level in that paper, although I had only some small difficulties with your fuller treatment in [1973]. I think we ought to leave room for *branching* where three different levels, say, species, biological ecosystem and human society, may be said to emerge from the same level, say, organisms or populations, but none emerges from any of the others, and I think we ought to leave room for artefacts which "emerge" from civilization but are not higher forms of being than man. Secondly, I do not agree that interactionism cannot be diagrammed (p. 507), unless of course *no* diagrams are possible for *any* interactions between the parts of a whole. And I do not think monism scores over dualism on any of the six counts you elaborate (p. 508). But it would take us too far afield to go into those difficulties in any detail, except for the fifth of your six scoring points, concerning jibing with evolutionary biology, which I shall take up later. Let me return to the main topic.

As I have said, I am not much concerned whether minds are thought to be substantial (though I prefer not) so long as they are thought real. On the other hand, I want to insist that objects such as scientific theories, philosophical problems, musical symphonies, poems and plays are *not material* but still are definitely *real*. Now, your way with 'real' is to say that what is not substantial is not real. In particular, you assign no reality to constructs independently of brain processes. You specifically say this in one place ([1977b], p. 116) but it is implicit in many others, especially when you say "Everything else (than substantial individuals) is fiction." You are quite explicit to deny the reality of what Popper calls World 3: "There are conceptual objects to be sure but not as constituents of the world, much less constituting a realm of their own. Constructs, whether busy or idle, are fictions" ([1977b], p. 162; compare also p. 34 and p. 158). I have a lot of difficulty with this view and I want to advance several arguments why it cannot be right. I suppose the shortest way would be to say I accept completely what Karl says about World 3 in his [1977] and refer you to his arguments, but I am going to be longer-winded in an effort to hang specific arguments against your view on the things you say and thus, as it were, carry the attack right to you. But I must digress to say how impressed I continue to be, as I read and re-read your [1977b], with your immense good sense and with the excellent way in which you are able to deal, I'd say almost commonsensically, with problems philosophers have vainly toiled at, caught in the toils of wrong-headed assumptions. Perhaps it is just the resonance of my feebly remembered upbringing in physics, but I frequently found myself cheering (even, occasionally, out loud) at what you achieved. On the other hand, every now and then I cried, "Ow!" as your energy in cutting away the bad philosophy hurt the good. It is your commitment to materialism that does the damage. I do not object to materialism as a methodological assumption for some branches of science, physics and chemistry especially, though I object to it even as an 'as if' for zoology and the social sciences. *A fortiori*, I think it wrong, as an ontological assumption. It would be foolish of me to indulge in psychoanalysis, but I can't help worrying at the roots of the viciousness you display in some of your stabs at various non-materialist philosophies (and philosophers). A useful place to start my criticism of materialism is with arguments about the reality of mental constructs.

My first argument fastens on your not assuming that constructs exist independently of brain processes. If you were right in this, then you could never communicate a construct to me since your brain does not act directly on mine (I assume we agree to rule out telepathy as the basis for communication between philosophers). The construct would have to exist in the gap between us – presumably encoded in sound or in writing but perhaps in a visual display (body language!) – for me to get what it was you thought. That what you have to say gets encoded on the way to my apprehending it not only shows the independence of constructs from brain processes but challenges your radical dichotomy between things and constructs (p. 117). I should want to leave room for an object to be in some way both a thing and a construct. I stress "in some way" because of the oddity of having constructs substantial (I sound as if I believed in substance!). But I think an ontology has to leave room for things to be, in addition to their other properties, the embodiment of constructs.

Now, if things embody constructs, they may do so faithfully or otherwise. For example, an orchestra may render a symphony more or less accurately or a theatrical company, a play. Let us suppose that the composer or the playwright is dead (Mozart, say, or Congreve), so as not to get ensnared in the dependence of the work on his brain. I, or other members of the audience including professional critics, may judge there to have been a good or bad performance of the symphony or play, not just as to technical skill but also as to interpretation. And in doing so, we should be employing standards that were objective in some way. We should not be intending merely to say how well or otherwise the performance had satisfied our taste or aroused or lulled us. (Compare your idea of the aim of the arts, [1974] p. 84.) We may of course want to refer to our taste or to our emotions, but we would be able to distinguish between those on the one hand and our judgment of the work against the objective standard on the other. You may want to say that each of us can function only with his own apprehension of the so-called standard (the symphony or the play), his own reading of the score or the text, and so his judgment is thoroughly subjective and the construct quite dependent on his brain. But the *effort* of each of the critics to have his own sense of the work conform to the work itself, so that he could be moved by an argument of another critic to see how he might have misapprehended it, belies thorough subjectivity.

It is a not uncommon experience for a poet to see more in what he wrote than he intended at the moment of writing, and a common understanding of poetry readers, that a poem is not necessarily identical to the author's interpretation of the text. This is a corollary in art of the commonplace in mathematics and science that a theory may have unintended consequences which may be discovered after its articulation and publication. None of the discussions as to what a given text means or whether this or that follows from a theory, make sense, it seems to me, if poems and theories have no independence from brain processes. The same applies to the standards of criticism (including logic). (Compare Karl's restatement of Haldane's argument against materialism [1977] pp. 76–81.)

My next argument concerns the way constructs influence physical objects. If you are not going to allow the reality of constructs then I think you have to deny that any construct ever affects the physical world. And this seems so contrary to experience. For example, a theory that you have about the mind, which you have embodied in the text of papers you sent me has caused me to spend many hours preparing and writing a reply. A second example: a novel (*Three Men in a Boat*) originally a construct in the mind (brain, to you!) of Jerome K. Jerome caused my son to roll about (literally) in uncontrollable laughter on reading it. Incidentally, I should like to put in a plug for literature: not all fiction is mere fiction! Nothing has yet come out of science that matches *King Lear* or *The Brothers Karamazov* or even *The Wind in the Willows* in its perception of what it is to be human. What's moving about such works is their truth! More obvious examples are perhaps the influence scientific theories have on the design and construction of works of engineering, or the influence information has on the medium that carries it.

You may want to stand by that version of your causal principles which requires a transfer of energy between a cause and its effect and in any case defines causation as a relation between two events in different things, ([1977b], pp. 326–7) and say that your theory about mind could not possibly cause me to spend time working on this reply nor the content of a book cause John to laugh because neither the theory nor the content of a book is an event and neither has any energy to transfer. Then what events would you pick out as the causes? And what relationship would the constructs in question have to the action of those events on their effects? What your theory says, and that it is interesting and challenging, and that *Three Men in a Boat* is very funny, seem to me extremely relevant. How would these relevant factors show up in your explanation? It seems to me a broadening of the idea of cause is needed, though there is no obvious reason why the concept you have defined should not be entirely adequate for *physical* causation.

I want to take up now point 5 in your chart of how monism beats dualism. Remember, I do not think monism beats dualism on any of the six points. On this one, I think dualism wins hands down.

> 5. Unlike dualism, which digs an unbridgeable chasm between man and beast, emergentist materialism *jibes with evolutionary biology*, which—by exhibiting the gradual development of the mental faculties along certain lineages—refutes the superstition that only Man has been endowed with a mind ([1977a], p. 508).

It is plain to me that dualists are not alone in digging "an unbridgeable chasm" between Man and Beast (at least all known species of Beast) because monists do not show any more signs than dualists do of wanting to attribute self-conscious thought, descriptive and argumentative language, and the capacity for creating mental constructs (such as the General Theory of Relativity, the design for the Concorde, or *Die Schöne Müllerin* – or even Boyle's Law, the design for the steam engine, or 'Greensleaves') to any known Beast, while dualists seem to me just as likely as monists to allow Beasts to have minds. I think you may be confusing dualism with popular Christianity, which usually has no dualist theory for animals and thinks it a matter of sentimentality if some dog-, cat-, or horse-lovers do.

You take me by surprise in claiming that the *gradual* development of *mental* faculties had been exhibited along certain lineages, by evolutionary biology. I thought that the most evolutionary biology could come up with by way of exhibition was a spotty record of different sizes of cranium or (for humanoid settlements) different tools and artefacts and architecture or (for living species) different behaviour patterns; that the idea of all development being gradual was a metaphysical postulate in the theory of evolution and not something that could be anywhere near exhibited for any developments in the remote past; and that calling any development a development of *mental* faculties is a daring hypothesis projected from human subjective experience—the *only* source any of us has of direct evidence of the mind. I assure you that I accept (in the usual tentative manner, open to criticism and so on) the gradualism of evolution theory and the extrapolation from subjective experience, but I should hardly call either of these *exhibited*, nor could I think either would refute the hypothesis that only Man (so far) has a self-conscious mind.

My argument for dualism faring better than monism at the job of jibing with evolutionary biology sits squarely on the shoulders of the argument Karl uses decisively against epiphenomenalism ([1977], pp. 72–5), part of which I shall briefly rehearse. From a Darwinian point of view, "mental processes are to be regarded (and, if possible, to be explained) as the produce of evolution by natural selection" (p. 72). The benefits which intelligent action confers on an animal need no rehearsing. The superior mental capabilities of the human being, especially his capacity for conceptual thought, for devising theories and testing them out (and letting them die in his stead, as Karl is fond of putting it) are plainly adaptive. A person's subjective experiences thus impinge upon the physical world to his advantage in something like what D. T. Campbell called "downward causation": "Where natural selection operates through life and death at a higher level of organization, the laws of the higher-level selective system determine in part the distribution of lower-level events and substances" ([1974], p. 180). (This would need extending beyond what Campbell had in mind at the time to allow for the causal influence of a higher level thing behaving idiosyncratically, to meet the purpose here.) Interactionism very easily and readily squares with the Darwinian point of view, since it, but not epiphenomenalism (which implies the causal ineffectiveness of subjective experiences) can explain the impact of the development of human mental capabilities on the physical world.

My argument is that *all* the evidence we have about the evolutionary advantage of Man's mind over Beast's mind relates to what we subjectively experience, to self-consciousness, to growing linguistic skills, to rationality, to scientific theory building and so on. There is not a shred of independent evidence that picks out the evolutionary advantage of Man's *brain* over Beast's *brain*, so far as mental activity is concerned. Rather, the only reason why any of the evidence about the advantages that accrue from intelligent behaviour can be applied to explain the puzzle of man's larger brain is that we have succeeded in linking mental processes to the brain, rather than to some other part of the anatomy, and there is, of course, overwhelming *empirical* evidence for that, though I should certainly challenge your claim that "there is definite evidence that every mental event is an event in the brain of some animal" ([1977b], p. 272). In the face of an alternative hypothesis that mental events ocucr in the minds of animals which interact causally with their brains, I should think you could count as "definite evidence" for one or other of the hypotheses only crucial evidence that favoured it over its rival. If we take human freedom seriously, as I think you and I both do, then interactionism requires that certain key neural systems be plastic. It does not require that these plastic systems initiate chains of events in the brain, though it does not rule it out. Monism, by contrast, would, I think, require that the initiation of chains of events leading to the *exercise* of the human advantage in survival in the physical world be accomplished by plastic neural systems assembled, we both agree, from parts which obey physical, chemical and microbiological laws of interaction (some deterministic, some indeterministic). And this leads to problems dualism does not face.

First, as Karl puts it

> the question arises whether the mental activities of the brain (on your view) are just part of its many physical activities or whether there is an important difference; and if so, what we can say about the difference ([1977], p. 96).

Secondly, I think you make it hard to sustain your claim not to deny mental activities, if you keep to Rule 7 in your *regulae philosophandi*:

"*Keep clear of subjectivism*—e.g., avoid definitions in terms of personal experience ... Every ontological elucidation ... should be couched in purely objective (subject-free) terms." I am sure that it is a sound guideline to go as far as you can without invoking subjective language. But when what you are trying to account for is the evolutionary advantage of subjective experience, even if you want to do so in part by linking cortical events with mental events you will be driven to use the terms of personal experience to make contact with your *explicanda*.

Allied to this difficulty is the one of giving anything like a satisfactory neurophysiological account of mental autonomy, and rational and artistically valid idiosyncrasy, when it is agreed on all hands that *chance* in the physical world is useless as the source of human freedom, though it is a necessary condition if any lower level of organisation is to be causally influenced downwards, for if the things that were assembled to make up a certain whole all behaved in a way that was completely explained by a set of deterministic laws, there would be no action of the whole on the parts, of the kind we need, to explain the evolution of man. You often choose 'self-assembly' as the appropriate term ([1973), p. 172, [1977a], pp. 502, 504), but this seems awkward. It suggests that a whole puts itself together, which would mean it had causally effective properties before it existed. It also suggests that the assembly of a whole is uninfluenced by an environment, which I am sure you would not endorse, especially in the case of the development of the brain, since you want the story of that development to accord with Darwinism in which selection pressure from the environment (including competitors for scarce resources) plays such a role. Assembly into a whole, surely, is either *required* of the parts given their properties and those of the environment, or it is accidental, though of course permitted by the laws governing the parts and their environment. Incidentally, you may want to rule out the emergence of mind as an *entity* because there is not at present any hypothesis as to how it was assembled, but I think you should rather pose that as a problem for interactionism which no one has yet addressed, arising out of assembly theory, as you have proposed it.

On the other hand, you have a problem too, arising out of assembly theory. How will you explain the unity of the self when all you have "self-assembling" is thousands of diverse neurons? A pretty thing if your neurons have solved the problem of democracy (or perhaps disproved Arrow's Impossibility Theorem). We know well how *un*unified every other herd of even slightly independent things is. The migration of birds or caribou would hardly serve as an analogue for neurons assembled in the brain since it is likely that migratory success depends largely on each animal knowing roughly where to go. Where no such consensus is enjoyed, it would seem that democracy works (for human beings) only if there is agreement to let certain individuals make certain crucial decisions for the group, and that would not do as an analogy either.

I want to stress the importance of the unity of the autonomous self. It is not just a question of people's sense of self-unity, though I think that alone would be important. There is this, too: the unity (integrity) of the human person, as well as his competence to enter into communion with other unified selves without abrogating their integrity and autonomy or his own, is an enormously important component of the theory of man in four major civilized traditions: Greek, Hebrew, Islamic, and Chinese. This metaphysical hypothesis of unity also has its place in science, since it is presupposed in the explanation we give of man's evolutionary advantage. The manner of your explanation here has to be unique. What is commonly to be explained in evolution is how a certain behaviour pattern is *forced* upon an organism by selection pressure and inherited gene structures following an accident in offspring variation. The pattern is to offer law statements which express the constraints upon the organism and thus explain a spectrum of behaviour traits. With a *plastic* neural system on the other hand, you have a reversal: the *removal* of constraint. This certainly would show how you leave room for idiosyncrasy but it does not offer much hope of showing how the resultant whole is unified, let alone self-consciously unified.

In other words, another "unbridgeable chasm" for the monist lies between the plain facts of everybody's personal experience and the plausibly conjectured "ficitions" (plastic neural systems) of neurophysiology. Everybody can contribute evidence for at least one centred, integrated, autonomous self whose plans, grasp of theory and creative ingenuity can cause changes in the physical world. (And there is a good argument for the existence of *other* selves in the use people can make of mental constructs they are told of. (Eccles uses this argument, I think.)) Interactionism tries to bridge the chasm by postulating a mind which "is actively engaged in reading out from the multitude of active centres at the higher level of brain activity" and which "'acts upon these neural centres, modifying the dynamic spatio-temporal patterns of the neural events" (Eccles [1977], p. 495). It is the obvious properties of the mind rather than the conjectured properties of the brain which afford support to the scientific theory of evolution, and monists who care to support that theory need to capture the properties of mind for the brain. But this can hardly be done simply by announcement, any more than a putsch which captures only the radio station can be called a coup d'état, though it may create the illusion of one for a while.

I'd like to invite you to set aside for a moment any commitment you may have to the presupposition of materialism, perhaps treating it temporarily as a useful methodological principle in science, especially the physical sciences, so that dualism is not disqualified before the competition begins. And then see if interactionism does not fare better than monism in jibing with biological evolutionary theory. I think it may be only the assumption of materialism which makes the *reality* of constructs and of minds seem questionable. The success of an interactionism uncommitted to the concept of substance at appending itself so naturally to evolutionary biology should register as a strong argument against materialism. I am well aware that your particular materialist solution gives more room than is usual to the plasticity required of the physical world if it is to be a home fit for human beings, and in this way displays more than the usual sense of the responsibility to take seriously the problem of accounting for the autonomous, intelligent, construct-building aspects of mind which focus man's evolutionary advantage. So I partly aim with this letter to tug you even further away from those theories of the mind which conflict with evolution by denying the mind any biological function towards the one that seems to me to jibe the best.

I cannot resist raising an objection to another of your rules:

> R3 *Try and formalize everything*: whatever is worth saying in any theoretical discipline, including metaphysics, can be said with the help of mathematics and ought so to be said for the sake of clarity and systematizability ([1977b], p. 8).

We have differed before on the extent to which formalization should be carried. You complain about trying to state philosophical theories in a natural language because of bad philosophy written that way, and I share many of your objections. But on the other hand, I have been appalled by the vast amount of pointless work that has been done and is being done formalizing what does not deserve the effort. Vast bodies of literature have developed, and heaven only knows how much scholarly energy wasted in efforts to solve riddles of induction that originate from mistaken assumptions or inadequate formalisms, or to find a formal way of stating the set of necessary and sufficient conditions for an explanation to be scientific –

as if there were any more hope of locating that animal than photographing the Loch Ness Monster. Well, perhaps there is some hope for the Monster. But it seems to me so plain that unformalizable factors are involved in what correctly counts as a scientific explanation that the search for those conditions was doomed from the start.

The primary misgiving I have about the insistence on formalizing *everything* – remember how much I admire your formalizing work in some fields – is not that it is a misuse of effort but that it pulls science away from art. I am immensely impressed by the search for truth that is at the core of the finest art – I'm conceiving art broadly, you understand, to include literature and (even) philosophy. I would rather work towards a unification of knowledge than suffer its bifurcation into, on the one hand, a science that more or less masters, by rigorous and formal systematics, the art of modelling the physical aspects of the world and then falls flat on its face when it comes to inquiring into the inquirers, and, on the other, an art that aims to understand the human condition and to display its apprehension, only to be robbed of credibility because of its failure (or refusal?) to command that grammar of mathematics which alone you allow to be the touchstone of truth. You'll not get anywhere near an adequate statement of the truth about the human being in any language as impoverished as all formalizations must be. Even natural languages have to be pushed and pulled and strained in fresh directions to yield expression of fresh insights. Such a unification of science and art could lead to science's learning to accommodate values and individuality. The main lesson for art (and even then much art knows it already) is to emulate science's openness to criticism. I don't mean simply aesthetic or technical criticism, in which of course the arts already abound, but rather appraisal of the perception in a work, its nearness to truth.

Could we at last turn to the difference between my view and Karl's? It concerns self-identity and it turns on the result of a thought experiment, "the flawless transplantation of a brain" (Popper, [1977], p. 117). The experiment is the heart of Karl's argument for giving up "Strawson's theory that the *person*, with its physical properties (of the whole human body) and its personal properties (those with a mental component) must be taken as *logically* primitive," (*ibid.*, pp. 117–8). I do not want to discuss Strawson's own case for the theory ([1959], chapter 3) since it would take me far afield into questions of the logical adequacy of criteria for the ascription of different kinds of predicates, though I should confess that within the terms of that kind of inquiry I think Strawson's case persuasive.

What I'd rather do is give reasons for thinking that Karl has drawn the wrong conclusions from his thought experiment. He argues that the transplantation of the brain "would amount to a transference of the mind, of the self," that we must then say that "a person's body no longer provides the unfailing basis for his personal identity" and that "we should have to predict ... that, after the transplantation, the person will claim identity with the donor of the brain, and that he will be able to "prove" this identity (by means like that used by Odysseus to prove his identity to Penelope)" (Popper [1977], p. 118).

I should say at once that I agree with Karl "that the mind is essential to the person" and that in discussing our expectations as to the outcome of the experiment we should discuss "the mind or the self" and "its conjectured liaison with the brain" (*ibid.*, p. 118). But I do not go along with his intuition as to the only possible outcome of this (thought) experiment. A preliminary, but not very important reason for rejecting his intuition, is that the conscious mind is not *attached* to the brain in his version of interactionism, though he does not reject the conjecture that *unconscious* mental states and processes may be identified with brain states and brain processes (*ibid.*, p. 94). In fact, Karl cautions Eccles not to pin down the location of the liaison with the brain. He says,

> I think that in view of the fact that after certain operations or injuries the liaison brain will, as far as we know, actually change its position ... I look at the very location of the liaison brain as being the result of interaction between the brain and the self-conscious mind (*ibid.*, p. 495).

Of course, I regard the idea of a flawless brain transplantation as very far fetched, but if I go along with the idea, I do not see why I should not conjecture that, if it's A's brain going into B's body, B's mind might try to contact the newly arrived brain, searching for its liaison areas and fumbling somewhat because of the differences. B's mind might be able to be comfortable with the body it was familiar with, but the signals from it would hardly be processed in quite the same way. But would B's mind *remember* what it had been like to deal with B's brain? It seems rather confusing. I do not say this is what would happen, but it is not obvious why such an operation should have such a straightforward result as Karl proposes, if his view avoids the presupposition of *attachment* of mind to brain.

Your view, by contrast, goes beyond mere attachments, identifying mental activity with brain activity and would be untouched by the challenge I raise to Karl's conclusion. If you were to spare such a far-fetched experiment a few moments' thought, I wonder how you would answer the question what identity the person assembled from A's brain and B's body (let's call him 'C', for 'composite') would claim. Karl's conclusion, that C would claim to be A despite having B's body, would be as consistent with monism as it is with dualism, I suppose, as would my different conclusion, based on considerations of the role of the body in the formation of personality, of who a person thinks himself to be. I shall preface my argument with a slightly different point, however. To be sure, Odysseus identified himself to Penelope by telling her a secret (how the great bed in their private chamber had been constructed and why it could not easily be moved) which only he, she and the maid by rights should know. But a little earlier that day he had been identified against his will by his old nurse who recognized a childhood scar on his knee as she gave him a footbath. Not only the brain banks "memories" by which we may be identified. And when Odysseus wished to prove himself to his father next day, he used both the scar and a childhood recollection. How easy it is for identity to be feigned is suggested by such stories as *The Man in the Iron Mask*. But this was hardly Karl's point. He wanted to argue that identity was *sufficiently* established by the unfeigned recollection of what a mind-brain would store. And I want to grant him his point that a recitation of such recollection might seem enough, and not quibble too much about how skeptical hearers of such recollections might be when all other evidence, voice, appearance and so on, pointed to C's being B. If the bodies of A and B had both been nondescript or even similar, the recollections might suffice. But suppose neither body had been nondescript. Suppose A had been a corner line-backer with the Calgary Stampeders and B a gorgeous, long-limbed and extraordinarily talented ballet dancer with the National Ballet of Canada. People would be very confused. C himself would be very confused, or is it 'herself'?

My argument starts with complete agreement with Karl about learning to be a self – leave aside for a moment the question whether a self is a mind or a person. I agree to reject the theory of the "pure self," the ego there prior to experience, and accept the suggestion "that being a self is partly the result of inborn dispositions and partly the result of experience, especially social experience" (Popper [1977], p. 111). The difference is, that to all Karl writes about the mind and the brain developing together in interaction I would add that the mind-brain and the body also develop together in interaction. I will take only slightly out of the ordinary examples to make my point. Consider a member of a University track team who has been running in competitions for some years and was one of his high school's best athletes. His mind will of course store memories of training and of meets, of days that went badly and of days when he won, or ran a personal best. My point is that who he is, his character, will have been formed by not only the capabilities of his body but also its demands upon him. Every social event in his ordinary life will have been coloured by his peers' knowledge of what he was, how well he could run, how well he could take losing, how well he could take winning. He and they will be accustomed to his choosing not to smoke and to his being the only one at a party to prefer apple juice to beer. He will know that no one in the school could pull a gag on him and run away. And so will they. By about sixteen, he will be pleased to know that no woman in the world could outrun him. And so would she. His uptake of oxygen with each breath will be perhaps four times that of his friends, his resting pulse lower, his walk more active, his relaxation more immediate, the obedience of his body to his will, less grudging, the knowledge of what he can call on it to do, more intimate.

Or consider a plain girl in school. The personality experience is indelible that she will not ever, ever be valued, in a society of cover-stories and cover-girls, for her cover, that she will not be a cheerleader, she will not make an evening dress look good, nor a pair of jeans, nor a photograph, that as she sits next to the boy-friend who has managed to take her to see a Jane Fonda movie, she and he are busy measuring the gap. I am not suggesting that she is irremediably harmed by this, on the contrary; she may be immeasurably enriched. I am maintaining that she is changed by it, trained by it, who "she" is is someone who knows this. (If, peculiarly, like colour-blindness, she possessed an inability to see it, it would be seen of her, and then "she" would have to cope with her reception in the world that does see it.) Transplant the brain of either of our examples into a somewhat contrasting body – an ordinary (unfit) student who smoked and drank, or a girl marked by glamour – or vice versa, and disorientation would be severe. But the immediate and continuous tensions a flawless transplant would produce are not confined to the self-conscious sense of identity: plainly physiological problems present themselves, their range depending on how far down into the nervous system the cut is made. Where does "the brain" end? One might say the entire spinal cord is brain. Certainly, in the sense in which animals without self-consciousness learn, human beings "learn" along the whole nervous system. Control of responses to external stimuli may go down to a single muscle fibre. I took Karl to mean, in his experiment, only to transplant the cerebrum, but even if he took the spinal cord, he would have to leave behind all the peripheral learning. Add the range of behaviour that may be triggered by the body's highly individualized chemistry, and the result of a transplant would be a ghastly amplification of the disorientation of commisurotomy. B's body would go about much of its business out of control of A's alien, self-conscious mind-brain. And the mind-brain would lack the memory to "understand" the signals it received and the motions the body went through of its own accord, learnt while in tandem with another mind.

Of course, I should not want to say that there was any chance C would self-consciously recollect being B. On the contrary, I accept that he would recall items that would have identified him as A if only he had not switched bodies. Now he has no clues from his perceptions of his body that any of his recollections have foundation. We might try *telling* him he was A (but that would beg the question of what he'd claim). Each person he met would tell him he was B unless he overrode their judgment with "proofs" of being A. He would not need to marshal proofs of being B, his voice, his appearance, his gait, many, many little habits would so attest. And who is to say who he "really" is? On my view, he would be neither A nor B, though partially identifiable as either. He could become "C." If he were ever to overcome the shocking tensions of the operation, what was A's mind-brain would have to interact with what was B's body, compromising handicaps and exploiting capabilities to begin the new life: a new person with strange lingering memories of another existence, and lacking a sense of roots. But that's enough of speculation. I have tried to show that considerations Karl did not discuss rather undermine the easiness with which a transplant patient might be thought to adopt the identity of the main donor, and thus to remove the only objection Karl raised to the logical primitiveness of person.

The arguments I have mustered concerning the interactions of mind-brain, body and society go to suggest that the person, the integrated compound of mind, brain and body, whose properties include his social relationships, is the appropriate unit of both public and private identification. In short, *the self is the person.* I think I might have you on my side in this matter despite our disagreement over the dualism of mind and brain.

My agreement with Karl on so many issues, like my very wide agreement with you, is not unquestioning, as you see. Nor are my views final, though many of them are firm. I should be delighted to know how you would answer the arguments I have raised against features of your philosophy and just as pleased to hear what arguments you would marshal against my way of trying to solve the problem of personal freedom in an orderly world (alias the mind-body problem), which at long last I have set out as clearly as I can in contrast to your own.

My very best wishes,

University of Guelph, Ontario, Canada

Tom

REFERENCES

Bunge, M.: 1967, *Scientific Research II, The Search for Truth*, Springer, New York.

Bunge, M.: 1973, *Method, Model and Matter*, D. Reidel, Dordrecht.

Bunge, M.: 1974, *Treatise on Basic Philosophy*, vol. I, D. Reidel, Dordrecht.

Bunge, M.: 1977a, 'Emergence and the Mind', *Neuroscience* **2**, 501–9.

Bunge, M.: 1977b, *Treatise on Basic Philosophy*, vol. III, D. Reidel, Dordrecht.

Bunge, M.: 1980, 'The Psychoneural Identity Theory', in Bindra (ed.), *The Brain's Mind*, Gardner Press, New York, pp. 89–108.

Campbell, D.T.: 1974, ' "Downward Causation" in Hierarchically Organised Biological Systems', in *Studies in the Philosophy of Biology*, F. J. Ayala and T. Dobzhansky (eds.), University of California Press, Berkeley and Los Angeles.

Popper, K.R., and J.C. Eccles: 1977, *The Self and Its Brain*, Springer International, London.

Settle, T.W.: 1977a, 'In Defence of Plain Metaphysics', in *Science and Metaphysics* Proceedings of the 1973 Fribourg conference of l'Académie Internationale de Philosophie des Sciences in *Archives de l'Institut International des Sciences Théoretiques*, Office International de Librairie, Brussels.

Settle, T.W.: 1977b, Review of Bunge 1973, *British Journal for the Philosophy of Science* **28**, 86–94.

Strawson, P.F.: 1959, *Individuals*, Methuen, London.

WILLIAM SHEA

THE YOUNG HEGEL'S QUEST FOR A PHILOSOPHY OF SCIENCE, OR PITTING KEPLER AGAINST NEWTON*

When G.W.F. Hegel defended the theses related to his *Habilitationsschrift* on *The Orbits of the Planets* on his thirty-first birthday on 27 August 1801 he had not achieved anything very remarkable in the world, and no one could have foreseen his later eminence. He had arrived at the University of Jena the year before at the invitation of his friend, the boy-wonder Schelling, who had been given a Chair at the recommendation of Schiller and Goethe in 1797. Shortly after his arrival Hegel began writing the *Habilitationsschrift* that would secure him the right to lecture, and an extended essay on *The Difference Between Fichte's and Schelling's System of Philosophy* that was published at the end of 1801. His first major work, *The Phenomenology of the Spirit*, did not appear until 1807.

The prolific and versatile Schelling had, by 1801, already written several books in which he outlined what he called a Philosophy of Identity, an attempt to bring together in a single vision God, nature and self-consciousness. The system consisted of two parts: the Philosophy of Nature and the Transcendental Philosophy. The Philosophy of Nature traces the process whereby an unconscious God reveals himself in the ever ascending levels of nature until self-consciousness emerges in man. The Transcendental Philosophy describes the stages of God's coming to know himself, a sequence that culminates in art, according to Schelling, or in religion, as the young Hegel believed. Hegel's philosophy subsequently developed and became more complex and subtle but in 1801 he was still very close to Schelling. (Their ways were to part a few years later when, in the Preface to his *Phenomenology of the Spirit*, Hegel spoke of Schelling's philosophy as "the night in which all cows are black".) Hegel's philosphy of nature, like Schelling's, is meant, if not to replace the natural sciences altogether, at least to provide them with a basic framework without which they remain little more than collections of unrelated information, data banks that serve no genuine purpose.

In his introduction to *The Difference between Fichte's and Schelling's System of Philosophy*, published in the same year as *The Orbits of the Planets*, Hegel writes that Philosophy becomes a need in times when the simple and beautiful harmony of existence is sundered by the awareness

J. Agassi and R.S. Cohen (eds.), Scientific Philosophy Today, 381–397.
Copyright © 1981 by D. Reidel Publishing Company.

of fundamental dichotomies and antinomies, when the believers become alienated from the gods, man from nature, the individual from his community. Out of this climate of crisis is born a philosophy that will eventually enable man to overcome society's many-dimensional alienation into a higher cultural synthesis, a genuinely integrated culture. This higher synthesis is described by Hegel in metaphors taken from the romantic conception of nature: a living whole of which the individuals are organs rather than atoms. This vital or organic philosophy is called "speculative" philosophy to distinguish it from "reflective" philosophy, a brush that Hegel uses to tar both British Empiricism and Kant's Transcendental Idealism. These reflective philosophies fall short of the true conception of philosophy because they try to solve intellectual riddles rather than give a meaningful vision of the whole. They pick up their knowledge of nature piecemeal from the natural sciences and unwittingly fall into two snares: first, they commit themselves to a concept of experience limited to sense perception, and, secondly, they profess a pervasive atomism that reduces the whole to the sum of its parts, and to a mechanism that excludes teleology from playing its essential role in cognition.

This is, of course, a very strong claim since it assumes that what is wrong with reflective philosophy is simply that it is not genuine philosophy. Under these conditions, one can hardly blame reflective philosophers (that is practically everyone who is not a Hegelian) for registering something like annoyance at the exalted status that Hegel bestows upon his own system. In a sense, he has elevated it above the possibility of rational attack by declaring that all criticism is, at the best, misinformed and, at the worse, perverse. Hegel took his philosophy too seriously to compare it to a game but if we could venture a comparison it would not be with tennis or football in which there are many different ways of scoring, but rather with fitting a jigsaw puzzle together where only one solution is possible. The Hegelian's decisive advantage lies in a feeling for the whole that is akin to remembering the jigsaw puzzle as it was before it was taken apart.

It is within the context of this larger philosophical program that we must consider Hegel's attempt to put the natural sciences on a proper philosophical footing. I shall draw mainly from *The Orbits of the Planets* but I shall also refer to some of the developments that Hegel made when he wrote his *Philosophy of Nature*, the second part of his *Encyclopaedia of the Philosophical Sciences* that appeared in 1817.

What is wrong with astronomy and physics? The short answer is:

Newton! for Newton is but the embodiment of British Empiricism. This would be melancholy enough but what makes it worse is the fact that the laws of physics and astronomy were much more felicitously expressed by a scientist who is far greater than Newton (one is tempted to say who exceeds Newton as much as Hegel towers over Locke). This man is Johann Kepler, and we may doubt whether Hegel considered it a mere coincidence that Kepler should have been born not only in Germany but near Stuttgart, Hegel's own native city, and that he should have studied, like Hegel, at the Seminary in Tübingen. Hegel writes: "After the happy genius of our great compatriot Kepler discovered the laws by which the planets move in their orbits, Newton is alleged to have demonstrated them with geometrical rather than physical arguments" [Hegel 1928, p. 350]. Hegel never wavered in this disparagement of Newton. If anything, his feeling that Kepler had been wronged grew with the years, and in his *Philosophy of Nature* he stated: "The honour of having discovered the law of universal gravitation has been attributed to Newton, who, by catching the popular imagination, has won the greatest applause, and obscured the glory of Kepler." But lest he be suspected of pettiness, Hegel adds in the same breath of voice: "It is often envy that motivates the debunking of great men, but on the other hand, it is a kind of superstition to regard their accomplishments as unsurpassable."[1]

Hegel feels, therefore, that in order to vindicate Kepler, he has to put Newton down. In the light of the holistic and vitalistic philosophy to which he subscribes, it comes as no surprise that he should accuse Newton of identifying mathematical construction with physical reality. "Care should be taken," he says, "not to confuse mathematical relations with physical ones by rashly assuming that the lines used in geometry to construct the proofs of its theorems are forces" [Hegel 1928, p. 350]. This disquiet about the inroads of mathematical physics is, of course, common to all the German Romantics. We find analogous statements in Schelling, and Goethe, in a celebrated passage, compared mathematicians to Frenchmen: "Talk to them and they will translate it into their own tongue where it at once becomes something altogether different."[2] Hegel quotes the passage in the *Principia* where Newton states that he uses

> the words attraction, impulse or propensity of any sort towards the centre, promiscuously and indifferently, and one for another, considering the forces not physically, but mathematically: wherefore the reader is not to imagine that by those words I anywhere take upon me to define the kind, or the manner of any action, the causes or the physical reason thereof, or that I attribute forces, in a true or physical sense, to certain centres

(which are only mathematical points); when at any time I happen to speak of centres as attracting or as endued with attractive powers [Newton 1971, vol. I, pp. 5–6].

But Hegel argues that although Newton expressly and repeatedly makes the distinction between the mathematical representations of the problems he is dealing with and the physical reality of the situation he is describing, it often appears to be quite absent from his expositions. In Hegel's days, gravity or attraction had indeed become for many physicists a purely factual question, but Newton himself had consistently maintained the distinction he established. The problem was that his interpreters chose to ignore it. For instance, we find Newton frequently protesting that he was ignorant of the physical nature of gravity. To Bishop Bentley who was theologizing the *Principia* or at least using Newton's work to pad his Boyle lectures, Newton wrote in 1693: "You some times speak of gravity as essential and inherent to matter. Pray, do not ascribe that notion to me; for the cause of gravity I do not pretend to know" [Turnbull 1961, p. 240]. But by the eighteenth century, Newton's distinction between the mathematical and the physical aspects of his exposition was overlooked with the result, as Alexandre Koyré expressed it, that scientists "became reconciled to the ununderstandable" [1965].

According to Hegel, mathematics can only deal with the *quantitative* aspect of the phenomenon, never with the underlying *qualitative* reality, because "geometry abstracts from time and arithmetic abstracts from space" [Hegel 1928, p. 350]. Newton's attractions or impulses are quantifiable because they belong to the first or mechanical level of inorganic nature that must be carefully distinguished from the second or physical level. At the first level, that of the science of mechanics, the determination of motion is completely external and is ascribed to shocks, pushes and impulses. At the second level, that of genuine physics, matter is recognized as being moved "from within," and the prototype of this motion is the spontaneous action of the magnet. By failing to distinguish mechanics from physics, Newton erected a "composite structure" in which it is difficult to recognize what is physically, and not merely mathematically, true. Kepler, on the contrary, did not – as he could easily have done – "cloak the pure and mathematical expression of the laws he discovered with a physical shape," in other words, he did not appeal to forces. Hegel adds that Kepler could have transformed his law that the area swept out by the radius vector is proportional to the time

into the following semblance of a physical law by saying that gravity is proportional to the arcs of equal sectors, and since the total area of the circles A and a are as the square of their radii, then $1/A : 1/a$ will be as $r^2 : R^2$. Here $1/A$ and $1/a$ would represent the quantity of motion or, if one wishes, the quantity of centripetal force, and Kepler could have said that the force of gravity or the centripetal force is inversely proportional to the radii or the distances [Hegel 1928, p. 354].

I have quoted this extended passage because it is a fair sample of Hegel's historiography as well as his science. His interpretation of Kepler is not only misleading but completely erroneous. Kepler does not provide a merely kinematical description of the motion of the planets. His approach is essentially dynamical and the concept of force plays a key role in his exposition. When Hegel proceeds to show that Kepler could have derived the law that the force of gravity is inversely proportional to the distances but did not, and hence deserves to be commended for not contaminating the purity of the mathematical formulation of his law, he is again wrong. Kepler did, in fact, maintain that force is inversely proportional to the distance. More serious than these historical errors, however, is Hegel's inept scientific reasoning whereby he does not realize that the conclusion of the argument (that he maintains Kepler could but did not pursue) is not identical with Newton's inverse square law since the force is said to vary as R^{-1} not R^{-2}!

Hegel finds the root of Newton's error in the decomposition of forces or the resolution of vector quantities into their components. This is a purely mathematical device, according to Hegel, and "is completely alien to a living force", by which he means a *natural* force. Hegel adds that this should come as no surprise since Newton even "takes apart light that is naturally simple" [Hegel 1928, p. 356]. We have here an echo of Goethe's campaign against Newton's *Opticks*. In 1801, Goethe had only published two short papers on the nature of light (his *Farbenlehre* appeared in 1808–1810), but Hegel and Schelling were well apprised of Goethe's views and they had the opportunity of discussing them with him. Goethe may even have encouraged the young Hegel to criticize Newton's astronomy and physics to herald the attack he was about to launch on his optics.

Hegel interprets the centripetal and the centrifugal forces as the components of the path of the planets: "For an infinitely small part of a circle is inscribed in a parallelogram as its diagonal so that the sides are, on the one hand, the tangent and the chord or the sine that is ultimately equal to it, and, on the other hand, the versed sine and the secant that is ultimately equal to it. Physical reality is ascribed to these lines so that one stands for the centrifugal force and the other the centripetal" [Hegel 1928, p. 358]. Unfortunately, Hegel fails to grasp the meaning of the fundamental proposition of Book I of the *Principia*. The following diagram illustrates how Newton actually proceeded in his resolution of the path of a planet attracted by a central body (see Figure 1). In a small interval of

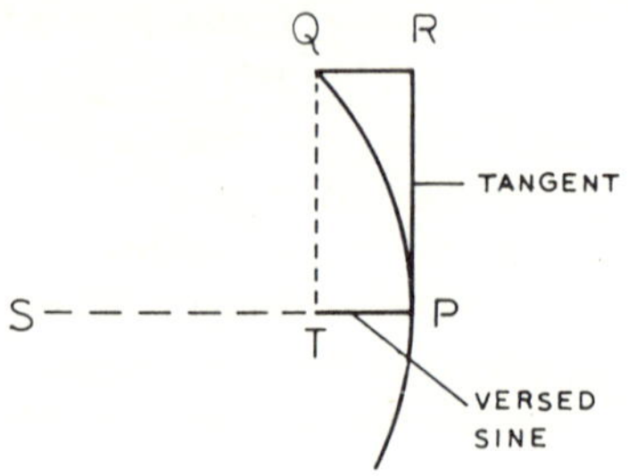

Fig. 1.

time, the planet moves along the arc PQ. The versed sine PT (we would use today 1-cosine) is the measure of the deflection of the trajectory and can be considered as the segment of the rectilinear path that the planet would have covered under the exclusive action of the central body. The tangent PR is the measure of the distance that the planet would have traversed had it continued to move inertially (i.e., without the application of any force). While the rate of increase of the versed sine PT is proportional to the square of the time, that of the tangent PR is simply proportional to the time. In the case of non-circular orbits, the line PS is no longer perpendicular to the tangent, and Newton no longer speaks of the versed sine but of the "arrow" (*sagitta*). Hegel ignores this distinction and proceeds on the assumption that Newton makes no difference between circular and elliptical paths.[3]

Newton never says that the tangent represents the centrifugal force. It corresponds to the inertial path that the planet would have followed without the intervening action of the central body. On Hegel's reading of the *Principia*, however, the tangent stands for the centrifugal force and the versed sine for the centripetal. This leads Hegel to believe that Newton attempted to account for the orbit of the planets and their varying velocities by combining the two "forces" in different ratios.

As we shall see later, Hegel's misunderstanding stems from his inability, or his unwillingness, to accept the principle of inertia or Newton's First Law which states that a body continues to move in a rectilinear path and is diverted from it only by the action of some other body. A body moving in a circle is rotating because of the existence of an external constraint. In other words, a centripetal force must be applied to constrain a body in circular motion. The situation was often obscured in textbooks by the survival of a terminology that was coined by Newton's immediate predecessors in the seventeenth century, for whom the task was not to measure the external constraint or force necessary to divert a body from its inertial path but the internal effort or force exerted by the body thus constrained, what they called the centrifugal force. But even in this conceptual scheme, the tangent was never identified with the centrifugal force.

Hegel's interpretation gives rise to problems that are indeed perplexing and about which he seems to believe that Newton was at a loss. If the centrifugal force is completely independent of the centripetal, what law does it obey? Or if it is not independent but always of the same magnitude as the centripetal force, how can it determine the variations in the velocity of each planet? Or, again, if these two forces are supposed to increase and decrease gradually, there must be a point where they are equal and in equilibrium; and this being the case, they will always continue equal, for there will be no reason for them to lose their equilibrium. Hence they cannot explain the variations of velocity. Hegel repeats the same charge with even greater forcefulness, if not greater clarity, in the *Philosophy of Nature*:

It is in the so-called explanation of uniformly accelerated and retarded motion by means of an alternating decrease and increase in the magnitude of the centripetal and centrifugal forces, that the confusion caused by the postulation of such independent forces is greatest. According to this explanation, the centrifugal force is less than the centripetal force in the movement of a planet from aphelion to perihelion; at perihelion itself however, the centrifugal force suddenly becomes greater than the centripetal force, and in the movement from perihelion to aphelion, the forces are supposed to work in the inverse relation.[4]

It says much about the philosophical climate in Berlin that Hegel was able to lecture on this topic repeatedly without anyone rising to point out his error. Three editions of the *Philosophy of Nature* appeared in Hegel's lifetime (in 1817, 1827 and 1830), but I have been unable to find any criticism of his interpretation of Newton prior to an article by William Whewell in the *Transactions of the Cambridge Philosophical Society* of 1849. Whewell's rebuttal of Hegel is in itself instructive. After citing the passage just quoted from the *Philosophy of Nature*, Whewell replies:

The most elementary knowledge of mechanics shows us that when a body is moving *obliquely* to the distance, it approaches to or recedes from the center in virtue of this obliquity, even if no forces at all act. And the total approach to the center is the approach due to this cause, *plus* the approach due to the centripetal force, *minus* the recess due to the centrifugal force. At the aphelion, the centripetal is greater than the centrifugal; and *hence* the motion becomes oblique; and *then*, the body approaches to the center on *both* accounts, and approaches on account of the obliquity of the path even when the centrifugal has become greater than centripetal force, which it becomes before the body reaches the perihelion. [Whewell 1849, p. 698]

What is interesting about this argument is that it appears to treat centrifugal force as a real not a fictitious force. Whewell himself recognized this incongruity and two paragraphs later he virtually repudiates the explanation he has just proferred:

But these modes of explanation, by means of the centripetal and centrifugal forces and their relation, are not necessary to Newton's doctrine, and are nowhere used by Newton; and undoubtedly much confusion has been produced in other minds, as well as Hegel's, by speaking of the centrifugal force, which is a mere intrinsic geometrical result of a body's curvilinear motion round a center, in conjunction with centripetal force, which is an extrinsic force, in the acting upon the body and urging it to the center. Neither Newton, nor any intelligent Newtonian, ever spoke of the centripetal and centrifugal forces as two distinct forces both extrinsic to the motion, as Hegel accuses them of doing. [Whewell 1849, p. 698]

Whewell is correct in assuming that Newton saw centrifugal force simply as a reaction to the centripetal force deflecting a body into a curved path. But a clear statement of this is not to be found in the *Principia* and many Newtonians, as Whewell did himself in an unguarded moment, spoke of the motion of the planets as a combination of a circulation and a motion along the rotating radius vector joining the planet to the sun. Whewell is wrong, however, in claiming that Newton does not occasionally offer an explanation in terms of centrifugal force. The most striking instance (and one that Hegel was to pounce upon) is his explanation of the slowing down of a pendulum at the equator. The problem of the rotation of the earth had been discussed in the Cartesian framework by Huygens who had shown that if the earth turned 17 times more quickly, the centrifugal force at the equator would equal the weight of the body and, hence, that the diurnal motion must take away a fraction of the weight. As the centrifugal force was proportional to the square of the speed, the fraction was 1/289. Newton continues to discuss the problem in the same terms, albeit in a different conceptual framework: "But if the diurnal motion was accelerated or retarded in any proportion, the centrifugal force would be augmented or diminished nearly in the same proportion squared" [Newton 1971, Bk. III. Prop. XIX, p. 427].

A fraction of the earth's attraction (1/289 at the equator) is used in keeping terrestrial objects from flying off by making them undergo an instantaneous deflection. This can be interpreted as though this fraction of the weight was used to counterbalance the centrifugal force. Because the centrifugal or the centripetal force may be employed in making the mathematical computations, Hegel feels justified in claiming that the entire explanation is merely a mathematical device, and in suggesting that it makes just as good physical sense to say that the oscillations of the pendulum are slower at the equator because of an increase in the force of gravity holding the pendulum more firmly in the perpendicular or line of rest. Flushed by his success at thus reinterpreting Newton, Hegel even claims that the earth is elongated at the poles: "let me add that this agrees perfectly with the shape of the earth which is greater at the equator where its diameter is shorter than its axis" [Hegel 1928, p. 372]. Jacques Cassini, misled by inaccurate geodesic measurements taken over a comparatively short meridian, had advanced this hypothesis at the beginning of the eighteenth century. In 1735 the Académie des Sciences decided to send an expedition headed by Maupertuis and Clairault to Lapland, and the new measurements confirmed that the earth was flattened at the poles as required by Newton's theory. After Maupertuis published his book in 1738, the issue was considered settled. It is strange that Hegel should have considered reviving it since even if the centrifugal and the centripetal forces were inverted this would not be an argument for a return to Cassini's position. The expedition of 1736 determined the length of the meridian and this bears no relation to the nature of gravity. In the *Philosophy of Nature* Hegel did not repeat this claim but he continued to contend that gravity is increased at the equator.

Having shown the inadequacies of the mathematical approach, Hegel warns us as we turn

> to the physical reality of the centrifugal force not to expect a philosophical construction of force from the experimental philosophy which Newton and the English in general have always considered not only to be far and away the best but the one and only philosophy [Hegel 1928, pp. 358–360].

Hegel gives two instances of Newton's "pitiable" experiments to illustrate centrifugal force. Both are taken from Definition V of Book I of the *Principia*. The first is a stone that is whirled about in a sling and endeavours to recede from the hand that turns it. The second is the famous thought-experiments of firing a leaden ball from the top of a mountain with increasing momentum until it acquires escape velocity and goes into orbit around the earth.

Although Hegel claims that the nature of centrifugal force is illustrated in both these instances, Newton actually used them to define *centripetal* force. Hegel quotes the passage describing the stone whirled in a sling but he omits the next sentence that reads: "The force which opposes itself to this endeavour, and by which the sling continually draws back the stone towards the hand, and retains it in its orbit... I call the centripetal force" [Newton 1971, pp. 2–3]. But Hegel, even at this early stage of his philosophical career, was not fettered by facts. The reason is that "philosophy can perhaps deduce *a priori* what the experimental method, that calls itself philosophy, wrongly and unsuccessfully tried to know from experiments" [Hegel 1928, p. 360].

The *a priori* deduction that Hegel has in mind is the one that Kant attempted in the section on "Dynamics" in his *Metaphysical Foundations of Natural Science* where he argues that only two moving forces of matter can be conceived. Kant assumes that a particle of matter is considered as a point, and that all action between two points occurs along a straight line. But along this straight line only two kinds of motion are possible, one, by which the points *recede* from one another, and a second by which they *approach* one another. Kant calls the causes of these motions the *attractive* and *repulsive* forces and to these he reduces all other forces in nature [Kant 1957, p. 50]. Hegel does not mention Kant by name in *The Orbits of the Planets* but he refers to him explicitly in *The Philosophy of Nature*:

> It is to be regarded as one of the many merits of Kant that in his *Metaphysical Foundations of Natural Science* he made an attempt at a so-called *construction* of matter, and by establishing a notion of matter revived the concept of a *philosophy of nature*.[5]

Hegel then proceeds to praise Kant for recognizing the forces of attraction and repulsion, but he criticizes him for assuming that what is to be attracted or repelled is already fully constituted matter. In the wake of Schelling's doctrine that attraction and repulsion constitute gravity, Hegel maintained from the outset that "gravity, namely identity itself, was the condition of forces." The same position is reaffirmed in his mature work: "By offering resistance matter repels itself from itself, and so constitutes repulsion, through which it posits its reality and fills space."[6] This is such a fundamental truth that even "the experimental philosophy, without knowing it, caught a glimpse of the opposition of attracting and repelling forces" [Hegel 1928, p. 360]. What blinded Newton to the significance of what he could not but see was his commitment to the spurious concept of inert matter and his uncritical use of mathematical limits. Let us examine this second alleged shortcoming first.

It is well known that although Newton presents his proofs in the *Principia* in geometric form, he actually arrived at his results with the aid of the calculus that he had invented several years earlier. The use of the calculus is thinly disguised, at least for those who read the *Principia* with the advantage of hindsight. For instance in Book I, Lemma VIII, Newton argues with the aid of Figure 2, that

If the right lines *AR*, *BR*, with the

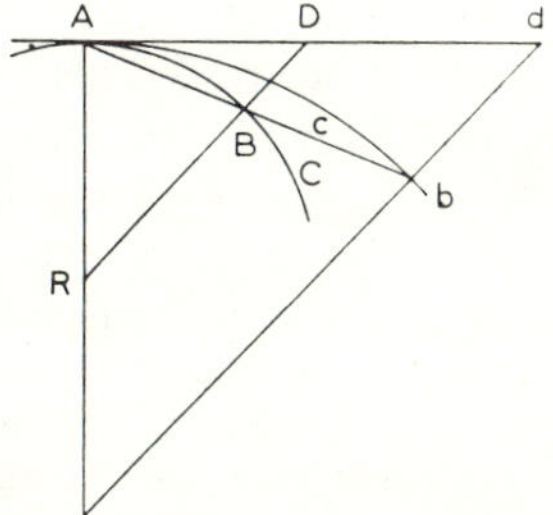

arc *ACB*, the chord *AB*, and the tangent *AD*, constitute three triangles *RAB*, *RACB*, *RAD*, and the points *A* and *B* approach and meet: I say, that the ultimate form of these evanescent triangles is that of similitude, and their ultimate ratio that of equality.

Fig. 2.

Hegel did not penetrate very deeply into the foundations of the calculus but what struck him was the use of "evanescent triangles." The conception of a limit is indeed problematic in the *Method of Fluxions* as Newton called his new mathematical method. The magnitude of the limiting ratio is ambiguous for as the arc and the chord are taken to be equal not before or after but only when they vanish, they have no being at the very moment when they should be caught and equated. It is for this reason that Bishop Berkeley in *The Analyst* criticizes Newton's fluxions as being "the ghosts of departed quantities", and argues that the fundamental idea of supposing a finite ratio to exist between two absolutely evanescent terms was absurd and unintelligible. Hegel was probably unaware of Berkeley's criticism but he was able to draw from analogous remarks in Joseph Louis Lagrange's *Théorie des fonctions analytiques*:

That method has the great inconvenience of considering quantities in the state in which they cease, so to speak, to be quantities; for though we can always conceive the ratios of two quantities as long as they remain finite, that ratio offers to the mind no clear and precise idea as soon as both of its terms become nothing at the same time.[7]

Lagrange's strategy to overcome the difficulty was to use Taylor's series to define the differential coefficient of $f(x)$ with respect to x as the coefficient of h in the expansion of $f(x + h)$, and so to avoid all reference to limits. But Lagrange was not fully aware of the limitations of the expansion of $f(x + h)$ and his proof, which was initially applauded, soon faced serious criticism. The conception of a limit was not put on a firm footing until the work of Cauchy in the nineteenth century. What Hegel saw in Lagrange's analysis was a clear distinction between a purely abstract (i.e. mathematical) mode of regarding functions, and their application to problems of physical motion. It buttressed his conviction that Newtonian physics had gone astray by following an artificial resolution of forces. Fortunately, from Hegel's viewpoint, there is another geometry, for

this method of constructing the phenomenon from absolute opposites should not be taken as the method of geometry. Geometry does not seek to construct a circle or any other curve from lines intersecting at a right angle or any other way; on the contrary, it takes as given the circle or any curve under consideration and from it establishes the resulting relations of the rest of the line [Hegel 1928, pp. 362–364].

This remarkable passage reveals that Hegel's understanding of geometry owes more to Renaissance Platonism than to the work of contemporary mathematicians. He assumes that if one grasps, intuitively, the correct shape or form of the universe, the meaning of all the parts will become obvious: "This is the true method that physical science should have followed faithfully, namely positing the whole and deducing from it the relations of the parts" [Hegel 1928, p. 364]. When we recollect that Kepler explained the nature of the solar system by pointing out that it fitted the geometry of the five regular solids, we begin to see another reason why Hegel saw Kepler as a "pure" genius. In the Prologue to his *Harmonices Mundi* Kepler praises Proclus for considering as one of the aims of geometry: "the teaching of how the heavens receives forms appropriate to its parts" [1940, p. 18]. On this view, Euclid began with the point and the line and ended with the five regular solids because these are the architectonic ideas of the world. Hence one passes naturally and without any break of continuity from the thirteenth book of Euclid's *Elements* to the study of physics.

This conception of mathematical forms is initimately linked to the notion of form, a concept derived from Aristotle whom Hegel was always to regard as his great forerunner among Greek philosophers. Weight or gravity is considered an essential property of matter and in the very first paragraph of *The Orbits of the Planets*, Hegel affirms that "gravity is the first force of nature". He never abandoned this Aristotelian tenet so completely demolished by scientists since the day of Galileo, and in the *Philosophy of Nature* he thrusts it to the fore: "the primary essence of matter is that it has weight. This is not an external property which may be separated from it. Gravity constitutes the substantiality of matter, which itself consists of a tendency towards a centre which falls outside it."[8] Hegel even retains Aristotle's distinction between heavenly and terrestrial bodies. The *Orbits of the Planets* begins with the statement that while terrestrial bodies "are not self-sufficient in respect to gravity", the heavenly bodies "carry their centre of gravity more perfectly and in the manner of gods progress through the aether". The same expression recurs in *The Philosophy of Nature*: "The movement of the heavenly bodies is not a pulling hither and thither, but free motion; as the ancients say, they go their ways like the blessed gods."[9] The key word here is "free" for Hegel considers the solar system as a living being. The Newtonians are in bondage to a science that is "alien to the life of nature, and the only primitive notion they can apply to matter is death, what they call the force of inertia, namely indifference to rest and motion" [Hegel 1928, p. 380]. As a result, they have to appeal to God to give the first impulse to matter and to wind up the cosmic clockwork when it becomes run down.

It is not that mechanics is not useful if one recognizes it for the abstract science that it is. Motion as conceived by the Newtonians is of considerable utility in dealing with practical problems involving levers, pulleys and winches, but at the level of rational analysis it is seen to be a mathematical device rather than a physical explanation. A rational philosophy cannot fail to take cognizance of the living forces in nature. The phenomena of light, heat, electricity and magnetism are sufficient in themselves to disclose the fact that matter is not dead. Nature is alive and what we witness as we ascend the ladder of being is a gradual increase of vital and autonomous activity. Hegel implicitly subscribes to the doctrine of the microcosm and the macrocosm when he claims that every natural thing "expresses the structure of the whole" [Hegel 1928, p. 348]. We find the same pattern recurring at a higher level of organization. For instance, matter spontaneously divides itself into two poles and between these poles there lies a "point of indifference". This is the case of the magnet but it can also be seen "in the level that imitates the natural line of magnetism in dead matter" or "in the focuses of the eclipse whose main axis is a true line of magnetism". Indeed "the line of magnetism passes into the form of a natural pendulum", the only difference being the loss "of the second pole". In the case of the solar system, "one and the same body produces both poles" and, hence, we have an empty focus. The justification of this truncated polarity is that "the true centre of force is necessarily the source of light" [Hegel 1928, p. 386].

Hegel calls the distinction of the poles a "real difference" in contrast to the "ideal difference" of the powers. This second difference is even more remarkable than the first for "when the mind, which is one of these powers, produces itself in abstraction from space, it is time, but inasmuch as it refers this production of itself to space, it constitutes the line". Having produced the line, the mind gives rise, apparently with equal ease, to the square and the cube. All this unfolds by virtue of an organic necessity that Hegel considers uncontroversial.

The dissertation ends with a critique of scientific research that is guided by mere empirical data. The example that Hegel chooses is Bode's law, an empirical correlation for the distances of the planets in the solar system whereby the number 4 is added to each term of the series 0, 3, 6, 12, 24, 48, 96, 192 and the results are divided by ten. The outcome is surprisingly good for most planets as the following table shows:

PLANETS	ACTUAL	CALCULATED	BODE'S LAW
Mercury	0.39	0.4	4 + 0
Venus	0.72	0.7	4 + 3
Earth	1.00	1.00	4 + 6
Mars	1.52	1.6	4 + 12
		2.8	4 + 24
Jupiter	5.20	5.2	4 + 48
Saturn	9.54	10.0	4 + 96
Uranus	19.19	19.6	4 + 192
Neptune	30.07	—	—
Pluto	39.57	38.8	4 + 384

When Uranus was discovered by Herschel in 1781 it was found to occupy the eighth position in the series, and this encouraged astronomers to look for the "missing" planet between Mars and Jupiter. In 1800, the German astronomer von Zach organized a systematic quest for the planet that should have been observed at 2.8 AU from the sun. Twenty-four astronomers joined in the search and it is this collective endeavour that Hegel ridicules on the grounds that an accidental sequence of numbers has nothing to do with the rational order of the universe. The astronomers would do better to follow the lead of Plato and the Pythagoreans, and Hegel quotes the series in the *Timaeus*, 1, 2, 3, 4, 9, 8, 27 (which he alters to read: 1, 2, 3, 4, 9, 16, 27 where 16 replaces 8) as a much more felicitous ordering. It has the two following advantages: first, it is not a mere arithmetical progression, and, secondly, the fourth and the fifth terms are separated by a wider distance, thus clearly indicating that there is no planet to be sought between Mars and Jupiter.

Hegel may have been indulging in heavy-handed irony, but the quest for the missing planet had begun to yield results even before he started writing his dissertation. On 1 January 1801, Piazzi, in Palermo, discovered a new star in the constellation of Taurus and on subsequent evenings he noticed that it changed its position. He imparted this information to Bode in Berlin who immediately conjectured that it was the elusive planet. The young mathematician Gauss set himself the task of computing its orbit and on the strength of his computations Olbers tracked it down on 1 January 1802, exactly a year after Piazzi. A second similar object was discovered only three months later. Olbers advanced the hypothesis that they were the fragments of a common planet since their neighbouring orbits were at 2.67 AU from the sun, close to the value of 2.8 AU determined by Bode's Law. Two more such bodies, christened asteroids by Herschel were discovered in 1804 and 1807 respectively. Hegel was an assiduous newspaper reader and he did not fail to come across the surprising news. He does not seem, however, to have considered it a rebuttal of his speculative philosophy of nature. When he published his *Philosophy of Nature* in 1817, he graciously admitted the existence of all four known asteroids, but he could not resist another – if somewhat more tentative – speculative fling. Abandoning his Pythagorean series, he merely classified the planets into three groups: the four inner planets which have no satellites or, like the Earth, only one, form the first group, the asteroids form the second, while the third group is made up of the three outer planets, all of which have many satellites or rings like Saturn. This new classification was to meet the same fate as Hegel's first attempt when Asaph Hall sighted two satellites of Mars in 1877. Hegel had died in 1831 and was thus spared the shock of this discovery. But would it have been a shock? We can doubt it. Hegel had constructed for himself a system that was immune to empirical falsification. When he met Bode several years after the publication of *The Orbits of the Planets* he betrayed no embarrassment. He merely welcomed the opportunity of stating his belief in the flimsiness of empirical generalizations:

I once made Mr. Bode sigh by saying that we know from experience that comets are followed by good vintage years, as was the case in 1811 and 1819, and that this double experience was just as good, if not better, than that regarding the return of the comets. What makes the comet-wine so good is that the water-process abandons the Earth, and so brings about an alteration in the state of the planet [Petry 1970, vol. III, pp. 29–30].

Well may Bode have sighed! We should not, however, assume that Hegel's interpretation of physics as the manifestation of an underlying organic reality has altogether vanished from science. The paradoxes of quantum mechanics have led some scientists to embrace an hypothesis that is strangely akin to Hegel's conjectures. In a modern journal, *Foundations of Science*, we can read as recently as 1971 that:

The assignment of a rudimentary degree of consciousness to atoms and the fundamental particles of matter not only offers an explanation of the quantum wave properties and strange behaviour of those particles, but it also explains how living organisms can result from aggregation of such particles, thus relating living organisms to so-called inanimate matter. Previous concepts explain neither the existence of living organisms nor the peculiar features of quantum physics.

Just as the concept of atoms and particles with no degree of consciousness was consistent with the basic features of classical physics, the new hypotheses are consistent with the basic features of quantum mechanics. The quantum laws are strange only by classical standards; when considered in a biological context, their strangeness finds explanation [Cochran 1970–1971, p. 248].

Whatever one may think of the tenability of such an hypothesis, it has the advantage of disclosing what was at the root of Hegel's dissatisfaction with mechanism, namely its inability to relate the inorganic world to the larger phenomenon of life. Fortunately, we need not be driven to Hegel's or Cochran's animism if we will but turn to volume 4 of Mario Bunge's *Treatise on Basic Philosophy* where a systemic world view is shown to answer most of the rational questions that can be raised about the emergence of vital properties from non-conscious matter.

McGill University

NOTES

* I wish to thank Prof. François de Gandt who has recently published an excellent translation of Hegel's dissertation (G.W.F. Hegel, *Les orbites des planètes*, Vrin, Paris,

1979), and Prof. Jonathan Robinson, with whom I am preparing an English translation of this work, for their scholarly advice. I am also grateful to the Social Sciences and Humanities Research Council for their generous support of my research.

[1] Petry [1970], vol. I, p. 272 (par. 270 of the *Enzyklopädie der philosophischen Wissenschaften*). This excellent translation is accompanied by a set of very useful notes about works that Hegel consulted.

[2] Goethe [1910–1932], vol. XXXIX, p. 92, cited in Jaki [1970], p. 45.

[3] This important distinction is lost in the Motte-Cajori translation where the Latin runs: "Eorundem quadrata sunt etiam ultimo ut sunt arcuum sagittae, quae chordas bisecant et ad datum punctum convergunt. Nam sagittae illae sunt ut subtensae *BD, bd*" (Cor. II of Lemma XI), the English has: "Their squares are also ultimately as the versed sines of the arcs bisecting the chords, and converging to a given point. For those versed sines are as the subtenses BD, bd". I am grateful to Prof. De Gandt for pointing this out to me.

[4] Petry [1970], pp. 266–267 (par. 270 of the *Enzyklopädie*).

[5] Petry [1970], p. 241 (par. 262 of the *Enzyklopädie*).

[6] Hegel [1928], pp. 360–362; Petry [1970], p. 243 (par. 262 of the *Enzyklopädie*).

[7] Lagrange [1797], cited by Petry [1970], p. 337.

[8] Hegel [1928], p. 348; Petry [1970], p. 242 (par. 262 of the *Enzyklopädie*).

[9] Hegel [1928], p. 348; Petry [1970], p. 262 (par 269 of the *Enzyklopädie*).

REFERENCES

Bunge, M.: 1979, *Ontology II: A World of Systems*. (His *Treatise on Basic Philosophy*, vol. 4). D. Reidel, Dordrecht.

Cochran, A. A.: 1970–71, 'Relationships between Quantum Physics and Biology', *Found. Phys.*

Goethe, J. W. von: 1909–1932, *Sämtliche Werke*, Müller, Munich.

Hegel, G. W. F.: 1928, *Erste Druckschriften*, G. Lasson (ed.), F. Meiner, Leipzig.

Jaki, S. L.: 1970, *The Relevance of Physics*, Chicago University Press, Chicago.

Kant, I.: 1957, *Metaphysische Anfangsgründe der Naturwissenschaft*, zweites Hauptstück, Erklärung 2, Zusatz, in Immanuel Kant, *Werke*, vol. V, W. Weischedel, ed., Insel, Wiesbaden.

Kepler, J.: 1940, *Harmonices Mundi*, in Kepler, *Gesammelte Werke*, vol. VI, M. Caspar (ed.), C. H. Beck, Munich.

Koyré, A.: 1965, *Newtonian Studies*, Harvard University Press, Cambridge.

Lagrange, J. L.: 1797, *Théorie des fonctions analytiques*, Paris.

Newton, I.: 1971, *Principia*, Motte's tr., rev. by F. Cajori, University of California Press, Berkeley.

Petry, M. J. (ed.): 1970, *Hegel's Philosophy of Nature*, vol. I, George Allen and Unwin, London.

Turnbull, H. W. (ed.): 1961, *The Correspondence of Isaac Newton*, vol. III, Cambridge University Press, Cambridge.

Whewell, W.: 1849, 'On Hegel's Criticism of Newton's *Principia*', in *Trans. Cambridge Phil. Soc. VIII* 698. (Reprinted in his *Philosophy of Discovery*, London 1860, as Appendix H.)

ROBERTO TORRETTI

THREE KINDS OF MATHEMATICAL FICTIONALISM*

> Dios sabe si hay Dulcinea o no en el mundo, o si es fantástica, o no es fantástica; y éstas no son de las cosas cuya averiguación se ha de llevar hasta el cabo. Ni yo engendré ni parí a mi señora, puesto que la contemplo como conviene que sea una dama que contenga en sí las partes que puedan hacerla famosa en todas las del mundo.
>
> *Quijote, II, xxxii.***

Mario Bunge's *Ontology*[1] has revived the program of seventeenth century continental European philosophy for furthering the study of being *qua* being with the aid of exact mathematical concept building. Undeterred by Kant's warnings against the use of the mathematical method in philosophy,[2] Bunge feels free to resort to the entire established repertory of mathematical structures in his search for an adequate representation of the basic structures of reality. Abstract algebra turns out to be particularly helpful – a fact that will not surprise us if we recall that Leibniz's pioneering work in the field was done primarily in the service of ontology (and not merely of logic, as some twentieth-century commentators would have it).[3] But while in Leibniz's philosophy mathematics can be regarded as a part of ontology, for, as the abstract theory of *possibilia*, it spells out the general framework with which every theory of *realia* must agree, Bunge deliberately excludes mathematics from ontology and subscribes to the thesis that the subject matter of pure mathematics is totally fictitious [1974–, I, 27]. Bunge writes:

> We feign that there are constructs, i.e., creations of the human mind to be distinguished not only from things (e.g., words) but also from individual brain processes. (Only, we do not assume that constructs exist independently of brain processes.) We distinguish four basic kinds of constructs: concepts, propositions, contexts and theories. (. . .) Concepts are the building blocks of propositions. (. . .) A concept is either an individual (. . .) or a set (. . .) or a relation (. . .). The most interesting relations are the functions. (. . .) A context is a set of propositions sharing a reference class, and a theory is a context closed under the operation of deduction. [1974–, III, 116 f]

All constructs revolve therefore about propositions, of which Bunge says, in another passage:

J. Agassi and R.S. Cohen (eds.), Scientific Philosophy Today, 399–414.
Copyright © 1981 *by D. Reidel Publishing Company.*

We do not claim that they exist in themselves but only that it is often convenient (for example in mathematics but not in metaphysics) to feign or pretend that they do. We do not assert that the Pythagorean theorem exists anywhere except in the world of phantasy called 'mathematics', a world that will go down with the last mathematician. [1974–, II, 85]

Thus Bunge, a staunch realist in the philosophy of physics, embraces fictionalism in the philosophy of mathematics. This combination is probably unprecedented, though Henri Poincaré may have been groping after it.[4] It is understandable in someone who, like Bunge, looks on Platonism as a "weird metaphysical hypothesis" [1974–, II, 85] and rejects or doubts the contention of objective idealism, that "concrete objects (things) have (...) intrinsic conceptual properties, in particular (...) mathematical features" [1974–, III, 118]. But mathematical fictionalism can be variously understood and, as I intend to show, not every conceivable version of it is equally at variance with Platonism. For greater precision I shall propose three interpretations of the thesis that mathematics is concerned with fictions. I shall then try to ascertain which of the three agrees better with the actual workings of mathematics, pure and applied. We shall see that the form of mathematical fictionalism that meets this condition, and which I take to be the one actually held by Bunge, does not differ from, say, Gödel's mathematical Platonism [1964] by much more than a change of emphasis. In my discussion, I shall take physical realism for granted; that is, I shall *call* real the things we handle, eat, perform physical experiments with, etc., and whatever can be said to interact with them, and I shall *profess* that mathematical physics has much to teach us concerning real things. In agreement with recent formulations of physical realism [cf. Armstrong 1978; also Bunge 1974–, III, 104ff], I shall assume that the properties and relations of physical things are no less real than the things themselves.

If someone says that we feign that there are constructs, to which our mathematical statements refer, he will most naturally be understood to mean that we produce objects in our fancy, which thereafter constitute definite, stable, albeit ghostly referents of our discourse. It is in this sense that some atheists maintain that there is a fantastic being, created by men in their own likeness, to whom the majority of mankind refer in their persistent talk of God. It is in this same sense that literary authors are usually said to create the characters about which they write in their books. However, the idea of creation, even if only of fictions, is wrought with such ontological difficulties that to many philosophers the ordinary understanding of fiction has seemed untenable. Thus, with regard to God, they would have us choose between Descartes' argument that God must exist, given that we can think of Him, and the positivist tenet that theology is nonsense. Likewise, they find it difficult to accept that such phrases as 'Pegasus, the flying horse' or 'the Mad Hatter's favourite tea blend' might refer to anything, fictitious or not. They prefer therefore to understand the feigning of referents for these and other kinds of discourse on the analogy of the feigning of feelings and actions. Just as one can feign sorrow while inwardly rejoicing, or feign selling a property one is merely placing under someone else's name, surely one can feign that he is referring to something when there is nothing in fact he could be speaking about. An understanding of mathematical discourse along some such lines has had many adherents in the twentieth century. Bunge, however, is not one of them. He dismisses them as "the literalists," who will not accept what to everybody else is clear, namely, "that mathematical symbols designate mathematical constructs" [1974–, II, 166]. There is still a third conceivable style of fictitious reference, illustrated by the familiar case where one actually refers to something real, but feigns it is other than it is. Thus, in fleet maneuvers the artillery officer commands his men to shoot at the enemy position, which is actually an empty beach. Under a dictatorship, a writer barely disguises his criticism of the government by telling a tale seemingly about a Babylonian king. When brought to bear on mathematical discourse these three senses of feigning yield three versions of fictionalism, which I shall call fictionalism$_1$, fictionalism$_2$, and fictionalism$_3$, respectively.

Let us first consider fictionalism$_3$. This appears to be a very suitable framework for describing the common procedures of applied mathematics. When asked to calculate the period of a given pendulum at a given place we feign that the pendulum hangs from a weightless, inextensible string, and that the sinus of the angle of displacement is equal to the angle itself, and rounding up the values of the pendulum's length and the local acceleration of gravity, we compute 2π times the square root of their quotient to an agreed decimal. From our fictitious assumptions we derive a result which is admittedly false, but which will differ from the measured values of the period by less than a prescribed number. At a less modest level, the cosmologist feigns that matter is homogeneously distributed in the universe and figures out the broad patterns of its evolution, although he knows all too well that matter is not equally distributed inside a pulsar and in intergalactic space. As the saying goes: one 'idealizes' reality in order to obtain an approximate yet manageable picture of it. Turning to pure mathematics, I find that fictionalism$_3$ may have inspired some of the mathematicians who in the nineteenth century carried out the arithmetization of analysis. As a result of their labors anyone speaking, say, of a complex number could henceforth be regarded as actually referring not to some new-fangled esoteric individual entity, but 'merely' to an ordered pair of infinite classes of classes of classes of natural numbers. But then, of course, a truly hardnosed fictionalist would observe that even the natural numbers do not really exist, at any rate, not the larger ones, nor the so-called set of them. Fictionalism$_3$ was probably also at the back of Hilbert's mind when he proposed to show that the mathematical infinite was just a manner of speaking, something to which the mathematician merely seemed to refer, when he was really reasoning only about the finite. On the other hand, the actual execution of his program for proving that this is so suggests rather a fictionalist$_2$ approach to classical mathematics – which is viewed as a lawful system of strings of meaningless symbols – coupled with a downright realist view of the new discipline of metamathematics, the theory of such lawful systems. Anyway, the impossibility of Hilbert's program makes the exact classification of his position less important to us than it would otherwise be. I have mentioned him only because, as I hope to make clear in the sequel, that very impossibility implies that fictionalism$_3$ is not a viable philosophy of classical mathematics.

The only passage I have found where Bunge appears to countenance fictionalism$_3$ is not in the main text of his *Treatise* but in a table of the chief views on signs and constructs [1974–, I, 31]. In the line summarizing his own position he enters the following equation:

Construct = Equivalence class of brain processes.

I take it to mean that this is what constructs, and hence mathematical objects, *really are*, though we *feign* them to be all sorts of things, quaternions, ultrafilters, Lie groups, inaccessible cardinals, and what not. Now, in the light of what we know about the structure of matter at the level relevant to brain processes, it should be evident that the number of significantly distinct brain states, including *possible* brain states and even making allowance for genetic mutations, must be less than some very large but finite integer. Since each kind of brain process is a transition or a finite sequence of transitions between brain states, their number is finite too. This means that all the mathematics that can be built by taking equivalence classes of brain processes – whether one feigns or not that they are something else – is comprised within combinatorics. Hence the undeniable philosophical importance of Hilbert's attempt to recreate Cantor's Paradise in a combinatorial setting, so to speak as a wall-paper garden or *coulisse*. Hence too the importance of Gödel's proof that not even elementary (i.e., non-analytic) number theory will fit into such a narrow setting. Generally speaking, it should be clear that not only the different kinds of human brain processes, but every kind of physical object available to man, will not suffice to provide all the objects that classical mathematics appears to speak about. Therefore fictionalism$_3$ is not a reasonable account of classical mathematical discourse. Consequently, if its objects do not exist of themselves as Platonic *noetà* – as Gödel himself believed – or in the mind of God, then *either* they are feigned through and through by the creative powers of the human intellect, which, to achieve this, must indeed be formidable, *or* mathematical discourse actually lacks referents and is a case of feigned referring. These are the theses I labelled fictionalism$_1$ and fictionalism$_2$, respectively. Let us take them up in this order.

Before going further into the matter I wish to remark that if mathematical objects are fictitious throughout, as fictionalism$_1$ maintains, there is no need to feign besides that they are brain processes, especially if one will end up by feigning that they are not. If we can feign them at all, then we can feign them to be what we want them to be, namely the elements, properties or relations of this or that mathematical structure, altogether clean of every psychic or encephalic connotation. This is not to deny that to think of them is indeed a mental process. But one must not confuse the act of feigning with what is feigned by it. The latter not only outlives the former, but may also be arrived at in many different ways. Thus, though in the present state of neurology we can say very little about such things, it does not seem at all likely that there was a great resemblance between, say, the brain processes of Ricci-Curbastro and those of Elie Cartan, when they worked on the development of modern differential geometry. Moreover, in stark contrast with the objects of ordinary imagination, which, according to some psychologists, can exhibit no more features than the act of imagining them bestows on them [cf. Sartre 1940], the fruits of mathematical fiction usually harbor surprises their creators never dreamt of. Suppose, for instance, that we feign – as indeed, according to fictionalism$_1$, someone must once have feigned – a standard model of first order arithmetic. For greater precision, let us assume that the model is ω, the set of finite ordinals, with addition and multiplication defined in the usual way. (The successor Sn of a finite ordinal n is $n \cup \{n\}$. The empty set $\emptyset$ is the only finite ordinal which is not the successor of another one, and may be denoted by 0. If m and n are finite ordinals, $m + 0 = m$; $m + Sn = S(m + n)$; $m \cdot 0 = 0$; $m \cdot Sn = (m \cdot n) + m$.) Given ω, one can build out of it a first order language, some subset of whose sentences is the axiomatic theory of first order arithmetic. (Take any system of Gödelization of first order syntax and identify the Gödel numbers with the expressions they stand for.) It can then be shown that if this theory has any model at all – and ω itself was contrived just for this role, in our hypothesis – it possesses a denumerable non-standard model, that is to say, a denumerable model not isomorphic with ω. (Let $\langle M, 0, S, +, \cdot \rangle$ be such a model; there are then many bijective mappings of ω onto M, but none of them are structure-preserving.) Thus by the very act of feigning our intended standard model of arithmetic we conjure up an unintended model. There is no magic in this, for the unintended model is denumerable and hence can be drawn entirely from the domain of the intended one. But it does confirm what I remarked about the undreamt-of surprises concealed in mathematical fiction, for, as is well known, non-standard models of arithmetic exhibit some shocking features. Let us say that an element a stands in *the line of succession* of an element b if a can be reached from b by compounding the successor function a finite number of times. (I.e., if $a = SS \ldots Sb = S^n b$.) Two elements will be said to belong to the same *dynasty* if they are identical or if one stands in the line of succession of the other. It is clear that belonging to the same dynasty is an equivalence which partitions any model of arithmetic into equivalence classes or dynasties. But while a standard model consists of single dynasty, a denumerable non-standard model comprises infinitely many of them, each of which has infinitely many members. Only one of them, the dynasty of zero, is a submodel isomorphic with ω, and hence a standard model. Every model of arithmetic is totally ordered by the relation '$<$' ($a < b$ if and only if $(Ex)(a + Sx = b)$). This order induces a total order in the set of dynasties of a denumerable non-standard model: one dynasty is less than another if a representative of the former is less than a representative of the latter. (It can be shown that this relation does not depend on the choice of the representatives.) The set of dynasties is dense in this order: if F and G are dynasties such that $F < G$, there is a dynasty H such that $F < H < G$. If fictionalism$_1$ is right, whoever first devised a standard model of arithmetic unwittingly feigned this monster as well.[5]

A different example may give us a firmer grasp of what I dare to call the substantive objectivity of mathematical structures. I propose to consider the structure underlying the game of chess. We shall view it as a finite directed graph,[6] the Graph of Chess, which can be constructed as follows. First take all injections of the 32 pieces of chess or of any subset thereof into the 64-square chessboard. Identify any two injections that can be obtained from one another by interchanging the values (i.e., the positions) of two equal pieces of the same color. The objects thus obtained we call *points*. In particular, the injection that yields the standard initial position of the game we call the starting-point. Take then every conceivable legal chess move. Each can be said to join a pair of points, namely, the point yielding the position at which the move begins and the point yielding the position at which the move ends. We call each such move an *edge*, which we regard as directed from the first point towards the second. We call an edge 'white' if it is a move of the whites; otherwise we call it 'black'. A path consisting of alternately black and white edges will be called a chequered path. The graph thus obtained is certainly finite, but it contains much more than we need. Thus, the injections of sets of pieces that do not contain both kings are isolated points of the graph which we may just as well ignore. But the Graph of Chess we are looking for is a subgraph of the graph we have built. It consists of (i) the starting-point and every point that can be reached from it by chequered paths whose first edge is white; and (ii) all the edges included in such paths. It is plain that this subgraph underlies the practice of the game and provides the necessary and sufficient basis for any theoretical study of it. The Graph of Chess is a definite, substantive entity. It can be embodied in an intricate fishing-net, with strings painted black and white and with the direction along each string given by the sense in which its threads are twisted. Indeed the graph is quite independent of the game itself. Except for the coloring of the edges and the privileged position of the starting-point, we may forget all about the game and consider the graph on its own. (In fact, it might even be unnecessary to distinguish the starting-point, if, as it is likely, it can be singled out by its very location in the graph - but I am not sure about this.) The different distributions of chess pieces on the chessboard can then be regarded as code-names for the points of this graph. As is often the case with code-names, they do not merely label the points but store information about them, specifically about their connections in the graph. Now, we will probably all agree that chess is a human invention. But was the Graph of Chess *created* by the inventor of the game? Should we not rather say that he *discovered* it? The Indian mathematician who, according to the legend, charged his king $2^{64} - 1$ grains of wheat for the invention of chess, probably had no inkling of graph theory. Yet he lighted on the marvelous idea of turning a fixed mathematical structure, which we now know can be aptly conceived as a graph, but of which nobody has ever had a complete view, into the setting of a game, in which two players, exploring the structure's labyrinthine paths, choosing by turns at each crossing which way to follow, seek to drive one another into a blind alley. Since the Graph of Chess, like every finite graph, is isomorphic with a graph in space, it is the structure of a possible physical object – such as the fishing-net I mentioned earlier – and it is therefore a substantial universal, in Bunge's sense [1974-, III, 105]. But if, as fictionalism$_1$ maintains, classical mathematical discourse actually succeeds in referring to the structures it appears to speak about, then these structures are not a whit less objective and substantive than the Graph of Chess, and they are comprised, together with it, in the total furniture of the universe.[7] Whether they are Platonic *eidoi*, or divine thoughts, or figments of human fancy, is but of little consequence in this regard. For the fictionalist$_1$ who upholds the last alternative thereby simply ascribes to men the power to bring forth and sustain precisely the same kind of determinate, changeless, invisible and untouchable, yet thoroughly intelligible entities that Plato placed in the *hyperouranios topos* and his medieval disciples in God's understanding. The metaphysical problems generated by the admission of such entities are not lessened but rather heightened by the fictionalist's choice, for his philosophy must not only account for the existence of the Platonic realm but must also trace its origin right down to us humble creatures. It looks therefore as if the only way of avoiding these difficulties and yet retaining classical mathematics were that of fictionalism$_2$. No wonder then that so many have followed it in our un-Platonic century.

Some of the strongest pronouncements is favor of fictionalism$_2$ are to be found in Wittgenstein's posthumous *Bemerkungen über die Grundlagen der Mathematik* [1964]. Thus, one could hardly think of a more uncompromising statement of this position than the following 'explication' of the term *ideal object*:

'The sign *a* designates an ideal object' should obviously say something about the meaning and hence about the use of *a*. And it says of course that this use is in a certain respect similar to that of a sign that has an object, *and* that it does not designate an object. [Wittgenstein 1964, p. 136]

Coupled with Wittgenstein's radical conventionalism in logic, his conception of mathematics as a "motley of techniques of proof" [1964, p. 84], to be learned by acquiring "conditioned calculating reflexes" [1964, p. 96] implies that mathematical discourse is a parrot's chatter, that not merely lacks reference, but is also empty of all thought. Indeed, in his opinion, the latter property must go with the former, for, as he puts it,

> What does it mean to obtain a new concept of the surface of a sphere? How is it then a concept of the surface of a *sphere*? Only in so far as it can be applied to real spheres. [1964, p. 134]

Wittgenstein's vision is, to my mind, so far removed from the reality of mathematics, except perhaps as one might encounter it in some dreary high-school or freshman course, that I do not think one ought to center upon this vision a discussion of fictionalism$_2$. I turn therefore to a more sensible version of the latter, which is sometimes disparagingly referred to as 'if-thenism' [see Musgrave 1977]. I take it that Rudolf Carnap adopted some such view after meeting Tarski in 1935. The if-thenist philosopher acknowledges the absolute value of logic and conceives of mathematical statements as logically true implication statements. E.g., 'If an object exemplifies such and such a structure, then it possesses this or that feature'. There need not be any such object for the statement to be true and interesting. Take a specific example. If in Hilbert's axioms of geometry we let X, Y and Z stand for the three arbitrary sets Hilbert calls 'points', 'lines' and 'planes', and we allow the letters R_1, R_2, R_3, R_4, R_5, to represent the five primitive relations of incidence (2), betweenness and congruence (2), we may define a Euclidean space as an octuple $\langle X, Y, Z, R_1, R_2, R_3, R_4, R_5\rangle$ that satisfies Hilbert's axioms. Then, the statement that, if S is a Euclidean space, the sum of the interior angles formed by three lines in S that mutually intersect by pairs is equal to two right angles, is true and significant, even if no Euclidean space exists at all and the terms 'S', 'line', 'angle', lack a referent. Indeed, the truth of the statement would be beyond all question if no Euclidean space could possibly exist due to the inconsistency of Hilbert's axioms, though in such a case we would not regard the statement itself as being particularly significant. This last remark suffices to explain what Wittgenstein somewhat irresponsibly called "the superstitious fear and awe of mathematicians in face of contradiction" [1964, p. 53]. They are not afraid of being wrong but they are afraid of being trifling. Unfortunately, if-thenism is not in a very good position to allay those fears. For, barring the simpler mathematical theories that can be translated into sound and complete formal systems in which it is syntactically impossible to derive a contradiction, most mathematical conceptions can only be proved consistent by producing a model of them, which, of course, if it is not a physical object, must be either a Platonic entity or a substantive fiction, in the sense of fictionalism$_1$. (Alternatively, one may produce a model for another, more likely conception, with the aid of which then a formal system spelling out the conception being discussed can be shown to be syntactically consistent.) This circumstance, however, though it may make if-thenism unpopular with working mathematicians, does not constitute a refutation of it. Pure mathematics can be carried on while the researcher blindly trusts that the theory he is busy with is satisfiable, even if there are no real or ideal or fictitious entities presently available to satisfy it. If the mathematician's faith turns out to be wrong, that will not ruin the truth of his conclusions, but only their relevance. The following difficulty is harder to overcome from the if-thenist point of view. The property of logical truth and the relation of logical consequence, which, if if-thenism is right, are all that mathematics is really about, have so far been explained satisfactorily, with the breadth and scope required by classical mathematics, only within the framework of Tarskian semantics. Now Tarskian semantics resorts to a naive understanding of the set-theoretical predicates 'to be a set' and 'to be a member of' which can only be substantiated by providing a model of a sufficiently rich version of set theory. In the particular case of these basic concepts, on which the meaningfulness of all further mathematical concept-building is made to rest by if-thenism, one ought therefore to say after Wittgenstein: "How are they then concepts of a *set*, of *set-membership*? Only in so far as they can be applied to real sets." It could also be argued against the if-thenist version of fictionalism$_2$ that the methods of proving universal statements of implication in mathematics normally involve instantiation, whereby reference is actually made to some mathematical object, either real, ideal, or fictitious. But I shall not press these points, for I find that the insufficiency of if-thenist fictionalism$_2$ can be laid bare once and for all by reflecting on some well-known methods of mathematical physics.

For the if-thenist it is only in applied mathematics that the whole enterprise of mathematicians finally comes to fruition. The antecedents of some of the implications proved here at last acquire referents of which they, and hence also the respective consequents, are true. On our assumption of physical realism – which, by the way, Carnap did not share – the statements of applied mathematics must be regarded as fully interpreted, at least in so far as they are concerned with reality. What matters to us now is that not all of them are exclusively concerned with it. In the cases I have in mind, which include the familiar applications of the calculus of variations in classical mechanics and in other branches of physics, a real physical object – a thing, state, process, etc. – is considered together with a multitude of unreal physical objects of the same kind and incorporated with the latter into a mathematical structure, the intrinsic properties of which are then used to single out the real object among the unreal ones (e.g., because it is the only path in the structure on which a certain integral takes an extremal value, etc.). I contend that if-thenist fictionalism$_2$ cannot account for cases like these. For suppose the if-thenist philosopher tried the usual gambit and claimed that the applied mathematician is not actually referring to the structure in question, consisting of a unique real and a multitude of unreal elements, but is merely stating the following logical true implication, which does not involve such a reference: '*If* such-and-such a structure existed, *then* it would contain those objects and the real one would hold such-and-such a peculiar position amidst the unreal ones'. In claiming this, the if-thenist would plainly misconstrue the applied mathematician's argument, the very upshot of which is that, on his assumptions, only the real object can actually exist, while all its near relatives within the structure are in effect impossible. The applied mathematician cannot therefore be understood to mean that if the structure considered by him did exist such and such consequences would ensue, for he is saying at the same time that the said structure cannot really exist at all. But in order to say so much he must succeed in referring to that structure and to the whole bunch of its unreal elements. He cannot just feign that he is referring to them. We may readily grant, indeed, that he does not actually refer to the real object, or, at any rate, that he does not refer to it directly, but only through an ideal or fictitious representative of it, which, as such, finds its natural place in the neighborhood of the unreal elements of the structure. (Thus the actual motions of a complex mechanical system can be represented by a simple path in configuration space.) But to that representative he must perforce refer, for it is only by being referred to that it can come to stand for anything. Here we face, therefore, a family of cases in which either there is no sense at all or actual reference is being made knowingly to unreal objects. It is clear that such cases can only be accounted for by Platonism – which bestows a meta-physical reality on the physically unreal objects referred to – or by fictionalism$_1$, and that they lie altogether beyond the reach of fictionalism$_2$.

The following example, drawn from S.W. Hawking and G.F.R. Ellis' book, *The Large-Scale Structure of Space-Time* [1973], will illustrate the point I want to make. It will also throw light on the relationship between this procedure of referring, in mathematical physics, to acknowledgedly fictional or ideal objects, and the procedure we described earlier, when talking about fictionalism$_3$, by which a real physical object is feigned to be other than it is and is pictured in an 'idealized' form. Hawking and Ellis assume that the physical world is or can to a good approximation be represented by a real four-dimensional non-compact Hausdorff C^∞-differentiable manifold, endowed with a Lorentz metric and with further geometrical objects corresponding to its diverse physical properties. Such a manifold is called a *spacetime*. A Lorentz metric on a spacetime M partitions all the tangent vectors at each point of M into the three mutually exclusive classes of spacelike, null and timelike vectors. A curve in M is said to be timelike if all vectors tangent to it are timelike; null, if all vectors tangent to it are non-zero and null; and non-spacelike if all its tangent vectors are either timelike or null and non-zero. Any spacetime is supposed to admit an everywhere timelike vector field, that makes it possible to classify all non-spacelike vectors at each point into future-directed and past-directed vectors. This classification applies also to the curves to which such vectors are tangent. It is also assumed that all transfers of energy and momentum take place along non-spacelike curves. In other words, all signals, all causal actions propagate along such curves. The authors assume moreover that the Lorentz world metric is bound to the distribution of matter by the Einstein field equations. Now, Gödel has shown that a spacetime whose metric is governed by the Einstein field equations can under certain conditions contain closed timelike curves [1949]. Hawking and Ellis regard the existence of such curves in the spacetime representing the physical world as paradoxical,

for one could imagine that with a suitable rocketship one could travel round such a curve and, arriving back before one's departure, one could prevent oneself from setting out in the first place. Of course there is a contradiction only if one assumes a simple notion of free will; but this is not something which can be dropped lightly since the whole of our philosophy of science is based on the assumption that one is free to perform any experiment. [1973, p.189].

To avoid such paradoxes the authors assume that the world fulfils three further conditions, namely, the *chronology condition*, that there are no closed timelike curves; the *causality condition*, that there are no closed non-spacelike curves; and the *strong causality condition*, which holds in a spacetime M if at every point P in M, every neighborhood of P contains a neighborhood of P which is not traversed by any given non-spacelike curve more than once. However, even the strong causality condition does not completely rule out all causal paradoxes, for a spacetime can be, so to speak, on the verge of violating the chronology condition in that the slightest variation of the metric would generate closed timelike curves. Hawking and Ellis do not countenance such a situation as physically viable because Einstein's Theory of Gravitation is "presumably the classical limit of some, as yet unknown, quantum theory of space-time and in such a theory the Uncertainty Principle would prevent the metric from having an exact value at every point" [1973, p. 197]. This difficulty can be overcome by assuming that the chronology condition is *stable*, in the sense that it is shared by all spacetimes that differ but slightly from the real one, and are, so to speak, its neighbors. In order to give an exact meaning to this notion of neighboring spacetimes one must define a topology on the set of all spacetimes. The authors do not tackle the problem of uniting in one connected topological space spacetimes with incompatible topologies – which, according to them, can be done – but they only consider the definition of a specific topology on any set of diffeomorphic spacetimes that differ only in their Lorentz metrics. We need not worry about the details of their definition [1973, p. 198]. It is enough to mention that a given spacetime is said to meet the *stable causality condition* if it has an open neighborhood in this topology such that every spacetime belonging to that neighborhood fulfils the chronology condition. The authors prove the following, cosmologically significant, theorem: A spacetime M meets the stable causality condition if and only if there is a real-valued function on M whose gradient is everywhere timelike. Any such function can be viewed as a cosmic time function, for either it or its product by minus one increases monotonically along every future-directed non-spacelike curve. On the other hand, any reasonable cosmic time function must have an everywhere timelike gradient. The authors have therefore succeeded in determining the necessary and sufficient condition for the existence of a cosmic time by means of a notion, the condition of stable causality, which can only be conceived of by inserting the real world, or if you please, its ideal or fictional mathematical representative, into a non-denumerable set of unreal worlds that resemble it, and by regarding them all as the points of a connected topological space. It is plain that this construction can only make sense if one actually succeeds in referring to the space of spacetimes and to its several points. It will be noticed, moreover, that Hawking and Ellis' move looks a good deal bolder that the familiar talk of possible worlds, say, in formal semantics, since the several spacetimes are not just collected in a set but are tightly knit together in a topological structure. And yet a short reflection should persuade us that some such move is implicit in the trite method of idealization we discussed at the outset, in connection with fictionalism$_3$. In that method one 'approximates' a real situation by an unreal one that is mathematically more manageable. Now the concept of approximation can only be given a clear meaning in a topological context. To view a given ideal physical system A as an approximate representation of the real system B is tantamount to conceiving of A and B as neighboring points in a suitable topological space. Such a conception is always tacitly but quite definitely implied in the mathematical handling of even the simplest physical situations. In the more familiar cases, one can approximate the real B by the ideal A only in as much as both systems are viewed as determined by the same parameters $x_1, \ldots, x_n$, which take the definite real values $a_1, \ldots, a_n$ at A, while taking at B other, indefinite, but neighboring values $b_1, \ldots, b_n$ (with $a_i - \sigma_i < b_i < a_i + \sigma_i$ for some small positive real number σ_i; $1 \leq i \leq n$). A and B are thereby treated as neighboring points in $\mathbf{R}^n$. In other, less simple cases, the topological space involved is more unusual—as in the above example from relativistic cosmology—but no less essential. Since approximation by idealization has been up to now the surest, indeed the only unquestionably successful way of increasing our understanding of the world, one might venture to conclude that the intellection of real things can only be achieved by finding the right place for them in a setting of *intelligibilia*. Whether the latter subsist eternally, as Platonists teach, or are man-made, as fictionalism$_1$ would have it, is not, I dare say, a question of much genuine philosophical moment. Indeed, shifting from one of these two views to the other does not imply or demand a greater change in the experienced content of the matter than does, say, a *Gestalt* switch. The foregoing discussion suggests a final remark regarding the claim, apparently made by Bunge, that real things do not actually possess 'mathematical features'. (See the quotation from [1974–, III, 118] on p. 400 above.) For aught I know, this claim may be correct, but unless it is wrong one cannot meaningfully contend that mathematical physics provides an approximate representation of physical truth, let alone that it converges to it; for, as we have seen, a mathematical object A can be said to approximate an object B only if the latter belongs together with the former in the same mathematical structure.

Universidad de Puerto Rico

NOTES

* A slightly different version of this paper was read, under the title 'Mathematics, Fictionalism and Ontology', at the meeting of the Society for Exact Philosophy held in Montreal on June 4–5, 1979.

** "God knows whether or not there is a Dulcinea in this world or if she is a fanciful creation. This is not one of those cases where you can prove a thing conclusively. I have not begotten or given birth to my lady, although I contemplate her as she needs must be, seeing that she is a damsel who possesses all those qualities that may render her famous in all parts of the world . . ." (Cervantes 1949, p. 723).

[1] I refer specifically to volume 3 of his [1974–], 7 vols., of which the first three had appeared when this was written.

[2] Kant, *Untersuchungen über die Deutlichkeit der Grundsätze der natürlichen Theologie und der Moral* (1764); *Kritik der reinen Vernunft*, Transzendentale Methodenlehre, I. Die Disziplin der reinen Vernunft (i) im dogmatischen Gebrauche, (iv) in Ansehung ihrer Beweise.

[3] See, in particular, the logico-ontological calculus developed in Leibniz [1890], pp. 236–247.

[4] On the meaning and purpose of Poincaré's conventionalism, see my [1978], pp. 320ff.

[5] For a proof of the above statements on non-standard models of arithmetic, see Boolos and Jeffrey [1974], pp. 192ff.

[6] We define a *directed graph* to be a pair $\langle V, E\rangle$, where V is any set and E is a subset of $V \times V$. The elements of V are the *points* or *vertices* of the graph; the elements of E are the *edges* of the graph. If $k = \langle x, y\rangle$ is an edge, k is said to *join* the points x and y, which are called the *endpoints* of k. A graph $G' = \langle V', E'\rangle$ is a subgraph of the graph $G = \langle V, E\rangle$ if $V' \subset V$ and $E' \subset E$. A *path* in a graph G is a sequence $k_1, k_2, \ldots, k_n$ of edges of G, such that, for every integer i $(1 \leq i < n)$, the second endpoint of k_i equals the first endpoint of k_{i+1}.

[7] For a forceful defense of fictionalism$_1$ from a working mathematician's standpoint, see R. Hersh [1979].

REFERENCES

Armstrong, D. M.: 1978, *Universals and Scientific Realism*, 2 vols., Cambridge University Press, Cambridge.

Boolos, G. F., and R. C. Jeffrey: 1974, *Computability and Logic*, Cambridge University Press, Cambridge.

Bunge, M.: 1974–, *Treatise on Basic Philosophy*, D. Reidel, Dordrecht.

Cervantes, M. de: 1949, *The Ingenious Gentleman Don Quixote de la Mancha*, trans. S. Putnam, Viking, New York.

Gödel, K.: 1949, 'An Example of a New Type of Cosmological Solution of Einstein's Field Equations of Gravitation', *Rev. Mod. Phys.* **21**, 447–450.

Gödel, K.: 1964, 'What is Cantor's Continuum Problem?', in *Philosophy of Mathematics. Selected Readings*, P. Benacerraf and H. Putnam (eds.), Prentice-Hall, Englewood Cliffs, N. J.

Hawking, S. W., and G. F. R. Ellis: 1973, *The Large-Scale Structure of Space-Time*, Cambridge University Press, Cambridge.

Hersh, R.: 1979, 'Some Proposals for Reviving the Philosophy of Mathematics', *Advances in Maths.* **31**, 31–50.

Kant, I.: 1764, *Untersuchungen über die Deutlichkeit der Grundsätze der natürlichen Theologie und der Moral*, Berlin.

Kant, I.: 1781, *Kritik der reinen Vernunft*, Hartknoch, Riga.

Leibniz, G. W.: 1890, *Die philosophischen Schriften von G. W. Leibniz*, vol. 7, C. I. Gerhardt (ed.), Weidmannsche Buchhandlung, Berlin.

Musgrave, A.: 1977, 'Logicism Revisited', *Brit. J. Phil. Sci.* **28**, 99–127.

Sartre, J.-P.: 1940, *L' Imaginaire*, Gallimard, Paris.

Torretti, R.: 1978, *Philosophy of Geometry from Riemann to Poincaré*, D. Reidel, Dordrecht.

Wittgenstein, L.: 1964, *Bemerkungen über die Grundlagen der Mathematik*, Blackwell, Oxford.

C. TRUESDELL

THE DISASTROUS EFFECTS OF EXPERIMENT UPON THE EARLY DEVELOPMENT OF THERMODYNAMICS

Nobody needs to be told that theories of nature must grow from experience, that experiment can sharpen, refine, and extend experience and correct our conception of it, that a scientific theory of an aspect of nature cannot be accepted until it has been somehow 'confirmed' by experiment. It does not follow that theory and experiment climb hand in hand up Jacob's ladder. It does not follow that experiments as an end in themselves are necessarily beneficial to anyone except, it may be hoped, those who perform them. It does not even follow that the theorist should scrupulously respect all such experimental data as may bear upon the branch of natural science he is trying to develop.

I offer these remarks not as lemmas of a philosophy of science, not in regard to ideal programs, but as inductions from the history of science as I have read it in the old way, searching and probing the sources in detail, case by case, problem by problem, line by line, equation by equation. Here I present an informal array of examples drawn from my *Tragicomical History of Thermodynamics.* That work contains full and explicit references to the sources, which I omit in this essay, since I intend it only as a crib for browsing giraffes of science, its history, and its philosophy.

1. Early experiment. Here there is no disaster. It is plain enough that without the experiments of the seventeenth and eighteenth centuries on thermometers and calorimetry, the basis on which we can think about relations between heat and work had wanted altogether. Experience with steam engines was central to the creation of thermodynamics, though the value of the numerous experiments performed upon them, often with scant comprehension of the physical processes that went on within them, is debatable.

2. Delaroche & Bérard (1812). Few experiments have had so great an influence on the history of physics as Delaroche & Bérard's regarding the specific heats of gases. Fox tells the story:

In order to investigate [the dependence of the specific heat of a gas upon its density], Delaroche and Bérard modified their first experiment so as to allow air to pass through

J. Agassi and R.S. Cohen (eds.), Scientific Philosophy Today, 415–423.
Copyright © 1981 *by D. Reidel Publishing Company.*

the apparatus at a pressure of 100.58 cm of mercury, as well as at atmospheric pressure, the initial temperature being very nearly the same in both cases. The result, that the volume specific heat at this higher pressure was to that at the ordinary pressure of 74.05 cm as 1.2396 to 1, was decisively, though not seriously, in error. Since the ratio of the pressures was 1.3583:1.0000, it followed that the specific heat by weight had decreased in the ratio 0.9126:1.000 as a result of the pressure increase. We know now, of course, that no variation at all should have been observed in the specific heat by weight, but in 1812 a decrease in specific heat with increasing pressure was expected and the quite unfounded confidence which Delaroche and Bérard placed in their result almost certainly owed a great deal to this fact. They based their conclusion on only two experiments conducted on air at the single higher pressure of 100.58 cm, the steady variation of pressure being impossible with their apparatus. The discrepancy between the volume specific heats deduced from the two experiments (1.2127 and 1.2665, of which 1.2396 was the mean) should in itself have made them suspicious, but without further examination they proceeded to extrapolate the results to other pressures and confidently assumed that they applied equally well to all gases. Their error, although of less than 10 per cent, was to prove one of the most influential in the whole history of the study of heat. Backed by the prestige associated with victory in the Institute's competition, the result quickly became standard and . . . was to mislead many calorists [Fox 1971, pp. 139–140].

The greatest of the calorists so misled was Carnot (1824). He was misled to the point that he not only set aside the lead offered by Laplace's theory of the speed of sound, regarded as an adiabatic process in an ideal gas, but also turned his back upon the whole corpus of experiment related to it. That corpus is compatible with the theory of calorimetry if and only if the ratio γ of the specific heats of an ideal gas is constant. Carnot refused to allow that possibility, for according to his theory it would have made the specific heat at constant pressure K_p a function of temperature alone, while Delaroche & Bérard's claim made both specific heats K_p and K_v increasing functions of volume at constant temperature. Carnot's esteem for Delaroche & Bérard's experiment above everything else is all the more remarkable when set against the fact, not noticed until recently, that Carnot's theory is not invariant under change of the unit of temperature *unless* γ is constant for ideal gases. In this statement the specific heats are of course defined with respect to the scale of temperature used by Carnot.

(The reader not at home in the byways of thermodynamics may need to be reminded that Clausius' theory makes constant the *difference*, not necessarily the *ratio*, of the specific heats of an ideal gas, and to be told that the Caloric Theory, whether restricted by Carnot's thermodynamic axioms or not, allows us to take the ratio *or* the difference constant but *not both*.)

Laplace himself was equally misled. While he stood fast by his theory and the experiments on the speed of sound, he let Delaroche & Bérard's result lull him into accepting the preposterous theory of specific heats that the Caloric Theory of heat delivers for an ideal gas when their ratio is assumed constant.

3. Regnault. Regnault's experiments excited the admiration of all who studied them in their day, especially Kelvin. Comparison with data accepted now indicates that his results were very accurate. His ponderous, subsidized program of "determining the principal laws and the numerical data that enter the calculation of steam engines" began in 1842; just about then appeared the first experiments of Joule and Mayer's Assertion; Holtzmann's Assertion, which was to play a key role in the thermodynamics of Clausius, lay three years in the future. The time was perfect for experiment to take the lead. Carnot's and Clapeyron's publications had shown the nature of the specific heats of gases to be *of the essence* to the motive power of heat engines. Regnault, not mentioning anything shown by Carnot or Clapeyron, set about making the most accurate thermometers and determining deviations from the ideal gas laws!

Fox tells the story:

In a history of the caloric theory of gases..., it must surely be the sterile experimenting of Victor Regnault that has pride of place as evidence of the decline; indeed, by any standards Regnault's failure to play a significant part in the development of thermodynamics deserves more than the passing comment it has received in earlier studies of our problem.

Regnault, we should recall, was a man of outstanding ability, and by the early 1840s he had the familiarity with steam-engine operation that seems to have been so important to most of the pioneers of energy conservation. Moreover, thanks to the French Government, he had been provided with assistants, equipment, and a laboratory that would have been the envy of his contemporaries both in France and elsewhere. With all this, how could he have failed? [Fox 1971, p. 315]

Regnault did finally get to the specific heats of gases, but only in 1853, three years after Clausius had conjectured, on the basis of theory alone, that they would be found constant. Regnault did find, indeed, that K_p for several gases was very nearly constant. Those who accepted Laplace's explanation of the speed of sound had to agree that K_p/K_v was sensibly constant. Thus Regnault's work and the data on the speed of sound could be taken as confirming Clausius' conjecture. Clausius, unlike Kelvin, had not waited for new data. *He had dared to go ahead with the theory*, so by the time experiment got going, thermodynamics had been standing in print for nearly three years. Not only that, also a good value of K_p at ordinary conditions had been obtained. Maxwell tells the story, not without a tinge of gentle sarcasm [1871, pp. 177–178; pp. 179–180 of later editions]:

Hence the determinations of the specific heat of gases were generally very inaccurate, till M. Regnault brought all the resources of his experimental skill to bear on the investigation, and, by making the gas pass in a continuous current and in large quantities through the tube of his calorimeter, deduced results which cannot be far from the truth.

These results, however, were not published till 1853, but in the meantime Rankine, by the application of the principles of thermodynamics to facts already known, determined theoretically a value of the specific heat of air, which he published in 1850. The value which he obtained differed from that which was then received as the best result of direct experiment, but when Regnault's result was published it agreed exactly with Rankine's calculation.

4. Joule. The hero of experimentation on the Interconvertibility of Heat and Work is surely Joule. Surely he had the idea that heat could always be specified in units of work, and he claimed that his experiments supported it. His results published before 1850, however, were so inaccurate as to cast doubt upon what he claimed for them.

Poincaré remarks that the early experiments of Joule and others were conceived as "*verifying* an established principle" – established, that is, on the basis of preconceptions molecular and otherwise – while later students, abandoning the historical order of things, preferred to regard those same experiments as themselves "*establishing experimentally* the principle of equivalence" [1892, §60]. However, the experiments in question do not suffice to discharge the responsibility the tradition of physics had laid upon them. The unacceptable spread of Joule's early results is summarised thus by Meyerson [1926, pp. 194–195]:

The numbers of the English physicist vary within extraordinarily large limits; the average at which he arrives is 838 foot-pounds (for the quantity of heat capable of increasing the temperature of a pound of water by 1° F, which is about equivalent to 460 kilogrammeters to 1°C.); but the different experiments from which this average is drawn furnish results varying from 742 to 1,040 foot-pounds (or from 407 to 561 kilogrammeters)—that is, by more than a third of the lowest value—and he even notes an experiment which gives 587 lb. (322 kilogrammeters) without seeing in it any source of particularly grave experimental errors. It is only in the postscript of this work that Joule tells of a series of experiments yielding as a result 770 foot-pounds (423 kilogrammeters), which approximates our present estimations. Moreover, if one considers that at this same moment Sadi Carnot and J. R. Mayer had already, each one for himself, calculated the equivalent of heat and had arrived at the figures of 370 and of 365 kilogrammeters (which is more than an eighth lower than Joule's value), it becomes really difficult to suppose that a conscientious scientist, relying solely on experimental data, could have been able to arrive at the conclusion that the equivalent must constitute, under all conditions, an invariable datum.

Rowland, writing earlier but still long after the fact, levels weightier charges [1880, p. 150]:

> One very serious defect in Joule's experiments is the small range of temperature used, this being only about half a degree Fahrenheit, or about six divisions on his thermometer. It would seem almost impossible to calibrate a thermometer so accurately that six divisions should be accurate to one per cent, and it would certainly need a very skilful observer to read to that degree of accuracy. Further, the same thermometer "A" was used throughout the whole experiment with water, and so the error of calibration was hardly eliminated, the temperature of the water being nearly the same.

Indeed, Joule obtained his final numbers in each experiment by simple averaging over many trials, and he carried out all his measurements at room temperatures, which anyone who has lived in Britain will expect to lie in a small, chilly interval, pretty close to freezing.

It is no wonder that Joule's contemporaries, even his friend Kelvin, were reluctant to accept his early results. Helmholtz (1847) rejected them at the very moment when accurate measurements over a broad range of temperatures could have exerted a decisive influence upon the course of thermodynamics. In a paper published in 1851 Kelvin tells us that Joule in a letter dated December 9, 1848, had put forward as a conjecture the celebrated evaluation of Carnot's function μ:

$$\mu = J/\theta,$$

J being the mechanical equivalent of a unit of heat and θ the centigrade temperature measured from absolute cold. A good folklorist would expect that Joule had induced this formula from his experiments, but that is not so, as those did not allow the temperature to vary much. It was Kelvin, as Kelvin himself tells us, who communicated to Joule the relation

$$\frac{\text{Work done in an isothermal process at temperature } \theta}{\text{Heat gained in that process}} = \theta\mu(\theta).$$

This formula follows easily from Carnot's general ideas, irrespective of any commitment for or against the Caloric Theory and of any particular equivalence of heat and work. By applying to it his general belief in a uniform and universal interconvertibility of heat and work in all circumstances, Joule concluded that $\mu = J/\theta$, J being his uniform and universal mechanical equivalent of a unit of heat. *Joule's conjecture was theoretical.* By his good luck, the simple and direct interconvertibility of heat and work in isothermal processes turned out to be consistent with the later, precise thermodynamics of ideal gases, although his vague expressions of a more general equivalence remain inassessable.

In his 'Account of Carnot's theory,' 1849, Kelvin without bluntly saying so undertook to test Joule's conjecture by experiment. He tabulated values of $\theta\mu(\theta)$. To do so, of course he used his table of μ, based upon Regnault's data, 'completed' by his own conjecture. He found that $\theta\mu(\theta)$ was not constant. One value agreed well with one of Joule's determinations, another agreed well with Carnot's calculation for a particular circumstance. In effect, Kelvin informs the reader that experiment controverts Joule's conjecture.

This is irony indeed. Kelvin was reluctant to accept a hypothesis unsupported by experiment. He appealed to the best data he could get. It was insufficient. Then, as physicists will do, to supply wanting data he *appealed to authority* to make what could be little better than *another conjecture*, one that seemed to him secondary and innocuous: that saturated steam obeyed the laws of ideal gases, even at high temperatures. In itself such a guess merely reflects the belief, current in earlier periods and lingering on even in 1849, that nearly all gases are nearly ideal. Nevertheless, conjecture it was – if anything, still *less* supported by experiment than was the one Kelvin was trying to test – and it was fatal. Not only did it break the chain of his calculation allegedly based upon the results of experiment, not only did it vacate his claim to determine μ by appeal to nature itself, but also it tied his hands. In possession of all the elements out of which classical thermodynamics was just about to be formed, Kelvin was not the man destined to form it. His failure should serve as a classic warning to theorists of two kinds: those who bind themselves too tightly to the results of experiment, and those who make plausible conjectures about the results of unperformed experiments. In 1849 Kelvin was of both these kinds.

It is only hindsight, sharpened by the far greater definition of Joule's later work, that makes the modern student expect Joule to have been the hero of the new science, under whose banner all physicists ought to have rallied.

True, by the end of 1850 two great theorists accepted Joule's results: Clausius and Rankine. Both of these theorists, however, were predisposed to them; each was guided by a molecular picture, in which he seems to have believed firmly,[1] and each sought from Joule's work not direction toward a theory but experimental evaluation of a constant, the existence of which he had already assumed. The decisive position that both Rankine and Clausius took was to *reject*, at last, the experimental data provided by Delaroche & Bérard. On the other hand, Rankine even so late as February 1850 rejected all the numerical values of J that Joule had obtained up to then. Clausius took the double risk of accepting Joule's work in general terms and of going on to conjecture that real gases insofar as not sensibly ideal would also fail to have sensibly constant specific heats.

5. *Fantasy*. We may imagine how different the history of thermodynamics would have been, had Delaroche & Bérard concluded that K_p was nearly constant for most gases, air included. Rough confirmation of Laplace's formula for the speed of sound would have suggested that γ = const. for air. Thus any competent theorist would have taken *the ideal gas with constant specific heats*, not as the embodiment of nature but as the most natural special case to consider. Attempting to apply the Caloric Theory, he would have derived one or both of the relations

$$K_p - K_v + p\frac{\partial K_p}{\partial p} + \rho\frac{\partial K_v}{\partial_\rho} = 0,$$

$$K_v = \frac{M}{\gamma\theta}\left[\frac{p^{1/\gamma}}{\rho}\psi'\left(\frac{p^{1/\gamma}}{\rho}\right)\right],$$

for both were latent in Laplace's work. Because neither of these equations allows an ideal gas to have constant and distinct specific heats, *our imaginary theorist would have had to reject the Caloric Theory*! The date is any time from 1812 onward. The theorist in question could easily have been Carnot, who had all the necessary apparatus.

That is not all. The same theorist could have calculated the motive power of a Carnot cycle *without use of any theory relating heat and work*. For an ideal gas with constant specific heats no general theory need be called upon. There is no reason why this hypothetical theorist – again, he could well have been Carnot – would have failed then to get the entirely elementary theorem Hoppe was to publish in 1856: An ideal gas with constant specific heats interconverts heat and work uniformly in Carnot cycles. This special case points the way to the general idea. If, as Carnot claimed, "the motive power of heat is independent of the agents used to realize it," then *all bodies* must interconvert heat and work uniformly in Carnot cycles because the ideal gas with constant specific heats does so! Clausius's Axiom would have lain in Carnot's hands, had Delaroche & Bérard's work been done well, or had Carnot simply rejected it. Again the fact was too much experiment, or too little for Delaroche & Bérard's data forbade those who accepted them from making the most natural first guess about the specific heats of gases.

Well then, suppose we forget Delaroche & Bérard, accept (as in historical fact we must) Carnot's dilemma as inevitable in 1824. What about Regnault? Had he chosen to determine K_p in 1840, he would certainly have done so correctly, and, had he published his result, Kelvin could scarcely have failed then to accept Joule's conclusions, inaccurate as was the experimentation from which Joule had drawn them.

The tragicomedy of classical thermodynamics has an experimental counterpart: the tragicomedy of early experiment on the specific heats of gases.

The Johns Hopkins University

ACKNOWLEDGEMENTS

I am grateful to Mr. C.-S. Man for his help in analysis of Kelvin's tables and Joule's experiments through 1850 and for locating earlier published criticism of the latter; to Dr. Fox and the Oxford University Press for permission to reprint two passages from his *The Caloric Theory of Gases*, 1971. The research reported here was carried out with the partial support of the U.S. National Science Foundation's Program in the History and Philosophy of Science.

The text of this article is revised and rearranged from passages in my *Tragicomical History of Thermodynamics*, Springer, New York, 1980, copyright © 1980 by Springer-Verlag, New York, reprinted here by permission.

NOTES

[1] Rankine in fact obscured his phenomenal theory by entwining it with his theory of molecular vortices, doomed to rejection. Clausius withheld publication of his kinetic theory for some years, but in his paper of 1850 he alluded to it immediately after his

reference to "the careful investigations of Joule" at the beginning: "To this it must be added that other facts have lately become known which support the view, that heat is not a substance, but consists in a motion of the least parts of bodies."

REFERENCES

Fox, R.: 1971, *The Caloric Theory of Gases from Lavoisier to Regnault*, Clarendon Press, Oxford.

Maxwell, J. C.: 1871, *Theory of Heat*, Longmans Green, London.

Meyerson, E.: 1926, *Identity and Reality*, 3rd ed., K. Loewenberg, tr., George Allen and Unwin, London. See also 2nd ed., 1912.

Poincaré, J. H.: 1892, *Thermodynamique* (1888/9), Georges Carré, Paris.

Rowland, H. A.: 1880, *Proceedings of the American Academy of Arts and Sciences* **15** = (2) **7**, 75–200.

RAIMO TUOMELA

INDIVIDUALISM AND CONCEPT FORMATION IN THE SOCIAL SCIENCES

1. We shall below sketch a moderately individualistic program for concept formation in the social sciences. This program is basically functionalist: concepts are to be characterized in terms of their functions or roles in the theories where they are introduced and employed. Our view is individualistic in the sense that we accept a principle of *conceptual individualism* which entails *partial* reduction of holistic social concepts into individualistic ones. The account to be given below is somewhat sketchy and general – detailed functionalist analyses of single social concepts will not be presented in this context.

My starting point is Sellars' functionalist program for concept formation in psychology (see, e.g., Sellars [1956] and [1968]), which I have explicated and elaborated to some extent in Tuomela [1977] . In that approach the idea was to *conceptually* introduce 'inner' mental concepts, such as the concepts of want, intention, belief, perception, construed 'adverbially' in terms of 'social practice' (overt intelligent, including especially verbal, behavior and dispositions so to behave, in the first place) and thus out of 'public means', so to speak. Yet, even if mental concepts *conceptually* depend on social practice, *ontologically* (i.e., in the order of being) mental states and episodes (i.e., what the mental concepts stand for) are prior to overt behavior items. Thus, indeed, they may cause behavior and hence mental concepts may play a central role in the (causal) explanations of behavior. In fact, the main motivation for introducing such inner mental concepts is just their explanatory import. (It is not paradoxical to say that 'in principle' we could get along rather well by just having a 'Rylean' language of overt actions and the like without any Cartesian remnants such as inner mental acts.)

Sellars' account of concept formation in psychology is stated by him in terms of a myth. Suppose we had Rylean ancestors, to use Sellars' terminology, who employed a somewhat limited behavioristic language in their communication. The descriptive predicates of the language concern only public properties of public objects located in space and enduring through time. This language permits the standard logical operations, and it also allows for subjunctive conditionals. The only way our Rylean ancestors could speak of anything resembling our notion of thinking was

J. Agassi and R.S. Cohen (eds.), Scientific Philosophy Today, 425–438.
Copyright © 1981 *by D. Reidel Publishing Company.*

in terms of overt verbal episodes, most centrally "thinking-out-loud". Similarly, e.g., perceiving could only be approached through "perceiving-out-loud", and so on. No inner episodes, conceptual nor non-conceptual, are admitted by this Rylean language.

On the other hand, it should be noticed that *action*-talk (e.g., about speech acts) is permitted. That is, the language does not consider behavior only as movements and responses, etc. Also theoretical and practical reasoning (and at least the rudiments of normative discourse) is possible in our Rylean community.

Given this Rylean language, Sellars' claim now is that if we add to it the resources of (1) *semantical* and (2) *theoretical* (i.e., concerning unobservable inner episodes) discourse, that suffices to enable the language users to come to acquire the concept of mental episode (event, state, process) as we *now* have it.

Semantical (or metalinguistic) discourse will enable the members of the Rylean community to characterize overt verbal episodes in semantical terms and thus to give these episodes characteristics analogous to the intentional characteristics we now give to mental episodes. Thus, it becomes possible to characterize an agent's verbal utterance, e.g., 'The earth is flat' as a *saying* (thinking-out-loud) that the earth is flat (roughly because 'The earth is flat' – utterance construed as a linguistic statement *means* that the earth is flat). Let us call Rylean discourse cum semantical discourse pre-Jonesian discourse. The addition of a theoretical discourse about inner episodes then gives a new kind of entities, which share the aboutness aspect with verbal episodes but which still are distinct from the latter. The resulting discourse will be called Jonesian.

This introduction of inner mental episodes in the mythical community is due to a genius, Jones by name. According to our philosophical myth he develops the hypothesis (or theory) that people's propensities to think-out-loud change during periods of silence as they would have changed if they had, during the interval, been engaged in a steady stream of thinkings-out-loud of various kinds. This is because they are the subjects of imperceptible episodic periods which (a) are analogous to thinkings-out-loud; (b) culminate, in candid speech, in thinkings-out-loud of the kind to which they are specifically analogous; (c) are correlated with the verbal propensities which, when actualized, are actualized in such thinkings-out-loud; (d) occur not only when one is silent but in candid speech, as the initial stage of a process which comes "into the open", so to speak, as overt speech (or as sub-vocal speech), but which can occur without this culmination, and does so when we acquire the ability to keep our thoughts to ourselves. [Cf. Sellars 1968, p. 159.] Jones then goes on to teach his fellow members to use mental language (according to his hypothesis) not only in the third-person case but also in the first-person case.

According to this so-called analogy theory of thinking (i.e., of representational mental episodes) (inner) mental concepts derive their meanings primarily from overt intelligent (especially verbal) behavior and thus from public culturally laden facts. I have elsewhere (in Tuomela [1977], Chapter 4) disucssed in some detail a reconstruction of the psychological theory introduced by Jones in the above myth. Here I will not be so much interested in that theory but rather with some new ways of forming concepts out of the fundamentum of overt intelligent behavior. This means modifying the above Jones-myth in some respects. Let us see how.

2. We spoke above about the Rylean language employed by the mythical community. Let us now refine our description in one respect and think that the extralogical predicates of this language may contain not only individualistic (psychological and social) predicates but also holistic social predicates as long as these holistic predicates are not clearly explanatory predicates introduced by some social theory. It would seem that, e.g., 'group' and 'school' are examples of such 'observational' (viz., non-explanatory) holistic concepts. (Whether there indeed are such 'observational' holistic social concepts depends on one's theoretical-observational dichotomy. It is not so important for our present concerns.)

Let λ be the set of extralogical predicates of our antecedently understood Rylean language (or, better, pre-Jonesian language, as the presence of semantical, metalinguistic talk may be assumed). Let now λ_i be the subset of λ which consists of predicates expressing properties of individuals (persons) and relations between individuals and which do not conceptually presuppose holistic social notions such as, e.g., 'cashing a cheque' does presuppose the concept of bank, etc. In fact, λ_i may be taken to consist of predicates expressing social actions (and capacities for social action) as clarified in detail in Tuomela [1979]. Let $\lambda_h = \lambda - \lambda_i$ be the set of non-explanatory social predicates applying to or at least conceptually presupposing collectives of individuals or to properties of such collectives (or relations between them).

Next we introduce our explanatory predicates. We first let μ_p be the set of inner mental predicates introduced by Jones' theory, call it $T_p(\lambda \cup \mu_p)$. This theory may be taken to comprise all the mutually consistent common sense psychological truths (and approximate truths) which we (presently) possess and which are constitutive of the predicates in μ_p. (We shall below consider what this constitutiveness means.) Secondly, we introduce a set μ_s which is the set of such holistic (and possibly other) explanatory predicates as a best explanatory social theory requires. (This talk about the best explanatory social theory is a piece of science fiction, but we cannot here try to clarify the notion.) Now we let $T_s(\lambda \cup \mu_s)$ be the conjunction of postulates which are constitutive of the predicates in μ_s.

Whether or not T_s coincides with the just mentioned best explanatory social theory does not matter. Presumably it does not, for the following reason. Our theory T_s, which is a counterpart to the Jonesian theory T_p in the psychological case, is meant to be *conceptually* adequate (viz., adequate in the order of conceiving) in a sense to be explicated. But even if it were conceptually adequate it need not yet be causally (and explanatorily) adequate (viz., adequate in the order of being). For T_s lacks the explanatory mental concepts of μ_p, and surely they are needed in the best explanatory social theory if there is any truth at all in explanatory methodological individualism.

To be a little more concrete let us consider some examples of holistic social concepts (members of μ_s and λ_h). Some familiar holistic predicates are: 'institution', 'organization', 'group', 'state', 'government', 'church', 'battalion', 'working class', 'free market', 'role', 'stratification', 'cohesion', 'status crystallization'. Of these predicates presumably at least those appearing towards the end of the list would seem to qualify as members of μ_s. But, as said, what matters is not which of the above and other predicates belong to λ_h and which to μ_s but rather which are members of $\lambda_h \cup \mu_s$ (call this set η_s) and which of λ_i. For our idea, soon to be discussed, is to construe the meanings of the predicates in η_s on the basis of those in λ_i, given the postulate conjunction T_s (= $T_s(\lambda \cup \mu_s)$ = $T_s(\lambda_i \cup \eta_s)$).

Let me here still make one more remark on holistic social concepts. It is often claimed that such concepts are or may be *normative* (or at least prescriptive, viz., concerned with ought-and may-rules). Consider, for instance, the concept of role. It is definable in terms of norms, in terms of what the role bearer *ought to do* or *may do* in various circumstances (see, e.g., Pörn [1977], pp. 62–63 for this kind of treatment of roles). It is also easy to see that, e.g., organizations and institutions embody norms in analogous ways. As this matter is a familiar and an acknowledged one I shall not here further discuss it nor argue for it in detail. Let me just put the matter in general terms and say that the social realm is through and through normative. The point I am concerned to make here is that it is arguable that the concept of a norm cannot be expressed in the impoverished, individualistic pre-Jonesean language and thus in terms of the vocabulary λ_i even if we allow the logic of this language to contain normative operators (e.g., 'Ought', 'Shall'). This is because norms are *intersubjective* and this intersubjectivity cannot be accounted for in individualistic terms. More specifically, as will be clarified in more detail below, I will tentatively accept an analysis of ought-and may-rules in terms of *we-intentions*; and the social concept of a we-intention (perhaps) is not strictly reducible to individualistic concepts [cf. Sellars 1963, pp. 202–205]. If this approach is correct, in the ultimate analysis of holistic social concepts *at most* an irreducible concept of we-intention is needed in addition to the concepts in λ_i.

There are also holistic concepts which are less explicitly normative than those mentioned above (cf. role). Thus, it might seem that, e.g., group cohesion is not a normative concept. Much of the discussion concerned with methodological individualism has been concerned with concepts like it. It is not completely implausible to suggest that properties of systems of objects, such as social psychological groups, be reducible to properties of its constituents and relations between them. Consider thus the following familiar reduction principle:

(R) If an object is a system of objects, then every property of the object must consist in the fact that its constituents have such and such properties and stand in such and such relations.

We may perhaps agree that this strict reductionist principle indeed holds in the case of some systems. Yet I think it cannot *a priori* be required to be true of, e.g., social psychological groups. In any case, much hangs on the term 'consists'. What if, speaking of a social psychological group and its cohesion, the left hand side is a normative concept whereas the analysans is non-normative? (The concept of group may be regarded as involving the concept of norm.)

Although I am inclined to accept ontological individualism, a better characterization of it than that given by (R) is needed. But in this paper we shall not really be concerned with problems of ontology (and, in any case, ontology is ultimately to be determined by scientific research, not philosophizing), and we shall leave (R) without revision here.

3. What is often called the doctrine methodological individualism (as

opposed to methodological holism) involves three ingredients which should be distinguished from each other. The first ingredient is *ontology*: What is the ontological nature of social objects and properties (cf. our (R))? The second is *meaning* (semantics): How are social concepts, such as those in η_s, to be semantically understood? The third ingredient is *explanation*: How are social phenomena and regularities best explained? Do individualistic theories together with suitable correspondence principles suffice to explain (and thus reduce) holistic theories?

I will below be almost exclusively concerned with the second problem and advocate what I call the doctrine (or thesis) of *Conceptual Individualism*. To state this thesis, we employ the constitutive theory $T_s(\lambda_i \cup \eta_s)$ and rely on our above discussion of it:

(CI) The meanings of the social predicates in η_s depend entirely upon their usage with the predicates in λ_i, assumed antecedently understood, within the theory $T_s(\lambda_i \cup \eta_s)$.

(CI) requires several comments. First, we note that this principle does not entail that the predicates of η_s be *explicitly* (or in any other strict sense) definable in terms of those in λ_i within T_s. It only says that any predicate in η_s gets whatever (inferential) meaning it has from the predicates in λ_i on the basis of the comprehensive theory T_s. Thus (CI) is not reductionistic in a strong sense, but it only, as we may say, functionally and partially determines the meanings of the η_s-predicates in terms of the λ_i-predicates within T_s. It allows for 'open' social predicates and for multiply denoting predicates. Multiple denotation means roughly the following. Suppose that T_s is a theory formalized in first-order predicate logic. Then, if for some fixed interpretations of the λ_i-predicates it is possible to model-theoretically satisfy T_s by means of two or more interpretations (viz., two or more sets of ordered sequences of individuals or – if you are a strong intensionalist – two or more non-extensional properties) of some predicate $P \in \eta_s$, then P is said to multiply denote these interpretations. But it should be noted that as such (CI) says nothing about ontology; it does not even claim that the predicates in $\lambda_i \cup \eta_s$ are referring ones. As to the logical form of T_s in general, let me here just say that it can be taken to be formalized in first-order logic or some second-order extension of first-order logic with the addition of some (sentential) intensional operators, e.g., 'Ought', 'Shall'. Suitable metalinguistic discourse is of course assumed available. (As to the logical features of (CI), see my detailed discussion of its exact counterpart (CF) for the psychological case in Tuomela [1977], pp. 76–88.)

We have been requiring all along that T_s be a theory *constitutive* of the predicates in η_s. (Note that the λ_i-predicates are assumed antecedently understood in (CI).) By our notion of constitutiveness roughly the following is meant. The meanings of the extralogical predicates of a scientific theory are determined, in general, by the whole theory (or theories) in which the predicates occur (cf. the meaning thesis (M) defended in Tuomela [1973], pp. 122–123). This is the constitutive theory as meant in this paper. But on pain of making the theory true *a priori* there must, it seems, be a way of separating the *meaning-giving* (or analytic) component from its *factual* (or synthetic component). I have in Tuomela [1973], Chapter V, and Tuomela [1977], Chapter 4, discussed this much debated question and argued for a modified version of Carnap's approach to this problem. I shall not here go into any details except for pointing out one thing. It is the analytic component that matters here, and what the factual component is is not so central. Also note that practically any postulate set introduced as constitutive for the meanings of some scientific predicates will turn out to have some factual content (see Tuomela [1977], Chapter 4).

There is also another, and perhaps better, way of specifying the meanings of scientific predicates. It goes in terms of the inferential network in which these predicates occur. We may, at least for this purpose, think of scientific laws and theories as rules of inference licensing singular inferences, which in a familiar example would state that if *a* is a piece of copper which is heated then one may infer that it will expand. Thus, e.g., 'copper' would get its meaning from all the inferential contexts (or some specified subset of them) resembling the mentioned one in which it is used. As, however, the inferential approach can at least in standard cases be translated back to the postulational approach I will not here discuss it further.

Now some remarks on the philosophical content of (CI) are due. First, we may connect (CI) to Sellars' Jones-myth and tell a new myth. The content of the new myth is the following. We suppose an individualistic Rylean community speaking only the language of the λ_i-predicates. Then we enrich this restricted individualistic Rylean discourse by adding semantical discourse (and thus metalinguistic categories) to it. In fact we now also get (the logical aspects) of *prescriptive* discourse, if we assume (as earlier) the presence of (individualistic) practical reasoning in our original community. For, it may be argued, practical thinking has deontic conclusions, viz., conclusions specifying what an agent should or ought to do. So we need ought- and may-rules from the beginning, and metalinguistic talk specifies the meaning of such prescriptive talk and enables the agents of our individualistic community to formulate oughts at least for themselves. (The concept of a norm perhaps still involves more, as we shall see, and therefore we may not yet have got normative discourse, viz., prescriptive discourse in the fullest contentual sense.)

Given the individualistic Rylean discourse as well as semantical and prescriptive discourse we still have to add theoretical (viz., explanatory) discourse for new types entities and properties. After that we may accept (CI) and introduce the predicates in η_s by specifying the functional connections these predicates have to the λ_i-predicates. For instance, would we have the concept of norm available, we could specify the concept of church (let 'church' $\in \eta_s$) by specifying the oughts concerning priests, church members, and so on and, perhaps, further by functionally characterizing this concept as what (in a sense) explains various religious activities of people. Similarly, e.g., 'cohesion' may be partially and functionally characterized as what accounts for or, at least involves, the group-members' likings of each other (and suitable other attitudes) and, e.g., their contact frequencies. Group cohesion may also be functionally characterized as what "keeps the group together" (or something like that).

In all, we get a social myth, corresponding to the Jones-myth in the psychological case, to functionalistically account for concept formation in the social sciences. As an extra bonus we get a strengthening of the Jones-myth: If the social myth is acceptable, then Jones may start with the individualistic Rylean predicate set λ_i (to which possibly the concept of we-intention, if regarded as belonging to λ_h, must be added) rather than having to employ the full Rylean set λ, for the latter can be conceptually founded on the former.

What our program for concept formation in the social sciences amounts to can be summarized by means of some abbreviatory formulas in the following way. We assume that our task is to formulate for the social sciences a scientific language or discourse which gives an adequate conceptual foundation, as it were, for all theorizing. Let us call the fullest such language Socialese. Then what we have said above entails this abbreviatory slogan, using 'd.' for 'discourse':

(S1) Socialese = individualistic Rylean d. + semantical d. +

normative d. + holistic Rylean d. + psychological d. + social theoretical d.

A conceptually adequate but (perhaps) causally inadequate conceptual framework is given by

(S2) Socialese* = individualistic Rylean d. + semantical d. + normative d. + holistic Rylean d. + social theoretical d.

We may now also put our basic problem here as follows: What are the conceptual fundamentals of Socialese and Socialese*? To this question our (S1) and (S2) give their respective answers by listing the various conceptual components needed. We may also say, equivalently, that given (a) individualistic Rylean discourse, (b) semantical discourse, (c) normative discourse and (d) the conceptual resources for theoretical discourse on the whole (without regard to its substantive nature), then (i) the use of the principle (CI) (given that it is acceptable, of course) gives us Socialese* and (ii) the use of (CI) plus the use of its analogue for the psychological case (called (CF) in Tuomela [1977], p. 76) give us Socialese.

It is worth noting in the present context that when employing Socialese there is the somewhat attractive possibility of defining the predicates of η_s *explicitly* in terms of the predicates in $\lambda_i \cup \mu_p$ even if such definitions are not possible in terms of only λ_i. I take it for granted that the predicates in μ_p cannot be explicitly defined in terms of the predicates in λ_i (this follows from the failure of strict logical behaviorism). Accordingly, there is no hope of explicitly defining the predicates on η_s in terms of the predicates in λ_i no matter how individualistic one wants to be, for surely individualistic accounts of holistic social concepts must refer to such mental features as wants, intentions, perceptions, beliefs and so on, viz., mental features expressed by μ_p-predicates.

In this paper I have not presented nor will I present strong arguments against the explicit definability of η_s-predicates in terms of predicates in $\lambda_i \cup \mu_p$. As will be seen below my argument does not at all rely on the alleged possibility of the notion of we-intention (representable by a predicate in η_s) not being explicitly definable by means of I-intentions and so-called loop beliefs and thus by means of predicates in $\lambda_i \cup \mu_p$. There are yet several methodological arguments that can be involved here for the openness (viz. non-definability) of the η_s-predicates occurring in social theories. (For a detailed discussion of these arguments, relying on various methodological gains accruing from the use of open explanatory predicates, see Tuomela [1973], especially Chapter VI.)

4. There is one remaining problem to be discussed. It is the problem of what normative discourse really amounts to. (The other components of Socialese should be clear enough.) As most social concepts presuppose the concept of a norm, the question becomes how much content is conceptually needed in the concept of normative discourse. For instance, if strict naturalism were acceptable, then we would not need irreducible normative discourse at all. But if we do need some kind of irreducible oughts and mays, how rich a conceptual framework does that require? We shall below briefly consider this problem from a certain angle, viz. from the point of view of practical reasoning, which I find congenial.

Sellars [1968] attempts to give a fairly naturalistic analysis of norms in terms of intentions and practical thinking. As I find his analysis a good starting point let me briefly sketch some of its main features. Let us here then tentatively accept that practical reasoning gives a good basis for an analysis of (moral and other) norms. Now it is an acknowledged feature of norms that they are *intersubjective*: The same norm should be applicable to different agents. But if we now think that practical reasoning is somehow intimately connected with norms, e.g., in the sense of entailing conclusions as to what one shall do, viz., conclusions which are expressions of intentions to do something, and if this is taken to be what connects practical reasoning with norms (e.g., what one ought *overridingly* to do), then the problem of intersubjectivity arises. That it does so arise is basically because an agent cannot really directly intend "for" another agent. To be sure, e.g., I can in a sense intend that you get quickly and comfortably home. But this only means that I commit myself to do whatever *I* regard as necessary to get you quickly and comfortably home. Thus, as one can strictly speaking only intend for oneself, the problem of intersubjectivity arises. Sellars [1963, pp. 196–205] emphasizes this problem and introduces the concept of *we-intention* to handle it. Given it, both you and I can intend in the we-mode (expressed in Sellars' analysis by locutions of the form 'We shall do *X*'), and a real comparison between we-intentions can be performed in the sense required by intersubjectivity.

Sellars [1963] also offers other, broader arguments for the need of we-intentions. One is that we-intending involves a special sort of consciousness ("we-consciousness", yet utterly different from traditional 'group-mind') which is the internalization of the concept of group. A third argument in effect reconstruable from Sellars [1974], pp. 41–42, is that the notion of *akrasia* involves a conflict between we-intention and 'I-intention' and that therefore we need a concept of we-intention which is not in any sense reducible to I-intentions. A fourth, and very general argument by Sellars [1963], p. 205, is that having the notion of we-intention indeed amounts to having moral discourse and hence normative discourse, more broadly (recall here our problem about the content of normative discourse). Thus we may also say that what one *ought to* do really amounts to what *we* will *insist* upon and what one *may do* to what *we* will *allow*. This broad argument has the expected technical counterpart in Sellars' system so that we-intentions come to appropriately necessarily co-vary with oughts.[1]

A fifth argument which may be considered here is that to have a notion of a rational agent acting for reasons we need a 'non-vacuous' 'I', a notion of self which contrasts with and is not characterizable without the concept of we-ness [cf. Sellars 1968, pp. 223–226]. 'I' in this sense is concerned with 'one of us' and thus with what it is to be a member of *our* community. The need for this kind of non-vacuous 'I' can also be seen from the plausibility of certain types of practical inference. Thus if we as a group do something, and I perform my part of this joint action, then we may want to make the concept pair I-we satisfy the following schema which is supposed to represent my practical reasoning and where (i) is meant to be an expression of we-intention by me:

(W1) (i) We will do *X*.
(ii) I am one of us.
(iii) I will do whatever I regard as necessary for our doing *X*.

Given this, I do what I do (say *Y*; I haven't extended (W1) to that) at least in part *because* I am one of us.

Not only does the notion of we figure in simple reasonings of the above kind but also in more complex ones. Thus we may consider the following practical inference schema to represent my reasoning. In it my being one of us serves as a partial reason for my doing one or more other actions, such as *Y* below, less directly connected to our doing *X*:

(W2) (i) We will do *X*.
(ii) *A* is one of us.
(iii) I intend that *A* will do whatever I regard as necessary for him to do in order for us to do *X*.
(iv) I will do whatever I regard as necessary to bring it about

that A does whatever I regard as necessary for him to do in order for us to do X.

(v) My doing Y (e.g., teaching A to do something) is such a thing.

(vi) I will do Y.

We shall not here elaborate the schemas (W1) and (W2). Their role in the present context is to indicate that one may regard the concept of we-intention as a functionally introduced notion, which inference schemas such as (W1) and (W2) partially characterize. We may even bring this closer to our postulational method of concept formation and our (CI). For we may translate these schemas into postulates. Consider thus the simpler (W1). Corresponding to it we may, for instance, describe an agent's reasoning directly in terms of (i)-(iii) of (W1) or we may (not completely unproblematically) employ an indirect (and simplified) third-person translation such as the following:

(P) (A) (G) (X) (If an agent A we-intends, or group G-intends, to do X and thinks he is a one of us, or a member of group G, then he will intend to do whatever he regards as necessary for our (the group G's) doing X.)

Whether or not our above discussion has convinced the reader of the necessity of at least some kind of we-discourse (at least we-intention), we shall take that need to be established. What remains is to lay out the possibilities that we have for an exact characterization of the concept of we-intention. First, we may follow Sellars and regard the concept of we-intention as an irreducible primitive concept which, furthermore, conceptually serves to give us normative discourse. If we opt for this alternative, then in our (S1) and (S2) of Section 3 we may substitute for 'normative discourse' the phrase 'we-discourse' (meaning we-intention talk primarily). But we must also at the same time say, as, and in so far as, the notion of we-attitude can be taken to belong to λ_h and thus to have some descriptive or non-normative content over and above that of I-attitude, that λ_i (together with (CI) does not create a sufficient descriptive (or factual) basis for even Socialese*, still less Socialese.

Secondly, we may accept the concept of we-intention as a primitive concept and accept a Sellarsian analysis of normative discourse in terms of it but introduce we-intentions only functionally in terms of (CI). Then, contrary to the first alternative, λ_i and (CI) serve as a sufficient descriptive (or factual) basis for Socialese and Socialese*.

Thirdly, we may disregard Sellars' third argument related to the weakness of will and change the preceding alternatives by claiming that the concept of we-intention is *explicitly definable* in terms of I-intentions. Without taking up fine points, the definition could roughly be as follows: A member M of a group G we-intends to do X if and only if (a) M I-intends to do X and (b) M believes that for all other members M' in G, M' I-intends to do X and M' believes that all other members M'' in G I-intend to do X.

Fourthly, we have a group of alternatives which use the notion of we-intention in one of the above senses (and employ it, e.g., in defining the concept of group) but which analyze normative discourse differently.[2] For instance, a completely naturalistic analysis of normative discourse would drop this component from our (S1) and (S2) altogether.

Although I am inclined to accept either the second or, perhaps, even the third of the above alternatives I shall not here discuss the matter further. After all, the purpose of the present section of this paper has been mainly to illustrate and elucidate our general doctrine of Conceptual Individualism and much less to argue for some specific details related to it.[3]

University of Helsinki

NOTES

[1] In the technical terminology Sellars uses we, accordingly, get the necessary equivalence of the following two statements (a) and (b) [see Sellars 1968, p. 219]:

(a) It shall_{we} be the case that each of us does A_i in C_i;

(b) If any of us is in C_i he ought to do A_i.

Of these (a) is an expression of a we-intention and (b) states a norm.

Castañeda and Aune have presented closely similar analysis of normative statements in terms of practical reasoning and intentions. (See, e.g., Aune [1977], Chapter IV for a lucid discussion.)

[2] In this connection we may note an interesting parallel between semantical and normative discourse. Sellars, for one, bases semantical discourse on Rylean facts. (Still he regards semantical discourse as irreducible and *sui generis*; cf. Sellars [1974], pp. 274–275.) We may now ask what this factual basis of semantical discourse really amounts to. Does λ_i suffice or do we, in addition, need λ_h? And, especially, do we need we-intentions? These questions are appropriate also in the case of normative discourse, as we have, in fact, just above been discussing the factual basis of norms. I shall not here try to answer

the above questions in the case of semantical discourse but only suggest that the proper answers are closely parallel with the normative case.

[3] In Tuomela (1982) I have presented a more elaborate account of we-intentions and in fact argued that they are explicitly definable almost in the manner of our third alternative above.

REFERENCES

Aune, B.: 1977, *Reason and Action*, D. Reidel, Dordrecht and Boston.

Pörn, I.: 1977, *Action Theory and Social Science*, D. Reidel, Dordrecht and Boston.

Sellars, W.: 1956, 'Empiricism and the Philosophy of Mind', in *Minnesota Studies in the Philosophy of Science*, vol. 1, H. Feigl and M. Scriven (eds.), University of Minnesota Press, Minneapolis, pp. 258–329.

Sellars, W.: 1963, 'Imperatives, Intentions, and the Logic of "Ought" ', in *Morality and the Language of Conduct*, H.-N. Castañeda and G. Nakhnikian (eds.), Wayne State University Press, Detroit, pp. 159–218.

Sellars, W.: 1968, *Science and Metaphysics*, Routledge and Kegan Paul, London.

Sellars, W.: 1974, *Essays in Philosophy and Its History*, Reidel, Dordrecht and Boston.

Tuomela, R.: 1973, *Theoretical Concepts*, Springer, Vienna and New York.

Tuomela, R.: 1977, *Human Action and Its Explanation: A Study on the Philosophical Foundations of Psychology*, D. Reidel, Dordrecht and Boston.

Tuomela, R.: 1979, *The Structure of Social Action*, Reports from the Department of Philosophy, University of Helsinki, no. 2.

Tuomela, R.: 1982, *A Theory of Social Action* (in press).

PAUL WEINGARTNER

A NEW THEORY OF INTENSION

1. INTRODUCTION

Even a survey study of the history of logic and philosophy shows that a great number of philosophers and logicians were concerned with extension and intension. Under these the following had contributed to it in more detail:

Aristotle[1], the Stoics,[2] Porphyrius,[3] Boethius,[4] Albert the Great,[5] Thomas Aquinas,[6] Duns Scotus,[7] Cajetanus,[8] Ockham,[9] Gassendus,[10] Leibniz,[11] Arnauld-Nicole,[12] Kant,[13] Bolzano,[14] J. S. Mill,[15] De Morgan,[16] Brentano,[17] Frege,[18] Meinong,[19] Mally,[20] Russell,[21] Carnap,[22] Church,[23] H. S. Leonard,[24] Quine,[25] Montague,[26] Bunge.[27]

Already about ten years ago when I made a closer study of the doctrines of these philosophers and logicians I got the impression that there is a frequent traditional view of extension and intension which has the following properties:

1.1 The extensions are understood in the same way (or in a very similar way) as they are understood in modern predicate logic of first order and in the theory of classes which can be developed in it, i.e., the theory of "virtual classes" (according to Quine's terminology) where the addition of the abstraction-operator is just a conservative extension.

1.2 The intensions are non-modal. They cannot therefore be interpreted correctly by a possible-world-semantics.

1.3 both extensions and intensions are connected with a very simple and straightforward ontological thesis about the possibilities of characterizing (ontological) entities (cf. section 2).

1.4 The relation between extension and intension is directly dependent on the traditional principle of the inverse relation between extension and intension (cf. section 3).

The philosophers and logicians who are representatives of a doctrine of

J. Agassi and R.S. Cohen (eds.), Scientific Philosophy Today, 439–464.
Copyright © 1981 *by D. Reidel Publishing Company.*

extension and intension with the above four properties are Aristotle, Porphyrius, Boethius, Albert the Great, Thomas Aquinas, Duns Scotus, Cajetanus, Leibniz, De Morgan, and to some extent H.S. Leonard and Bunge. It is also important to notice that for instance Frege, Carnap, and Montague are not in that tradition. This is so mainly because they understand 'intension' differently, especially if their concept of intension is a modal concept.

It took me quite a time to discover that the traditional view on extension and intension characterized by 1.1–1.4 can be introduced easily into the standard theory of virtual classes by a few definitions – to which the key-concept was that of intensional inclusion (cf. 3.5) – and that the resulting theory has the following additional properties:

1.5 All intensions are reducible to or definable in terms of extensions and all extensions are reducible to and definable in intensions.

1.6 To every law in the standard theory of virtual classes (which contains extensional concepts) there is a corresponding dual law which contains exclusively intensional concepts.

1.7 The only identity needed is extensional identity, i.e., the intensions of this dual theory do not require a different or additional kind of identity.

One may very well question whether a theory satisfying conditions 1.5–1.7 can or should be called a theory of intension. And in fact it took me some time to accept these 'intensions' as reasonable intensions at all. A strong case in point was that I saw that at least some of the problems which are usually said to be solvable only with the help of such intensions which are non-extensional (for instance which are said to require a possible world-semantics) have a quite reasonable and much simpler answer with the intensions proposed here. For these reasons I see the whole situation roughly in the following way: It is a mistake to presuppose that there is only one kind of "correct intension". Rather there are different kinds suitable for the solution of different problems and sometimes also for the (simpler or more complicated) solution of the same problem. I do not deny that it is necessary to introduce other kinds of intensions (especially modal ones) for certain purposes and for the solution of certain problems. The claim of this paper is only that there is also a very simple theory of intension which is characterized by 1.1–1.7 (and described in section 4 below) and which is able to solve some of the problems involved in intensional logics at the same time being an interpretation of a frequent traditional doctrine.

1.8 This paper is divided into 4 parts. In section 2 the ontological thesis is described (2.1), more explicitly stated (2.2), and connected and supported by citations of traditional and contemporary doctrines (2.3). In section 3 the principle of the *inverse ratio*, i.e., of the opposite relation between extension and intension is formulated (3.1, 3.2) and explained (3.3) and passages from the history of philosophy and logic are quoted in which this principle is described (3.4). In section 3.5 the key-concept for the theory – intensional inclusion – is defined (3.52, 3.53) and the definitions of intensional intersection and union are added. Section 4 gives a short presentation of the theory with the most important theorems and hints for proofs.

2. THE ONTOLOGICAL THESIS

2.1 *Informal Characterization of the Thesis*

Entities are twofold: Those which can be predicated of other entities and those which cannot. This distinction is type-theoretical and is due to Aristotle. It was again used in a more sophisticated way in Whitehead and Russell's *Principia Mathematica.* According to this distinction one may separate individuals in a type-theoretical sense[28] – those entities which cannot be predicated of other entities, they can be just identical with other individuals – from properties or classes or sets. Moreover one can weaken the concept 'individual' still further to 'individual in the relative sense'; in this sense 'individuals' are just those entities which are of lowest type among the entities considered. Thus the individuals in ontology may be individuals in the proper sense (of type 0) whereas the individuals to which the individual variables of the standard first order predicate calculus refer need not be individuals in the proper sense, but they are certainly individuals in the relative sense; a similar remark holds for instance for the 'individuals' in mathematics, i.e., for the natural numbers; they are in a sense closer to the ontological individuals than the individuals of a logical calculus since both (the ontological and the mathematical individuals) meet two necessary conditions: the type-theoretical one considered above and the uniqueness condition. Entities which are not individuals (relative to a system) can be described or characterized (in that system) in a twofold way: First by means of entities of equal or lower type than the entity to be described. Second by means of entities of equal or higher type than the entity to be described. Since identity can be defined using either lower-type entities – $a = b \leftrightarrow (z)(z \in a \leftrightarrow z \in b)$ – or higher type entities – $a = b \leftrightarrow (z)(a \in z \leftrightarrow b \in z)$ – the characterization of entities by other entities which are identical to them can be traced back to the above mentioned twofold way. The only two exceptions are individuals (proper or relative) and classes of the highest type. The former can only be described by entities of a type higher than 0, i.e., in the second way, the latter can only be characterized by lower-type entities, i.e., in the first way.

2.2 *Different Versions of the Thesis*

The thesis can be formulated in different (not necessarily equivalent) ways:

THESIS 1. To characterize an entity *a* by entities of lower type than the type of *a* is to characterize it extensionally; whereas to characterize *a* by entities of higher type than the type of *a* is to characterize it intensionally.

THESIS 2. To describe an entity *a* by the individuals which are its elements, by the entities of which *a* can be predicated or by its subsets is to describe it extensionally, while to describe *a* by its properties or by the supersets of which it is an element or a subclass is to describe it intensionally.

THESIS 3. (1) The extension$_1$ of an entity *a* of type τ is (defined as) the class of those entities *x* of type $\tau - 1$ such that $x \in a$ (*x* is an element of *a*, or: *x* is an *a*, or: *a* is predicated of *x*).[29]

(2) The extension$_2$ of an entity *a* of type τ is (defined as) the class of those entities *x* of the same type (τ) such that $x \subseteq a$ (*x* is a subset of *a*) or $x \subset a$ (*x* is a proper subset of *a*).

(3) The intension$_1$ of an entity *a* of type τ is (defined as) the set of those characteristics (properties) or sets *z* of type $\tau + 1$ such that $a \in z$.

(4) The intension$_2$ of an entity *a* of type τ is (defined as) the class of those entities *z* of the same type (τ) such that $a \subseteq z$ (*a* is subset of *z*) or $a \subset z$ (*a* is proper subset of *z*).

Here $a \subseteq b$ and $a \subset b$ are defined in the usual way: $a \subseteq b = df(x)(x \in a \rightarrow x \in b)$; $a \subset b = df(x)(x \in a \rightarrow x \in b) \wedge \sim(a = b)$

THESIS 4. (1) $\text{Ext}_1(a) = df\{x: x \in a\}$
(2) $\text{Ext}_2(a) = df\{x: x \subseteq a\}$ or $\text{Ext}_2'(a) = df\{x: x \subset a\}$
(3) $\text{Int}_1(a) = df\{z: a \in z\}$
(4) $\text{Int}_2(a) = df\{z: a \subseteq z\}$ or $\text{Int}_2'(a) = df\{z: a \subset z\}$

Thesis 4 is thesis 3 in a more symbolic notation, but it lacks the reference to the type. If this is wanted one could rewrite Th 4 (1) in the following way (analogously for (2)–(4)): $\text{Ext}_1(a^\tau) = df\{x^{\tau-1} : x^{\tau-1} \in a^\tau\}$

2.3 *The Thesis in the History of Logic and Philosophy*

In this section passages and references of philosophers and logicians are given in which the thesis is stated. It is not claimed however that each of these passages is literally or syntactically identical with at least one of the four formulations of the thesis. On the other hand it is claimed that the passages and the references state sometimes explicitly, sometimes implicitly the content of at least one of the formulations of the thesis. To be more accurate: Since thesis 1 is very general, its content underlies many doctrines of extension and intension. Thesis 2 is more explicit. But in one of its alternatives it is the gist of many theories of extension and intension. Thesis 3 is most explicit: it defines two kinds of extension and two kinds of intension. There is hardly a doctrine of extension and intension which claims all these four forms of thesis 3. But it may be interesting to observe that those doctrines which speak about the relations of species and genus can be interpreted with thesis 3 (2) and (4) or (1), (2) and (4). Very roughly speaking, this holds for Aristotle and the Middle Ages, Leibniz and De Morgan. From the *Logique du Port Royal* on one may also interpret some doctrines with thesis 3 (1) and (3), even if the principle of the inverse relation between extension and intension can be correctly interpreted only with thesis 3 (2) and (4) or (1), (2) and (4).

Aristotle: "In all such cases the basic principle is that the genus has a wider application than the species and its differentia; for the differentia also has a narrower application than the genus."[30]

Porphyrius: "In another sense genus means that to which the species is subordinate, a meaning asserted as a definition of genus, perhaps, because of a similarity of this sense with the two former senses."[31]

Boethius: "Omne quod genus est, plures sub se species continet, omne quod species, plures in se differentias habet. Genus enim, id est animal, in hoc homine, id est specie, superabundat et superest, quod homo solum homo est, animal vero non solum homo, sed etiam bos vel avis vel alia huiusmodi. Species vero in eo superant genera sua, quod eas differentias quas species in actu habent, eas genera non habent."[32]

Albert the Great: "Genera quidem abundant de speciebus in continentia earum quae sub genere sunt specierum ... Species vero constitutae abundant a generibus continentia propriarum differentiarum, per quas constituuntur ..."[33]

Thomas Aquinas: "A more general universal is related to the less general both as whole and as part. As whole, in so far as the more universal virtually contains both the less universal and other things – for instance, *animal* contains not only *man* but also *horse*. As part, on the other hand, in so far as the less general contains in its definition both the more general and other things; for instance, *man* contains not only *animal* but also *rational*."[34]

Duns Scotus: "Dico enim quod in fundamentis est majus Universale Genus quam Species. Similiter in intentionibus, quia praedicatur de differentibus specie, et numero: Species autem de differentibus numero tantum: et ita habet plura supposita, non tamen in actu exercitio, imo pauciora."[35]

Cajetanus: "Ad hoc breviter dicitur, quod esse magis collectivum multorum potest intelligi dupliciter. Uno modo intensive; et sic species est magis collectiva, quia magis unit adunata . . . Alio modo extensive; et sic genus est magis collectivum, quia multo plura sub sua adunatione cadunt, quam sub speciei ambitu. Unde species et genus se habent sicut duo duces, quorum alter habet exercitum parvum, sed valde unanimem, alter exercitum magnum, sed diversarum factionum. Ille enim magis colligit intensive, hic extensive."[36]

Gassendus Petrus: "Quidquid convenit Generi, Speciei etiam convenit; ut, quia convenit Animali esse sensu praeditum, id Homini etiam convenit."[37]

Leibniz: "La maniere d'enoncer vulgaire regarde plustost les individus, mais celle d'Aristote a plus d'egard aux idées ou universaux. Car disant *Tout homme est animal*, je veux dire que tous les hommes sont compris dans tous les animaux; mais j'entends en même temps que l'idée de l'animal est comprise dans l'idée de l'homme. L'animal comprend plus d'individus que l'homme, mais l'homme comprend plus d'idées ou plus de formalités; l'un a plus d'exemples, l'autre plus de degrés de realité; l'un a plus d'extension, l'autre plus d'intension."[38]

Arnauld A. – Nicole P.: "Mais quand nous parlons ici de mots généraux, nous entendons les univoques qui sont joints à des idées universelles et générales. Or, dans ces idées universelles, il y a deux choses qu'il est très important de bien distinguer, *la compréhension et l'étendue.*

J'appelle *comprehension* de l'idée, les attributs qu'elle renferme en soi, et qu'on ne peut lui ôter sans la détruire, comme la compréhension de l'idée du triangle enferme extension, figure, trois lignes, trois angles, et l'égalité de ces trois angles à deux droits, etc.

J'appelle *étendue* de l'idée les sujets à qui cette idée convient; ce qu'on appelle aussi les inférieurs d'un terme général, qui a leur égard, est appelé supérieur, comme l'idée du triangle en général s'étend à toutes les diverses espèces de triangle."[39]

Kant: "Inhalt und Umfang der Begriffe. Ein jeder Begriff, als Teilbegriff, ist in der Vorstellung der Dinge enthalten; als *Erkenntnisgrund, d. i. als Merkmal* sind diese Dinge *unter ihm* enthalten. – In der erstern Rücksicht hat jeder Begriff einen *Inhalt*; in der anderen einen *Umfang*. Inhalt und Umfang eines Begriffes stehen gegen einander in umgekehrtem Verhältnisse. Je mehr nämlich ein Begriff *unter* sich enthält, desto weniger enthält er *in sich* und umgekehrt."[40]

Mill, J. S.: "The word white, denotes all white things, as snow, paper, the foam of the sea, etc., and implies, or in the language of the schoolmen, *connotes* the attribute *whiteness*."

"The name, therefore, is said to signify the subjects *directly* the attributes *indirectly*, it *denotes* the subjects, and implies, or involves, or indicates, or as we shall say henceforth *connotes* the attributes."[41]

Meinong, A.: "Zieht man, um hierüber klar zu werden, ein beliebiges kategorisches Urteil von der Form '*M* ist *N*' in Betracht . . . Nun bemerkt man aber leicht, daß hier zum Meinen des Subjektgegenstandes auch das *N* herangezogen werden kann . . . Noch deutlicher tritt der Umstand, daß auch das *N* dem Meinen dienen kann, dann hervor, wenn man dafür sorgt, daß diese Funktion nicht, wie in der Wendung '*M*, das *N* ist' geschieht, durch den Anteil des *M* verdunkelt wird. Nimmt man dem *M* alle Bestimmungen, so daß an deren Stelle gleichsam nur der Raum für Bestimmungen übrigbleibt – sagt man also 'etwas, das *N* ist' oder 'das, was *N* ist', so wird alles Bestimmen einzig und allein durch *N* besorgt."

This passage, quoted from §45, is concerned with extension; 'those things which are *N*' ('das, was *N* ist') is just a formulation of Ext_1 (Thesis

4). The next passage (from the same paragraph) is concerned with intension; 'that which is the genus of *M* ('etwas, das das Genus zu *M* abgibt') seems to be a formulation of Int_1 (Thesis 4):

"Die Charakteristik des Soseinsmeinens sei noch durch den Hinweis darauf vervollständigt, daß nicht nur das Prädikat eines gegebenen Satzes, mehr kurz als genau gesprochen, dazu dienen kann, das Subjekt, sondern auch umgekehrt das Subjekt dazu, das Prädikat zu meinen. Wieder wird das Bestimmen, das sich diesmal vom Subjekte her vollzieht, deutlicher, wenn man statt des Prädikates *N* sozusagen den leeren Platz dafür in Gestalt jenes "etwas" setzt, was zunächst ergibt: "*M* ist etwas". Wenden wir dies nun so: "etwas, das *M* ist", oder "etwas, das das Genus zu *M* abgibt" oder dgl. so ist die bestimmende Funktion des *M* und damit die Tatsache klar, daß hier das Soseinsmeinen sich in ähnlicher Weise des *M* bedient, wie oben des *N*, wobei die eine Weise des Soseinsmeinens immerhin eine Art Inversion der anderen darstellt."[42]

Russell, B.: "Class may be defined either extensionally or intensionally. That is to say, we may define the kind of object which is a class, or the kind of concept which denotes a class: this is the precise meaning of the opposition of extension and intension in this connection. But although the general notion can be defined in this two-fold manner, particular classes, except when they happen to be finite, can only be defined intensionally, i.e., as the objects denoted by such and such concepts. I believe this distinction to be purely psychological: logically, the extensional definition appears to be equally applicable to infinite classes, but practically, if we were to attempt it, Death would cut short our laudable endeavour before it had attained its goal. Logically, therefore extension and intension seem to be on a par."[43]

Leonard, H. S.: "By the extension of a term is meant the group or class of all those objects to each of which the term refers or applies." "More generally, the total contingent intension of any term is the set of all characteristics common to the extension of that term."[44]

This is an informal way of stating the definitions (1) and (4) of Thesis 4 (cf. 2.2). According to the second reference the total contingent intension of $A = \{F: (x)(x \in A \rightarrow x \in F)\}$ which is $\{F: A \subseteq F\}$ which is the same as Thesis 4(4).

Bunge, M.: "The extension of a concept is the set of all objects, real or unreal, to which the concept can apply." – "The intension $I(C)$ of a concept C is, then, the set of properties and relations P_i subsumed under the concept or which the concept, so to speak, synthesizes."[45]

According to Frisch similar views have been taken by John of St. Thomas, De Morgan, W. S. Jevons, A. Bain, J. M. Keynes, H. Lotze, T. Fowler, J. Rickaby, T. Pesch, J. Gredt, G. Frege (concerning extension), F. Gonseth, L. S. Stebbing, M. Black, H. L. Searles, L. Couturat.[46]

3. THE PRINCIPLE OF THE INVERSE RATIO

This principle is also called principle of the opposite (or inverse) relation between extension and intension. The principle can be formulated in different (not necessarily equivalent) ways. In all the different forms the principle claims that there are two types of inclusion or two logical meanings of 'something is in something else'. But whereas the first type of inclusion seems to have the same content in all different formulations of the principle, the second type is described mainly in two ways; therefore the different forms in which the principle is presented may be divided into two groups.

3.1 *First Group:*

P1 Class a is part of class b iff the set of characteristics of b is part of the set of characteristics of a.

P2 The species is part of the genus, class in class. In another sense the genus is part of the species, notion in notion.

P3 a is extensionally included in b iff b is intensionally included in a.

P4 Class a is extensionally included in class b iff the set of properties (characteristics) of the class b are intensionally included in the set of properties (characteristics) of the class a.

3.2 *Second Group:*

P5 Class a is part of class b iff class b is part of the definiens in the definition of a.

P6 The species is part of the genus, class in class. In another sense the genus is part of the species: insofar the parts of the definition (of a notion) are (or correspond to) parts of the whole notion.

3.3 *Explanation*

It suffices to give examples for the first group: man is an animal, class in class; animal is in man in the sense that all the characteristics (properties) which are possessed by animals are possessed by man.

The explanation of the second group presupposes the traditional (Aristotelian) view of proper definition (as distinguished by him and philosophers of the Middle Ages from description, tautological necessity and purely meaning declaration)[47]:

A proper definition in which a species is defined consists of genus and differentia. In modern terms: The definiendum consists of a class concept (the species), the definiens consists of the intersection of two class concepts (the genus and the differentia).[48] Example: man = animal $\cap$ rational. Thus it is understandable that the genus 'animal' is part of the definiens (of the definition) or of the notion 'man' (of the species) being defined; the same holds for the differentia 'rational'.

3.4 *The Principle of the Inverse Ratio in the History of Logic and Philosophy*

As in section 2.3 passages and references of philosophers and logicians who state the principle of the inverse ratio are given. These passages show the strong connection and interdependence of the different forms of the thesis (sections 2.1 – 2.3) and the principle of the inverse ratio.[49]

Aristotle: "Again, those divisions into which the form, apart from quantity, can be divided, are also called parts of the form. Hence species are called parts of their genus." This passage is concerned with extension, the next one with intension: "The elements in the definition of each thing are also called parts of the whole. Hence the genus is even called a part of the species, whereas in another sense the species is part of the genus."[50]

Porphyrius: "Whatever is predicated of the genus as genus is also predicated of the species under it; and whatever is predicated of the difference as difference will also be predicated of the species formed from it."[51] "Thus the individual is contained by the species and the species by the genus, for the genus is a kind of whole, the individual a part. The species is both a whole and a part, a part of another and a whole, not of another but in others. The whole is in the parts."[52]

Boethius: "Genera supervadunt species suas . . ., species vero genera differentiarum pluralitate. Animal enim, quod est genus, supervadit hominem, quod est species, quia non hominem solum continet, verum etiam bovem, equum aliasque species. . . Species vero, ut homo, supervadit genus, ut animal, multitudine differentiarum".[53]

Thomas Aquinas: "He posits eight modes in which something is said to be in something ... The third mode is as man is said to be in animal, or as any species is in its genus. The fourth mode is as a genus is said to be in the species. And lest this mode seem extraneous, he signifies the reason why he says this. The genus as well as the differentia is part of the definition of the species. Hence in a certain way both the genus and the differentia are said to be in the species as parts in the whole."[54]

"He now gives four senses in which something is said to be a *part* ... In a second sense parts mean those things into which something is divided irrespective of quantity; and it is in this sense that species are said to be parts of a genus. For a genus is divided into species, but not as a quantity is divided into quantitative parts. For a whole quantity is not in each one of its parts, but a genus is in each one of its species ... In a fourth sense parts mean those things which are placed in the definition of anything, and these are parts of its intelligible structure; for example, animal and two-footed are parts of man. From this it is clear that a genus is part of a species in this fourth sense, but that a species is part of a genus in a different sense, i.e., in the second sense."[55]

Cajetanus: Cf. the citation in 2.3.

Leibniz: "Hactenus quantitates ex individuis terminorum aestimavimus. Et cum dictum est omnis homo est animal, consideratum est omnia individua humana esse partem individuorum animalis. Sed inversa plane est ratio aestimandi secundum ideas. Nam uti homines sunt pars animalium, ita contra notio animalis est pars notionis quae homini competit, homo enim est animal rationale."[56] "Two terms which contain each other but do not coincide are commonly called 'genus' and 'species'. These, in so far as they compose concepts or terms (which is how I regard them here) differ as part and whole, in such a way that the concept of the genus is a part and that of the species is a whole, since it is composed of genus and differentia. For example, the concept of gold and the concept of metal differ as part and whole; for in the concept of gold there is contained the concept of metal and something else – e.g., the concept of the heaviest among metals. Consequently, the concept of gold is greater than the concept of metal."[57]

Kant: "Inhalt und Umfang eines Begriffs stehen gegen einander in umgekehrtem Verhältnisse. Je mehr nämlich ein Begriff *unter* sich enthält, desto weniger enthält er *in sich* und umgekehrt."[58]

De Morgan: "The logicians who have *recently* introduced the distinction of extension and comprehension, have altogether missed this opposition of the quantities, and have imagined that the quantities remain the same.

Thus, according to Sir W. Hamilton 'All *X* is some *Y*' is a proposition of comprehension, but 'Some *Y* is all *X*' is a proposition of extension. In this the logicians have abandoned both Aristotle and the laws of thought from which he drew the few clear words of his dictum: 'the genus is said to be part of the species; but in another point of view (αλλως) the species is part of the genus'. *All animal* is in *man*, notion in notion: *all man* is in *animal*, class in class. In the first, all the notion *animal* part of the notion *man*: in the second, all the class *man* part of the class *animal*. Here is the opposition of the quantities."[59]

Bunge, M.: "The intension of concepts behaves inversely to their extension: the more properties are assembled the less individuals will share them. In other words, the intension of general concepts is included in or is at most identical with the intension of the corresponding specific concepts. Symbolically, if $(x)(Sx \rightarrow Gx)$, then $I(G) \subseteq I(S)$, where '*S*' designates a species, '*G*' a genus, and '$\subseteq$' the relation of class inclusion."[60]

According to Frisch similar views have been taken by John of St. Thomas, Petrus Gassendus, the Port Royalists, H. Lotze, J. Rickaby, L. Couturat.[61]

3.5 *Intensional Inclusion*

3.51 There is hardly a problem to interpret the traditional concept of inclusion, if it is expressed by saying: species is contained (included) in genus like a part is in its whole. What is meant is the standard class-inclusion in the two forms:

$$a \subseteq b \leftrightarrow (x)(x \in a \rightarrow x \in b)$$
$$a \subset b \leftrightarrow (x)(x \in a \rightarrow x \in b) \wedge \sim(a = b) \text{ where}$$
$$a = b \leftrightarrow (x)(x \in a \leftrightarrow x \in b)$$

Since by Thesis 4 $\{x\colon x \in a\}$ is the Ext_1 of a, it is clear that $a \subseteq b$ and $a \subset b$ are called *extensional inclusions*.

3.52 There is however a problem how to interpret the type of inclusion which is expressed by saying: genus is contained in species, insofar as the set of properties (characteristics) of the genus are included in the set of properties (characteristics) of the species; or since the genus-term is a part of the whole definiens by which species is defined. Or more generally and independent of an interpretation of the tradition: How can one define the kind of inclusion which is meant when I say: the set of properties which are common to all elements of class a are included in the set of properties which are common to all elements of class b? It was clear to me from the very beginning that an adequate answer to this question will provide a key-concept for a new interpretation of intensions. The solution to this question is the following definition of intensional inclusion:[62]

3.53 Set a is intensionally included in set b (symbolically: $a \sqsubseteq b$) iff for all sets Z: if a is a subset of Z then b is a subset of Z. Symbolically:

$$a \sqsubseteq b \leftrightarrow (Z)\,(a \subseteq Z \rightarrow b \subseteq Z)$$

The concept of intensional inclusion just defined is the dual concept to that of the familiar extensional inclusion $\subseteq$ usually called set-or class-inclusion.

That the given definition is in fact an adequate interpretation of the kind of inclusion involved in the principle of the inverse relation between extension and intension is proved by the fact that this principle follows as a theorem from the dual theory, i.e., $a \subseteq b \leftrightarrow b \sqsubseteq a$ is a theorem of the theory of (virtual) classes (cf. Theorem 3 in section 4).

3.54 In order to complete the duality, intensional intersection and union are defined analogously:

$$(a \sqcap b) = c \leftrightarrow (Z)\,[(a \subseteq Z \wedge b \subseteq Z) \leftrightarrow c \subseteq Z]$$
$$(a \sqcup b) = c \leftrightarrow (Z)\,[(a \subseteq Z \vee b \subseteq Z) \leftrightarrow c \subseteq Z]$$

Adding the three definitions of intensional inclusion, intersection and union to the standard theory of (virtual) classes results in a dual theory of (virtual) classes which is both extensional and intensional. Since the extension is conservative and the definitions are noncreative the dual part which is intensional is already implicitly contained in the standard theory we know. It only was not made explicit and so a reasonable and simple way of having some kind of intension was not made use of so far.

4. THEORY OF EXTENSION AND INTENSION

4.1 *Notation*

'a', 'b', 'c', 'x', 'y', 'z', 'X', 'Y', 'Z' are class-variables or set-variables. Atomic sentences are formed with the usual membership-relation. Sentential connectives $\neg$, $\vee$, $\wedge$, $\rightarrow$, $\leftrightarrow$, are used in the usual way. Well-formed formulas are defined in the usual way. '*P*' for 'Propositional calculus'. '$\underset{\cdot}{=}$' stands for extensional identity as defined by D1 T4.1. '$\doteq$' stands for intensional identity as defined by D8.

4.2 *Definitions:*

$$a \underset{\cdot}{=} b \leftrightarrow (z)(z \in a \leftrightarrow z \in b) \text{ extensional identity} \qquad \text{D1}$$
$$a \subseteq b \leftrightarrow (z)(z \in a \rightarrow z \in b) \text{ extensional inclusion} \qquad \text{D2}$$
$$a \sqsubseteq b \leftrightarrow (Z)\ (a \subseteq Z \rightarrow b \subseteq Z) \text{ intensional inclusion} \qquad \text{D3}$$
$$a \cap b \underset{\cdot}{=} c \leftrightarrow (x)\ [(x \in a \wedge x \in b) \leftrightarrow x \in c] \qquad \text{D4}$$
$$a \cup b \underset{\cdot}{=} c \leftrightarrow (x)\ [(x \in a \vee x \in b) \leftrightarrow x \in c] \qquad \text{D5}$$
$$a \sqcap b \underset{\cdot}{=} c \leftrightarrow (Z)\ [(a \subseteq Z \wedge b \subseteq Z) \leftrightarrow c \subseteq Z] \qquad \text{D6}$$
$$a \sqcup b \underset{\cdot}{=} c \leftrightarrow (Z)\ [(a \subseteq Z \vee b \subseteq Z) \leftrightarrow c \subseteq Z] \qquad \text{D7}$$

D1, D2, D4 and D5 are well-known definitions of the standard theory of classes and of set theory. The new definitions are D3 (*intensional inclusion*), D6 (*intensional intersection*) and D7 (*intensional union*). With the help of the additional definitions D3, D6 and D7, a *theory of extension and intension* can be introduced into the theory of classes and into set theory. For the proof of the theorems of the theory of extension and intension only the principles of first order predicate logic are needed, nothing else; i.e., no special axiom of set theory.

4.3 *Theorems:*

$$a \subseteq b \rightarrow b \sqsubseteq a \qquad \text{T1}$$

Proof: $(Z)[a \subseteq b \rightarrow (b \subseteq Z \rightarrow a \subseteq Z)]$ is a theorem (transitivity of implication). Therefore $a \subseteq b \rightarrow (Z)(b \subseteq Z \rightarrow a \subseteq Z)$ and so by D3 T1.

$$b \sqsubseteq a \rightarrow a \subseteq b \qquad \text{T2}$$

Proof: $b \sqsubseteq a \rightarrow (Z)(b \subseteq Z \rightarrow a \subseteq Z)$ from D3; by universal instantiation: $b \sqsubseteq a \rightarrow (b \subseteq b \rightarrow a \subseteq b)$ and therefore $b \subseteq b \rightarrow (b \sqsubseteq a \rightarrow a \subseteq b)$ by P. Since $b \subseteq b$ is a theorem, T2.

$$a \subseteq b \leftrightarrow b \sqsubseteq a \qquad \text{T3}$$

Proof: From T1 and T2.

T3 is the principle of the opposite relation between extension and intension. It says: *a is extensionally included in b* iff *b is intensionally included in a*. Or: class *a* is included in class *b* iff notion *b* is included in notion *a*.

Or: the class *a* is extensionally included in the class *b* iff the set of properties (characteristics) of the class *b* is intensionally included in the set of properties (characteristics) of the class *a*.

EXAMPLES: The class of men is extensionally included in the class of animals iff the set of characteristics (properties) of animals is intensionally included in the set of characteristics of men. The class of cardinal numbers is extensionally included in the class of ordinal numbers iff the set of characteristics of ordinal numbers is intensionally included in the set of characteristics of cardinal numbers.[63]

Another example raises a new question: Can we say that the class of natural numbers is included in the class of rational numbers iff the set of properties of rational numbers is intensionally included in the set of properties of natural numbers? Since there is a one-one-correspondence, this seems dubious. On the other hand, there are rational numbers which are not natural numbers. An unambiguous formulation of the relation between natural and rational numbers, according to the principle T3, seems to be this: The class *A* of (classes of) numbers which are described by an axiom-system for natural numbers is extensionally included in the class *B* of (classes of) numbers which are described by an axiom-system for rational numbers, iff the set of properties which is attributed to *B* by the axiom-system of rational numbers is intensionally included in the set of properties which is attributed to *A* by the axiom-system of natural numbers.

Finally the following two questions have to be distinguished here:

(1) *Is* T3 *valid?* (2) *Is* T3 *a good interpretation of the principle of the opposite relation between extension and intension?*

The answer to the first question is clearly: Yes, if we accept the usual theory of classes and First-Order Predicate Logic, T3 is simply a theorem of the theory of classes, if one replaces '$b \sqsubseteq a$' by its definiens '$(Z)(b \subseteq Z \rightarrow a \subseteq Z)$'. The answer to the second question seems to me to be 'Yes' with some restrictions. T3 seems to be a quite good interpretation of the traditional principle of the opposite relation between extension and intension. The reason is that in the relevant philosophical and logical contexts this principle is almost exclusively applied to *genus* and *species* but not generally to classes (in the modern sense of the word). This is a strong restriction if one realizes that *species*, in the proper sense, for Aristotle is a class of substances in the proper sense (i.e., real existing individuals) or a first order predicate of individuals with the additional restriction that it expresses an essence. Thus animals are a *species* (there is an essential difference between plants and animals), but tables and houses are not (since artificial things have only accidental properties, no essence according to Aristotle). If on the other hand, T3 is understood as the principle of the opposite relation between extension and intension which is generally applicable to classes (in the modern sense) one may have to choose certain unambiguous (and therefore often complicated) formulations and to exclude other formulations in order to recognize the old principle in the new instance at all.

$$(a \subseteq b \wedge b \subseteq a) \leftrightarrow (b \sqsubseteq a \wedge a \sqsubseteq b) \qquad \text{T4}$$

Proof: From T3 by P.

$$a \doteqdot b \leftrightarrow (Z)(a \subseteq Z \leftrightarrow b \subseteq Z) \qquad \text{T4.1}$$

Proof: From D1, D2, D3 and T4.

$$x \in (a \cap b) \leftrightarrow x \in a \wedge x \in b \qquad \text{T5}$$

Proof: Replace 'c' by '$a \cap b$' in D4.

$$(a \sqcap b) \subseteq Z \leftrightarrow a \subseteq Z \wedge b \subseteq Z \qquad \text{T6}$$

Proof: Replace 'c' by '$a \sqcap b$' in D6.

$$a \subseteq (b \cap c) \leftrightarrow (x)\,[x \in a \rightarrow (x \in b \wedge x \in c)] \qquad \text{T7}$$

Proof: Use D2 and T5.

$$a \sqsubseteq (b \sqcap c) \leftrightarrow (Z)[a \subseteq Z \rightarrow (b \subseteq Z \wedge c \subseteq Z)] \qquad \text{T8}$$

Proof: Use D3 and T6.

$$x \in (a \cup b) \leftrightarrow (x \in a \vee x \in b) \qquad \text{T9}$$

Proof: Replace 'c' by '$a \cup b$' in D5.

$$(a \sqcup b) \subseteq Z \leftrightarrow a \subseteq Z \vee b \subseteq Z \qquad \text{T10}$$

Proof: Replace 'c' by '$a \sqcup b$' in D7.

$$a \subseteq (b \cup c) \leftrightarrow (x)[x \in a \rightarrow (x \in b \vee x \in c)] \qquad \text{T11}$$

Proof: Use D2 and T9.

$$a \sqsubseteq (b \sqcup c) \leftrightarrow (Z)[a \subseteq Z \rightarrow (b \subseteq Z \vee c \subseteq Z)] \qquad \text{T12}$$

Proof: Use D3 and T10.

$$(a \cap b) \subseteq c \leftrightarrow (x)[(x \in a \wedge x \in b) \rightarrow x \in c] \qquad \text{T13}$$

Proof: Use D2 and T5.

$$(a \sqcap b) \sqsubseteq c \leftrightarrow (Z)[(a \subseteq Z \wedge b \subseteq Z) \rightarrow c \subseteq Z] \qquad \text{T14}$$

Proof: Use D3 and T6.

$$(a \cup b) \subseteq c \leftrightarrow (x)[(x \in a \vee x \in b) \rightarrow x \in c] \qquad \text{T15}$$

Proof: Use D2 and T9.

$$(a \sqcup b) \sqsubseteq c \leftrightarrow (Z)[(a \subseteq Z \vee b \subseteq Z) \rightarrow c \subseteq Z] \qquad \text{T16}$$

Proof: Use D3 and T10.

$$a \sqcap b \doteqdot a \cup b \qquad \text{T17}$$

Proof: $[(a \sqcap b) \subseteq Z] \leftrightarrow [(a \cup b) \subseteq Z]$ by T4.1; transform the left part of the equivalence by T6, D2, P and T15 into $(a \cup b) \subseteq Z$.

$$a \sqcup b \sqsubseteq a \cap b \qquad \text{T18}$$

Proof: Use D3, transform T10 and T13 into:
$(Z)[[(x)(x \in a \rightarrow x \in Z) \vee (x)(x \in b \rightarrow x \in Z)] \rightarrow (x)[(x \in a \rightarrow x \in Z) \vee (x \in b \rightarrow x \in Z)]]$

$$a \cap b \subseteq a \sqcup b \qquad \text{T19}$$

Proof: Use T18 and T3.

The theorems T17, T18, and T19 are – besides the theorems T3 and T4.1 – the most important of the theory. They explain the relations between intersection and union. T17 says: *the intensional intersection of a and b is (extensionally) identical with the extensional union of a and b.* T18 says: *The intensional union of a and b is intensionally included in the extensional intersection of a and b.* T19 says: *The extensional intersection of a and b is extensionally included in the intensional union of a and b.*

EXAMPLES: T17: The intensional intersection of the sets of characteristics of animals and men is extensionally identical with the extensional union of the sets of animals and men. T18: The intensional union of the sets of characteristics of living things and of things without senses [this is just the set of common characteristics of things without senses] is intensionally included in the set of characteristics of the extensional intersection of living things and things without senses [this intersection is just the class of plants]. T19: The extensional intersection of living things and things without senses (i.e., the class of plants) is extensionally included in the class of things which have as their characteristics the intensional union of the characteristics of living things and things without senses.

T20 Analogous to the theorems of the logic of classes, the following are theorems:

$$(a \sqcap b) \sqsubseteq a,\ a \sqsubseteq (b \sqcup a),\ a \sqsubseteq (b \sqcap a) \leftrightarrow a \sqsubseteq b,$$
$$(a \sqcup b) \sqsubseteq a \leftrightarrow b \sqsubseteq a,\ a \sqcap b \doteqdot b \sqcap a,\ a \sqcup b \doteqdot b \sqcup a,$$
$$[(a \sqcap b) \sqcap c] \doteqdot [a \sqcap (b \sqcap c)],\ [(a \sqcup b) \sqcup c]$$
$$\doteqdot [a \sqcup (b \sqcup c)],\ a \sqcap a \doteqdot a \sqcup a \doteqdot a.$$

$$a \cap \Lambda \doteqdot \Lambda \qquad a \cap V \doteqdot a$$ T21

Proof: Replace 'b' and 'c' by 'Λ' or 'V' in D4.

$$a \sqcap \Lambda \doteqdot a \qquad a \sqcap V \doteqdot V$$ T22

Proof: Replace 'b' by 'Λ' and 'c' by 'a' or 'b' and 'c' by 'V' in D6.

$$a \cup \Lambda \doteqdot a \qquad a \cup V \doteqdot V$$ T13

Proof: Replace 'b' by 'Λ' and 'c' by 'a' or 'b' and 'c' by 'V' in D5.

$$a \sqcup \Lambda \doteqdot \Lambda \qquad a \sqcup V \doteqdot a$$ T24

Proof: Replace 'b' and 'c' by 'Λ' or 'b' by 'V' and 'c' by 'a' in D7.

T25 The distribution laws are analogous to those of the theory of classes:

$$(a \sqcup b) \sqcap c \doteqdot (a \sqcap c) \sqcup (b \sqcap c)$$

Proof: Use T6 and T10.

$$(a \sqcap b) \sqcup c \doteqdot (a \sqcup c) \sqcap (b \sqcup c)$$

Proof: Use T10 and T6.

T26 The following laws are analogous to those of the theory of classes:

$$a \sqsubseteq (b \sqcap c) \rightarrow a \sqsubseteq (b \sqcup c)$$

Proof: Use T8 and T12.

$$(a \sqcup b) \sqsubseteq c \rightarrow (a \sqcap b) \sqsubseteq c$$

Proof: Use T14 and T16.

T27 The following three theorems state interrelations between extensional and intensional functors:

$$a \sqsubseteq (b \sqcap c) \leftrightarrow (b \cup c) \subseteq a$$

Proof: Use T3 and T17.

$$(a \sqcap b) \sqsubseteq c \leftrightarrow c \subseteq (a \cup b)$$

Proof: Use T3 and T17.

$$(a \sqcup b) \sqsubseteq c \leftrightarrow c \subseteq (a \cap b)$$

Proof: Use T16, T3, D3, T7.

$$(a \sqcup b) \subseteq c \rightarrow (a \cap b) \subseteq c \qquad \text{T28}$$

Proof: Use T18 and D2.

$$a \sqsubseteq (b \sqcup c) \rightarrow (b \cap c) \subseteq a$$

Proof: Use T3, T28.

Universität Salzburg and *Institut für Wissenschaftstheorie, Internationales Forschungszentrum, Salzburg*

NOTES

1 Cf. (Top) 121b; (Cat) 3b21–23; (AnPost) 96a24–26. (Met) 1023b17–25.

2 Sextus Empiricus (AMt) VIII, 11. Translated in Bochenski (HFL) §19.21

3 (Isg), especially chapters 2, 3, and 8.

4 (IPC), p. 112. Cf. *ibid.* Lib. II, pp. 92-94; Lib. IV, pp. 261-263; Lib. V, p. 306.

5 (CPr) Tr. VI, 5, p. 181. Cf. *ibid.* Tr. IV, 6, p. 100.

6 (APC) IV, 4(435); (AMC) V, 21 (1094, 1097); (STh) I, 85, 3 ad 2.

7 (ULQ) questio 17.

8 (CPI) I, p. 25. According to William Hamilton and Thomas Spencer Baynes it is Cajetanus who uses the expressions 'extensive' and 'intensive' for the first time.

9 (SLg) I, 20. Ockham, though, having the usual view on extension and extensional inclusion refuses to say that "in another sense the genus is part of the species."

10 (OpO) Vol. I, Pars II, Canon XVI, p. 105.

11 (OFI) p. 300.

12 (LAP), p. 31.

13 (Lg) A 147, 148.

14 (WSL), §56, 64, 66. Like Ockham he is exceptional in his view on intensions: they are not properties of objects belonging to the extension.

15 (SLR) I, ch. 2, 5, p. 19. Mill deviates from the traditional understanding of extension and intension and is criticized for that by De Morgan.

16 (SPS) p. 201. Cf. the passage quoted in 3.4, note 59.
17 (LRU) p. 80.
18 (SBd), (GGA), §§2, 27, 32.
19 (Ann), p. 268 s.
20 (LgS), p. 93 s.
21 (PMt), p. 69.
22 (MgN), p. 23 ss.
23 (IML), p. 3–31, (FLS), (ORF) I and II.
24 (PpR), p. 214–268.
25 (QPA), (LPV) especially chapters VII and VIII. Quine is listed here, not because he has proposed any theory of intension but because he has given important and severe criticism of a number of proposals to clarify or define intensions, especially of all those approaches in which intensions are not dual counterparts of extensions in the sense that they are not mutually interdefinable or reducible. From this it does not follow that such approaches are impossible but Quine's criticism showed indirectly a number of conditions which have to be met and difficulties which have to be overcome. To the theory of intension proposed in this paper (cf. chapters 3.5 and 4) Quine's criticism is not applicable since these intensions can be incorporated into the standard theory of (virtual) classes since they satisfy the conditions 1.1 to 1.7 given below.
26 (Prg) (PIL).
27 (SRI) p. 65 ss. (TBP) I, chapters 2, 3, and 4.
28 "Individuals in a type-theoretical sense" include a much wider class of entities than individuals in the proper sense; for the latter the type-theoretical condition is but one necessary condition, though not sufficient. In his *Categories* Aristotle requires two additional conditions for real individuals in space and time: the second condition is the uniqueness condition (i.e., that there is at least one and at most one). But so far the two conditions are also satisfied by mathematical entities like natural numbers. The third condition for individual substances (substances in the primary sense) is – according to Aristotle – that they are bearers of contingent and contrary properties in the sequence of time; i.e., they have contingent property P_1 at t_1 and contingent property P_2 at t_2 such that P_2 is contrary to P_1. *Categories*, chapter 5, 2a, 3b and 4a. Cf. Weingartner (CEM) and (PUD).
29 In order to be able to express some of the formulations of the thesis more generally the epsilon $\in$ used here may be on one hand the elementhood-relation of the standard theory of (virtual) classes and of set theory but on the other hand it can also be interpreted as the primitive relation of predication. Thus '$x \in y$' means "x is an element of y" if y is a class or set and '$x \in y$' means "x is a y" or "x has the property y" if y is a property, attribute or characteristic.
30 Aristotle, (Top) IV, 1, 121b. Observe that in cases where special sets (like species and genus) are used Ext_2 rather than Ext_1 is the correct interpretation, although Ext_2 (i.e., the class of subsets) is much more general than the class of all species falling under one genus.
31 Porphyrius, (Isg) II, 2b.
32 [Everything that is a genus contains within it many species; everything that is a species has within it many *differentiae*. For a genus, let us say animal in this man, as species, exceeds and excels, because man is only man, while animal is not only man, but also ox or bird or other things of this kind. Species on the other hand exceed their genera be-

cause those *differentiae* which species have in actuality genera do not have.] Boethius, (IPC), p. 112.

[33] [Indeed, genera exceed species in terms of the things contained in those species under the genus ... Constituted species, on the other hand, exceed genera in terms of the things contained in the *differentiae* constituting them ...] Albert the Great, (CPr), Tr. VIII, 5, 131.

[34] Thomas Aquinas, (STh) I, 85, 3 ad 2.

[35] [For I say that, on the level of things, a genus is more universal than a species. Similarly, on the level of intentions, because a genus is predicated on entities that are different in both species and number while a species is predicated on entities that are different in number only, it follows that a species stands for a greater number of things, not however in reality, because there it stands for fewer.] Duns Scotus, (ULQ), in *Opera Omnia*, Vol. I, Quaestio 17, p. 248.

[36] [With regard to this it can be said that that which is more collective of many things can be understood in two ways. In one way intensively; thus the species is more collective because it more successfully unites things ... In another way extensively; thus the genus is more collective because many more things fall under its unifying than under the range of [one of its] species. Hence the species and genus are related as two leaders, one of which has a small but very united army, the other a large army of different factions. For the former collects things better intensively, the latter extensively.] Cajetanus (Cardinalis, Thomas de Vio), (CPI) Cap. I, p. 25.

[37] [Whatever is consistent with the genus is also consistent with the species; for example, since it is consistent with animal to be endowed with sense, so it is also consistent with man.] Gassendus, P., (OpO), Vol. I, Pars II, Canon XVI, p. 105.

[38] Leibniz, G.W., (NEE) IV, 17, 8. ["The common manner of statement concerns individuals, whereas Aristotle's refers rather to ideas or universals. For when I say *Every man is an animal* I mean that all the men are included amongst all the animals; but at the same time I mean that the idea of animal is included in the idea of man. 'Animal' comprises more individuals than 'man' does, but 'man' comprises more ideas or more attributes: one has more instances, the other more degrees of reality; one has the greater extension, the other the greater intension." *New Essays on Human Understanding*, trans. and ed. by P. Remnant and J. Bennett, Cambridge University Press, Cambridge, 1981.]

[39] Arnauld, A.-Nicole, P., (LAP) p. 55. ["Though general words can be equivocal or univocal, when we speak of general words in this text we refer to univocal words which are linked to universal ideas.

We distinguish the comprehension of a universal idea from the extension of a universal idea.

The comprehension of an idea is the constituent parts which make up the idea, none of which can be removed without destroying the idea. For example, the idea of a triangle is made up of the idea of having three sides, the idea of having three angles, and the idea of having angles whose sum is equal to two right angles, and so on.

The extension of an idea is the objects to which the word expressing the idea can be applied. The objects which belong to the extension of an idea are called the inferiors of that idea, which with respect to them is called the superior. Thus, the general idea of triangle has in its extension triangles of all kinds whatsoever." *The Art of Thinking: Port-Royal Logic*, trans. by J. Dickoff and P. James. Bobbs-Merrill, Indianaoplis, 1964, p. 51.]

[40] Kant, (Lg) A 147, 148, §7. ["*Intension and Extension of Concepts.* Every concept, *as a partial concept*, is contained *in* the presentation of things; as a *ground of cognition*, i.e., *as a characteristic*, it has these things contained *under it*. In the former regard, every concept has an *intension* [content]; in the latter, it has an *extension*.

Intension and extension of a concept have an inverse relation to each other. The more a concept contains *under* it, the less it contains *in* it." *Logic*, trans. by R.S. Hartman and W. Schwarz, Bobbs-Merrill, Indianapolis, 1974, pp. 101–102.]

[41] Mill, J. S., (SLR) I, chapt. 2, 5, p. 19 and 20.

[42] [In order to be clear about this, let us consider an arbitrary categorical judgment of the form '*M* is *N*' ... But now it is easily noted that here *N* can also be referred to the meaning of the subject-object ... The case that *N* can also serve as the meaning stands out even more clearly if one takes care that this function is not obscured through the *M* portion, as happens in the phrase, '*M*, which is *N*'. If all determinations are taken from *M* so that in their place, as it were, only space for determinations remains – thus if one says, 'something, which is *N*' or 'those things which are *N*' – then all determination is taken care of through and only through *N* alone.]

[The characteristics of '*Soseinsmeinen*' would be completed through the indication that not only can the predicate of a given sentence, shorter than exactly as it is spoken, serve to mean the subject for that purpose, but also vice versa, the subject can serve to mean the predicate. Again the determining which now takes place from the subject becomes clearer if one, so to speak, substitutes the empty place for the predicate *N*, in the form of any 'something' which then produces: '*M* is something'. If we now turn this around: 'something which is *M*' or 'something which transfers the genus to *M*' or something similar, then the determining function of *M*, and thereby the fact, is clear that here the '*Soseinsmeinen*' makes use of *M* as it did of *N* above, whereby one way of '*Soseinsmeinen*' still describes a kind of inversion of the other.] Meinong, (Ann) p. 268 and 270.

[43] Russell, B., (PMt), p. 69. Cf. *ibid.* Appendix A, 'The Logical and Arithmetical Doctrines of Frege,' p. 501–522.

[44] Leonard, H. S., (PpR) p. 214 and 241.

[45] Bunge, M., (SRI) p. 67 and 66.

[46] Cf. Frisch, J. C., (ECL), p. 183–214.

[47] Aristotle, (Met) 1029b 31–1030a 18. Cf. Thomas Aquinas (AMC) VII, 3, 1327–1330.

[48] Of course, though species, differentia and genus are class concepts not every class concept is some species or differentia or genus. Every proper species must be of first order, i.e., predicated of proper individuals, that is of individuals which satisfy the three conditions mentioned in note 28. This predication must be an essential predication (which implies but is not equivalent to necessary predication). Moreover the predication-relation between individual, species and genus is transitive which is not true for classes in general. For short: species, differentia and genus are very special kinds of classes. Cf. Porphyrius (Isg), chapter 2.

[49] Concerning the question whether the cited passages are identical with one of the formulations P1-P6 what has been said at the beginning of section 2.3 holds here too: it is only claimed that the content of one of these principles is stated more or less explicitly in the historical texts.

[50] Aristotle (Met) 1023b17 and 1023b23.

[51] Porphyrius, (Isg), chapter 8.

[52] *Ibid.*, chapter 3.
[53] [Genera surpass their species . . ., species surpass genera in their plurality of *differentiae*. For animal as genus surpasses man as species, because the genus contains not only man, but also ox, horse, and other species . . . But a species, such as man, surpasses a genus, such as animal, because of its plurality of *differentiae*.] (IPC), p. 306.
[54] (APC), IV, 4, (435).
[55] Thomas Aquinas, (AMC), V, 21, (1093, 1094, 1096, 1097).
[56] Leibniz, G.W., 'De formae logicae comprobatione per linearum ductus', in Leibniz (OFI), p. 292–321, p. 300. [Up to now we have judged quantities according to individuals belonging to terms. And when it is said that every man is an animal, one considers all human individuals to be a part of the individuals of animal. But, judging things on the basis of ideas, the converse is true. For just as men are a part of animals, so, conversely, the notion of animal is part of the notion which coincides with man, for man is a rational aninal.]
[57] Leibniz, 'Elements of a Calculus', in Leibniz (LgP), p.17–24, p. 20 (= OFI, p. 52 s.).
[58] Kant, (Lg), A 148. ["Intension and extension of a concept have an inverse relation to each other. The more a concept contains *under* it, the less it contains *in* it." *Logic*, trans. by R.S. Hartman and W. Schwarz, Bobbs-Merrill, Indianapolis, 1974, pp. 101–102.]
[59] De Morgan, (SPS), p. 201, note 2.
[60] Bunge, (SRI), p. 66.
[61] Cf. Frisch, (ECL), p. 183–209.
[62] I have proposed this definition for the first time in a paper on the thesis of extensionality delivered in 1970 at the International Meinong Colloquium in Graz (cf. (FEP)). The claim in footnote 129 on page 178 of the Proceedings (cf. Haller (JSN)) saying that this definition was also proposed by Bas v. Fraassen in (MRP) is wrong, since his definition is different in two respects: it is modal and it has also a different structure without the modal impact. The theory was first presented in (PCI), p. 264 ss. being a consequence of the set-theoretic predicate calculus developed there. That it can be easily incorporated into first order predicate calculus (or better: into the theory of virtual classes) – independently of this set-theoretic predicate calculus – I discovered in 1973, cf. the publications (IIS) and (SPI). For a more complete report of the respective historical doctrines see my (WTh) II, 1, chapters 3.41–3.45 (p.117–157) and (FEP).
[63] Several objections against the principle interpreted by T3 have been made by philosophers: Cf. F. Brentano, (LRU), p. 80, B. Bolzano, (WSL) I, §120 (also referred to in Kneale W. – Kneale M., (DLg), p. 364 f.). These objections show that without a precise definition of 'intensional inclusion' the interpretation of T3 leads to difficulties. The main source for the construction of counterexamples seems to depend on the question of what an 'additional property' is, when sets of properties (characteristics) are compared with one another. It seems that all 'counterexamples' to the principle make use of properties which are, on one hand, already included in the original set of properties (by normal use of their meaning) but are, on the other hand, used as "additional properties."

REFERENCES

Albert the Great
(CPr) *Commentaria in De Praedicabilibus* (in *Opera Omnia*), L. Vivès, Paris, 1890.

Aristotle
(An Post) *Analytica Posteriora*, in *Prior and Posterior Analytics* (vol. 1 of *Works*), intr. comm. W.D. Ross, Clarendon Press, Oxford 1957.
(Cat) *Categories* (vol. 1 of *Works*), J.A. Smith and W.D. Ross, eds. Clarendon Press, Oxford.
(Met) *Metaphysics*, intr. comm. W.D. Ross (vol. 8 of *Works*), Clarendon Press, Oxford, 1958.
(Top) *Topics*, (vol. 1 of *Works*), J.A. Smith and W.D. Ross, eds. Clarendon Press, Oxford.
Arnauld A. – Nicole P.
(LAP) *La logique ou l'art de penser*, Paris 1662 and 1752, (known as "Logique du Port Royal").
Bochenski, I. J. M.
(HFL) *A History of Formal Logic*, Notre Dame University Press, Notre Dame, 1961.
Boethius
(IPC) *In Isagogen Porphyrii Commenta*, ed. Samuel Brandt. Vienna, Leipzig, 1906. (*Corpus Scriptorum Ecclesiasticorum Latinorum*, ed. Academiae Litterarum Caesareae Vindobonensis, vol. 48).
Bolzano, B.
(WSL) *Wissenschaftslehre*, 4 vols., ed. A. Höfler – W. Schultz, Meiner, Leipzig I-II 1929, III 1930, IV 1931.
Brentano, F.
(LRU) *Die Lehre vom richtigen Urteil*, F. Mayer-Hillebrandt, ed., Francke, Bern, 1956.
Bunge, M.
(SRI) *Scientific Research I*, Springer, Berlin, 1967.
(TBP) I *Treatise on Basic Philosophy*, vol. I, Reidel, Dordrecht, 1974.
Cajetanus
(CPI) *Commentaria in Porphyrii Isagogen ad Praedicamenta Aristotelis*, P. Insardus – M. Marega, ed., Rome, 1934.
Carnap, R.
(MgN) *Meaning and Necessity*, 2nd ed., enl., Chicago University Press, Chicago, 1956.
Church, A.
(FLS) 'A Formulation of the Logic of Sense and Denotation', in P. Henle (ed.), *Structure, Method and Meaning, Essays in Honour of H.M. Sheffer*, Liberal Arts Press, New York, 1951, p. 3–24.
(IML) *Introduction to Mathematical Logic*, vol. I, Princeton University Press, Princeton, 1956.
(ORF) 'Outline of a Revised Formulation of the Logic of Sense and Denotation', Part I and II, *Noûs* **7** (1973), 24–33 and **8** (1974), 135–156.
Couturat, L.
(LgL) *La Logique de Leibniz* (1901), Olms, Hildesheim, 1961.
Duns Scotus
(ULQ) 'In Universam Logicam Questiones', in *Opera Omnia*, Vivès, Paris, 1891, Vol. I.
Fraassen, B. v.
(MRP) 'Meaning Relations among Predicates', *Noûs* **1**(1967), 161–179.

Frege, G.

(GGA) *Grundgesetze der Arithmetik I*, Pohle, Jena, 1893; Darmstadt, 1962.

(SBd) 'Über Sinn und Bedeutung', *Zeitschrift für Philosophie und philosophische Kritik* **100**(1892), 25–50. English Translation in Frege, *Translations from the Philosophical Writings*, P. Geach and M. Black, ed., Blackwell, Oxford, 1952.

Frisch, J. C.

(ECL) *Extension and Comprehension in Logic*, Philosophical Library, New York, 1969.

Gassendus Petrus

(OpO) *Opera Omnia* (Lugduni: Sumptivus Laurentii, Anisson & Joan. Bapt. Devent, 1658).

Haller, R. ed.

(JSN) *Jenseits von Sein und Nichtsein, Beiträge zur Meinong-Forschung*, Academische Druck-u. Verlagsanstalt, Graz, 1972.

Kant, I.

(Lg) *Logik* (Immanuel Kants Logik, ein Handbuch zu Vorlesungen), Königsberg, 1800.

Kneale, W. and Kneale, M.

(DLg) *Development of Logic*, Clarendon Press, Oxford, 1962.

Leibniz, G.W.

(NEE) 'Nouveaux Essais sur l'Entendement Humain', in *Die philosophischen Schriften von G. W. Leibniz*, vol.V, ed., C. I. Gerhardt, Weidmann, Berlin, 1875–1890.

(LgP) *Logical Papers*, G. H. R. Parkinson, ed., Clarendon Press, Oxford, 1966.

(OFI) *Opuscules et fragments inédits de Leibniz*, L. Couturat, ed., Alcan, Paris, 1903.

Leonard, H. S.

(PpR) *Principles of Reasoning* (1957), Dover, New York, 1967.

Mally, E.

(LgS) *Logische Schriften*, K. Wolf and P. Weingartner, eds., Reidel, Dordrecht, 1971.

Meinong, A.

(Ann) *Über Annahmen*, Barth, Leipzig, 1910.

Mill, J. S.

(SLR) *A System of Logic, Ratiocinative and Inductive*, Parker, London, 1843 (1960).

Montague, R.

(FPh) *Formal Philosophy, Selected Papers of R. Montague*, R. H. Thomason, ed., Yale U. P., New Haven, 1974.

(Prg) 'Pragmatics' in: *Contemporary Philosophy*, vol. 1 R. Klibansky, ed., La nuova Italia, Florence, 1968, pp. 102–122. (Reprint in Montague (Fph)).

(PIL) 'Pragmatics and Intensional Logic', *Synthese* **22**(1970), 68–94. (Reprint in Montague (FPh)).

Morgan de, A.

(SPS) 'Syllabus of a Proposed System of Logic', in De Morgan, (SLW), pp. 147–207.

(SLW) *On the Syllogism and Other Logical Writings*, P. Heath, ed., Routledge and Kegan Paul, London, 1966.

Ockham.

(SLg) *Summa Logicae*, Ph. Boehner, ed., Franciscan Institute, St. Bonaventure, N.Y., 1951–.

Porphyrius

(Isg) *Isagoge*, Translation, Introduction and Notes by E. W. Warren, The Pontifical Institute of Mediaeval Studies, Toronto, 1975.

Quine, W. V. O.

(LPV) *From a Logical Point of View*, Harvard University Press, Cambridge, 1961.

(QPA) 'Quantifiers and Propositional Attitudes', *The Journal of Philosophy* **53**(1956) 177–187.

Russell, B.

(PMt) *The Principles of Mathematics* (1903), Allen and Unwin, London, 1937.

Sextus Empiricus

(AMt) *Adversos Mathematicos*, in *Sextus Empiricus*, R. G. Bury, ed., Harvard University Press, Cambridge, Mass., 1961.

Thomas Aquinas

(AMC) *Commentary on the Metaphysics of Aristotle*, J. P. Rowan, tr., Regnery, Chicago, 1961.

(APC) *Commentary on Aristotle's Physics*, R. J. Blackwell, R. J. Spath, W. E. Thirkel, tr., Yale University Press, New Haven, 1963.

(STh) *Summa Theologica*, P. Caramello, ed., Marietti, Rome, 1952.

Weingartner, P.

(CEM) 'On the Characterization of Entities by Means of Individuals and Properties', in Weingartner, ed., *Logic and Ontology*, (separate issue of the) *Journal of Philosophical Logic* **3** (1974).

(FEP) 'Die Fraglichkeit der Extensionalitätsthese und die Probleme einer intensionalen Logik', in R. Haller, ed., (JSN), pp. 127–178.

(IIS) 'On the Introduction of Intensions into Set Theory', in M. Przełęcki, K. Szaniawski, R. Wójcicki, eds., *Proceedings of the Conference of Formal Methods in the Methodology of Empirical Sciences*, Reidel, Dordrecht, 1975.

(PCI) 'A Predicate Calculus for Intensional Logic', *Journal of Philosophical Logic* **2**(1973) 220–303.

(PUD) 'The Problem of the Universe of Discourse of Metaphysics', in *Archives de l'Institut International des Sciences Théoretiques*, Actes du Colloque "Science et Métaphysique" de l'Académie Internationale de Philosophie des Sciences, Fribourg, 1973; Office International de Librairie, Brussels, 1976, pp. 207–254.

(SPI) 'A Simple Proposal for Introducing Intensions into Science' in '*Archives de l'Institut International des Sciences Théoretiques*, Actes du Colloque "La Semantique dans les Sciences" de l'Académie Internationale de Philosophie des Sciences, Rixensaart, 1974; Office International de Librairie, Brussels, 1978, pp. 193–209.

(WTh) *Wissenschaftstheorie I, Einführung in die Hauptprobleme*, Frommann-Holzboog, Stuttgart, 1971; 2nd edition 1978; *II*, 1 *Grundlagenprobleme der Logik und Mathematik*, Frommann-Holzboog, Stuttgart, 1976.

JOHN WETTERSTEN

THE PLACE OF MARIO BUNGE

INTRODUCTION

The name of Mario Bunge conjures unease amongst those who are in the know. He is today one of the most well known of living philosophers, his numerous publications are well known, and his positions leave sharp impressions; nevertheless, there has been relatively little critical appraisal of his work. This seems a pity, since his work is quite intriguing and recognized as such by many who look for challenges yet are frustrated by him. The current situation of a well-known, controversial and slightly read author seems paradoxical enough to call for some explanation. My own conjecture, for what it is worth, is that Bunge's work is not presented as the obvious alternative to the work of the philosophers who are in the public eye, and therefore it does not generate the public discussion it profitably might generate, because it is not clearly seen as constituting such an alternative. Now why is an alternative seen, yet not as an alternative? In order to answer this we need know what, in general, makes one view one alternative to the other. Tentatively, let me say, it is a question: two views are alternatives to each other when viewed as competing answers to a given question. We can now answer my question thus. Perhaps Bunge's view is not publicly debated because it is not seen as an alternative to publicly debated views, because the question that might make them competing is not obvious. The question, which can make Bunge's view a competitor to publicly debated answers to it, to conclude my conjecture, is this. Under what conditions can two views be made competing by devising a question to which both views may be deemed answers, and when are these competing and when not? This question, I may note in passing, is a problem that I need to solve, however partially, in writing this essay, as well as the subject of the essay. To this question we have some classical answers. First comes Collingwood's answer. He is the first, because it is a question he has invented. After Collingwood comes Quine, Popper, and of course, Kuhn and Feyerabend who say it cannot be answered, and whose views will not be discussed here. Finally, in my opinion, comes Bunge. In this paper, then, I will discuss Collingwood's questions, and

J. Agassi and R.S. Cohen (eds.), Scientific Philosophy Today, 465–486.
Copyright © 1981 *by D. Reidel Publishing Company.*

answers by Collingwood, Quine, Popper and Bunge.

My assumption here is that people's views about the possibility of contrasts between views from different frameworks are various and have a deep impact on how they perceive alternatives and how they propose to appraise alternatives once they have been described. I hope here merely to make some of this background knowledge concerning alternatives explicit in hopes this will further the dialectic.

I hope here to present the problem interestingly. It is no doubt not the only one that can do the job at hand. And – I hope Professor Bunge will not mind – some sacrifice of precision is needed in order to gain an overview. For me, and I hope for my audience as well, this attempt to see the forest before examining the trees may be useful. Even if subsequent examination may refute an initial conjecture about Bunge's place, this conjecture may help us to come to an improved appraisal of his philosophy.

In presenting my view of Bunge's philosophy I have found it useful to adopt a new perspective. Such a procedure is naturally suspect since it seems more reasonable to try traditional problems and ways of placing alternatives before inventing new ones. In Bunge's case, however, there is, to repeat, a blatant paradox which I feel it useful to explain and which I have felt forces me to a very broad and somewhat new framework for placing some contemporary philosophers. I hope this placing is useful though it admittedly is not unique.

Before coming to the problem at hand, I should observe a general feature of Bunge's philosophy which I find puzzling. He is deeply involved with two modes in contemporary philosophy which are normally seen to be at odds with each other. One of these modes is analytic philosophy, which deems the primary task of philosophy to be the analysis of concepts, and for many the tools for this are from logic, especially set theory. The other mode is fallibilism or the philosophy that views all knowledge as conjectural and the primary tool of its growth as criticism and conjecture. Bunge's analytic and fallibilistic biases together place him in no-man's land: one group objects to his analysis as excessive, the other to his fallibilism and metaphysics as mere conjectures. In this situation we cannot deal with his philosophy quite adequately if we deem him merely in one school or the other. Nor can we deal with his philosophy if we fail to understand why these strains are combined, what problems they solve, and whether there are alternatives.

There are various explanations of this phenomenon. We might explain

Bunge's analytic bias away as merely an outmoded position which is grafted onto a fallibilist philosophical position. In that case we might expect our effort to come to terms with Bunge's philosophy problematic since we would not then view parts of it as intended to solve any problem and could not unify various older solutions, but wish to explain them away.

When faced with a difficulty of interpretation, however, it seems wiser to seek a real problem before explaining a part of a text away. We may then require some problems which these various threads in Bunge's philosophy respond to. In this way we may raise two problems, at least. The first is, whether the problem we ascribe to an author is worth solving, and the second is whether his proffered solution is the best available.

In this essay, then, I want to propose a problem to which Bunge's combination of analytic techniques and fallibilist philosophy may respond. I would like to describe in broad, even superficial terms how this problem has arisen and alternatives to it. I wish to suggest that it has been an influential problem in the shaping of many of our views even though the study of it has nevertheless been neglected: it has not itself been well appraised nor have alternative solutions to it been explicitly juxtaposed. I suggest that in this way we may see Bunge's philosophy as part and parcel of an ongoing discussion of an influential problem with several alternative answers. In doing so we may see that Bunge's position is not merely on two sides of a divide but offers solutions to problems caused by attempts to reconcile agreeable aspects of each side of the divide. This would present Bunge's combination as not eclectic but a tour de force.

In my attempt to formulate the problem, however, I am not endorsing the view of it as serious. The difficulty posed by putative serious problems requires that their claims for seriousness themselves be evaluated. I will here present one in order to set the stage for such understanding and evaluation. Some philosophers, for example Popper (whose views I will discuss below), do not deem the problem serious. Whether serious or not, however, its impact is largely due to its influence on leading twentieth century philosophers.

Let me now finally come to the problem, explain it in broad terms, and then turn to Collingwood's early and powerful portrayal of it. I will then present in broad terms the views of how it should be solved of Collingwood, Quine, Popper and Bunge. In doing so my main concern is merely to describe these positions in broad, even superficial terms, in hopes this may set the stage for some further appraisal.

1. TWO TRENDS IN CONTEMPORARY PHILOSOPHY

There is a traditional problem, the problem of how to explain how we have knowledge, without falling quite unintentionally into either a relativist or dogmatist position. Relativism declares the choice of frameworks arbitrary; it results from the acceptance of many frameworks. Dogmatism declares only one framework acceptable, even without sufficient warrant. Both, it is agreed, are to be avoided as undesirable; neither, it seems, can be avoided without incurring the other.

This difficulty of avoiding either horn of the dilemma is due to the difficulty of forming a theory of how frameworks may be rationally appraised. It is thought, on the one hand, that selection should be unique: we should find proof; while, on the other hand, the very existence of alternative frameworks seems to block such proof. If we seek proof and recognize the role of frameworks, we thus seem pushed to dogmatism by declaring only proofs in one framework acceptable, or relativism by declaring proofs in any framework acceptable each within its own framework, especially since it appears we must use a framework to provide a proof. So, it seems reasonable to set aside temporarily, perhaps even permanently, the desire to find proof and pose instead a seemingly more narrow problem: how else can we rationally appraise frameworks? How can we rationally appraise frameworks even though proof is not possible? How can we be reasonable about the choice, that is, of a proper background or context to a given problem? Now, logic has traditionally been designed to yield proof. We are proposing, then, a lower aim for our use of logic. This would seem to be a reasonable move, since the lowering of standards should make it easier to employ logic. This has turned out annoyingly not to be the case: the use of logic in appraisals of frameworks or contexts seems problematic because it in no way recognizes the peculiar features of frameworks or contexts. Standard logics, since they are designed for absolute proof, do not themselves recognize contexts, but must treat all propositions as discrete entities whose structure, meaning, and truth value, are absolute, i.e., depend on nothing but the structure of language and the facts, neither of which poses a problem for proof or for rationality as proof. When we introduce frameworks and rational appraisal rather than proof, however, this situation changes somewhat drastically. This is so because we now accept in argument propositions whose meanings depend on contexts. We want also to use our traditional logic which recognizes only absolute propositions to supply rules for our arguments. But are the rules which may be quite good if propositions are meaningfully independent of context, good enough even if the meaning of propositions does depend on context? As we shall shortly see, there are reasons for thinking the answer is 'no'. If it is 'no', then we have indeed a serious problem.

For, our problem is, how can we avoid both relativism and dogmatism; since when we seek proof and recognize the existence of frameworks we seem to fall into one or the other, we attempt to solve the problem by lowering our standards in some way; yet we then find that our very ability to employ our basic rules of logic becomes questionable.

These are problems involved in attempts to get out of this dilemma. One: Can there be context-free meaning of propositions? Two: Can we translate a context-dependent proposition to a context-free one? Three: Are there translation rules for that? Evidently, options are limited, since having context-independent meanings offers both the possibility and the rules for translation, context dependence permits the possibility and impossibility of translation, and possibility of translation permits rules to exist or not, but impossibility of translation excludes the existence of rules. The popular public opinion is straightforward. Most philosophers see here no problem, for they view meaning as context-free anyway and we only have a problem if the meaning of propositions is context-dependent. Now each of the other live options which includes the theory that meaning is context-dependent, has its representatives. Quine and Bunge seem to claim that even though meanings of propositions are context-dependent we can translate in accord with rules; I am a bit leery here since much depends on the interpretation of the word 'rules'. Popper claims we can translate but rules of translation must be very limited or not exist at all. Kuhn and Feyerabend abandon all possibility of translation (this is their incommensurability thesis). I will not discuss their view here. Collingwood holds we have no rules of translation but aspires to create them all the same.

Now, except for the extremes, those who see no problem at all here and those who block all possible solution – each thinker engaged in our problem has a theory of how we can engage in discussion between contexts. Collingwood says we cannot now translate but he holds we shall save the day when we obtain a new logic of questions. Popper says we can translate without rules and solve problems which translations incur as we meet them. This leads to a piecemeal approach; we translate, we are critical of our translations, and deal with difficulties due to poor translation, when criticism reveals such poverty. Quine says, we can translate and we can use rules, but only in a fixed, even highly explicit, context. More on this later. Bunge says, we can translate only if we provide systematic though tentative theories of how translation is possible.

2. COLLINGWOOD'S PROBLEM

The problem just introduced provides a framework for my survey. I dub this problem "Collingwood's problem." It was not unique to him, yet he was one of the first to perceive it and the first to deem it important and face it explicitly. Even while trying to solve it, others failed to present it so forcefully. I may thus use a discussion of Collingwood's problem as he conceived it to set my present framework. Assuming there is a context-free system of logic, Collingwood's problem, how can we translate between frameworks, reduces into: how can we translate between context-dependent theory and logic?

During his lifetime Collingwood's position in the philosophical community was not enviable. His colleagues deemed his theories and problems to be out of date and did not see the point of his objections to their views: communication was thus cut to a minimum. This perception of Collingwood as a back-slider had two very close causes. The first of these was his maintenance of a theory that thought occurred in contexts. His view came so close to idealism as to be identified with it. He was deemed a mere variant of Bradley, Bosanquet and Green with an Italian flavor due to a smidgen of Croce.

This misapprehension of Collingwood has recently been remedied. Yet the remedy still leaves Collingwood in limbo, as long as it consists merely in acknowledging Collingwood's realism. This is admittedly progress, since Collingwood was not only a realist but one whose realism is a crucial part of his philosophy: it provides a primary motive force for its development and a criterion for its success or failure.

Yet, this very recognition of Collingwood's realism also leads to a rather negative perspective on Collingwood's work, as long as it is merely grafted onto the earlier view of Collingwood as a fellow traveler of the idealists: it makes him possibly inconsistent, and inconsistent for sure if we assume that he had hoped to transform an idealistic perspective into a realist philosophy and failed. Thus Collingwood is only saved from being an outdated idealist by the view that he is a failed realist due to inability to jettison early idealist sympathies.

These readings of Collingwood can be explained by a commonly used historiography which views philosophers primarily as system builders and evaluates their success or failure as philosophers by their approximation of their philosophies to the ideal of a system. It was this very idealistic historiography of Collingwood's (realistic) critics which damned Collingwood as an idealist. Without saying that such an historiography is utterly useless, we can acknowledge that following it narrowly we can ignore important developments. This indeed has happened in the case of Collingwood.

An alternative approach to the history of thought is to include as valuable the work of those philosophers who pose new and challenging problems, whether or not they themselves solve these problems and whether or not their solutions are satisfactory. If we adopt this point of view we can see how Collingwood was indeed more than a mere failure at system building, even though his philosophy never succeeded in achieving its aim of explaining how thought constrained to stay within frameworks could be realistic. It is his analysis of a problem and his proposal of an incomplete program for solving it in which his originality lay.

The fundamental problem Collingwood faced was a product of his early affinity for both the idealists and the realism of Cook Wilson. As a student he found the realist criticism of Bradley and his friends convincing. Yet soon after he found Cook Wilson's realist views empty and refuted by his own techniques. Since he rejected both Bradley's idealism and the realism of Cook Wilson he needed some third view. In developing his view he hoped to maintain the view that rationality required frameworks. Yet he did not want to be a relativist. He was sympathetic therefore to Cook Wilson's criticism even while rejecting it. In order to adopt the proper critical and objectivist stance, while rejecting the errors of Cook Wilson's theory of criticism, Collingwood felt he had to develop a theory which could allow for proper critical and rational discussion, i.e., critical and rational discussion which recognized importance of context for rationality.

Collingwood sought a theory of rational thought as bound by frameworks, and a theory of science which is, in some sense realist. It is not clear to me what were for Collingwood (and for succeeding thinkers) sufficient conditions for a realist view; but some necessary conditions for this, and the problems these conditions pose, may be identified. In broad terms any realist theory must avoid dogmatism and relativism as seemingly easy and final solutions to the problem of how knowledge is possible.

Even while attempting to avoid them, it is easy to lapse into them.

Collingwood views Hegel as the best example of a relativist due to the severity of the limitations his theory of frameworks places on rationality. This is a standard view and needs no explanation. More surprisingly he views Russell as the best example of dogmatism. Collingwood views Russell's position as dogmatic because it ignores frameworks entirely. The failure to take into account frameworks leads in Collingwood's view to an irrationalism which is evident in Russell's use of logic. This use of logic becomes arbitrary due to arbitrary removal of propositions from their context. The combination of the use of arbitrary techniques combined with the appearance of proof, indeed the appearance of the highest degree of rationality possible, leads to dogmatism. Arbitrary moves and spurious proof leads to dogmatic assurance. The analysis of this dogmatic position leads to a problem which Collingwood deems it necessary to solve in order to maintain realism with an adequate theory of frameworks.

The problem Collingwood saw was how to avoid the arbitrariness which resulted from the use of Russell's logic. This arbitrariness was due to the translation of propositions which could only be properly understood in some context into a language (logic) which was blind to context and then proceeding to engage in manipulations with these propositions while forgetting the context from which they came. It was this mistranslation into the language of logic which caused the difficulty.

Collingwood could have merely dispensed with logic and translation as impossible. But he could not take this course and maintain his realism: such a course would have blocked his aims to have a critical discussion which aimed at the discovery of truth. Collingwood thus accepted both the theory that all (rational) thought was bound by frameworks. He also held that in order for thought to be objective or realist we must employ logic and critical discussion. Yet all logic presumes, from Collingwood's point of view, a dogmatic posture. It does so because it presumes that propositions can be true or false quite independent of context. This leads to dogmatism because it leads to erroneous arguments which use arbitrary techniques with spuriously firm conclusions.

The severity of the difficulty may be increased. If we translate from propositions within a given framework into a logic whose framework is the supposition that there is no framework, we will be bound to make mistakes. We cannot make such a translation and preserve meaning and thus the validity of arguments, as they are translated from one language or context to another: the change of context which occurs in translation will change meaning. As we cannot change meaning without changing validity, the validity of logic is thus spurious and based on an error.

3. COLLINGWOOD'S PROGRAM

Collingwood thought that some proper translation between the propositions occurring within the frameworks of science and logic was needed, as a minimum requirement for objective or rational thought. Now this problem of translation may be posed as a general problem of how we may translate between any two languages. The translation between informal language and logic would be merely a special case. This broad problem poses a special difficulty for Collingwood, however. This difficulty is that Collingwood sees some translations as impossible, as he sees no way of translating from one context to another unless we can devise some broader context in which we may place the two. Some frameworks can and some cannot be placed in a broader context. In order to set the broader context one has to specify presuppositions. But this leads to a regress – possibly an infinite regress. To stop the regress by accepting the truth of some presuppositions on faith is dogmatism. So Collingwood thought that the presuppositions of some frameworks are absolute for those frameworks in the sense that they are not questioned or reduced to further ones, so as to block infinite regress, yet they had no truth value, and so are not dogmas: they are simply taken for granted and not questioned – and for want of a context in which to question them. The question then is: how can there be presuppositions with no truth value. To this we need Collingwood's idea of truth-value which differs from that of ordinary logic. Truth, he said, depends on context, hence truth value without context is impossible. Now two contexts including their different presuppositions can merge into one broader context, but if we have propositions from two contexts, each with different absolute (unquestioned) presuppositions, then we cannot place these in a unified context unless both contexts are transcended. Absolute presuppositions are adopted merely as a device, as a way to set up a context. They are, at least in this capacity, beyond the scope of rational discussion. There may be some ideal, broad, unique, unifying context but such a framework is, and must be, beyond human knowledge. This means that we can have no theory of translation between frameworks. Such a translation requires some further unifying framework. Yet no such unifying framework is possible unless we wait until a new one is created that transcends them both. What then?

Let us now accept the view that any discussion of any question takes place within some unquestioned presuppositions which constitute a context or framework. We may ask now, how is reasoned discussion of any question, even within a framework, at all possible? Collingwood held that no propositional logic could be adequate even for this purpose, because the logic must itself account for the contextual features of each individual proposition. Collingwood is thus led to a more narrow program than he might have adopted because it does not provide a broad theory of translation. Nevertheless it is super ambitious because Collingwood set himself the task of providing an entirely new logic which could take into account the contextual features of each and every proposition.

Collingwood's idea is that a formal logic of contexts, the logic of questions, can formalize the contextual features of informal language and thus offer contexts for questions and answers and rational debates. In that case, a translation between the informal and formal language will not be a translation from one context or framework to another, but rather translation into logic – indeed a mere explication, a mere heightening of the awareness of what one takes unquestioningly for granted – a translation which will preserve the context and the framework within which rational questioning occurs. We will thus translate not only individual propositions but also their context when we translate statements from the informal or poor awareness mode of some framework into a more formal mode of heightened awareness in the same framework. And the ability to translate from different contexts into the formal system would then, perhaps, offer a common ground for rational discussion, but on this Collingwood is systematically very vague and seemingly uncommitted.

Collingwood is forced by the logic of his position to abandon hope for any theory of proper translation between frameworks and to the view that an entirely new logic was needed. All he could show was that, looking at metaphysics as a historical science, we can see how absolute presuppositions do provide frameworks for discussions, and if we want to understand a dead discussion afresh we must revive its context and the structure of that context. Yet he could neither supply the alternative logic which he needed for the completion of his program nor could he offer a theory of translation within that new logic. The logic of questions then, the logic of contexts, makes all discussions events in history.

Collingwood's rationalism, his theory of contexts as problems, and his theory of the need for a logic of contexts as problems, cannot be overlooked. Since Collingwood has no logic and there were no students to carry on the program, he has been almost completely ignored. Alan Donagan still uses the fact that his program is unexecuted in order to dismiss it. Yet the inner logic of the program is retained in his problem and the problem keeps popping up in others' programs.

4. QUINE

Perhaps the most famous discussion of translation of all is Quine's. Quine's problem of translation has the same root source as Collingwood: Quine sees the meaning of all propositions as context-dependent. And even though he is quite pessimistic about the possibility of good translation he still wants translation to be as good as it can be. Quine's pessimism is based in his theory of the indeterminacy of translation. This theory is that any proposition may be translated in any number of ways without our being capable of seeing the mistranslation. We may only discover it, perhaps to our surprise or peril, when some new situation arises unexpectedly. This possibility is deemed real because of Quine's view that context is always important to translation. When we translate some simple sentence we do so in a context. We see if our translation is correct by seeing if the person who uses the proposition we wish to translate responds in appropriate ways. This test is the best we have, says Quine, but is still quite weak. It is quite weak because the context in which we test the response of other individuals might be perceived differently by them than by us, even though they respond appropriately. To put it his way, their response is occasional, and the occasion is the context, our perception is not occasional and so we must provide a context. As a result when we extend the use of our translation to further contexts we may do so in different ways, even if our grasp of the context is quite adequate. So how can we explain the possibility of translation, and how can we extend it?

Quine explains the possibility of translation in two ways. On the one hand we have the possibility of translating occasional sentences, i.e., sentences which are specific to contexts in which humans interact most intimately. This translation is based on our ability to build up with other people expected reactions to the use of particular sentences. These expected reactions and the translation of occasion sentences become his goal: only here can translation become as good as we can make it: the transition from the occasional to the context-dependent is a kind of extrapolation.

The translation of occasional sentences leaves open many possibilities of translation of other types of sentences. This poses a problem in itself since we need a policy toward translation of those sentences as well. But this is not all. The successful translation of occasional sentences depends on the establishment of common contexts since only this will enable us to successfully project reactions. The context of the use of occasion sentences depends on the theoretical framework in which they occur. We thus have an interest in the use of our framework for the successful translation of occasion sentences.

Now Quine believes that we can translate sentences at a higher level. This translation is much farther from the highly rich texture of the translation of occasional sentences. It is too far from specific contexts to approach being really good. Quine therefore disregards almost entirely the aim of establishing good translations at this level in the sense of accurately translating meaning. He adopts instead Duhem's policy: we must establish rules which will refine our system to make the translation of occasional sentences to the richer languages more effective, but also permit translating them into the less rich ones. For Quine occasional sentences occur in everyday life whereas for Duhem they occur in the laboratory.

For example, translation – I should say paraphrase – into the language of modern logic does not translate meaning but refines our propositions to improve the utility of our framework. It so happens that the use of paraphrase is highly useful, i.e., it leads to an increased ability to translate occasional sentences. Thus, Quine wishes to make our framework, i.e., context, increasingly definite by refining our theoretical statements in such a way as to serve to improve our translation of occasional sentences: our ability to feel at home with each other, or to have expected relations.

Now this reading of Quine as a mere conventionalist seeking social unity seems to overlook his Platonism. Indeed it does not seem entirely reconcilable or at least I do not see how. Yet it seems true anyway and for this reason: even his ontology is evaluated on its usefulness. Existence claims per se are uninteresting: they are themselves a type of tool we can use to provide a context for translation of occasional sentences. Perhaps, however, the connection, tenuous or not, is this. The core of logic – including the law of contradiction but excluding the law of the excluded middle – is universal, he says. Yet, one cannot hope to translate from a sufficiently rich language to the core, since the core is analytic yet not clearly demarcated, since the distinction between the analytic and the rest is not clear cut. The Platonic ontology belongs to the core and is thus outside our discourse.

Like Collingwood, Quine abjures any possibility of a context that can unify various contexts. The best we can do, he thinks, is to improve the context in which we find ourselves. Quine's philosophy, just as Hegel's, is ambiguous as to whether it describes merely the current stage of social developments, the current and highest stage, or a universal theory. On the one hand Quine only wishes to refine our framework, no transcendent task is possible; yet he appears to describe this task as if it were transcendent, i.e., the same for all contexts.

Quine thus faces a problem similar to Collingwood's when he tries to explain the selection of frameworks or rather the framework. They both wish to avoid relativism yet neither can hold that a single framework can be justified. Quine's solution is that even though truth is not equal to what works, and everything that works is not true, we may conjecture that whatever works is true. And he recommends this policy. This allows Quine to deem some existing framework true and to fix it as the framework even though he cannot justify it: he confesses to be a non-justificationist. This move can furthermore permit him to provide a pragmatist justification proper for various aspects of his theory of logic: to the extent that he can devise a theory of how logic may work within the context of a single framework, he may deem this theory the true theory of logic.

Quine's position has not been challenged by anyone until quite recently. Gellner has sharply criticized his views. I explain this by the fact that Quine's program of seeking to fix a framework and to operate entirely within it in order to allow for translation of occasion sentences was never deemed the important part of Quine's philosophy. Instead it was thought by many that it was desirable to solve problems of logic or language piecemeal within the framework. This was so because most philosophers of logic did not take seriously the problem of rationality at large, and its translation into the problem of choice of framework, and thus of translation between frameworks. Those who did, were often influenced by Quine so as not to discuss frameworks. Thus even though Quine set the tenor for much of the discussion of logic and language he did not influence it in such a way as to lead logicians and philosophers of language to notice his central problem or his radical conservative pragmatism. Indeed, the same criticism levelled against Collingwood above holds against Quine; is his view of translation translatable? Can there be argument between frameworks? Here translation breaks down.

Gellner became the first critic of Quine's broad program because he is an outsider with an eye for frameworks and especially an eye for unstated ones. Gellner's critique of Quine may be seen as a product of Gellner's social perspective. Whereas Quine hopes to devise a theory – which turns out to be social – to preserve and improve the standards of translation – Gellner reverses the order of things and first looks at the social theory and examines its viability. This is quite readily done since Gellner is a student of the social roles of frameworks and a critic of Collingwood. Can we dismiss a Gellner style criticism? For most philosophers social theory has not been deemed central and so perhaps they may dismiss it; yet sociology is important for Quine's philosophy – even for his philosophy of logic. Gellner's criticism thus shows the severe limitation of Quine's philosophy. If we do not agree on a framework what do we do? Quine does not want to enforce his rules and refrains (with difficulty), as Gellner has already noted, from doing so. Yet he observes acceptance – by persuasion or by force – of some rules and not of others. And the rules that he accepts with profit are accepted by societies that exist longer. Thus Quine's only hope reduces to social Darwinism. Those societies which do enforce or achieve uniform rules of translation will succeed; perhaps they will be rational but more importantly social expectations – a feeling of being at home will be maintained. Thus social Darwinism and pragmatism are joined (as in Dewey) to select those societies which may properly use logic for the conservative maintenance or radical imposition of a single framework. Surprisingly Quine's final concern is not his initial one, for reason or rationality, but for social unity. And this reason is a mere tool toward this non-rational goal.

5. POPPER

Karl Popper is important in this survey of alternatives because even though he accepts the basic presumption that no propositions have meaning outside of some context, he holds that translation of propositions poses no serious difficulty: he holds this even though he does not require or seek to form any translation rules which would render such translation viable. In order to explain why Popper deems it possible to thereby explain away Collingwood's problem it is useful to explain some of the background to this problem.

Collingwood held a position which in the new jargon is called justificationist. On this view we can only have knowledge when truth is established. We cannot, on this view, allow for error. Collingwood's problem becomes serious, given this background, because we can show in a general way that error is not precluded from our normal ways of proceeding. Now Popper, as is now well known, rejects justificationism. He holds instead that we make conjectures as a matter of course. Hence error is unavoidable. As long as we test and improve our conjectures we are proceeding well enough. This posture applied to the present case, suggests the view that we may translate as we deem appropriate and take it from there.

Translations are always possible, says Popper, if we translate in accord with any rule we care to. We may, for example, choose the rule: translate all the words of one language into a single word of another. We may be more restrictive and require a correspondence of a sentence in one language with some sentence of another which will allow us to translate in either direction. We may also specify some translation of meaning. In this way translation is always possible. This way translation, however poor, is guaranteed, and as we go along we may increase the restrictions to whatever degree we deem desirable within the range of feasibility. All this will hardly be contested by Quine. Indeed, he is clearly aware of it. The question is, is what we want at all feasible? This depends on what we want. Quine says he does not want certainty, yet his thesis of untranslatability is proven by the proof that we cannot prove the adequacy of translation. By distinction, Popper claims that the degree to which translation is possible is quite adequate for the purposes of rational discussion – including scientific discussion – even though it is never absolutely precise and even though this impression may at times cause difficulty. Popper sees the existence of mistranslation as endemic but differs in both his appraisal of the problem and in the remedy. The justificationist Collingwood seeks a general solution which will forestall improper translation. Quine seeks to make translation as good as possible. Popper seeks merely to obtain critical tools which will enable us to uncover mistranslations when they cause us trouble. Popper holds that if we can uncover mistakes in translation and correct them to a sufficient degree to solve our problems we needn't worry about Collingwood's problem posing a block to the successful and effective use of logic to promote rational discussion. What this means is the proposal that we assume we understand each other well enough until we hit a snare. Rather than criticize this eminently fallibilist view, let me contrast it with another fallibilist view and return to it later.

6. BUNGE

Bunge's voluminous writings provide almost no discussion whatsoever of translation proper. Yet in this paper I am trying to explain Bunge's place in contemporary philosophy by his solution to this problem. This seems rather quixotic: surely if this problem were central Bunge would have explicit and detailed views on this as he does on most other issues. This need not be the case, however, if his problem is concealed in his system as a recurrent problem which his system attempts to overcome in detailed ways even without a general solution, or, if the problem is a skeleton in Bunge's closet. I do not have preference for explaining Bunge's relative silence on this. But I would like to explain why the problem seems endemic to his philosophy and why it appears we can read him as entertaining two possible solutions.

Bunge's central motivating force in all his philosophical work is to increase exactness as a way of furthering the growth of knowledge. This aim would lead us to expect *a priori* that if Collingwood's problem is any problem at all it must occur for Bunge. This is so because Bunge emphasizes even to the extreme both attitudes which lead to the problem. He pursues on the one hand the context-dependent character of propositions and thus of theories and on the other hand, the exact analysis of theories as a sine qua non of heightened rationality. This raises a severe problem in reading Bunge, whether or not it turns out to be serious or not for Bunge's philosophy. This problem is: how does Bunge deem exactness possible when theories and propositions are conjectural, context-dependent and changing? There seems to be no existing theory to remove this unease – the one this essay starts with – concerning the suspicion that Bunge merely makes eclectic combinations leading to a rather arbitrary idea of exactness. Can we remove this unease? Does Bunge have a unified theory, including a theory of exactness, or not? I think yes.

In order to seek a resolution of this unease, then, I have posed the problem of whether Bunge solves the problem of translation. I do so because Bunge's program of using formal analysis in an exact way can only succeed if he can provide an adequate theory of how we may translate any semi-formal scientific theory into a more formal version, how we can describe their relations, and how this translation improves science. *A priori*, I can see three ways he may solve this problem. He may seek to solve the problem given some exact and universal standards of analytic philosophy, or he may seek to rest his case on a view such as Popper's which is satisfied with a mere piecemeal approach, or he may wish to do better than either and incorporate some virtue of each. This type of solution *a priori* would best resolve the unease, referred to in the beginning of this essay. Since Bunge emphasizes exactness to such a high degree and Popper's view could seem to render this emphasis rather empty, I will not seek this type of answer from Bunge. Instead I will suggest first, that he has a solution to the problem which is very much like Collingwood's. Secondly, that this solution seems to be too strong for his purpose of seeking growth, but that he may have a second solution which beats all options by having exact rules of translation, growth, systematicity and may still allow for piecemeal criticism.

Let me turn now to the first, Collingwood-like solution. Collingwood thought that rationality was threatened because we could only be fully rational if we used logic and we could not properly use existing logic because if we translated statements whose meaning is context-dependent in, for example, science to the language of modern logic, where the meaning of statements is presumed independent of context, we will destroy rational argument. This could only be remedied, says Collingwood, by a new logic. Now Bunge's version of this problem arises because he wishes to provide formal analyses of scientific theory. He recognizes that the meaning of propositions in science are context-dependent. Logic alone, then, is on his view insufficient. Whether his views are quite the same as Collingwood's I cannot say. He does however require theories of analysis which are especially designed for science. My first conjecture would be that he uses his theory of analysis of scientific theory to provide an alternative to Collingwood's theory, an alternative which would accomplish the same task of providing a formal language which allows for proper, exact translation.

This view of Bunge's philosophy is rendered plausible because we may explain how Bunge's proposals for the use of systems theory may solve the problem of translation as Collingwood conceived it. By this I mean that the problem may be deemed solved if we can provide a formal language into which we may translate scientific theories in such a way as to preserve context. On Collingwood's view the preservation of context in translation would be sufficient to guarantee exactness. Notice I speak of science alone since Bunge equates science and rationality and is impatient with the rest. Bunge may be read as sharing the view that exactness in translation is a result of translation of context. Furthermore we may deem the only block to such translation the construction of a context-sensitive formal language.

Now this view of Bunge's philosophy still may leave us with unease. If we say that Bunge deems the use of systems theory just as powerful as

Collingwood deemed the use of his projected logic of problems we are faced with the following problem. Collingwood deemed translation to be so perfect as to merely heighten our awareness of the structure of debate. At the same time as it revealed structure it fixed context: there could be no or very little change or growth of context: once a context was fixed we could not change it: we could only place it in a more general context and this only to a limited degree since many frameworks were discrete and had to remain so.

This Collingwoodian reading of Bunge seems too strong then, because he wants to use the formalization of scientific theories for broader purposes. These broader purposes are the improvement of the statement of theories through formalization and the growth of the contexts of frameworks through criticism and replacement of theories of varying depth. This fact heightens our unease: how can Bunge have a theory of exact translation when exact translation may block his aims of pursuing growth of contexts or frameworks.

Let us try again to see if we can read Bunge in such a way as to provide for both exact translation as complementary to improvement and growth, thereby resolving our initial unease. My final conjecture as to how this is possible, and is indeed the central proposal of the present essay, will follow shortly. Instead of seeing Bunge as deeming the development of his theories of the analysis of scientific theory, as meeting the final conditions of translation, on this reading we may see them as providing a theory – indeed a systematic though tentative theory – of how such translation is possible. The following reading solves the problem with a weaker solution than Collingwood's or that sought by most analytic philosophers since his solution is conjectural, designed to serve specific purposes, and improvable. The following reading provides a stronger solution than Popper's since Bunge systemizes his rules and, while increasing the standards for exactness provides for their test, piecemeal or not, in a way quite in keeping with Popper's own standards.

The desideratum for my reading of Bunge is that he must provide a theory of exactness which is non-arbitrary. We need rules for proper translation which serve some purpose. In particular he needs to allow for improvement of theory and growth of context or framework. Yet we also need a theory which does not deem proper translation merely a function of the local standards in which this translation occurs. If so we have no resolution of our unease that Bunge's systematic emphasis on exactness is empty or arbitrary.

My proposal then for reading Bunge is this. Thesis One: Exactness is itself context-dependent. Two: Proper analysis reveals context. Three: There are degrees of propriety of analysis. Four: Techniques of analysis are improvable. Five: The best techniques available are Bunge's. They concern the degree of exactness we need to meet specific purposes. They can, for example, enable us to provide desiderata for how a new translation may be deemed appropriate both as a paraphrase and as an improvement due to increased precision and as a means of evaluating the application of these desiderata when they compete. Here then are rules that satisfy those desiderata and which Bunge seems to endorse. I do not intend them to be complete.

Let me proceed following Agassi's observation concerning frameworks ('Between Metaphysics and Methodology') by translating the five theses above into rules and then ask whether they fit Bunge's view. Thesis one is that exactness is context-dependent. We may translate this to a rule: be as exact as the context allows. This rule fits Bunge's appraisal of say physics and social science. Social science is now so confused that we cannot immediately achieve the same degree of exactness in any part of the social sciences. The second thesis is that proper analysis reveals context. This may be translated into the rule: start an analysis by formalizing a context. Observe that Bunge always does that. Thesis three says there are degrees of propriety of analysis. We may translate this to the rule: do not worry too much about the correctness of your formalization of the context until the end of the process, after which you can criticize yourself. Thesis four is that techniques of analysis are improvable. This can read: look for latest developments in mathematics in order to see if they can be used in analysis of scientific theories. Observe that Bunge does this. The fifth thesis is that Bunge's techniques are the best. This reads, compare his techniques with others and try to improve upon all of them.

A step towards the implementation of the fifth rule is making Bunge's rules explicit as done in the previous paragraph. Another step is extending his rules to new domains. An attempt in this direction is the next and last section.

7. ON TRANSLATION PROPER

In the foregoing I have discussed the problem of translation. The problem however has been a stand-in for the problem of rationality: translation is a requirement of argument and a standard for understanding. Bar-Hillel in a very remarkable essay starts quite simply with the problem of translation. He poses the problem: how can we translate between the Indian language Choctaw and quantum theory? Bar-Hillel explains that if we deem Choctaw closed, if we cannot enrich it with rules and concepts, we can trivially show that we cannot translate quantum theory into Choctaw. If, however, we agree to enrich Choctaw, then we may do so quite easily, i.e., we may simply add English to Choctaw. This, Bar-Hillel points out, is ridiculous. Yet, since either the extreme of a strictly closed language or the extreme of a strictly open one is unacceptable, we need some theory of a semi-open language: a language with some restrictions on enrichment which still allow for translation.

Bar-Hillel suggests that his problem may be solved in each case if the speakers of the languages can agree in some meta-language on what restrictions to invoke. This will do for each case but he wants more. He would like to have a universal meta-language. This is highly problematical. Since it is, we do not know if there is even a simple meta-language common to all language users and rich enough to provide the needed means of discussing the rules of language. I do not even know whether Bar-Hillel and I could discuss the rules of language.

Now since we have no theory of how to solve Bar-Hillel's problem, this problem of translation becomes a problem of rationality. It does so because we must now evaluate our restrictions on translation vis à vis their ability to promote understanding, argument, etc.

Chomsky and Quine represent two extreme solutions to Bar-Hillel's problem. Chomsky claims (or rather claimed) that the existence of various languages is only apparent. Since all human beings have the same deep structure we all share the same language: translation may be superseded by derivation. Whatever can be said in any surface structure can be understood by any human since it can be derivable from the deep structure: it can appear in various surface structures and our criterion for enrichment is clear: any statement may be incorporated merely by placing it in the universal context of our deep structure.

Quine takes an equally extreme position. On Quine's theory our framework is a mere tool to organize our social interaction. If we find a problem of translation, i.e., if we want to translate a sentence from one speech community into that of another we may do so only if we can come to agree on the proper behavioral reactions to the use of a sentence. We want our speech community to be as broad as possible, but if someone's reactions prove continually abhorrent we merely deem him outside our speech community. This is no problem since we may draw the boundaries of our speech community as wide or as narrow as we like.

Each of these theories seeks to avoid Bar-Hillel's problem. Chomsky seeks to reject the problem by rejecting the idea that there is more than one language. Quine seeks to reject it by finding standards for translation in our ability to establish agreed upon reactions.

These two positions are two of Bunge's *bêtes noires*. The first presumes a super-rationalism which rejects the possibility of improvement of our language. The second is a super-irrationalism which deems agreement and social uniformity the only criterion. The latter position is an extension of Duhem's theory of science to all of society. It must lead however to various societies which may deem other societies as beyond intercourse: membership in a society is determined by proper activity and relations between societies can only be by force. Perhaps not physical but social; but nevertheless, as Gellner has already noted concerning Quine's theory: coercion is always just present there.

Chomsky's theory is equally reactionary since it seeks to place all language in one innate framework. All disagreement, all newness in language, is mere appearance which can be explained away by the proper theory of language. Furthermore language and learning is a mere mechanism and this mechanism fails to explain anything that is specific to language. Thus Chomsky merely explains away all problems and reassures us about the unity of reason. Yet his program could never provide solutions to real and exciting problems such as that posed by Bar-Hillel.

Bunge's program as well as Popper's tries to respond to the problem posed by Bar-Hillel. They do so by allowing openness which Quine does not and growth by enrichment which Chomsky does not. Popper hopes to do this entirely piecemeal, whereas Bunge on my reading recommends the use of a framework or background to solve this problem for as long as the framework is accepted, thus combining piecemeal and systematicity which is his usual trait. This is the trait which evokes unease as it promotes both the critical and the analytic stances.

NOTES

* Joseph Agassi has commented on and criticized both in general and in detail, earlier drafts of this essay. This essay is a product of these conversations. I am grateful for his generous and insightful help.

REFERENCES

Agassi, J.: 1968–69, 'Changing Our Background Knowledge', Review of Mario Bunge *Scientific Research I, II*, *Synthese* **19**, 453–64.

Agassi, J.: 1975, 'Between Metaphysics and Methodology', *Poznań Studies* **1**, 1–8; reprinted in his *Science and Society* (*Boston Studies in the Philosophy of Science*, vol. 65), D. Reidel, Dordrecht, 1981.

Bar-Hillel, Y.: 1964, *Language and Information*, Addison-Wesley, Reading, Mass.

Bar-Hillel, Y.: 1970, *Aspects of Language*, Magnes Press, Jerusalem; North-Holland, Amsterdam.

Bunge, M.: 1959, *Metascientific Queries*, Charles C. Thomas, Springfield, Ill.

Bunge, M.: 1967a, *Scientific Research I, The Search for System*, Springer, New York.

Bunge, M.: 1967b, *Scientific Research II, The Search for Truth*, Springer, New York.

Bunge, M.: 1967c, *Foundations of Physics*, Springer, New York.

Bunge, M.: 1973, *Method, Model and Matter*, D. Reidel, Dordrecht, Holland, and Boston.

Bunge, M.: 1974–79, *Treatise on Basic Philosophy*, vols. I-IV, D. Reidel, Dordrecht, Holland, and Boston. *Semantics I, Sense and Reference*, 1974. *Semantics II, Interpretation and Truth*, 1974. *Ontology I, The Furniture of the World*, 1977. *Ontology II, A World of Systems*, 1979.

Bunge, M.: 1974, 'The Concept of Social Structure', in *Developments in the Methodology of Social Science*, W. Leinfellner and E Koehler (eds.), Reidel, Dordrecht and Boston, pp. 175–215.

Bunge, M.: 1979, *Causality, The Place of the Causal Principle in Modern Science*, rev. ed., Dover Publ., New York.

Collingwood, R. G. 1939, *An Autobiography*, Oxford University Press, London and New York.

Collingwood, R. G.: 1940, *An Essay on Metaphysics*, Clarendon Press, Oxford.

Davidson, D., and J. Hintikka: 1969, *Words and Objects, Essays on the Work of W. V. Quine*, D. Reidel, Dordrecht, Holland.

Donagan, A.: 1962, *The Later Philosophy of R. G. Collingwood*, Clarendon Press, Oxford.

Gellner, E.: 1973, 'Thought and Time, or, the Reluctant Relativist', *Times Literary Supplement*, 3708 (30 March 1973), 337–339; reprinted in his *The Devil in Modern Philosophy*, Routledge and Kegan Paul, London, Boston, 1974.

Gellner, E.: 1975, 'The Last Pragmatist', *Times Literary Supplement*, July 25.

Popper, K.: 1947. 'New Foundations for Logic', *Mind* **56**.

Popper, K.: 1961, *The Logic of Scientific Discovery*, Science Editions, New York.

Quine, W. V. O.: 1953, *From a Logical Point of View*, Harvard University Press, Cambridge.

Quine, W. V. O.: 1960, *Word and Object*, MIT Press, Cambridge.

Wettersten, J.: 1978a, 'Traditional Rationality vs. a Tradition of Criticism, a Criticism of Popper's Theory of the Objectivity of Science', *Erkenntnis* **12** 329–38.

Wettersten, J.: 1978b, 'Tasks without Purpose', review of W. Leinfellner and E. Koehler, eds., *Developments in the Methodology of Social Science*, *Phil. of Sci.* **8** 299–311.

CURRICULUM VITAE OF MARIO BUNGE

1. PERSONAL

Born September 21, 1919, Buenos Aires, Argentina. Canadian citizen.
Physics student, University of La Plata, 1938–44.
Doctor of physico-mathematical sciences. University of La Plata, 1952.

2. FELLOWSHIPS AND AWARDS

Fellow, Conselho Nacional de Pesquisas (Brazil), 1953.
Fellow, Fundación Ernesto Santamarina, Buenos Aires, 1954.
Research Fellow, Alexander von Humboldt-Stiftung, Freiburg i. Br., 1965–66.
Fellow, John Simon Guggenheim Memorial Foundation, Aarhus and Zürich, 1972–73.
Award of merit, University of Wisconsin (Green Bay), 1979.
Doctor of Laws, *honoris causa*, Simon Fraser University, 1981.
Guest of honor, symposium on 'Science and Philosophy in the Work of Mario Bunge', Peñíscola, Spain, 1981.

3. POSITIONS HELD

Founder and secretary general, later headmaster, Universidad Obrera Argentina, 1938–43.
Teaching assistant, Experimental Physics, University of La Plata, 1941.
Secretary general, Federación Argentina de Sociedades Populares de Educación, 1942–44.
Teaching assistant, Mathematical Physics, University of Buenos Aires, 1947–52.
Subdirector, Biophysics Laboratory, Dirección de Medicina Tecnológica, Ministerio de Salud Pública de la Nación, 1949.
Lecturer, Inter American Course on Modern Physics organized by UNESCO, Universidad Mayor de San Andrés, La Paz, Bolivia, 1955.
Lecturer, Physics Department and Instituto Pedagógico, Universidad de Chile, 1955.
Editor, Asociación Física Argentina, 1956–63.
Assistant Professor of Theoretical Physics, University of Buenos Aires, 1956.
Assistant Professor of Theoretical Physics, University of La Plata, 1956.
Professor of Theoretical Physics, University of Buenos Aires, 1956–58.
Professor of Theoretical Physics, University of La Plata, 1956–59.
Professor of Philosophy, University of Buenos Aires, 1957–62.
Councillor, Facultad de Filosofía y Letras, Universidad de Buenos Aires, 1958
Visiting Professor of Philosophy, University of Pennsylvania, 1960–61.
Visiting Lecturer, Universidad de la República, Montevideo, 1962.

J. Agassi and R.S. Cohen (eds.), Scientific Philosophy Today, 487–488.
Copyright © 1981 *by D. Reidel Publishing Company.*

Visiting Lecturer, Universidad Central, Quito, Ecuador, 1962.
Visiting Professor of Philosophy, University of Texas, Spring, 1963.
Visiting Professor of Philosophy and Physics, Temple University, 1963–64.
Distinguished Visiting Professor of Philosophy and Physics, University of Delaware, 1964–65.
Visiting Professor of Physics, Universität Freiburg, Summer semester, 1966.
Professor of Philosophy, McGill University, Montreal, 1966–1981.
Investigador especial, Universidad Nacional Autónoma de México, Summer, 1968.
Head, Foundations and Philosophy of Science Unit, McGill University, 1969–
Honorary Research Professor, Aarhus Universitet, Fall term, 1972.
Visiting Professor, ETH Zürich, Spring term, 1973.
Research professor, Instituto de Investigaciones Filosóficas, Universidad National Autónoma de Mexico, 1975–76.
UNESCO consultant, 1979.
Lecturer, Centro de Investigaciones Científicas, La Paz, México, 1979, 1981.
National Lecturer, Sigma Xi, The Scientific Research Society, 1980–82.
Frothingham Professor of Logic and Metaphysics, McGill University, 1981–.

4. LEARNED SOCIETIES AND JOURNALS

Charter member, and later editor, Asociación Física Argentina (1944–1963).
Editor, *Minerva* (Revista Continental de Filosofía), 1944–45.
Charter member, Agrupación Ríoplatense de Lógica y Filosofía Científica, President, 1960–63.
Editor, and partly translator, *Cuadernos de Epistemología* (Facultad de Filosofía y Letras, Univ. Buenos Aires, 1960–63), a collection of 50 booklets.
Member, Editorial Board, *Ciencia e Investigación*, 1963–
Member, Académie Internationale de la Philosophie des Sciences, 1965–
General Editor, *Studies in the Foundations, Methodology and Philosophy of Science*, Springer 1966–1971.
Member, Editorial Board, *International Journal of Theoretical Physics*, 1969–1978.
Member, Institut International de Philosophie, 1969–
General editor, *Library of Exact Philosophy*, Springer-Verlag, 1969–
Member, Governing Board, Philosophy of Science Association, 1971–73.
Consulting Editor, *Theory and Decision*, 1972–.
Member, National Council, Canadian Society for the History and Philosophy of Science, 1973–76.
Member, Editorial Board, *International Journal of General Systems*, 1973–
Editor, *Episteme Library*, 1974–79.
Member, Editorial Board, *Teorema*, 1974–
Member, Editorial Board, *Poznan Studies in the Philosophy of the Sciences and the Humanities*, 1975–78.
Charter member, Society for Exact Philosophy (1971–). V. P., 1974–76.
Charter member and first president, Asociación Mexicana de Epistemología (1976–).
Member, Intern. Editorial Board, *Applied Mathematical Modelling* (1977–).
Member, Advisory Editorial Board, *Epistemologia* (1978–).
Member, Editorial Board, *Technology in Society* (1979–).
Member, Scientific Council, World Future Studies Federation (1978–).
Editor, *Foundations and Philosophy of Science & Technology* series, Pergamon (1979–).

LIST OF PUBLICATIONS OF MARIO BUNGE

BOOKS

1. *Temas de educación popular*, El Ateneo, Buenos Aires, 1943.
2. *La edad del universo*, Laboratorio de Física Cósmica, La Paz, 1955.
3. *Causality: The Place of the Causal Principle in Modern Science*, Harvard University Press, Cambridge, Mass., 1959.
4. *Metascientific Queries*, Charles C. Thomas, Springfield, Ill., 1959.
5. *Ética y ciencia*, Siglo Veinte, Buenos Aires, 1960.
6. *La ciencia*, translation of three chapters of #4, Siglo Veinte, Buenos Aires, 1960.
7. *Antología semántica* (Editor), Nueva Visión, Buenos Aires, 1960.
8. *Cinemática del electrón relativista*, (Ph.D. dissertation), Universidad Nacional de Tucumán, Tucumán, 1960.
9. *Causalidad*, translation of #3. Editorial Universitaria de Buenos Aires, Buenos Aires, 1960.
10. *Intuition and Science*, Prentice-Hall, Englewood Cliffs, N. J., 1962.
11. *Pritchinost*, Translation of #3. Publishing House for Foreign Literature, Moscow, 1962.
12. *The Myth of Simplicity*, Prentice-Hall, Englewood Cliffs, N. J., 1963.
13. *Causality*, paperback edition of #3, with new Foreword and Appendix, The World Publishing Co., Cleveland and New York, 1963.
14. *La ciencia*, second, enlarged edition of #6, Siglo Veinte, Buenos Aires, 1963.
15. *The Critical Approach: Essays in Honor of Karl Popper* (Editor), includes a Preface, Free Press, Glencoe, 1964. (See article #71.)
16. *Intuición y ciencia*, translation of #10, Editorial Universitaria de Buenos Aires, Buenos Aires, 1964.
17. *Delaware Seminar in the Foundations of Physics* (Editor), includes an Introduction, Springer, Berlin, Heidelberg, New York, 1967.
18. *Scientific Research 1, The Search for System*, Springer, Berlin, Heidelberg, New York, 1967.
19. *Scientific Research 2, The Search for Truth*, Springer, Berlin, Heidelberg, New York, 1967.
20. *Foundations of Physics*, Springer, Berlin, Heidelberg, New York, 1967.
21. *Quantum Theory and Reality* (Editor), includes an Introduction, Springer, Berlin, Heidelberg, New York, 1967.
22. *Az oksag*, translation of #3. Gondolat Kiado, Budapest, 1967.
23. *Intuitsia i nauka*, translation of #10 with a study by V. G. Vinogradov, Progress, Moscow, 1967.

J. Agassi and R.S. Cohen (eds.), Scientific Philosophy Today, 489–500.
Copyright © 1981 *by D. Reidel Publishing Company.*

24. *O Przycznowosci*, translation of #3, Panstwowe Wydawnictwo Naukowe, Warsaw, 1968.
25. *La investigación científica*, translation of #18 and #19, Ediciones Ariel, Barcelona, 1969.
26. *La causalità*, Italian translation of #3, with a new Preface and a study by E. Panaitescu: 'La causalità e il determinismo secondo Mario Bunge', Boringhieri, Turin, 1970.
27. *Problems in the Foundations of Physics* (Editor), Springer, Berlin, Heidelberg, New York, 1971.
28. Japanese translation of #3, Iwanami, Tokyo, 1972.
29. *Ética y ciencia*, 2nd revised ed. of #5, Siglo Veinte, Buenos Aires, 1972.
30. *Teoría y realidad*. Ariel, Barcelona, 1973.
31. Reprint of #25 with Preface by Eramis Bueno, Instituto Cubano del Libro, Havana, 1972.
32. *Philosophy of Physics*, D. Reidel, Dordrecht, 1973.
33. *Method, Model and Matter*, D. Reidel, Dordrecht, 1973.
34. *Exact Philosophy* (Editor), D. Reidel, Dordrecht, 1973.
35. *The Methodological Unity of Science* (Editor), D. Reidel, Dordrecht, 1973.
36. *Filosofia fiziki*, Russian transl. of #32, Progress, Moscow 1974.
37. *Sense and Reference*, vol. I of *Treatise on Basic Philosophy*, D. Reidel, Dordrecht, Boston, 1974.
38. *Interpretation and Truth*, vol. II of *Treatise on Basic Philosophy*, D. Reidel, Dordrecht, Boston, 1974.
39. *Philosophie de la physique*. Transl. of #32. Paris. Ed. du Seuil, 1974.
40. *Teoria e realidade*, transl. of #30, Editora Perspectiva, São Paulo, 1974.
41. Reprint of #10, Greenwood Press, Westport, Conn. 1975.
42. *Tratado de Filosofía Básica*, Vol. 1, Portuguese transl. of #37, Editora da Universidade de São Paulo & Editora Pedagógica e Universitária, São Paulo, 1976.
43. *Tratado de Filosofía Básica*, Vol. 2, Portuguese transl. of #38, Ed. da Universidade de S. Paulo & Ed. Pedagógica e Universitária, São Paulo, 1976.
44. *Tecnología y filosofía*, Universidad Autónoma de Nuevo León, Monterrey, Mexico, 1976.
45. *Ética y ciencia*, 3rd. ed, new appendix: 'Por una tecnoética', Siglo Veinte, Buenos Aires, 1976.
46. *The Furniture of the World*, vol. III of *Treatise on Basic Philosophy*, D. Reidel, Dordrecht, Boston, 1977.
47. *Filosofía de la física*, Spanish transl. of #32, Ariel, Barcelona, 1978.
48. *A World of Systems*, vol. IV of *Treatise on Basic Philosophy*, D., Reidel, Dordrecht, Boston, 1979.
49. *Causality in Modern Science*, 3rd. ed. of #3, corrections and new Preface. Dover Publications, New York, 1979.
50. *Epistemología. Curso de actualización*, Ariel, Barcelona, 1980.
51. *The Mind-Body Problem*, Pergamon Press, Oxford, New York, 1980.
52. *Ciencia y desarollo*, Siglo Veinte, Buenos Aires, 1980.
53. *Epistemologia*, Portuguese translation of #50. Queiroz and Editoria da Universidade de S. Paulo, São Paulo, 1980.

54. *Ciência e desenvolvimento*, Portuguese translation of #52. Itaitia, Belo Horizonte; Editora da Universidade de S. Paulo, São Paulo 1980.
55. *Materialismo y ciencia*, Ariel, Barcelona, 1981.
56. *Scientific Materialism*, D. Reidel, Dordrecht, 1981

ARTICLES

['L' signifies a letter or brief note]

1. 'Introducción al estudio de los grandes pensadores', *Conferencias* (Buenos Aires) III, Nos. 3 & 4, 1939.
2. 'El tricentenario de Newton', Universidad Obrera Argentina, Buenos Aires, 1943.
3. 'Significado físico e histórico de la teoría de Maxwell', Buenos Aires, 1943.
4. 'La epistemología positivista', *Nosotros* (Buenos Aires) VIII, No. 93, 283, 1943.
5. 'A new representation of types of nuclear forces (L)', *Physical Review* **65**, 249, 1944.
6. 'Una nueva representación de los tipos de fuerzas nucleares', *Revista de la Facultad de Ciencias Físicomatemáticas* (La Plata), III, 221, 1944.
7. '¿Qué es la epistemología?', *Minerva* (Buenos Aires) **1**, 27, 1944.
8. 'Precursores, predecesores y predictores', *Minerva* (Buenos Aires) **1**, 61, 1944.
9. 'Una de las posibles metafísicas', *Minerva* (Buenos Aires) **1**, 167, 1944.
10. 'Auge y fracaso de la filosofía de la naturaleza', *Minerva* (Buenos Aires) **1**, 213, 1944.
11. 'Una nueva interpretación de Rousseau', *Minerva* (Buenos Aires) **1**, 274, 1944.
12. 'Nietzsche y la ciencia', *Minerva* (Buenos Aires) **2**, 44, 1944.
13. 'Ludwig Boltzmann', *Minerva* (Buenos Aires) **2**, 70, 1944.
14. 'Cómo veía el mundo Florentino Ameghino', *Minerva* (Buenos Aires) **2**, 184, 1945.
15. 'El spin total de un sistema de más de dos partículas', *Revista de la Unión Matemática Argentina* X, 13, 1945.
16. 'Neutron-proton scattering at 8.8 and 13 MeV', *Nature* **156**, 301, 1945.
17. 'Fenómenos de resonancia en la difusión de neutrones por protones', *Revista de la Unión Matemática Argentina* **XI**, 35, 1945.
18. 'La fenomenología y la ciencia', *Cuadernos Americanos* (México) No. 4, 108, 1951.
19. 'Bemerkung über den Massendefekt des Wasserstoffatoms', *Acta Physica Austriaca* **5**, 77, 1951.
20. 'Mach y la teoria atómica', *Boletin del Químico Peruano* **3**, No. 16, 12, 1951.
21. 'What is chance?', *Science and Society* **15**, 209, 1951.
22. 'New dialogues between Hylas and Philonous', *Philosophy and Phenomenological Research* **15**, 192, 1954.
23. 'Exposición y crítica del principio de complementaridad', Laboratorio de Física Cósmica, La Paz, 1955.
24. 'A picture of the elctron', *Nuovo Cimento*, ser. X, **1**, 977, 1955.
25. 'Strife about complementarity', *British Journal for the Philosophy of Science* **6**, 1; **6**, 141, 1955.
26. 'The philosophy of the space-time approach to the quantum theory', *Methodos* **7**, 295, 1955.

27. 'A critique of the frequentist theory of probability', *Congresso Internacional de Filosofia*, São Paulo (Brazil), III, 787, 1956.
28. 'La interpretación causal de la mecánica ondulatoria', *Ciencia e Investigación* (Buenos Aires) **12**, 272, 1956.
29. 'Nuevas constantes del movimiento del electrón', *Revista de la Unión Matemática Argentina y de la Asociación Física Argentina* **XVVIII**, 25, 1956.
30. 'La antimetafísica del empirismo lógico', *Anales de la Universidad de Chile* **CXIV**, No. 102, 43, 1956.
31. 'Do computers think?', *British Journal for the Philosophy of Science* **7**, 139; **7**, 212, 1956.
32. 'Beitrag zur Diskussion über philosophische Fragen der modernen Physik', *Deutsche Zeitschrift für Philosophie* **4**, 467, 1956.
33. 'A survey of the interpretations of quantum mechanics', *American Journal of Physics* **24**, 272, 1956.
34. '¿Ha progresado la filosofía en el siglo XX?', *Revista do Livro* (Rio de Janeiro) **I**, No. 3/4, 15, 1956.
35. 'El método científico', *Revista del Mar Dulce* (Buenos Aires) No. 3, 1956.
36. 'Ubicación de la física teórica', *Revista de la Universidad de Buenos Aires* **1**, 405, 1956.
37. 'Las ideas fundamentales de la mecánica ondulatoria', *Ciencia y técnica* **123**, No. 616, 3, 1957.
38. 'Lagrangian formulation and mechanical interpretation', *American Journal of Physics* **25**, 211, 1957.
39. 'Filosofar científicamente y encarar la ciencia filosóficamente', *Ciencia e Investigación* **13**, 244, 1957.
40. '¿Qué es la ciencia?', Facultad de Ingeniería, Buenos Aires, 1958.
41. 'Sobre la imagen física de las partículas de spin entero', *Ciencia e Investigación* **14**, 311, 1958.
42. 'On multi-dimensional time (L)', *British Journal for the Philosophy of Science* **9**, 39, 1958.
43. '¿Qué significa "ley científica"?', Universidad Nacional Autónoma, México, 1958.
44. 'A filosofia tem progredido durante o século XX?', translation of ♯34, *Revista Filosófica* (Coimbra) **8**, No. 22, 1959.
45. 'Análisis epistemológico del principio de Arquímedes', Facultad de Filosófia y Letras, Buenos Aires, 1959.
46. 'Comentario crítico de algunas ideas de Poincaré sobre la hipótesis; Facultad de Filosofía y Letras, Buenos Aires, 1959.
47. '¿Qué es un problema científico?', *Holmbergia* (Buenos Aires) **VI**, No. 15, 47, 1959.
48. '¿Cómo sabemos que existe la atmósfera?', *Revista de la Universidad de Buenos Aires* **IV**, No. 2, 246, 2959.
49. 'La axiomática de Peano', Centro de Estudiantes de Filosofía y Letras, Buenos Aires, 1959.
50. 'On the connections among levels', *Proceedings of the XIIth International Congress of Philosophy* **VI**, 63, 1960.
51. 'Levels: a semantical preliminary', *Review of Metaphysics* **13**, 396, 1960.

52. 'The place of induction in science', *Philosophy of Science* **27**, 262, 1960.
53. 'Probabilidad e inducción', *Ciencia y Técnica* (Buenos Aires) **129**, 240, 1960.
54. 'Are there timeless entities?', *Miscelanea de Estudos a Joaquim de Carvalho*, Figueira da Foz (Portugal) No. 3, 290, 1960.
55. 'Analyticity redefined', *Mind* **LXX**, 239, 1961.
56. 'The weight of simplicity in the construction and assaying of scientific theories', *Philosophy of Science* **28**, 129, 1961.
57. 'Kinds and criteria of scientific law', *Philosophy of Science* **28**, 260, 1961.
58. 'Laws of physical laws', *American Journal of Physics* **29**, 518, 1961.
59. 'Causality, chance, and law', *American Scientist* **49**, 432, 1961.
60. 'Ley y determinación', *Scientia* **55**, 1, 1961.
61. 'Ethics as a science', *Philosophy and Phenomenological Research* **XX**, 139, 1961.
62. 'Significación del humanismo en el mundo contemporáneo', *Revista de la Universidad de Buenos Aires* **VI**, 563, 1961.
63. 'The complexity of simplicity', *Journal of Philosophy* **LIX**, 113, 1962.
64. 'Causality: A rejoinder', *Philosophy of Science* **29**, 306, 1962.
65. 'La teoría del conocimiento en nuestro tiempo,' *Ciencia e Investigación* **18**, 60, 1962.
66. 'Cosmology and magic', *The Monist* **44**, 116, 1962.
67. 'An analysis of value', *Mathematicae Notae* XVIII, 95, 1962.
68. 'Bertrand Russell y la teoría del conocimiento', in *La filosofía del siglo XX y otros ensayos*, Alfa, Montevideo, 1962.
69. 'Tecnología, ciencia y filosofía', *Revista de la Universidad de Chile CXXI*, No. 126, 64, 1963.
70. 'A general black box theory', *Philosophy of Science* **30**, 346, 1963.
71. 'Phenomenological theories', in Book ♯15, pp. 234–254.
72. 'Physics and reality', *Dialectica* **19**, 195, 1965.
73. 'Technology as applied science', *Technology and Culture* **7**, 329, 1966.
74. 'Mach's critique of Newtonian mechanics', *American Journal of Physics* **34**, 585, 1966.
75. Reprint of ♯72. *Dialectica* **20**, 174, 1966.
76. 'Are there operational definitions of physical concepts?', (in Russian) *Voprosi filosofii* No. 11, 66, 1966.
77. 'On null individuals (L)', *Journal of Philosophy* **63**, 776, 1966.
78. Reprint of ♯56. In M. H. Foster and M. Martin, eds., *Probability, Confirmation, and Simplicity*, Odyssey Press, New York, 1966.
79. 'The structure and content of a physical theory', in Book ♯17, pp. 15–17.
80. 'A ghost free axiomatization of quantum mechanics', in Book ♯21, pp. 105–117.
81. 'Quanta and philosophy', *Proceedings of the 7th Inter-American Congress of Philosophy* I, Presses de l'Université Laval, Quebec, 1967.
82. 'Physical axiomatics', *Reviews of Modern Physics* **39**, 463, 1967.
83. 'Quanta y filosofía', translation of ♯81, *Crítica* (México) **1**, No. 3, 41, 1967.
84. 'Analogy in quantum mechanics: from insight to nonsense', *British Journal for the Philosophy of Science* **18**, 265, 1967.
85. 'Machs Kritik an der Newtonschen Mechnik', in *Symposium aus dem Anlass des 50. Todestages von Ernst Mach*, Ernst-Mach-Institut, Freiburg i. Br., 1967.

86. 'The maturation of science', in I. Lakatos and A. Musgrave, eds., *Problems in the Philosophy of Science*, North-Holland, Amsterdam, 1968.
87. 'The nature of science', in R. Klibansky, ed., *Contemporary Philosophy* II, La Nuova Italia Editrice, Florence, 1968.
88. 'Philosophy and physics', in R. Klibansky, ed., *Contemporary Philosophy* II, La Nuova Italia Editrice, Florence, 1968.
89. 'On Mach's nonconcept of mass (L)', *American Journal of Physics* **36**, 167, 1968.
90. 'Problems and games in the current philosophy of natural science', *Proceedings of the XIVth International Congress of Philosophy* **II**, 1968.
91. 'Physique et métaphysique du temps', *Proceedings of the XIVth International Congress of Philosophy* **II**, 1968.
92. 'Towards a philosophy of technology', reprint of #73, in S. Dockx, ed., *Civilisation technique et humanisme*, Office internationale de librairie, Brussels, 1968.
93. 'Conjunction, succession, determination, and causation', *International Journal of Theoretical Physics* **1**, 299, 1968.
94. 'Physical time: the objective and relational theory', *Philosophy of Science* **35**, 355, 1968.
95. 'La vérification des théories scientifiques', in *Démonstration, vérification, justification: Entretiens de l'Institut International de Philosophie*, Nauwelaerts, Louvain, Paris, 1968.
96. 'Les concepts de modèle', *L'âge de la science* **I**, 165, 1968.
97. 'Theory of partial truth: not proved inconsistent' (L), *Philosophy and Phenomenological Research* **29**, 297, 1968.
98. 'Arten und Kriterien wissenschaftlicher Gesetze', translation of #105, in G. Kröber, ed., *Der Gesetzbegriff in der Philosophie und den Einzelwissenschaften*, Akademie-Verlag, Berlin, 1968.
99. 'Filosofía de la investigación científica en los países en desarrollo', *Acta Científica Venezolana* **19**, No. 3, 118, 1968.
100. 'Corrections to *Foundations of Physics*: Correct and incorrect', *Synthese* **19**, 443, 1969.
101. Reprint of #99, *Mensurae* (Buenos Aires) **2**, No. 11, 1969.
102. 'Machs Beitrag zur Grundlegung der Mechanik', *Philosophia Naturalis* **11**, 189, 1969.
103. 'The metaphysics, epistemology and methodology of levels', in L. L. Whyte, A. G. Wilson, and D. Wilson, eds., *Hierarchical Structures*, American Elsevier, New York, 1969.
104. 'Alexander von Humboldt und die Philosophie', in H. Pfeiffer, ed., *Alexander von Humboldt: Werk und Weltgeltung*, Piper, Munich, 1969.
105. 'Analogy, simulation, representation', *Revue internationale de philosophie* **23**, 16, 1969.
106. 'What are physical theories about?', in N. Rescher, ed., *Studies in the Philosophy of Science: American Philosophical Quarterly Monograph* No. 3, 1969.
107. 'Azar, probabilidad y ley', *Diánoia* (México) **15**, 141, 1969.
108. 'Models in theoretical science', *Proceedings of the XIVth International Congress of Philosophy III*, 1969.

109. 'A covariant position operator for the relativistic electron' (with A. J. Kálnay), *Progress of Theoretical Physics* **42**, 1445, 1969.
110. 'Four models of human migration: An exercise in mathematical sociology', *Archiv für Rechts- und Sozialphilosophie* **55**, 451, 1969.
111. 'The arrow of time' (L), *International Journal of Theoretical Physics* **3**, 77–78, 1970.
112. 'Time asymmetry, time reversal, and irreversibility', *Studium Generale* **23**, 562–570, 1970.
113. Reprint of #99, *Folia humanística* (Barcelona) **8**, 141–154, 1970.
114. 'La ciencia ¿es éticamente neutral?', *Folia humanística* (Barcelona) **8**, 241– XX, 1970.
115. 'Alexander von Humboldt y la filosofía', Spanish translation of #104, *Folia humanística* **8**, 535–546, 1970.
116. 'The so-called fourth indeterminacy relation' (L), *Canadian Journal of Physics* **48**, 1410–1411, 1970.
117. 'Problems concerning intertheory relations', in P. Weingartner and G. Zecha (eds.), *Induction, Physics, and Ethics*, pp. 285–325. D. Reidel, Dordrecht, 1970.
118. 'Comments on Groenewold's paper', in volume mentioned in #114, pp. 202–207, 213–214.
119. 'Theory meets experience', in H. Kiefer and M. K. Munitz (eds.), *Contemporary Philosophic Thought*, Vol. 2, pp. 138–165, 1970.
120. 'The physicist and philosophy', *Zeitschrift für allgemeine Wissenschaftstheorie* **1**, 196–208, 1970.
121. Reprint of #84. Archives de l'Institut International des Sciences Théoriques No. 16: *La symétrie*. Office internationale de librairie, Bruxelles, 1970.
122. 'Physik und Wirklichkeit', German translation of #72, in L. Krüger, ed., *Erkenntnisprobleme der Naturwissenschaften*, Kiepenheuer and Witsch, Cologne, 1970.
123. Reprint of #105, *General Systems* **25**, 27–34, 1970.
124. 'Space and time in contemporary science' (in Russian), *Voprosi filosofii* No. 7, pp. 81–92, 1970.
125. 'Virtual processes and virtual particles: real or fictitious?' (L), *International Journal of Theoretical Physics* **3**, 507–508, 1970.
126. 'Conjonction, succession, détermination, causalité', in J. Piaget (ed.), *Les théories de la causalité*. Presses universitaires de France, Paris, 1971.
127. 'A philosophical obstacle to the rise of new theories in microphysics', in E. W. Bastin (ed.), *Quantum Theory and Beyond*, pp. 263–273. Cambridge University Press, 1971.
128. 'The paradox of addition and its dissolution', *Crítica* (México) **3**, 27–31, 1971.
129. 'Is scientific metaphysics possible?', *Journal of Philosophy* **68**, 507–520, 1971.
130. 'A mathematical theory of the dimensions and units of physical quantities', in Book # 27, pp. 1–16.
131. '"Scientific metaphysics": addenda et corrigenda', *Journal of Philosophy* **68**, 876, 1971.
132. 'A new look at definite descriptions', *Philosophy of Science* (Japan) **4**, 131–146, 1971.
133. Reprint of #110, *General Systems* **16**, 87–92, 1971.

134. 'On method in the philosophy of science', *Archives de philosophie* **34**, 551–574, 1971.
135. 'Space and time in modern science', *II Bienal de Ciência e Humanismo* (São Paulo), pp. 21–34, 1971.
136. 'Seudociencia y seudofilosofía: dos monólogos paralelos', *Ciencia nueva* (Buenos Aires) No. 15, pp. 41–43, 1972.
137. Reprint of #112, in J. T. Fraser et al. (eds.), *The Study of Time*, Springer, Berlin, Heidelberg, New York, 1972.
138. 'A program for the semantics of science', *Journal of Philosophical Logic* **1**, 317–328, 1972.
139. 'Metatheory', in *Scientific thought*, a UNESCO project, pp. 227–252. Mouton/UNESCO, Paris, The Hague, 1972.
140. 'Modelo del dilema electoral argentino', *Ciencia nueva* (Buenos Aires) No. 21, pp. 52–54, 1972.
141. Reprint of Book #19, Ch. 12, in Carl A. Mitcham and Robert Mackey, eds., *Philosophy and Technology*, The Free Press, Riverside, N. J., 1972.
142. Reprint of #72, in Edward A. Mackinnon, ed., *The Problem of Scientific Realism*, Appleton-Century-Crofts, New York, 1972.
143. 'Adevar', in *Mario Bunge* (Logicieni si filosofi contemporani, No. 2), pp. 65–115. Centrul de Informare si Documentare in Scintele sociale si politice, Bucharest, 1973.
144. 'Meaning in science', *Proc. XVth World Congress of Philosophy* **2**, 281–286, 1973.
145. 'Normative Wissenschaft ohne Normen – aber mit Werten', *Conceptus* **VII**, 57–64, 1973.
146. 'Bertrand Russell's *regulae philosophandi*', in Book #35, pp. 3–12.
147. 'On confusing "measurement" with "measure" in the methodology of the behavioral sciences', in Book #35, pp. 105–122.
148. 'The role of forecast in planning', *Theory and Decision* **3**, 207–221, 1973.
149. 'A decision theoretic model of the American War in Vietnam', *Theory and Decision* **3**, 323–338, 1973.
150. '¿Es posible una metafisica científica?', Spanish translation of #126, *Teorema* **III**, 435–454, 1973.
151. 'Conceptul de structura sociala', *Informatica si modele matematice in stiintele sociale* (Bucharest) II, No. 2, pp. 5–57, 1973.
152. Reprints of #56 and #73 in Alex C. Michalos (ed.), *Philosophical Problems of Science and Technology*, Allyn and Bacon, Boston, 1974.
153. 'The relations of logic and semantics to ontology', *Journal of Philosophical Logic* **3**, 195–210, 1974.
154. Reprint of #73, in Friedrich Rapp, ed., *Contributions to a Philosophy of Technology*, D. Reidel, Dordrecht, Boston, 1974.
155. 'The concept of social structure', in W. Leinfellner and E. Köhler (eds.), *Developments in the Methodology of Social Science*, pp. 175–215, D. Reidel, Dordrecht, Boston, 1974.
156. 'Les présupposés et les produits métaphysiques de la science et de la technique contemporaines', *Dialogue* **13**, 443–453, 1974.

157. 'Metaphysics and science', *General Systems* **19**, 15–18, 1974.
158. 'Things', *International Journal of General Systems* **1**, 229–236, 1974.
159. 'Teoria stiintifica', Rumanian translation of Ch. 7 of #18, *Epistemologie: Orientari contemporane*. Ed. Ilie Parvu. Editura politica, Bucharest, 1974.
160. 'The methodology of development indicators', UNESCO, Methods and Analysis Division, Dept. of Social Sciences, 1974.
161. 'Crítica de la noción fregeana de predicado', *Revista Latinoamericana de Filosofía* **1**, 5–8, 1975.
162. 'Entscheidungstheoretische Modelle in der Politik: Vietnam', German transl. of #149, in R. Simon-Schaefer and W. Ch. Zimmerli, eds., *Wissenschaftstheorie der Geisteswissenschaften*, Hoffmann and Campe, Hamburg, 1975.
163. 'El significado en ciencia', Spanish transl. of #144, *Teoría* (México) **1**, No. 1, 1975.
164. 'What is a quality of life indicator?', *Social Indicators Research* **2**, 65–80, 1975.
165. 'Towards a technoethics', *Philosophic Exchange* **2**, No. 1, pp. 69–79, 1975.
166. 'Welches sind die Besonderheiten der Quantenphysik gegenüber der klassischen Physik?' (with Andrés J. Kálnay) in R. Haller and J. Götschl, eds., *Philosophie und Physik*, Vieweg, Braunschweig, 1975.
167. 'La paradoja de la adición: respuesta al Maestro Margáin', *Crítica* **7**, 105–107, 1975.
168. 'Ontología y ciencia', *Diánoia* 50–59, 1975.
169. 'La representación conceptual de los hechos', *Teorema* **5**, 317–360, 1975.
170. Reprint of 144, *Poznán Studies in the Philosophy of the Sciences and the Humanities* **1**, No. 4, 1975, 56–64.
171. 'A critical examination of dialectics', in Ch. Perelman, ed., *Dialectics/Dialectique*, pp. 63–77. Martinus Nijhoff, The Hague. Commentary by I. Narsky, 'Bemerkungen über den Vortrag von Prof. Bunge', pp. 78–86, 1975.
172. Russian translation of #170, *Voprosi filosofii* No. 4, 1975.
173. '¿Hay proposiciones?' *Aspectos de la Filosofía de W. V. Quine*, Teorema, Valencia, 1975.
174. Reprint of #99, in Jorge A. Sábato (ed.), *El pensamiento latinoamericano en la problematica ciencia-tecnología-desarrollo-independencia*, Paidos, Buenos Aires, 1975.
175. French transl. of #166, *Fundamenta scientiae* No. 11, 1976.
176. 'Possibility and probability', in W. Harper and C. Hooker (eds.), *Foundations of Probability Theory, Statistical Inference, and Statistical Theories of Science*, Vol. III, pp. 17–33. D. Reidel, Dordrecht, Boston, 1976.
177. Reprint of #164 with slight changes, in J. King-Farlow and W. Shea (eds.), *Values and the Quality of Life*, Neale Watson Academic Publications, New York, 1976.
178. 'The relevance of philosophy to social science', in W. Shea (ed.), *Basic Issues in the Philosophy of Science*, pp. 136–155, Neale Watson Academic Publications, New York, 1976.
179. 'El método en la biología', *Naturaleza* (México) **7**, 70–81, 1976.
180. 'El ser no tiene sentido y el sentido no tiene ser', *Teorema* **VI**, 201–212, 1976.
181. 'Differentiation, participation and cohesion', (with Máximo García-Sucre), *Quality and Quantity* **10**, 171–178, 1976.

182. Reprint of ♯168, *La filosofía y la ciencia en nuestros días*, pp. 27–40. Ed. Grijalbo, Mexico, 1976.
183. Spanish transl. of ♯177, *La filosofía y las ciencias sociales*, pp. 43–69. Ed. Grijalbo, Mexico, 1976.
184. 'A model for processes combining competition with cooperation', *Applied Mathematical Modelling* **1**, 21–23, 1976.
185. Polish transl. of ♯144, *Poznanskie Studia z Filosofii Nauki*, Vol. 1, pp. 13–23. Panstwowe Nydawnictwo Naudowe, Warsaw-Poznan, 1976.
186. Reprint of ♯156, *Science et métaphysique*, pp. 193–206, Office international de librairie, Bruxelles, 1976.
187. 'Is science value-free and morally neutral?', *Philosophy and Social Action* **II**, No. 4, pp. 5–18, 1976.
188. 'Examen filosófico del vocabulario sociológico', *Diánoia* **XXII**, 56–75, 1976.
189. Greek transl. of ♯87, *Deukalion* **3**, 351–363, 1976.
190. 'Qué es y para qué sirve la epistemología?', *Revista de la Universidad de México* **XXXI**, No. 2, 1–7, 1976.
191. Reprint of ♯165, *The Monist* **60**, 96–107, 1977.
192. 'The interpretation of Heisenberg's inequalities', in H. Pfeiffer, ed., *Denken und Umdenken: zu Werk und Wirkung von Werner Heisenberg*, pp. 146–156. Piper, Munich, 1977.
193. 'Reply to van Rootselar's criticisms of my theory of things', *Intern. J. General Systems* **3**, 181–182, 1977.
194. 'A theory of properties and kinds', (with Arturo Sangalli), *Intern. J. General Systems* **3**, 183–190, 1977.
195. 'Tres politicas de desarrollo científico y una sola eficaz', *Interciencia* **2**, 76–80, 1977.
196. Reprint of ♯195, in Enrique Leff, ed., *Primer Simposio sobre Ecodesarrollo*, pp. 88–96. Asociación Méxicana de Epistemología, México, D.F., 1977.
197. Spanish transl. of ♯126, in *La teoría de la causalidad*, Ed. Sígueme, Salamanca, 1977.
198. 'A relational theory of physical space', (with A. Garcia Mâynez), *Intern. J. Theoretical Physics* **15**, 961–972, 1977.
199. 'Emergence and the mind', *Neuroscience* **2**, 501–509, 1977.
200. 'Levels and reduction', *Am. J. Physiology: Regulatory, Integrative and Compar. Physiol.* **2**, 75–82.
201. 'The philosophical richness of technology', in F. Suppe and P. D. Asquith, eds., *PSA* **2**, pp. 153–172, 1977.
202. 'The GST challenge to the classical philosophies of science', *Intern. J. General Systems* **4**, 29–37, 1977.
203. 'General systems and holism', *General Systems* **XXII**, 87–90, 1977.
204. 'States and events', in William E. Hartnett, ed., *Systems: Approaches, Theories, Applications*, pp. 71–95. D. Reidel, Boston, Dordrecht, 1977.
205. Reprint of ♯153, in Edgar Morscher, Johannes Czermak, and Paul Weingartner, eds., *Problems in Logic and Ontology*, pp. 29–43. Akademische Druck-und-Verlagsanstalt, Graz, 1977.
206. '¿Qué es y a qué puede aplicarse el método científico?', *Diánoia*, 88–101, 1977.

207. 'A systems concept of the international system', in Mario Bunge, Johan Galtung, and Mircea Malitza, eds., *Mathematical Approaches to International Relations*, pp. 291–305. Rumanian Academy of Social and Political Sciences, Bucharest, 1977.
208. 'Quantum mechanics and measurement', *International Journal of Quantum Chemistry* **12**, Suppl. 1, 1–13, 1977.
209. 'The mind-body problem in the light of contemporary biology', (with Rodolfo Llinás), *16th World Congress of Philosophy: Section Papers*, pp. 131–133, 1978.
210. 'Restricted applicability of the concept of command in neuroscience: dangers of metaphors' (with Rodolfo Llinás), *The Behavioral and Brain Sciences* **1**, 30–31, 1978.
211. 'Physical space', in M. Svilar and A. Mercier, eds., *Space*, pp. 133–148. Comments by H. Törnebohm, pp. 149–167. Other comments and author's replies, pp. 167–171. Peter Lang, Bern, Frankfurt, Las Vegas, 1978.
212. 'A model of evolution', *Applied Mathematical Modelling* **2**, 201–204, 1978.
213. 'The limits of science', *Epistemologia* **1**, 11–32, 1978.
214. 'Iatrofilosofía', in F. Alonso de Florida, ed., *Ensayos de Yatrofilosofía*, pp. 3–5. Academia Nacional de Medicina, Mexico, 1978.
215. 'La enfermedad como estado o proceso', in vol. cited in #214, pp. 65–69.
216. 'Conocimiento objetivo y mundos popperianos', *Semestre de Filosofía* **I**, No. 2, pp. 7–25, 1978.
217. 'A cultura como sistema concreto', *Ciência e Filosofía* **1**, No. 1, 1978.
218. 'A systems concept of society: beyond individualism and holism', *Theory and Decision* **10**, 13–30, 1979.
219. 'The five buds of technophilosophy', *Technology in Society* **1**, 67–74, 1979.
220. 'The mind-body problem, information theory, and Christian dogma', *Neuroscience* **4**, 453–454, 1979.
221. 'Philosophical inputs and outputs of technology', in George Bugliarello and Dean B. Doner, eds., *The History and Philosophy of Technology*, pp. 262–281. University of Illinois Press, Urbana, 1979.
222. Prefacio to Augusto Fernandez Guardiola, ed., *La cosciencia*, pp. 5–8. Ed. Trillas, Mexico, 1979.
223. 'La bancarrota del dualismo psiconeural', in #222, pp. 71–84. Comment by Carlos Pereda, *op.cit.*, pp. 85–87.
224. Russian transl. of #223, *Filosofskie Nauki* No. **2**, 77–87, 1979. Comments by D. I. Dubrovskii, *ibid.*, pp. 88–97.
225. 'A model of secrecy', *Journal of Irreproducible Results* **25**, 25–26, 1979.
226. 'The mind-body problem in an evolutionary perspective', in *Brain and Mind*, [Ciba Foundation Series 69], pp. 53–63, Excerpta Medica, Amsterdam, 1979.
227. 'Relativity and philosophy', in J. Bärmark (ed.), *Perspectives in Metascience*, pp. 75–88, Regiae Societas Scientiarum et Litterarum Gothoburgensis, [Interdisciplinaria 2], Göteborg, 1979.
228. Reprint of #229. *Physics in Canada* **35**, 105–111, 1979.
229. 'The Einstein-Bohr debate over quantum mechanics: Who was right about what?', *Lecture Notes in Physics* **100**, 204–219, 1979.

230. 'Some topical problems in biophilosophy', *J. Social and Biological Structures* **2**, 155–172, 1979.
231. 'Reply to Craig Dilworth's review of the *Treatise on Basic Philosophy*, vols. 1–4', *Epistemologia* **II**, 524–428, 1979.
232. Reprint of #213. *The Physiologist* **23**, 7–13, 1980.
233. 'From neuron to behavior and mentation: An exercise in levelmanship', in H. M. Pinsker and W. D. William (eds.), *Information Processing in the Nervous System*, pp. 1–16, Raven Press, New York, 1980.
234. 'Valor biológico y valor psicológico', in J. J. E. Gracia (ed.), *El hombre y su conducta/Man and His Conduct*, pp. 102–111, Editorial Universitaria, Rio Piedras, P. R., 1980.
235. 'Technoethics', in M. Kranzberg (ed.), *Ethics in an Age of Pervasive Technology*, pp. 139–142, Westview Press, Boulder, Colo., 1980.
236. Reprint of #199, in A. D. Smith, R. Llinás, and P. G. Kostyuk (eds.), *Commentaries in the Neurosciences*, pp. 633–641, Pergamon Press, Oxford, 1980.
237. 'Introduction: The mind-body problem', in Dalbir Bindra (ed.), *The Brain's Mind*, pp. 1–5, Gardner Press, New York, 1980.
238. 'The psychoneural identity theory', in Bindra (ed.), *The Brain's Mind*, pp. 89–108, Gardner Press, New York, 1980.
239. Reprint of #218 in *General Systems* **24**, 27–44, 1980.
240. German translation of #120. *Physik und Didaktik* **8**, 261–267, 1980.
241. 'The geometry of a quantal system' (with Máximo Garcia-Sucre)', *International Journal of Quantum Chemistry* **19**, 83–93, 1981.
242. 'Systems all the way', *Nature and System* **3**, 37–47, 1981.
243. 'Developmental Indicators', *Social Indicators Research* **9**, 369–385, 1981.

SELECTED REVIEWS OF BOOKS BY MARIO BUNGE

Causality: The Place of the Causal Principle in Modern Science

C. J. Ducasse, *Isis* **51** (1960), 88–90.
R. Harré, *Philosophical Books* **1**, No. 2 (1960), 5–7.
S. Issman, *Revue Internationale de Philosophie* **14** (1960), 454–455.
J. O. Wisdom, *Nature* **187** (1960), 92.
K. W. Rankin, *Mind* **70** (1961), 107–109.
R. Schlegel, *Philosophy of Science* **28** (1961), 72–82.
S. Morgenbesser, *Scientific American* **204**, No. 2 (1961), 175–178.
L. Ponticelli, *Rivista di Filosofia Neo-Scolastica* **65** (1971), 523–525.

Metascientific Queries

R. Harré, *Philosophical Books* **1**, No. 3 (1960), 2–3.
D. Riepe, *Philosophy and Phenomenological Research* **20** (1960), 552–553.
A. L. Leroy, *Revue philosophique* **151** (1961), 535.
P. K. Feyerabend, *Philosophical Review* **70** (1961), 396–405.
E. H. Hutten, *Mind* **71** (1962), 137.

Intuition and Science

J. A. Wojciechowski, *Dialogue* **2** (1963), 238–239.
M. Scriven, *American Mathematical Monthly* **70** (1963), 232–233.

The Myth of Simplicity

H. Leblanc, *Dialogue* **3** (1964), 201–203.
R. Ackermann, *Philosophy and Phenomenological Research* **24** (1964), 447–448.
G. Schlesinger, *Philosophical Review* **74** (1965), 402–404.
R. Harré, *Philosophical Quarterly* **16** (1966), 85–86.

Foundations of Physics

V. Frei, *Czechoslovak Journal of Physics* **B18** (1968), 1502–1503.
M. Strauss, *Deutsche Zeitschrift für Philosophie* **17** (1969), 1133–1140.
M. Strauss, *Synthese* **19** (1968–69), 433–442.
H. Freudenthal, *Synthese* **21** (1970), 93–106.

J. Agassi and R.S. Cohen (eds.), Scientific Philosophy Today, 501–503
Copyright ©1981 *by D. Reidel Publishing Company.*

Scientific Research

S. J. Schmidt, *Physikalische Blätter* **23** (1967).
R. Torretti, *Anales de la Universidad de Chile* **125** (1967), 346–350.
V. Frei, *Czechoslovak Journal of Physics* **B18** (1968), 1348–1350.
G. Klaus, *Deutsche Literaturzeitung* **89** (1968).
D. Wittich, *Deutsche Zeitschrift für Philosophie* **16** (1968), 96–104.
J. Agassi, *Synthese* **19** (1969), 453–464.
S. G. Edgerton, *History of Education Quarterly* **9** (1969), 492–496.
R. Haller, *Acta Physica Austriaca* **29** (1969), 278–282.
R. B. Lindsay, *J. Franklin Institute* **287** (1969), 79–80.
G. I. Ruzavin, *Voprosi filosofii* No. 1 (1971), 169–173.
R. Beneyto, *Teorema* **3** (1971), 140–142.
W. Schäfer, *Erasmus* **25** (1973), 1–5.
F. Russo, *Archives de philosophie* **36** (1973), 373–393.

Philosophy of Physics

S. Brush, *Physics Today* Sept. 1973, p. 61.
M. Robinson, *American Journal of Physics* **42** (1974), 797–798.
G. Vollmer, *Zeitschrift für allgemeine Wissenschaftstheorie* **4** (1973), 407–409.
Y. Gauthier, *Dialogue* **13** (1974), 206–209.
R. Torretti, *Diálogos* **9** (1974), 176–181.
J. Bub, *Philosophia* **5** (1975), 352–356.
P. Kirschenmann, *Philosophy of Science* **42** (1975), 94–96.

Method, Model and Matter

G. Vollmer, *Zeitschrift für allgemeine Wissenschaftstheorie* **4** (1973), 380 383.
R. Torretti, *Diálogos* **9** (1973), 156–167.
M. A. Quintanilla, *Teorema* (1974), 147–151.
R. Tuomela, *Philosophy of Science* **41** (1974), 429–430.
R. Harré, *Isis* **66** (1975), 264–266.
M. Ruse, *Philosophia* **5** (1975), 348-351.
T. Settle, *British Journal for the Philosophy of Science* **28** (1977), 86–94.
F. Russo, *Archives de philosophie* **43** (1980), 171.

Treatise on Basic Philosophy

J. Kekes, *Philosophy and Phenomenological Research* **37** (1977), 426–428.
I. G. McFetridge, *Mind* **87** (1978), 144–146.
R. Torretti, *Diálogos* **14** (1979), 151–156.
F. Russo, *Archives de philosophie* **43** (1980), 151–152.
G. Vollmer, *Zeitschrift für allgemeine Wissenschaftstheorie* **10** (1979), 405–407.
T. D. Bowler, *International Journal of General Systems* **6** (1980), 51–53.

A. Rapoport, *Behavioral Science* **25** (1980), 166–168.
R. Torretti, *Diálogos* **25** (1980), 211–212.
C. Dilworth, *Epistemologia* **2** (1979), 415–425.
M. A. Quintanilla, *Dialectica* **34** (1980), 295–297.
H. Barreau, *Revue de Métaphysique et de Morale*, No. 1 (1981), 112–117.

The Mind-Body Problem

D. M. MacKay, *Neuroscience* **5** (1980), 2329–2330.
A. Battro, *Revista Latinoamericana de Filosofía* **6** (1980), 284–286.
R. Palfree, *American Scientist* **69** (1981), 110.

INDEX OF NAMES

BOSTON STUDIES IN THE PHILOSOPHY OF SCIENCE

Editors:
ROBERT S. COHEN and MARX W. WARTOFSKY
(Boston University)

1. Marx W. Wartofsky (ed.), *Proceedings of the Boston Colloquium for the Philosophy of Science 1961-1962.* 1963.
2. Robert S. Cohen and Marx W. Wartofsky (eds.), *In Honor of Philipp Frank.* 1965.
3. Robert S. Cohen and Marx W. Wartofsky (eds.), *Proceedings of the Boston Colloquium for the Philosophy of Science 1964-1966. In Memory of Norwood Russell Hanson.* 1967.
4. Robert S. Cohen and Marx W. Wartofsky (eds.), *Proceedings of the Boston Colloquium for the Philosophy of Science 1966-1968.* 1969.
5. Robert S. Cohen and Marx W. Wartofsky (eds.), *Proceedings of the Boston Colloquium for the Philosophy of Science 1966-1968.* 1969.
6. Robert S. Cohen and Raymond J. Seeger (eds.), *Ernst Mach: Physicist and Philosopher.* 1970.
7. Milic Capek, *Bergson and Modern Physics.* 1971.
8. Roger C. Buck and Robert S. Cohen (eds.), *PSA 1970. In Memory of Rudolf Carnap.* 1971.
9. A. A. Zinov'ev, *Foundations of the Logical Theory of Scientific Knowledge (Complex Logic).* (Revised and enlarged English edition with an appendix by G. A. Smirnov, E. A. Sidorenka, A. M. Fedina, and L. A. Bobrova.) 1973.
10. Ladislav Tondl, *Scientific Procedures.* 1973.
11. R. J. Seeger and Robert S. Cohen (eds.), *Philosophical Foundations of Science.* 1974.
12. Adolf Grünbaum, *Philosophical Problems of Space and Time.* (Second, enlarged edition.) 1973.
13. Robert S. Cohen and Marx W. Wartofsky (eds.), *Logical and Epistemological Studies in Contemporary Physics.* 1973.
14. Robert S. Cohen and Marx W. Wartofsky (eds.), *Methodological and Historical Essays in the Natural and Social Sciences. Proceedings of the Boston Colloquium for the Philosophy of Science 1969-1972.* 1974.
15. Robert S. Cohen, J. J. Stachel and Marx W. Wartofsky (eds.), *For Dirk Struik. Scientific, Historical and Political Essays in Honor of Dirk Struik.* 1974.
16. Norman Geschwind, *Selected Papers on Language and the Brain.* 1974.
18. Peter Mittelstaedt, *Philosophical Problems of Modern Physics.* 1976.
19. Henry Mehlberg, *Time, Causality, and the Quantum Theory* (2 vols.). 1980.
20. Kenneth F. Schaffner and Robert S. Cohen (eds.), *Proceedings of the 1972 Biennial Meeting, Philosophy of Science Association.* 1974.
21. R. S. Cohen and J. J. Stachel (eds.), *Selected Papers of Léon Rosenfeld.* 1978.
22. Milic Capek (ed.), *The Concepts of Space and Time. Their Structure and Their Development.* 1976.
23. Marjorie Grene, *The Understanding of Nature. Essays in the Philosophy of Biology.* 1974.

24. Don Ihde, *Technics and Praxis. A Philosophy of Technology.* 1978.
25. Jaakko Hintikka and Unto Remes, *The Method of Analysis. Its Geometrical Origin and Its General Significance.* 1974.
26. John Emery Murdoch and Edith Dudley Sylla, *The Cultural Context of Medieval Learning.* 1975.
27. Marjorie Grene and Everett Mendelsohn (eds.), *Topics in the Philosophy of Biology.* 1976.
28. Joseph Agassi, *Science in Flux.* 1975.
29. Jerzy J. Wiatr (ed.), *Polish Essays in the Methodology of the Social Sciences.* 1979.
32. R. S. Cohen, C. A. Hooker, A. C. Michalos, and J. W. van Evra (eds.), *PSA 1974: Proceedings of the 1974 Biennial Meeting of the Philosophy of Science Association.* 1976.
33. Gerald Holton and William Blanpied (eds.), *Science and Its Public: The Changing Relationship.* 1976.
34. Mirko D. Grmek (ed.), *On Scientific Discovery.* 1980.
35. Stefan Amsterdamski, *Between Experience and Metaphysics. Philosophical Problems of the Evolution of Science.* 1975.
36. Mihailo Marković and Gajo Petrović (eds.), *Praxis. Yugoslav Essays in the Philosophy and Methodology of the Social Sciences.* 1979.
37. Hermann von Helmholtz: *Epistemological Writings. The Paul Hertz/Moritz Schlick Centenary Edition of 1921 with Notes and Commentary by the Editors.* (Newly translated by Malcolm F. Lowe Edited, with an Introduction and Bibliography, by Robert S. Cohen and Yehuda Elkana.) 1977.
38. R. M. Martin, *Pragmatics, Truth, and Language.* 1979.
39. R. S. Cohen, P. K. Feyerabend, and M. W. Wartofsky (eds.), *Essays in Memory of Imre Lakatos.* 1976.
42. Humberto R. Maturana and Francisco J. Varela, *Autopoiesis and Cognition. The Realization of the Living.* 1980.
43. A. Kasher (ed.), *Language in Focus: Foundations, Methods and Systems. Essays Dedicated to Yehoshua Bar-Hillel.* 1976.
46. Peter L. Kapitza, *Experiment, Theory, Practice.* 1980.
47. Maria L. Dalla Chiara (ed.), *Italian Studies in the Philosophy of Science.* 1980.
48. Marx W. Wartofsky, *Models: Representation and the Scientific Understanding.* 1979.
50. Yehuda Fried and Joseph Agassi, *Paranoia: A Study in Diagnosis.* 1976.
51. Kurt H. Wolff, *Surrender and Catch: Experience and Inquiry Today.* 1976.
52. Karel Kosík, *Dialectics of the Concrete.* 1976.
53. Nelson Goodman, *The Structure of Appearance.* (Third edition.) 1977.
54. Herbert A. Simon, *Models of Discovery and Other Topics in the Methods of Science.* 1977.
55. Morris Lazerowitz, *The Language of Philosophy. Freud and Wittgenstein.* 1977.
56. Thomas Nickles (ed.), *Scientific Discovery, Logic, and Rationality.* 1980.
57. Joseph Margolis, *Persons and Minds. The Prospects of Nonreductive Materialism.* 1977.
58. Gerard Radnitzky and Gunnar Andersson (eds.), *Progress and Rationality in Science.* 1978.

59. Gerard Radnitzky and Gunnar Andersson (eds.), *The Structure and Development of Science.* 1979.
60. Thomas Nickles (ed.), *Scientific Discovery: Case Studies.* 1980.
61. Maurice A. Finocchiaro, *Galileo and the Art of Reasoning.* 1980.
62. William A. Wallace, *Prelude to Galileo.* 1981.
63. Friedrich Rapp, *Analytical Philosophy of Technology.* 1981.
64. Robert S. Cohen and Marx W. Wartofsky (eds.), *Hegel and the Sciences.* (Forthcoming).
65. Joseph Agassi, *Science and Society.* 1981.
66. Ladislav Tondl, *Problems of Semantics.* 1981.
67. Joseph Agassi and Robert S. Cohen (eds.), *Scientific Philosophy Today.* 1981.
68. Władysław Krajewski (ed.), *Polish Essays in the Philosophy of the Natural Sciences.* (Forthcoming).
69. James H. Fetzer, *Scientific Knowledge.* (Forthcoming).

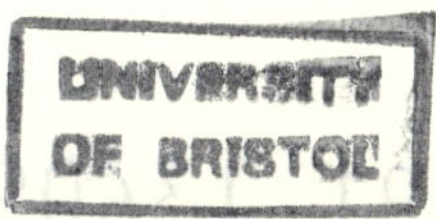
UNIVERSITY OF BRISTOL